任务型语码转换式双语教学系列教材

总主编 刘玉彬 副总主编 杜元虎 总主审 段晓东

数学与统计学

MATHEMATICS AND STATISTICS

主　编　王立冬　周文书
副主编　刘　满　张　友
主　审　袁学刚　卢　静

大连理工大学出版社

图书在版编目(CIP)数据

数学与统计学 / 王立冬，周文书主编. — 大连：
大连理工大学出版社，2014.8
任务型语码转换式双语教学系列教材
ISBN 978-7-5611-9442-3

Ⅰ. ①数… Ⅱ. ①王… ②周… Ⅲ. ①高等数学－高等学校－教材②统计学－高等学校－教材 Ⅳ. ①O13 ②C8

中国版本图书馆 CIP 数据核字(2014)第 180251 号

大连理工大学出版社出版

地址：大连市软件园路80号　邮政编码：116023
发行：0411-84708842　邮购：0411-84703636　传真：0411-84701466
E-mail：dutp@dutp.cn　URL：http://www.dutp.cn
大连业发印刷有限公司印制　大连理工大学出版社发行

幅面尺寸：183mm×233mm　印张：22.75　字数：740千字
2014年8月第1版　　　　　　　　2014年8月第1次印刷

责任编辑：邵　婉　　　　　　　　责任校对：齐　悦
封面设计：波　朗

ISBN 978-7-5611-9442-3　　　　　　定　价：40.00元

2014年的初夏,我们为广大师生奉上这套"任务型语码转换式双语教学系列教材"。

"任务型语码转换式双语教学"是双语教学内涵建设的成果,主要由两大模块构成:课上,以不影响学科授课进度为前提,根据学生实际、专业特点、学年变化及社会需求等,适时适量地渗透英语专业语汇、语句、语段或语篇,"润物细无声"般地扩大学生专业语汇量,提高学生专业英语能力;课外,可向学生提供多种选择的"用中学"平台,如英语科技文献翻译、英语实验报告、英语学术论文、英语小论文、英语课程设计报告、模拟国际研讨会、英语辩论、工作室英语讨论会等,使学生的专业英语实践及应用达到一定频度和数量,激活英语与学科知识的相互渗透,培养学生用英语学习、科研、工作的能力及适应教育国际化和经济一体化的能力。

为保证"任务型语码转换式双语教学"有计划、系统、高效、科学地持续运行,减少教学的随意性和盲目性,方便师生的教与学,我们编写了这套"任务型语码转换式双语教学系列教材"。

本套教材的全部内容均采用汉英双语编写。

教材按专业组册,涵盖所有主干专业课和专业基础课,力求较为全面地反映各学科领域的知识体系。

分册教材编写以中文版课程教材为单位,即一门课为分册教材的一章,每章内容以中文版教材章节为序,每门课以一本中文教材为蓝本,兼顾其他同类教材内容,蓝本教材绝大部分是面向21世纪的国家规划教材。

教材的词汇短语部分,注意体现学科发展的新词、新语,同时考虑课程需求及专业特点,在不同程度灵活渗透了各章节的重要概念、定义,概述了体现章节内容主旨的语句及语段。分册教材还编写了体现各自专业特点的渗透内容,如例题及解题方法,课程的发生、发展及前沿简介,图示,实验原理,合同文本,案例分析,法条,计算机操作错误提示等。

部分教材补充了中文教材未能体现的先进理论、先进工艺、先进材料或先进方法的核心内容,弥补了某些中文教材内容相对滞后的不足;部分教材概述了各自专业常用研究方法、最新研究成果及学术发展的趋势动态;部分教材还选择性地把编者的部分科研成果转化为教材内容，以期启发学生的创新思维，开阔学生的视野，丰富学生的知识结构，从教材角度支持学生参与科研活动。

本套教材大多数分册都编写了对"用中学"任务实施具有指导性的内容，应用性内容的设计及编写比例因专业而异。与专业紧密结合的应用性内容包括英语写作介绍，如英语实验报告写作，英语论文写作，英语论文摘要写作，英语产品、作品或项目的概要介绍写作等。应用性内容的编写旨在降低学生参与各种实践应用活动的难度，提高学生参与"用中学"活动的可实现性，帮助学生提高完成"用中学"任务的质量水平。

考虑学生英语写作和汉译英的方便，多数分册教材都编写了词汇与短语索引。

"任务型语码转换式双语教学系列教材"尚属尝试性首创，是多人辛勤耐心劳作的结果。尽管在编写过程中，我们一边使用一边修改，力求教材的实用性、知识性、先进性融为一体，希望教材能对学生专业语汇积累及专业资料阅读、英语写作、英汉互译能力的提高发挥作用；尽管编者在教材编写的同时也都在实践"任务型语码转换式双语教学"，但由于我们缺乏经验，学识水平和占有资料有限，加上为使学生尽早使用教材，编写时间仓促，在教材内容编写、译文处理、分类体系等方面存在缺点、疏忽和失误，恳请各方专家和广大师生对本套教材提出批评和建议，以期再版时更加完善。

在教材的编写过程中，大量中外出版物中的内容给了我们重要启示和权威性的参考帮助，在此，我们谨向有关资料的编著者致以诚挚的谢意！

编 者

2014 年 5 月

随着科学技术的飞速发展,数学不仅被广泛深入地应用于自然科学、信息技术和工程技术,而且已渗透到诸如生命科学、社会科学、环境科学、军事科学、经济科学等领域,它已成为表达严格科学思想的媒介,人们越来越深刻地认识到,没有数学就难以取得当代的科学成就。正是由于自然科学各学科数学化的趋势以及社会科学各部门定量化的要求,许多学科都或直接或间接、或先或后地经历着数学化的进程。现在已经没有哪一领域能够抵御得住数学的渗透,这正诠释了马克思所说"一门科学只有当它达到能够成功地运用数学时,才算真正发展了"的精辟论述。所以在科学王国中,数学占有极为特殊的地位,并作为一门独立科学存在于世。它既是一门专业领域,又是基础(思维)工具;既是语言,又是文化;既能与经管科学交叉,又能与理工结合,且能向文科渗透。

数学的这种特殊位置和应用的广泛性,加之英语作为信息交流的一种重要工具,确定了数学的语言英文表达有着极为重要的意义,它已成为科学技术交流和传播的重要基础工具之一。数学教学与外语的有机结合,有利于学生综合素质的全面提高,顺应时代发展方向。

因此,编写适合双语教学的,同时又与国内数学课程内容相适应的教材已势在必行。

目前,双语教学的教学模式基本有两方面的选择。关于教材,或直接采用原版教材,或采用中文版教材,加外语补充材料。关于授课,则采用全外语授课,或部分外语授课,或在使用原版教材的基础上采用全中文授课。各高校大多根据学生的外语水平及教师的外语特长在上述几种情况中选择。近年来,学生的外语水平有了明显的提高,师资的外语及专业能力也有了本质上的变化。因此,双语教学的模式也面临真正意义上的提升。

数学课程实施双语教学的目的在于提高数学教育教学质量。通过双语教学,学生可以学习国外先进的学科体系、教学理念和丰富的数学逻辑内涵以及数学在其他学科领域中的基本应用。本教材引文内容以高等学校数学课程教学大纲为依据,参考多种国外原版教材,以使语言表述准确、地道;书中内容涵盖了数学课程中的主要数学概念、常见的专业词汇等,前面加*的词汇表示重点词汇,必须掌握。

编写本书的直接目的是为讲授数学课程的广大教师和学习数学课程的学生提供掌

握相关内容的英文描述服务,进而使得学生学过本课程后,能够独立阅读相关的英文教材和文献,进而提升英语科技文献翻译以及使用英语撰写英语实验报告、英语课程设计报告以及英语学术论文的实际能力。它既可作为学生学习数学课程的配套教材和扩大知识领域的参考书,也可作为数学专业英语词汇查找的工具书。

参与本书编写的人员有(按姓氏拼音为序):楚振艳、丁淑妍、董莹、葛仁东、焦佳、李阳、刘恒、刘力军、刘满、李秀文、卢静、吕娜、马玉梅、牛大田、齐淑华、曲程远、孙雪莲、藤颖俏、王金芝、王书臣、徐毅、余军、张友、张誉铎。

大连民族学院开展"任务型语码转换式双语教学"已有多年,特别是结合本校师资及学生特点施行的"渗透式双语教学"工作已获得国家级教学成果二等奖。这种双语教学模式的特点是:根据不影响学生课程授课进度为前提,结合学生实际、专业特点、学年变化及社会需求等,适时适量地渗透英语专业词汇、语句、语段或语篇,"润物细无声"般地扩大学生专业词汇量,提高学生专业英语能力。本书是我们进行双语教学的又一次有益尝试,也是一个新的阶段总结,恳请各位专家、同行以及广大读者提出宝贵的建议和意见,我们会一直努力!

编者
2014 年 7 月

数学基础课程 /1

第一部分　数学分析 /1

引 言 / 1
第一章　实数集与函数 / 3
第二章　数列极限 / 5
例 题 / 6
第三章　函数极限 / 10
例 题 / 12
第四章　函数的连续性 / 13
例 题 / 14
第五章　导数与微分 / 15
例 题 / 17
第六章　微分中值定理及其应用 / 18
例 题 / 19
第七章　实数的完备性 / 20
例 题 / 21
第八章　不定积分 / 21
第九章　定积分 / 22
例 题 / 23
第十章　定积分的应用 / 25
第十一章　反常积分 / 25
例 题 / 26
第十二章　数项级数 / 27
例 题 / 28
第十三章　函数列与函数项级数 / 30
例 题 / 31
第十四章　幂级数 / 32
例 题 / 33
第十五章　傅里叶级数 / 35
例 题 / 35
第十六章　多元函数的极限与连续 / 36
例 题 / 37
第十七章　多元函数微分学 / 38
例 题 / 39
第十八章　隐函数定理及其应用 / 41
第十九章　含参量积分 / 42
例 题 / 42
第二十章　重积分 / 45
例 题 / 45
第二十一章　曲线积分 / 47
例 题 / 48
第二十二章　曲面积分 / 49
例 题 / 49
练 习 / 51

第二部分　高等代数 / 57

引 言 / 57
第一章　多项式 / 57
第二章　行列式 / 58
例 题 / 58
第三章　线性方程组 / 59
例 题 / 59
第四章　矩 阵 / 62
例 题 / 63
第五章　二次型 / 64
第六章　线性空间 / 65
第七章　线性变换 / 65
第八章　λ- 矩阵 / 66
第九章　欧几里得空间 / 67
例 题 / 67
第十章　双线性函数 / 68
练 习 / 68

第三部分　解析几何 / 71

引 言 / 71
第一章　矢量与坐标 / 71
例 题 / 73
第二章　轨迹与方程 / 74
例 题 / 75
第三章　平面与空间直线 / 76
例 题 / 77
第四章　柱面、锥面、旋转曲面与二次曲面 / 78
例 题 / 79
第五章　二次曲线的一般理论 / 81
例 题 / 82
第六章　二次曲面的一般理论 / 82
例 题 / 83
练 习 / 84

数学与应用数学专业课程 / 85

第一部分　数学模型 / 85

引 言 / 85
第一章　数学模型概论 / 86
第二章　最优化模型 / 87
例 题 / 88
第三章　微分方程模型 / 90
例 题 / 91

第四章　概率统计模型 / 93
例　题 / 94
第五章　基本算法工具包 / 96
第六章　最新算法 / 97
第七章　数学建模竞赛 / 97

>> 第二部分　复变函数与积分变换 / 98
引　言 / 98
第一章　复数与复变函数 / 99
例　题 / 100
第二章　解析函数 / 101
例　题 / 101
第三章　复变函数的积分 / 103
例　题 / 104
第四章　解析函数的级数表示 / 104
例　题 / 105
第五章　留数及其应用 / 107
例　题 / 108
第六章　共形映射 / 109
第七章　傅里叶变换 / 109
第八章　拉普拉斯变换 / 110
练　习 / 111

>> 第三部分　运筹与优化 / 116
引　言 / 116
第一章　线性规划模型 / 117
例　题 / 118
第二章　线性规划的解法 / 120
例　题 / 120
第三章　对偶理论与灵敏度分析 / 122
例　题 / 123
第四章　运输问题及其解法 / 124
例　题 / 126
第五章　多目标规划 / 127
例　题 / 129
第六章　整数规划 / 130
例　题 / 132
第七章　动态规划 / 134
例　题 / 135
第八章　图与网络分析 / 136
例　题 / 138
第九章　存储论 / 140
例　题 / 141
第十章　排队论 / 142
例　题 / 144
第十一章　对策论 / 145
例　题 / 148
第十二章　决策分析 / 148
例　题 / 150

>> 第四部分　常微分方程 / 152
引　言 / 152
第一章　基本概念 / 153
例　题 / 153
第二章　初等积分法 / 154
例　题 / 156
第三章　存在和唯一性定理 / 157
例　题 / 158
第四章　奇解 / 159
例　题 / 160
第五章　高阶微分方程 / 161
第六章　线性微分方程组 / 162
例　题 / 163
第七章　微分方程的幂级数解法 / 164
第八章　定性理论与分支理论 / 165
第九章　边值问题 / 166
练　习 / 167

>> 第五部分　实变函数与泛函分析 / 169
引　言 / 169
第一章　预备知识 / 170
例　题 / 171
第二章　点集的拓扑概念 / 172
例　题 / 172
第三章　测度论 / 173
第四章　可测函数 / 174
例　题 / 174
第五章　积分理论 / 175
例　题 / 175
第六章　抽象空间论 / 178
例　题 / 178
第七章　抽象空间之间的映射 / 179
例　题 / 180
练　习 / 181

>> 信息与计算科学专业课程 / 184

>> 第一部分　数值分析 / 184
引　言 / 184
第一章　数值计算中的误差分析 / 185
第二章　多项式插值方法 / 186
第三章　样条插值 / 188
第四章　最佳逼近 / 189
第五章　数值微分与数值积分 / 191
第六章　常微分方程(组)数值解 / 192

第七章　偏微分方程(组)数值解 / 193
练　习 / 194

>> 第二部分　信息论 / 197

引　言 / 197
第一章　介绍 / 198
第二章　信息理论 / 199
第三章　离散无记忆信道和容量成本方程 / 200
第四章　离散无记忆信源和扭曲率方程 / 200
第五章　高斯信道和信源 / 201
第六章　信源-信道编码理论 / 201
练　习 / 202

>> 第三部分　数据结构与算法 / 206

引　言 / 206
第一章　线性表 / 206
第二章　栈和队列 / 207
第三章　树和二叉树 / 207
第四章　图 / 208
第五章　内排序 / 209
第六章　查找 / 209
练　习 / 209

>> 第四部分　离散数学 / 211

引　言 / 211
第一章　命题逻辑 / 214
例　题 / 215
第二章　一阶逻辑 / 216
例　题 / 218
第三章　关系 / 219
例　题 / 220
第四章　函数 / 221
例　题 / 222
第五章　图论 / 223
例　题 / 226
第六章　树及其应用 / 227
例　题 / 228
第七章　代数系统 / 229
例　题 / 232

>> 第五部分　计算机组成原理 / 234

引　言 / 234
第一章　计算机系统概论 / 234
例　题 / 235
第二章　运算方法和运算器 / 235
例　题 / 236
第三章　存储系统 / 237
例　题 / 238
第四章　指令系统 / 239

例　题 / 239
第五章　中央处理器 / 241
例　题 / 241
第六章　系统总线 / 242
例　题 / 242
第七章　外围设备 / 244
例　题 / 244
第八章　输入输出系统 / 245
例　题 / 246

>> 保险精算专业课程 / 248

>> 第一部分　风险理论 / 248

第一章　效应理论与保险 / 248
练　习 / 249
第二章　个体风险模型 / 249
练　习 / 250
第三章　聚合风险模型 / 250
练　习 / 251
第四章　破产理论 / 252
练　习 / 253

>> 第二部分　复利数学 / 255

引　言 / 255
第一章　利息的基本概念 / 255
第二章　年金 / 257
例　题 / 257
第三章　收益率 / 258
例　题 / 258
第四章　债务偿还 / 260
例　题 / 260
第五章　债券及其定价理论 / 262
例　题 / 262
第六章　复利数学的应用 / 264
例　题 / 264
第七章　金融分析 / 265
练　习 / 266

>> 第三部分　寿险精算实务 / 268

引　言 / 268
第一章　人寿保险的主要类型 / 268
第二章　保单现金价值与红利 / 269
第三章　特殊年金与保险 / 269
第四章　寿险定价概述 / 270
第五章　资产份额定价法 / 270
第六章　资产份额法的进一步分析 / 271
第七章　准备金评估 I / 272
第八章　准备金评估 II / 272

第九章 寿险公司内含价值 / 273
第十章 偿付能力监管 / 274
第十一章 养老金概述 / 274
第十二章 养老金数理及实例 / 275
练习 / 276

》第四部分 多元统计分析 / 279
第一章 矩阵代数基本知识 / 279
第二章 统计分析 / 280
例题 / 281
第三章 主成分分析 / 283
例题 / 283
第四章 因子分析 / 287
例题 / 289
第五章 聚类分析 / 292
例题 / 294
第六章 判别分析 / 296
例题 / 298

》第五部分 概率论与数理统计 / 299
引言 / 299
第一章 概率论的基本概念 / 299
例题 / 300
第二章 随机变量及其分布 / 302
例题 / 303
第三章 多维随机变量及其分布 / 304
例题 / 305
第四章 随机变量的数字特征 / 306
例题 / 306
第五章 大数定律及中心极限定理 / 307
例题 / 308
第六章 样本及抽样分布 / 308
第七章 参数估计 / 309
例题 / 310
第八章 假设检验 / 311
例题 / 311
练习 / 312

》第六部分 统计学原理 / 313
引言 / 313
第一章 绪论 / 313
第二章 统计设计和统计调查 / 314
第三章 统计整理 / 315
第四章 总量指标和相对指标 / 315
第五章 平均指标和变异指标 / 316
第六章 动态数列 / 317
第七章 统计指数 / 317
例题 / 318

第八章 抽样调查 / 319
例题 / 320
第九章 相关与回归分析 / 320
例题 / 321

》第七部分 生命表基础 / 323
引言 / 323

》第一篇 生存模型及其应用 / 323
第一章 生存模型及其性质 / 323
例题 / 324
第二章 生命表 / 325
例题 / 325
第三章 完整样本数据情况下表格生存模型的估计 / 326
第四章 非完整样本数据情况下表格生存模型的估计 / 327
例题 / 328
第五章 参数生存模型的估计 / 328
例题 / 329
第六章 大样本数据下年龄的处理及暴露数的计算 / 331
例题 / 332

》第二篇 人口统计 / 333
第七章 死亡和生育测度 / 333
例题 / 333
第八章 人口模型 / 334
例题 / 335
第九章 人口规划及人口普查应用 / 335
例题 / 336

》第三篇 人口统计 / 337
第十章 表格数据修匀 / 337
例题 / 338
第十一章 参数修匀 / 340
例题 / 340

》第八部分 抽样调查 / 342
引言 / 342
第一章 简单随机抽样 / 343
第二章 分层抽样 / 344
第三章 系统抽样 / 346
第四章 整群抽样 / 347
第五章 多阶段抽样 / 348
第六章 不等概率抽样 / 349
第七章 二重抽样 / 350

》参考文献 / 352

数学基础课程

第一部分　数学分析
Part 1　Mathematical Analysis

引 言

数学分析的萌芽、发生与发展，经历了一个漫长的时期．萌芽时期是从古希腊数学家欧多克索斯（Eudoxus，约公元前408～355）提出穷竭法和阿基米德（Archimedes，公元前287～212）用穷竭法求出抛物线弓形的面积开始．公元263年，刘徽为《九章算术》作注时提出"割圆术"，以及1328年英国大主教布雷德沃丁（Bradwardine，1290～1349）在牛津发表著作中给出类似于均匀变化率和非均匀变化率的概念，这些都是极限思想的成功运用．到16世纪中叶，数学分析正式进入了酝酿阶段，其中有两部著作在当时有很大的影响：一是德国数学家开普勒（Kepler，1571～1630）的《新空间几何》；另一部是意大利数学家卡瓦列里（Cavalieri，1598～1647）的《不可分量几何》．

17世纪上半叶开始到中叶是数学分析的奠基性工作阶段．主要先驱有法国的帕斯卡（Pascal，1623～1662）和费马（Fermat，1601～1665），英国的沃利斯（Wallis，1616～1703）和巴罗（Barrow，1630～1677）．

17世纪下半叶，牛顿（Newton，1642～1727）和莱布尼茨（Leibniz，1646～1716）在总结前人工作的基础上分别独立地给出了微积分的概念．微积分诞生以后，曾就它是否严密及基础是否稳固爆发过一场大的争论，为此有许多数学家企图弥补出现的不严密性，如英国数学家麦克劳林（Maclaurin，1698～1746）、泰勒（Taylor，1685～1731），法国数学家达朗贝尔（D'Alembert，1717～1783）等，其中，达朗贝尔曾试图将微积分的基础归结为极限，但遗憾的是，他并未沿着这条路走到底．

与此同时，许多数学家在不严密的基础上对微积分创立了许多辉煌的成就：如瑞士数学家欧拉（Euler，1707～1783）以微积分为工具解决了大量的天文、物理、力学等问题，开创了微分方程、无穷级数、变分学等诸多新学科．1748年他出版了《无穷小分析引论》——世界上第一本完整的有系统的分析学用书．还有法国数学家拉格朗日（Lagrange，1736～1813）、拉普拉斯（Laplace，1749～1827）、勒让德（Legendre，1752～1833）、傅里叶（Fourier，1768～1830）等在分析学方面都作了重大的贡献，但在微积分基础上仍没有找到解决的办法．

进入19世纪以后，分析学的不严密性到了非解决不可的地步！但那时还没有变量和极限的严格定义，不知道什么是连续，不知道什么是级数的收敛性．定积分的存在性都是含糊不清的，这可从挪威数学家阿贝尔（Abel，1802～1829）在1826年所说的"在高等分析中仅有很少几个定理是用逻辑上站得住脚的形式证明，人们到处发现从特殊跳到一般的不可靠的推论方法"这句话中看出．为了解决分析的严密性问题，捷克数学家波尔查诺（Bolzano，1781～1848），挪威的阿贝尔和法国的柯西（Cauchy，1789～1857）作了大量的工作．1821年，法国理工大学教授柯西写了《分析教程》一书，将分析学奠定在极限的概念之上，把纷乱的概念理出了一个头绪．但是他的叙述仍然使用"无限趋近"之类的语言，仍不是严格的．因此遭到了一些数学家的反对，德国数学家魏尔斯特拉斯（Weierstrass，1815～1897）就是其中之一，他认为变量无非是一个字母，用来表示区间的数．这一想法导致了变量x在$(x_0-\delta, x_0+\delta)$取值时，$f(x)$在$(f(x_0)-\varepsilon, f(x_0)+\varepsilon)$取值的新方法．由此得到了如今广泛使用的"$\varepsilon\delta$"语言．

因为分析学使用的工具是极限，而极限又要用到实数．因此，分析学的严密性是建立在实数理论基础上的．而在这方面，法国数学家柯西、梅雷（Méray，1835～1911）以及德国数学家海涅（Heine，1821～1881）、康托（Cantor，1845～1918）、戴德金（Dedekind，1831～1916）等都为建立实数理论作出了贡献．

19世纪后半叶，数学分析在理论上有了很大进展，1870年海涅提出了一致连续的概念；1895年法国数学家波莱尔（Borel，1871～1956）给出了有限覆盖定理；1872年魏尔斯特拉斯给出了处处连续而不可微的例子；德国数学家黎曼（Riemann，1826～1866）和法国数学家达布（Darboux，1842～1917）分别于 1854 年和 1885 年给出了有界函数可积性的定义和充要条件．这些概念和例子构成了现今数学分析教科书的主要内容．现在，数学分析已根植于自然科学和社会科学的各学科分支之中．微积分作为数学分析的基础，不仅要为全部数学方法和算法工具提供方法论，同时还要为人们灌输逻辑思维方法．目前数学分析的主要内容已是高校数学专业必修课和理工管等学科的重要基础课之一．

数学分析已形成四大块结构：分析引论、微分学、积分学、无穷级数与广义积分．数学分析的立论数域是实数连续统，研究的主要对象是函数，研究问题使用的主要工具是极限．

Introduction

The germination, appearance and development of Mathematical Analysis went through a long period. The germination period started from the method of exhaustion put forward by the ancient Greek mathematician Eudoxus (about 408～355 BC) and Archimedes (about 287～212 BC) who worked out the area of parabolic arch. The idea of limits is well put into practice, such as, in 263 BC, Liu Hui raised "Cyclotomic Method" as he glossed for a book named *Nine Chapters of Arithmetic*; In 1328, the British archbishop Bradwardine (1290～1349) gave the definition for homogeneous rate of change and non-homogeneous of change in his book published in Oxford. By the middle of the 16th century, the preparing period of Mathematical Analysis really started. Two famous works made great influence at that time. One was *New Space Geometry* by the German mathematician Kepler (1571～1630), another was *Geometria Indivisibilibus Continuorum Nova Quadam Ratione Promota* by the Italian mathematician Cavalieri (1598～1647).

Great foundation of Mathematical Analysis had been laid from the early 17th century to the middle of 17th century. Among the pioneers were Pascal (1623～1662) and Fermat (1601～1665) from France, Wallis (1616～1703) and Barrow (1630～1677) from UK.

In the late 17th century, Newton (1642～1727) and Leibniz (1646～1716) founded Calculus based on the works of early mathematicians. Right after its birth, there was a heated debate over whether it was logically strict and fundamentally stable. Consequently, many mathematicians tried to remedy its loose foundation, among whom were Maclaurin (1698～1746) and Taylor (1685～1731) from UK, D'Alembert (1717～1783) from France. In particular, D'Alembert once tried to define the base of calculus to limit, but to our regret, abandoned the idea halfway.

Meanwhile, many mathematicians had made great achievement on the loose Calculus. For example, the Switzerland mathematician Euler (1707～1783), by using Calculus as a tool, solved many problems in the fields of astronomy, physics and mechanics, and also founded many new subjects such as differential equations, infinite series and calculus of variations. And the first systematically integrated book on analysis, *Introductio in Analysin Infinitorum*, was published in 1748. Moreover, Lagrange (1736～1813), Laplace (1749～1827), Legendre (1752～1833), Fourier (1768～1830) also contributed a lot to Mathematical Analysis. But no efficient solution to the loose base of Mathematical Analysis had been found.

Stepping into the 19th century, the loose foundation of Mathematical Analysis came up to the degree that had to be solved. But there were no strict definition for variable and limits. Terms such as continuity and the convergence of series were unknown. The existence of definite integral was still ambiguous, which could be seen from the statement of the Norwegian mathematician Abel (1802～1829) in 1826, "Only few proofs of the theorems in advanced analysis can logically hold water. Unreliable reasoning methods drawing conclusions of general cases from special ones can be found everywhere". In order to solve the loose foundation of Mathematical Analysis, the Czechic mathematician Bolzano (1781～1848), the Norway mathematician Abel and the French mathematician Cauchy(1789～1857) did great amount of work. In 1821, Prof. Cauchy, the Science and Engineering university of France, wrote the book *Analysis Course*, in which Mathematical Analysis was defined on the concept of limit, and thus got a major line out of the disorderly numerous concepts. But the language was still not strict enough to avoid the expressions such as "approach infinitely", thus met the opposition of some mathematicians, among whom was the German mathematician Weierstrass (1815~1897) who believed that the variable was not more than a letter, which is used to represent the number in an interval. This idea resulted in the new method that if x belongs to the interval $(x_0 - \varepsilon, x_0 + \varepsilon)$, then $f(x)$ must be a number of the interval $(f(x_0) - \varepsilon, f(x_0) + \varepsilon)$. Hence today's widely used "$\varepsilon\text{-}\delta$" language came into being.

Since the tool of Mathematical Analysis is limit, which is related to real numbers, the strictness of Mathematical Analysis is based on the real number theory. In this aspect, the French mathematician Cauchy, Méray (1835~1911), the German mathematician Heine (1821~1881), Cantor (1845~1918) and Dedekind (1831~1916) all made great contribution to the foundation of real number theory.

In the late 19th century, the theoretical developments of Mathematical Analysis are very rapid: In 1870, Heine put up the concept of uniform continuity. In 1895, Borel (1871~1956) gave the theorem of finite covering. In 1872, Weierstrass gave a function which is continuous at every point but not differentiable. In 1854 and 1885, the German mathematician Riemann (1826~1866) and the French mathematician Darboux (1842~1917) respectively gave the definition of bounded function, integrability and its necessary and sufficient conditions. All of these made up the major contents of Mathematical Analysis nowadays. At present, Mathematical Analysis is rooted in different subjects of natural science and social science. Calculus, as the base of Mathematical Analysis, not only supplies all mathematical methods and algorithms with methodology, but also cultivates people's thinking mode. Presently, the major contents of Mathematical Analysis have already become the compulsory course for math majors, and the selective course for science, engineering and management majors.

Mathematical Analysis has formed a structure composed of four major parts: analysis theory, differentials, integrals, infinite series and generalized integrals. Mathematical Analysis is founded on the continuum of real numbers, and the subject for study in Mathematical Analysis is functions. The major research tool in Mathematical Analysis is limits.

第一章 实数集与函数
Chapter 1 Set of Real Numbers and Functions

初等数学中研究的主要对象基本上是常量,而在数学分析中我们研究的是变量.变量的变化范围是实数集.变量之间的对应关系是函数.本章的内容主要包括实数、函数、复合函数、初等函数的基本概念及它们的一些性质.

The main object investigated in Elementary Mathematics is constant quantities, while it is variables that we investigate in Mathematical Analysis. The changeable domain of a variable is a set of real numbers and the correspondent relation between variables is called a function. The contents of this chapter mainly include some fundamental concepts such as real numbers, functions, composite functions, elementary functions and their properties.

单词和短语 Words and expressions

实数及其性质	real number and its properties	三角不等式	triangle inequality
有理数	rational number	反三角不等式	inverse triangle inequality
无理数	irrational number	伯努利不等式	Bernoulli inequality
定义	definition [defiˈniʃən]	确界原理	principles of supremum and infimum
命题	proposition [prɔpəˈziʃən]	开区间	open interval
加	plus	闭区间	closed interval
减	minus	半开区间	semi-open interval
乘	multiplied by / times	半闭区间	semi-closed interval
除	over / is to / divided by	有限区间	finite interval
绝对值与不等式	absolute value and inequality	无限区间	infinite interval

邻域	neighborhood
去心邻域	deleted neighborhood
和	sum
差	difference
积	product ['prɔdəkt]
商	quotient ['kwəuʃnt]
数轴	number axis / number line
封闭性	closeness
阿基米德性质	Archimedean property
稠密性	density
上界与下界	upper and lower bounds
有界集	bounded set
无界集	unbounded set
存在域	existence domain
上确界	supremum
下确界	infimum
有序完备集	order-complete set
实数的完备性	completeness of real numbers
全序域	complete ordered field
完备性公理	axiom of completeness
戴德金分割	Dedekind cut
戴德金性质	Dedekind property
常量与变量	constant and variable quantities
函数的定义	definition of function
定义域	domain
值域	range
自变量	independent variable
因变量	dependent variable
中间变量	intermediate variable
单调性	monotonicity
初等函数	elementary function
常量函数	constant function
幂函数	power function
指数函数	exponential function
对数函数	logarithmic function
三角函数	trigonometric function
反三角函数	inverse trigonometric function
反函数	inverse function
复合函数	compound function
映射	mapping
逆映射	inverse mapping
像	image
原像	primary image
分段函数	piecewise function
符号函数	sign function
狄利克雷函数	Dirichlet function
黎曼函数	Riemann function
有界函数	bounded function
单调函数	monotone function
单调增函数	monotone increasing function
严格单调函数	strictly monotone function
奇(偶)函数	odd (even) function
周期函数	periodic function
最小正周期	minimal positive period
绝对值函数	absolute value function
恒等函数	identity function
多项式函数	polynomial function
线性函数	linear function
二次函数	quadratic function
有理函数	rational function
双曲正弦	hyperbolic sine
双曲余弦	hyperbolic cosine
三角恒等式	trigonometric identity
奇偶恒等式	odd-even identity
余函数恒等式	cofunction identity
毕达哥拉斯恒等式	Pythagorean identity
半角恒等式	half-angle identity
积恒等式	product identity
和恒等式	sum identity
加法恒等式	addition identity
倍角恒等式	double-angle identity

基本概念和性质 Basic concepts and properties

1 非空实数集 S 称为有上界(下界),如果存在数 $M(L)$,使得对一切 $x \in S$,都有 $x \leqslant M (x \geqslant L)$. 数 $M(L)$ 称为 S 的一个上界(下界).

A nonempty set S of real numbers is said to have upper (lower) bound provided that there is a number $M(L)$ having the property that $x \leqslant M(x \geqslant L)$ for all x in S. Such number $M(L)$ is called an upper bound (a lower bound) for S.

2 集称为有界的,如果集既有上界又有下界.

A set is said to be bounded if it has not only upper bound but also lower bound.

3 如果集 S 的所有上界集合有最小元 M,则 M 称为集 S 的上确界(或最小上界).

If the set of all upper bounds of a set S has the smallest number M, then M is called the supremum (or the least upper bound) of S.

4 集 S 的上确界 M 有下列两个性质：(i) M 是集 S 的上界，即对任意 $x \in S$，有 $x \leqslant M$；(ii) 没有比 M 小的数是 S 的上界，即 $\forall \varepsilon > 0, \exists y \in S$，使得 $y > M - \varepsilon$.

The supremum M of a set S has the following two properties: (i) M is an upper bound of S, i. e., for any $x \in S$, we have $x \leqslant M$; (ii) No numbers less than M can be an upper bound, i. e., for any positive number ε, there exists a number $y \in S$, such that $y > M - \varepsilon$.

本章重点

因为在数学分析中一元函数微积分讨论问题的范围是实数。而函数是数学分析研究的主要对象。所以对于实数与函数必须掌握如下几点：

1 为什么要学习实数？
2 为什么要引入确界的概念？
3 为什么要学习绝对值不等式？
4 何谓函数？怎样确定函数的定义域？何谓函数的值域？
5 映射与函数的区别是什么？
6 何谓初等函数？
7 掌握复合函数概念，会将复合函数"分解"为基本的初等函数.

Key points of this chapter

Because the scope for problems discussed in unary calculus is real numbers, but functions are the main object in Mathematical Analysis, and thus the following points must be mastered for real numbers and functions.

1 Why are real numbers studied?
2 Why are the notions of supremum and infimum introduced?
3 Why is the absolute value inequality studied?
4 What is a function? How to determine the domain of a function? What is the region of a function?
5 What is the difference between mapping and function?
6 What is an elementary function?
7 Master the notion of composite functions and the method to "decompose" composite functions into basic elementary functions.

第二章　数列极限
Chapter 2　Limits of Sequences

数学分析中研究问题的主要工具是极限，而实数列是最简单也是最重要的函数之一．事实上，一般函数的许多性质都能由所了解的数列得到．所以本章的内容包括实数列的极限、收敛数列的性质、收敛数列的运算法则、数列极限存在的判别准则等．

The major research tool in Mathematical Analysis is limit, while sequences of real numbers are the most simple, but one of the most important functions. In fact, many properties of general functions can be deduced from the understanding of sequences. Accordingly, the contents of this chapter include limits of sequences of real numbers, properties of convergent sequences, operational rules of convergent sequences, existence criteria of limits of sequences, and so on.

单词和短语 Words and expressions

★ 数列极限　limit of sequence
发散数列　divergent sequence
无穷小数列　infinitesimal sequence
收敛数列　convergent sequence

★ 唯一性定理　uniqueness theorem
★ 有界性定理　boundedness theorem
保序性　inheriting order properties
保不等式性　inheriting inequality

子列　subsequence
严格递增　strictly increasing
单调递增数列　monotone increasing sequence
单调递减数列　monotone decreasing sequence
严格递减　strictly decreasing

必要条件　necessary condition
充分条件　sufficient condition
★ 夹逼定理　squeeze principle
★ 柯西收敛准则　Cauchy convergence criterion

基本概念和性质　Basic concepts and properties

1 收敛数列的和的极限等于极限的和.
The limit of the sum of convergent sequences is equal to the sum of the limits.

2 收敛数列的积的极限等于极限的积.
The limit of the product of convergent sequences is equal to the product of the limits.

3 收敛数列的商的极限等于极限的商.
The limit of the quotient of convergent sequences is equal to the quotient of the limits.

4 定义域为全体自然数集,值域为实数集的函数称为实数列,记为
$$f:N \to R \quad 或 \quad f(n), n \in N.$$
A function whose domain is the set of natural numbers and range is a set of real numbers is called a real sequence. Thus a real sequence is denoted symbolically by
$$f:N \to R \quad or \quad f(n), n \in N.$$

5 设 $\{a_n\}$ 为数列, a 为常数. 若对任给的正数 $\varepsilon > 0$, 总存在正整数 N, 使得当 $n > N$ 时, 有 $|a_n - a| < \varepsilon$, 则称数列 $\{a_n\}$ 收敛于 a, 常数 a 称为数列 $\{a_n\}$ 的极限, 并记作
$$\lim_{n \to \infty} a_n = a \quad 或 \quad a_n \to a(n \to \infty).$$
Let $\{a_n\}$ be a sequence and a be a constant, $\{a_n\}$ is said to be convergent to a and a is called the limit of $\{a_n\}$ if for any $\varepsilon > 0$, there exists a positive integer N, such that $|a_n - a| < \varepsilon$ for all $n > N$, and the limit is denoted by
$$\lim_{n \to \infty} a_n = a \text{ or } a_n \to a(n \to \infty).$$

6 单调数列收敛当且仅当数列有界.
A monotone sequence is convergent if and only if it is bounded.

例 题　Examples

例 1 用数列收敛的定义证明下列极限:

(1) $\lim\limits_{n \to \infty} q^n = 0, |q| < 1$;   (2) $\lim\limits_{n \to \infty} \dfrac{1+2+3+\cdots+n}{n^3} = 0.$

证明 (1) 若 $q = 0$, 则结果是显然的. 令 $\varepsilon > 0$, 我们要寻找一个自然数 N, 使得当 $n > N$ 时, 有
$$|q^n - 0| < \varepsilon.$$
现设 $0 < |q| < 1$. 记 $h = 1/|q| - 1$, 则有 $|q| = 1/(1+h)$ 和 $h > 0$. 因此, 对每个自然数 n, 由 $(1+h)^n \geq 1 + nh$ 得到, $|q^n| \leq \dfrac{1}{1+nh} < \dfrac{1}{nh}$.

选取自然数 N 使得 $N = 1/\varepsilon h$. 因而对所有的 $n > N$, 有
$$|q^n - 0| = |q^n| < \dfrac{1}{nh} < \dfrac{1}{Nh} < \varepsilon.$$

(2) 易证 $\dfrac{1+2+3+\cdots+n}{n^3} = \dfrac{n(n+1)}{2n^3} = \dfrac{n+1}{2n^2} < \dfrac{n+n}{2n^2} = \dfrac{1}{n}.$ 于是 $\forall \varepsilon > 0$, 取 $N = \dfrac{1}{\varepsilon}, \forall n > N$, 必有 $\left|\dfrac{1+2+3+\cdots+n}{n^3} - 0\right| < \dfrac{1}{n} < \varepsilon$, 所以
$$\lim_{n \to \infty} \dfrac{1+2+3+\cdots+n}{n^3} = 0.$$

Ex. 1 Prove the following limits by using the definition of convergent sequence:

(1) $\lim\limits_{n \to \infty} q^n = 0, |q| < 1$; (2) $\lim\limits_{n \to \infty} \dfrac{1+2+3+\cdots+n}{n^3} = 0$.

Proof (1) Obviously, if $q = 0$, the result is valid. Let $\varepsilon > 0$. We need to look for a natural number N such that $|q^n - 0| < \varepsilon$ is valid for all $n > N$.

Observe that since $0 < |q| < 1$, if we set $h = 1/|q| - 1$, it follows that $|q| = 1/(1+h)$ and $h > 0$. Hence, using the inequality $(1+h)^n \geqslant 1 + nh$, we have $|q^n| \leqslant \dfrac{1}{1+nh} < \dfrac{1}{nh}$ is valid for every natural number n.

We may choose a natural number N such that $N = 1/\varepsilon h$. Consequently, we have
$$|q^n - 0| = |q^n| < \frac{1}{nh} < \varepsilon \text{ is valid for all integers } n > N.$$

(2) It is easy to show that $\dfrac{1+2+3+\cdots+n}{n^3} = \dfrac{n(n+1)}{2n^3} = \dfrac{n+1}{2n^2} < \dfrac{n+n}{2n^2} = \dfrac{1}{n}$. Thus for any $\varepsilon > 0$, taking $N = \dfrac{1}{\varepsilon}$, we have that $\left|\dfrac{1+2+3+\cdots+n}{n^3} - 0\right| < \dfrac{1}{n} < \varepsilon$ as $n > N$. Consequently,
$$\lim_{n \to \infty} \frac{1+2+3+\cdots+n}{n^3} = 0.$$

例 2 设 $\lim\limits_{n \to \infty} a_n = a, \lim\limits_{n \to \infty} b_n = b$, 且 $a < b$. 证明:存在正数 N, 使得当 $n > N$ 时,有 $a_n < b_n$.

证明 由 $a < b$, 有 $a < \dfrac{a+b}{2} < b$. 因为 $\lim\limits_{n \to \infty} a_n = a < \dfrac{a+b}{2}$, 由保序性定理,存在 $N_1 > 0$, 使得当 $n > N_1$ 时有 $a_n < \dfrac{a+b}{2}$. 又因为 $\lim\limits_{n \to \infty} b_n = b > \dfrac{a+b}{2}$, 所以,又存在 $N_2 > 0$, 使得当 $n > N_2$ 时有 $b_n > \dfrac{a+b}{2}$. 于是取 $N = \max\{N_1, N_2\}$, 当 $n > N$ 时,有 $a_n < \dfrac{a+b}{2} < b_n$.

Ex. 2 If $\lim\limits_{n \to \infty} a_n = a$, $\lim\limits_{n \to \infty} b_n = b$, and $a < b$, then there exists a positive number N such that $a_n < b_n$ as $n > N$.

Proof From $a < b$, we obtain $a < \dfrac{a+b}{2} < b$, and thus $\lim\limits_{n \to \infty} a_n = a < \dfrac{a+b}{2}$. Using the theorem of inheriting order properties, we know that there exists $N_1 > 0$ such that $a_n < \dfrac{a+b}{2}$ as $n > N_1$. Similarly, from $\lim\limits_{n \to \infty} b_n = b > \dfrac{a+b}{2}$, we know that there exists $N_2 > 0$ such that $b_n > \dfrac{a+b}{2}$ as $n > N_2$. Taking $N = \max\{N_1, N_2\}$, we have
$$a_n < \frac{a+b}{2} < b_n \text{ as } n > N.$$

例 3 令 $a_n = 1 + \dfrac{1}{2^\alpha} + \dfrac{1}{3^\alpha} + \cdots + \dfrac{1}{n^\alpha}, n = 1, 2, \cdots$, 其中 $\alpha \geqslant 2$. 证明数列 $\{a_n\}$ 收敛.

证明 显然 $\{a_n\}$ 是单调递增的,下证 $\{a_n\}$ 有上界. 事实上,
$$a_n \leqslant 1 + \frac{1}{2^2} + \frac{1}{3^2} + \cdots + \frac{1}{n^2} \leqslant 1 + \frac{1}{1 \cdot 2} + \frac{1}{2 \cdot 3} + \cdots + \frac{1}{(n-1) \cdot n}$$
$$= 1 + \left(1 - \frac{1}{2}\right) + \left(\frac{1}{2} - \frac{1}{3}\right) + \cdots + \left(\frac{1}{n-1} - \frac{1}{n}\right)$$
$$= 2 - \frac{1}{n} < 2, n = 1, 2, \cdots.$$

于是由单调有界定理, $\{a_n\}$ 收敛.

Ex. 3 Let $a_n = 1 + \dfrac{1}{2^\alpha} + \dfrac{1}{3^\alpha} + \cdots + \dfrac{1}{n^\alpha}, n = 1, 2, \cdots$, where $\alpha \geqslant 2$. Prove that $\{a_n\}$ is convergent.

Proof Obviously, $\{a_n\}$ is a monotone increasing sequence. Next we prove $\{a_n\}$ has an upper bound. In fact,

$$a_n \leqslant 1 + \frac{1}{2^2} + \frac{1}{3^2} + \cdots + \frac{1}{n^2} \leqslant 1 + \frac{1}{1\cdot 2} + \frac{1}{2\cdot 3} + \cdots + \frac{1}{(n-1)\cdot n}$$

$$= 1 + \left(1 - \frac{1}{2}\right) + \left(\frac{1}{2} - \frac{1}{3}\right) + \cdots + \left(\frac{1}{n-1} - \frac{1}{n}\right)$$

$$= 2 - \frac{1}{n} < 2, \quad n = 1, 2, \cdots.$$

From the monotonic and bounded theorem, we know that $\{a_n\}$ is convergent.

例 4 (1) 设 $\lim\limits_{n\to\infty} a_n = a$,证明: $\lim\limits_{n\to\infty}\dfrac{a_1 + a_2 + \cdots + a_n}{n} = a$.

(2) 利用(1)的结论证明下列极限:

① $\lim\limits_{n\to\infty}\dfrac{1 + \frac{1}{2} + \frac{1}{3} + \cdots + \frac{1}{n}}{n} = 0$; ② $\lim\limits_{n\to\infty}\sqrt[n]{n} = 1$; ③ $\lim\limits_{n\to\infty}\dfrac{1}{\sqrt[n]{n!}} = 0$.

证明 (1) 因为 $\lim\limits_{n\to\infty} a_n = a$,于是有 $\forall \varepsilon > 0, \exists N_1 > 0, \forall n > N_1, |a_n - a| < \dfrac{\varepsilon}{2}$. 从而当 $n > N_1$ 时,有

$$\left|\frac{a_1 + a_2 + \cdots + a_n}{n} - a\right|$$

$$= \left|\frac{a_1 + a_2 + \cdots + a_n - na}{n}\right|$$

$$\leqslant \frac{|a_1 - a| + |a_2 - a| + \cdots + |a_{N_1} - a|}{n} + \frac{|a_{N_1+1} - a| + |a_{N_1+2} - a| + \cdots + |a_n - a|}{n}$$

$$\leqslant \frac{A}{n} + \frac{n - N_1}{n} \cdot \frac{\varepsilon}{2} \leqslant \frac{A}{n} + \frac{\varepsilon}{2},$$

其中 $A = |a_1 - a| + |a_2 - a| + \cdots + |a_{N_1} - a|$ 是一个定数. 再由 $\lim\limits_{n\to\infty}\dfrac{A}{n} = 0$,知存在 $N_2 > 0$,使得当 $n > N_2$ 时,$\dfrac{A}{n} < \dfrac{\varepsilon}{2}$. 因此取 $N = \max\{N_1, N_2\}$,当 $n > N$ 时,有

$$\left|\frac{a_1 + a_2 + \cdots + a_n}{n} - a\right| \leqslant \frac{A}{n} + \frac{\varepsilon}{2} < \frac{\varepsilon}{2} + \frac{\varepsilon}{2} = \varepsilon.$$

(2) ① 令 $a_n = \dfrac{1}{n}$,则 $\lim\limits_{n\to\infty} a_n = \lim\limits_{n\to\infty}\dfrac{1}{n} = 0$,所以 $\lim\limits_{n\to\infty}\dfrac{1 + \frac{1}{2} + \frac{1}{3} + \cdots + \frac{1}{n}}{n} = 0$.

② 令 $a_1 = 1, a_n = \dfrac{n}{n-1}, n = 2, 3, \cdots$,则 $\lim\limits_{n\to\infty} a_n = 1$,于是

$$\lim\limits_{n\to\infty}\sqrt[n]{n} = \lim\limits_{n\to\infty}\sqrt[n]{1 \cdot \frac{2}{1} \cdot \frac{3}{2} \cdot \frac{4}{3} \cdot \cdots \cdot \frac{n}{n-1}} = \lim\limits_{n\to\infty}\sqrt[n]{a_1 a_2 \cdots a_n} = \lim\limits_{n\to\infty} a_n = 1.$$

③ 令 $a_n = \dfrac{1}{n}, n = 1, 2, \cdots$,则 $\lim\limits_{n\to\infty} a_n = 0$,所以

$$\lim\limits_{n\to\infty}\frac{1}{\sqrt[n]{n!}} = \lim\limits_{n\to\infty}\sqrt[n]{\frac{1}{1 \cdot 2 \cdot 3 \cdot \cdots \cdot n}} = \lim\limits_{n\to\infty}\sqrt[n]{1 \cdot \frac{1}{2} \cdot \cdots \cdot \frac{1}{n}} = \lim\limits_{n\to\infty}\frac{1}{n} = 0.$$

Ex. 4 (1) Let $\lim\limits_{n\to\infty} a_n = a$, then we have $\lim\limits_{n\to\infty}\dfrac{a_1 + a_2 + \cdots + a_n}{n} = a$.

(2) Prove that the following limits are valid by using the result in (1):

① $\lim\limits_{n\to\infty}\dfrac{1 + \frac{1}{2} + \frac{1}{3} + \cdots + \frac{1}{n}}{n} = 0$; ② $\lim\limits_{n\to\infty}\sqrt[n]{n} = 1$; ③ $\lim\limits_{n\to\infty}\dfrac{1}{\sqrt[n]{n!}} = 0$.

Proof (1) From $\lim\limits_{n\to\infty} a_n = a$, we know that, for any $\varepsilon > 0$, $\exists N_1 > 0, \forall n > N_1$ such that $|a_n - a| < \frac{\varepsilon}{2}$. Consequently, as $n > N_1$, we have

$$\left|\frac{a_1 + a_2 + \cdots + a_n}{n} - a\right|$$
$$= \left|\frac{a_1 + a_2 + \cdots + a_n - na}{n}\right|$$
$$\leqslant \frac{|a_1 - a| + |a_2 - a| + \cdots + |a_{N_1} - a|}{n} + \frac{|a_{N_1+1} - a| + |a_{N_1+2} - a| + \cdots + |a_n - a|}{n}$$
$$\leqslant \frac{A}{n} + \frac{n - N_1}{n} \cdot \frac{\varepsilon}{2} \leqslant \frac{A}{n} + \frac{\varepsilon}{2},$$

where $A = |a_1 - a| + |a_2 - a| + \cdots + |a_{N_1} - a|$ is a determined number. Also, from $\lim\limits_{n\to\infty}\frac{A}{n} = 0$, we know that there exists $N_2 > 0$ such that $\frac{A}{n} < \frac{\varepsilon}{2}$ is valid as $n > N_2$. Taking $N = \max\{N_1, N_2\}$, as $n > N$, we have

$$\left|\frac{a_1 + a_2 + \cdots + a_n}{n} - a\right| \leqslant \frac{A}{n} + \frac{\varepsilon}{2} < \frac{\varepsilon}{2} + \frac{\varepsilon}{2} = \varepsilon.$$

(2) ① Let $a_n = \frac{1}{n}$, then we obtain $\lim\limits_{n\to\infty} a_n = \lim\limits_{n\to\infty}\frac{1}{n} = 0$, thus

$$\lim\limits_{n\to\infty}\frac{1 + \frac{1}{2} + \frac{1}{3} + \cdots + \frac{1}{n}}{n} = 0.$$

② Let $a_1 = 1$ and $a_n = \frac{n}{n-1}, n = 2, 3, \cdots$, then $\lim\limits_{n\to\infty} a_n = 1$, and thus

$$\lim\limits_{n\to\infty}\sqrt[n]{n} = \lim\limits_{n\to\infty}\sqrt[n]{1 \cdot \frac{2}{1} \cdot \frac{3}{2} \cdot \frac{4}{3} \cdot \cdots \cdot \frac{n}{n-1}} = \lim\limits_{n\to\infty}\sqrt[n]{a_1 a_2 \cdots a_n} = \lim\limits_{n\to\infty} a_n = 1.$$

③ Let $a_n = \frac{1}{n}, n = 1, 2, \cdots$, then $\lim\limits_{n\to\infty} a_n = 0$, and thus

$$\lim\limits_{n\to\infty}\frac{1}{\sqrt[n]{n!}} = \lim\limits_{n\to\infty}\sqrt[n]{\frac{1}{1 \cdot 2 \cdot 3 \cdot \cdots \cdot n}} = \lim\limits_{n\to\infty}\sqrt[n]{1 \cdot \frac{1}{2} \cdot \cdots \cdot \frac{1}{n}} = \lim\limits_{n\to\infty}\frac{1}{n} = 0.$$

本章重点

极限是研究数学分析的主要工具,故在数学分析中处于十分重要的地位.要求掌握:

1 学会用数学语言描述极限.

2 深刻理解和熟练书写数列收敛和发散的"ε-N"定义.

3 会用数列收敛定义证明下列数列的极限并记住以下常见数列的极限值:

(1) $\lim\limits_{n\to\infty} q^n = 0, |q| < 1$ (2) $\lim\limits_{n\to\infty}\frac{1}{n^\alpha} = 0, \alpha > 0$ (3) $\lim\limits_{n\to\infty}\sqrt[n]{a} = 1, a > 0$

(4) $\lim\limits_{n\to\infty}\sqrt[n]{n} = 1$ (5) $\lim\limits_{n\to\infty}\frac{a^n}{n!} = 0, a > 0$ (6) $\lim\limits_{n\to\infty}\frac{1}{\sqrt[n]{n!}} = 0$

4 写出数列 $\{a_n\}$ 极限是 a 的等价命题.

5 柯西收敛数列有什么意义?

6 掌握判定数列极限存在的判定定理和应用,判别哪种类型的数列的收敛问题用夹逼定理,哪种类型用单调有界必有极限定理.

Key points of this chapter

Limits are the main tools in studying Mathematical Analysis, and thus occupy an important status. The following need mastering:

1. Learn to describe limits by using mathematical language.

2. Understand and familiarize how to write the "$\varepsilon\text{-}N$" definitions of sequence convergence and sequence divergence.

3. Be able to prove the limits of the following sequences by using the definition of sequence convergence and remember:

(1) $\lim\limits_{n\to\infty} q^n = 0, |q|<1$ (2) $\lim\limits_{n\to\infty}\dfrac{1}{n^a} = 0, a>0$ (3) $\lim\limits_{n\to\infty}\sqrt[n]{a} = 1, a>0$

(4) $\lim\limits_{n\to\infty}\sqrt[n]{n} = 1$ (5) $\lim\limits_{n\to\infty}\dfrac{a^n}{n!} = 0, a>0$ (6) $\lim\limits_{n\to\infty}\dfrac{1}{\sqrt[n]{n!}} = 0$

4. Write out the equivalent proposition that the limit of sequence $\{a_n\}$ is a.

5. What sense does a Cauchy convergent sequence make?

6. Master the test theorem and its applications of existence of test sequence limits, and judge which kind of sequence convergence problems is suitable to squeeze theorem, which kind of sequence convergence problems is suitable to the theorem that monotone and bounded sequences must have limits.

第三章 函数极限
Chapter 3 Limits of Functions

在第二章,我们主要考虑定义域为自然数集的实值函数,即主要考虑实数列的情况。本章我们首先介绍函数极限的概念(右极限和左极限)、极限存在的条件以及一些关于极限的定理;然后给出两个重要极限以及无穷大量、无穷小量和无穷小阶的概念。

In Chapter 2, we have considered real-valued functions that have natural number set as their domain, i. e. , sequences of real numbers. In this chapter, we first introduce the concepts of limits of functions (left limit and right limit), conditions of existence of limits and some theorems about limits, respectively. Then we present two important limits and the concepts of infinities, infinitesimals and the order of infinitesimals.

单词和短语 Words and expressions

★ 函数极限 limit of function
无穷极限 infinite limit
单侧极限 one-sided limit
右(左)极限 right (left) limit / right (left) hand limit
★ 函数极限的性质 property of limit of function
★ 局部有界性 local boundedness
★ 海涅定理 Heine theorem
无穷大量 infinity

无穷小量的阶 order of infinitesimal
高(低)阶无穷小量 infinitesimal of higher (lower) order
同阶无穷小量 infinitesimal of same order
等价无穷小量 equivalent infinitesimal
k-阶无穷小量 k-order infinitesimal
垂直渐近线 vertical asymptote
斜渐近线 oblique asymptote
水平渐近线 horizontal asymptote

基本概念和性质 Basic concepts and properties

1. 和的极限等于极限的和.
The limit of a sum is equal to the sum of the limits.

2. 差的极限等于极限的差.
The limit of a difference is equal to the difference of the limits.

3. 积的极限等于极限的积.
The limit of a product is equal to the product of the limits.

4. 商的极限等于极限的商(分母极限不为零).
The limit of a quotient is equal to the quotient of the limits (provided that the limit of the denominator

5 常数因子可提到极限符号前面.
A constant factor can be taken out of the limit computation.

6 幂的极限等于极限的幂.
The limit of a power is equal to the power of the limit.

7 定义：设 $f(x)$ 为定义在 $(-\infty, +\infty)$ 上的函数，A 为常数. 若对于 $\forall \varepsilon > 0, \exists M(\geqslant 0)$，使得当 $|x| > M$ 时有 $|f(x)-A| < \varepsilon$，则称函数 $f(x)$ 当 $x \to \infty$ 时以 A 为极限，记作 $\lim\limits_{x \to \infty} f(x) = A$ 或 $f(x) \to A(x \to \infty)$.

Definition: Let $f(x)$ be a function defined in $(-\infty, +\infty)$, and A be a constant. For any $\varepsilon > 0$, if there exists a number $M(\geqslant 0)$ such that $|f(x)-A| < \varepsilon$ as $|x| > M$, then the number A is called the limit of $f(x)$ as $x \to \infty$. We denote this as $\lim\limits_{x \to \infty} f(x) = A$ or $f(x) \to A(x \to \infty)$.

8 定义：设函数 $f(x)$ 在点 x_0 的某去心邻域 $U^{\circ}(x_0; \delta')$ 内有定义，A 为常数. 若对于 $\forall \varepsilon > 0$，存在正数 $\delta(<\delta')$ 使得当 $0 < |x-x_0| < \delta$ 时有 $|f(x)-A| < \varepsilon$，则称函数 f 当 x 趋于 x_0 时以 A 为极限. 记作 $\lim\limits_{x \to x_0} f(x) = A$ 或 $f(x) \to A(x \to x_0)$.

Definition: Let $f(x)$ be a function defined in a certain deleted neighborhood of x_0, and A be a constant. For any $\varepsilon > 0$, if there exists a number $\delta > 0$ such that $|f(x)-A| < \varepsilon$ for all x satisfying $0 < |x-x_0| < \delta$, then the constant number A is called the limit of $f(x)$ as $x \to x_0$, and is denoted by $\lim\limits_{x \to x_0} f(x) = A$ or $f(x) \to A(x \to x_0)$.

9 柯西准则：设 $f(x)$ 在 $\mathring{U}(x_0; \delta')$ 内有定义. $\lim\limits_{x \to x_0} f(x)$ 存在的充要条件是：任给 $\varepsilon > 0$，总存在正数 $\delta(<\delta')$，使得对任何 $x', x'' \in \mathring{U}(x_0; \delta)$，有 $|f(x')-f(x'')| < \varepsilon$.

Cauchy principle: Let $f(x)$ be a function defined in $\mathring{U}(x_0; \delta')$. $\lim\limits_{x \to x_0} f(x)$ exists $\Leftrightarrow$ for any $\varepsilon > 0$, there exists a positive number $\delta(<\delta')$ such that for any $x', x'' \in \mathring{U}(x_0; \delta)$, we have $|f(x')-f(x'')| < \varepsilon$.

10 夹逼定理：设函数 f, g 和 h 满足：对所有在点 c 邻域内的 x（可能不包含 c）$f(x) \leqslant g(x) \leqslant h(x)$. 若 $\lim\limits_{x \to c} f(x) = \lim\limits_{x \to c} h(x) = L$，则 $\lim\limits_{x \to c} g(x) = L$.

Squeeze Theorem: Let f, g and h be functions satisfying $f(x) \leqslant g(x) \leqslant h(x)$ for all x in the neighborhood of the point c, except possibly at c. If $\lim\limits_{x \to c} f(x) = \lim\limits_{x \to c} h(x) = L$, then $\lim\limits_{x \to c} g(x) = L$.

11 (1) $\lim\limits_{x \to 0} \dfrac{\sin x}{x} = 1$（特殊三角极限）；(2) $\lim\limits_{x \to \infty} \left(1 + \dfrac{1}{x}\right)^x = e$（e 极限）.

这是两个重要的公式，许多包含三角函数的 $\dfrac{0}{0}$ 型极限可用公式(1)帮助解决，1^{∞} 型的极限可用公式(2)解决.

(1) $\lim\limits_{x \to 0} \dfrac{\sin x}{x} = 1$ (special trigonometric limit); (2) $\lim\limits_{x \to \infty} \left(1 + \dfrac{1}{x}\right)^x = e$ (e-limit).

The two formulae are very important, many limits of indeterminate forms of the type $\dfrac{0}{0}$ involving trigonometric functions can be computed with the help of the formula (1) and limits of indeterminate forms of the type 1^{∞} can be computed with the help of the formula (2).

12 无穷小量虽然都是以零为极限的，但是不同的无穷小量趋于零的速度是不同的. 它的应用是：(1) 用等价无穷小代替求函数的极限；(2) 判别正项级数的敛散性.

The speeds approaching to 0 of different infinitesimals may be different although their limits are all zero. Their applications are:

(1) We can find the limits of functions by using the method of substitution of equivalent infinitesimals into functions. (2) Determine convergence or divergence of positive term series.

例 题 Examples

例 1 用定义证明下列函数的极限:

(1) $\lim\limits_{x\to\infty}\dfrac{x^2-5}{x^2-1}=1$; (2) $\lim\limits_{x\to 1}\dfrac{x^2-1}{2x^2-x-1}=\dfrac{2}{3}$.

证明 (1) 对 $\forall \varepsilon>0$,设 $|x|>1$,要证

$$\left|\dfrac{x^2-5}{x^2-1}-1\right|=\dfrac{4}{x^2-1}\leqslant\dfrac{4}{(|x|-1)(|x|+1)}\leqslant\dfrac{4}{|x|-1}<\varepsilon,$$

只需 $|x|>1+\dfrac{4}{\varepsilon}$,于是对 $\forall \varepsilon>0$ 和 $\forall |x|>M$,其中 $M=1+\dfrac{4}{\varepsilon}$,有 $\left|\dfrac{x^2-5}{x^2-1}-1\right|\leqslant\dfrac{4}{|x|-1}<\varepsilon$,所以 $\lim\limits_{x\to\infty}\dfrac{x^2-5}{x^2-1}=1$.

(2) 当 $x\neq 1$ 时有

$$\left|\dfrac{x^2-1}{2x^2-x-1}-\dfrac{2}{3}\right|=\left|\dfrac{x+1}{2x+1}-\dfrac{2}{3}\right|=\left|\dfrac{x-1}{3(2x+1)}\right|,$$

若 x 满足 $0<|x-1|<1$,则有 $|2x+1|>1$. 于是,对 $\forall \varepsilon>0$,只要取 $\delta=\min\{3\varepsilon,1\}$,则当 $0<|x-1|<\delta$ 时,便有

$$\left|\dfrac{x^2-1}{2x^2-x-1}-\dfrac{2}{3}\right|<\dfrac{|x-1|}{3}<\varepsilon.$$

Ex. 1 Prove the following limits of functions by definition:

(1) $\lim\limits_{x\to\infty}\dfrac{x^2-5}{x^2-1}=1$; (2) $\lim\limits_{x\to 1}\dfrac{x^2-1}{2x^2-x-1}=\dfrac{2}{3}$.

Proof (1) For $\forall \varepsilon>0$ and for $|x|>1$, to prove

$$\left|\dfrac{x^2-5}{x^2-1}-1\right|=\dfrac{4}{x^2-1}\leqslant\dfrac{4}{(|x|-1)(|x|+1)}\leqslant\dfrac{4}{|x|-1}<\varepsilon,$$

the inequality $|x|>1+\dfrac{4}{\varepsilon}$ must be valid. Consequently, for $\forall \varepsilon>0$ and for $\forall |x|>M$, where $M=1+\dfrac{4}{\varepsilon}$, this leads to $\left|\dfrac{x^2-5}{x^2-1}-1\right|\leqslant\dfrac{4}{|x|-1}<\varepsilon$, and thus we have $\lim\limits_{x\to\infty}\dfrac{x^2-5}{x^2-1}=1$.

(2) If $x\neq 1$, then we have

$$\left|\dfrac{x^2-1}{2x^2-x-1}-\dfrac{2}{3}\right|=\left|\dfrac{x+1}{2x+1}-\dfrac{2}{3}\right|=\left|\dfrac{x-1}{3(2x+1)}\right|,$$

furthermore, if x satisfies $0<|x-1|<1$, we obtain $|2x+1|>1$. Consequently, for $\forall \varepsilon>0$, if $\delta=\min\{3\varepsilon,1\}$, then we have

$$\left|\dfrac{x^2-1}{2x^2-x-1}-\dfrac{2}{3}\right|<\dfrac{|x-1|}{3}<\varepsilon, \text{ as } 0<|x-1|<\delta.$$

例 2 (1) 设 $\lim\limits_{x\to 0}f(x^3)=A$,证明: $\lim\limits_{x\to 0}f(x)=A$;

(2) 设 $\lim\limits_{x\to 0}f(x^2)=A$,问是否成立 $\lim\limits_{x\to 0}f(x)=A$?

证明 (1) 设 $\lim\limits_{x\to 0}f(x^3)=A$,则 $\forall \varepsilon>0, \exists \delta'>0, \forall x(0<|x|<\delta')$(即 $0<|x^3|<\delta'^3$),有 $|f(x^3)-A|<\varepsilon$. 取 $\delta=\delta'^3>0$,则当 $0<|x|<\delta$ 时,有 $0<|\sqrt[3]{x}|<\delta'$,从而 $|f(x)-A|<\varepsilon$,这就说明了 $\lim\limits_{x\to 0}f(x)=A$.

(2) 当 $\lim\limits_{x\to 0}f(x^2)=A$ 时,不一定成立 $\lim\limits_{x\to 0}f(x)=A$. 例如: $f(x)=\begin{cases}1 & x>0 \\ -1 & x<0\end{cases}$,则 $\lim\limits_{x\to 0}f(x^2)=1$,但极限 $\lim\limits_{x\to 0}f(x)$ 不存在.

Ex. 2 (1) If $\lim\limits_{x\to 0} f(x^3) = A$, we then have $\lim\limits_{x\to 0} f(x) = A$;

(2) If $\lim\limits_{x\to 0} f(x^2) = A$, whether the limit $\lim\limits_{x\to 0} f(x) = A$ is valid or not?

Proof (1) From $\lim\limits_{x\to 0} f(x^3) = A$, we know that there exists $\delta' > 0$, for $\forall \varepsilon > 0$ and for $\forall x (0 < |x| < \delta')$ (namely, $0 < |x^3| < \delta'^3$), $|f(x^3) - A| < \varepsilon$ is valid. Take $\delta = \delta'^3 > 0$, we then have $0 < |\sqrt[3]{x}| < \delta'$ as $0 < |x| < \delta$, and thus $|f(x) - A| < \varepsilon$, this implies $\lim\limits_{x\to 0} f(x) = A$ is valid.

(2) However, as $\lim\limits_{x\to 0} f(x^2) = A$, $\lim\limits_{x\to 0} f(x) = A$ is not valid identically. For example, if $f(x) = \begin{cases} 1 & x > 0 \\ -1 & x < 0 \end{cases}$, we then have $\lim\limits_{x\to 0} f(x^2) = 1$, but $\lim\limits_{x\to 0} f(x)$ does not exist.

例 3 试确定 a 的值,使下列函数与 x^a 当 $x \to 0$ 时为同阶无穷小量:

(1) $\dfrac{1}{1+x} - (1-x)$;   (2) $\sqrt{1+\tan x} - \sqrt{1-\sin x}$.

解 (1) 因为 $\dfrac{1}{1+x} - (1-x) = \dfrac{x^2}{1+x} \sim x^2 (x \to 0)$,于是 $\lim\limits_{x\to 0}\dfrac{x^2}{x^2(1+x)} = 1$,所以 $a = 2$;

(2) 因为
$$\sqrt{1+\tan x} - \sqrt{1-\sin x} = \frac{\tan x + \sin x}{\sqrt{1+\tan x} + \sqrt{1-\sin x}}$$
$$= \frac{\sin x(1+\cos x)}{(\sqrt{1+\tan x} + \sqrt{1-\sin x})\cos x} \sim x \quad (x \to 0)$$

所以 $a = 1$.

Ex. 3 Try to determine the values of a such that the following functions are infinitesimals of the same order with respect to x^a as $x \to 0$:

(1) $\dfrac{1}{1+x} - (1-x)$;   (2) $\sqrt{1+\tan x} - \sqrt{1-\sin x}$.

Solution (1) Since $\dfrac{1}{1+x} - (1-x) = \dfrac{x^2}{1+x} \sim x^2 (x \to 0)$, we have
$$\lim_{x\to 0}\frac{x^2}{x^2(1+x)} = 1, \text{ and thus } a = 2;$$

(2) From $\sqrt{1+\tan x} - \sqrt{1-\sin x} = \dfrac{\tan x + \sin x}{\sqrt{1+\tan x} + \sqrt{1-\sin x}}$
$$= \frac{\sin x(1+\cos x)}{(\sqrt{1+\tan x} + \sqrt{1-\sin x})\cos x} \sim x \ (x \to 0)$$

we know that $a = 1$.

本章重点

1. 理解海涅定理的意义以及它在极限论中的作用.
2. 掌握两个重要极限的应用.

Key points of this chapter

1. Understand the sense of Heine theorem and its effect in the limit theory.
2. Master the applications of the two important limits.

第四章 函数的连续性
Chapter 4 Continuity of Functions

本章我们研究连续性的概念.

在数学中很多函数有下列特殊性质:如果函数的自变量变化不大时,函数的改变也不大. 这种函数常被认为是有特别好的行为,即铅笔不离开纸就能画出函数的图形,我们称该性质为函数的连续性.

首先给出的是连续函数、函数间断点和区间上连续函数的详细陈述;然后详细论述了连续函数的局部性质和连续函数的运算,包括闭区间上连续函数基本性质的讨论;最后将详细讨论初等函数的连续性.

In this chapter, we deal with the concept of continuity.

In mathematics, many functions have the following specific properties: If the change of the independent variable of a function is small, then the change of the dependent variable is also not large. This kind of functions is usually regarded as especially well behavior, that is, the graph of a function can be drawn without lifting your pencil from the paper. We call this property the continuity of functions.

First of all, we give a detail treatment of concepts of continuous functions, discontinuous points of functions, continuous functions on interval. Secondly, we present detail discussions of local properties of continuous functions and the operations of continuous functions, including discussion of properties of continuous functions over closed interval. Finally, continuity of elementary functions will be discussed in detail.

单词和短语 Words and expressions

自变量的增量　increment of independent variable
函数的增量　increment of function
右(左)连续　right (left) continuous
间断点及其分类　discontinuity point and its classification
可去间断点　removable discontinuity
跳跃间断点　jump discontinuity
★第一类间断点　discontinuity of the first kind
★第二类间断点　discontinuity of the second kind
连续函数的局部性质　local properties of continuous function
连续函数的复合性质　composition properties of continuous function
闭区间上连续函数的性质　properties of continuous function over closed interval
极值定理　extreme value theorem
★最大值和最小值定理　maximum and minimum value theorem
★介值性定理　intermediate value theorem
★零点定理　zero-point theorem
★一致连续性定理　uniform continuity theorem
反函数的连续性　continuity of inverse function
局部保号性　local inheriting order property
初等函数的连续性　continuity of elementary function

例 题　Examples

例 1　指出函数 $f(x)=\mathrm{sgn}(\cos x)$ 的间断点并说明其类型.

解　因为 $f(x)=\mathrm{sgn}(\cos x)=\begin{cases} 1 & 2n\pi-\dfrac{\pi}{2}<x<2n\pi+\dfrac{\pi}{2} \\ 0 & x=n\pi+\dfrac{\pi}{2} \\ -1 & 2n\pi+\dfrac{\pi}{2}<x<2n\pi+\dfrac{3\pi}{2} \end{cases}$,所以 f 在 $x=2n\pi\pm\dfrac{\pi}{2}(n=0,\pm 1,\pm 2,\cdots)$ 间断.

由于 $\lim\limits_{x\to 2n\pi+\frac{\pi}{2}^-}\mathrm{sgn}(\cos x)=-1$,$\lim\limits_{x\to 2n\pi+\frac{\pi}{2}^+}\mathrm{sgn}(\cos x)=1$,$\lim\limits_{x\to 2n\pi-\frac{\pi}{2}^+}\mathrm{sgn}(\cos x)=1$,$\lim\limits_{x\to 2n\pi-\frac{\pi}{2}^-}\mathrm{sgn}(\cos x)=-1$,故 $x=2n\pi\pm\dfrac{\pi}{2}(n=0,\pm 1,\pm 2,\cdots)$ 是 f 的跳跃间断点.

Ex. 1　Point out the discontinuous points and explain its classification of the function $f(x)=\mathrm{sgn}(\cos x)$.

Solution For $f(x) = \operatorname{sgn}(\cos x) = \begin{cases} 1 & 2n\pi - \frac{\pi}{2} < x < 2n\pi + \frac{\pi}{2} \\ 0 & x = n\pi + \frac{\pi}{2} \\ -1 & 2n\pi + \frac{\pi}{2} < x < 2n\pi + \frac{3\pi}{2} \end{cases}$, it is easy to show that $x = 2n\pi \pm \frac{\pi}{2}\,(n = 0, \pm 1, \pm 2, \cdots)$ are discontinuous points of f.

Moreover, $\lim\limits_{x \to 2n\pi + \frac{\pi}{2}^-} \operatorname{sgn}(\cos x) = -1$, $\lim\limits_{x \to 2n\pi + \frac{\pi}{2}^-} \operatorname{sgn}(\cos x) = 1$, $\lim\limits_{x \to 2n\pi - \frac{\pi}{2}^+} \operatorname{sgn}(\cos x) = 1$, $\lim\limits_{x \to 2n\pi - \frac{\pi}{2}^-} \operatorname{sgn}(\cos x) = -1$. So $x = 2n\pi \pm \frac{\pi}{2}\,(n = 0, \pm 1, \pm 2, \cdots)$ are jump discontinuous points of f.

例 2 设当 $x \neq 0$ 时 $f(x) \equiv g(x)$，而 $f(0) \neq g(0)$. 证明：f 与 g 两者中至多有一个在 $x = 0$ 连续.

证明 因为 $f(x) \equiv g(x)$，所以 $\lim\limits_{x \to 0} f(x) = \lim\limits_{x \to 0} g(x)$，假设 f 与 g 两个都在 $x = 0$ 连续，则 $f(0) = \lim\limits_{x \to 0} f(x) = \lim\limits_{x \to 0} g(x) = g(0)$. 与题设 $f(0) \neq g(0)$ 矛盾，所以 f 与 g 两者中至多有一个在 $x = 0$ 连续.

Ex. 2 Assume that $f(x) \equiv g(x)$ as $x \neq 0$ and $f(0) \neq g(0)$. Prove that at most one of the functions f and g is continuous at $x = 0$.

Proof Since $f(x) \equiv g(x)$ as $x \neq 0$, we know that $\lim\limits_{x \to 0} f(x) = \lim\limits_{x \to 0} g(x)$. Assume that f and g are all continuous at $x = 0$, then we have $f(0) = \lim\limits_{x \to 0} f(x) = \lim\limits_{x \to 0} g(x) = g(0)$, which contradicts with the known $f(0) \neq g(0)$, that is to say, at most one of the functions f and g is continuous at $x = 0$.

本章重点

1. 函数 $f(x)$ 在点 a 连续有哪些等价叙述？
2. 连续函数有什么优点？
3. 是否存在区间上任一点都不连续的函数？
4. 证明闭区间上连续函数的四个性质为什么要使用实数的连续性定理？
5. 掌握本章定理的内容和证明方法.
6. 初等函数的连续性是怎样证明的？
7. 证明方程根的存在性时，想到用本节的哪个定理？
8. 最大值和最小值定理的条件是什么？结论是什么？

Key points of this chapter

1. What equivalent depictions do the function $f(x)$ is continuous at point a have?
2. What advantages do continuous functions have?
3. Whether there exist discontinuous functions at any points over an interval or not?
4. Why is continuity theorem of real numbers used to prove the four properties of continuous functions over a closed interval?
5. Master the contents and proving methods of theorems of the chapter.
6. How to prove continuity of elementary functions?
7. Which theorem is reminded to prove the existence of roots of equations?
8. What are the conditions of the maximum and minimum theorem? What is the conclusion?

第五章 导数与微分
Chapter 5 Derivatives and Differentials

本章我们将致力于导数与微分的研究.

首先利用了第三、四章给出的概念，我们考虑曲线的切线，由极限方法确定切线导出函数导数的概念.

同时，证明了达布定理，从证明中得知，为了使定义在开区间上的实值函数是某一函数的导数，它的必要条件是给定的函数具有介值性．

然后，我们将根据公式

$$f'(x) = \lim_{\Delta x \to 0} \frac{f(x+\Delta x) - f(x)}{\Delta x}$$

计算一些我们熟悉的函数的导数并给出求导法则，并且给出参变函数的求导公式以及高阶函数的概念和几个初等函数的 n 阶导数公式．

最后，我们给出了微分的概念、微分的几何意义，并根据导数的公式建立微分公式和运算法则以及应用微分来说明计算函数值的问题．

We will devote this chapter to the study of derivatives and differentials.

First of all, using the concepts presented in Chapters 3 and 4, we consider the tangent lines of curves and define the derivative of a function by means of limits.

Meanwhile, Darboux Theorem is proved. It asserts that in order for a real-valued function that is defined on an open interval to be the derivate of another function, it is necessary that the given function possesses the intermediate value property.

Thereafter, we compute the derivatives of some familiar functions by the following formula

$$f'(x) = \lim_{\Delta x \to 0} \frac{f(x+\Delta x) - f(x)}{\Delta x}$$

and then give the rules for finding derivatives. We also present the formula for finding the derivatives of a function presented parametrically and show the notion of a higher derivative and the derivative formula of the n-th order of several elementary functions.

Finally, we give the concept and the geometric meaning of differentials, and construct the ordinary formulas and the operational rules of differentials according to the formulas of derivatives. As an application, we apply differentials to illustrate the problem of approximating value of functions.

单词和短语 Words and expressions

有限增量公式　finite increment formula
变化率　rate of change
差商　difference quotient
左导数　left derivative
右导数　right derivative
单侧导数　one-sided derivative
导函数　derivative function
★可导函数　derivable function
★导数的几何意义　geometric meaning of derivative
费马定理　Fermat theorem
达布定理　Darboux theorem
导函数的介值定理　intermediate value theorem of derivative function
导数的四则运算　algebra of derivatives
和的导数　derivative of sum
差的导数　derivative of difference
积的导数　derivative of product
商的导数　derivative of quotient
反函数的导数　derivative of inverse function
最大值　maximum value

最小值　minimum value
复合函数的导数　derivative of composite function
对数求导法　logarithmic derivative
圆的参数方程　parametric equation of circle
椭圆的参数方程　parametric equation of ellipse
摆线（旋轮线）的参数方程　parametric equation of cycloid
星形线的参数方程　parametric equation of asteroid
二阶导数　second derivative
三阶导数　third derivative
★ n 阶导数　n-th derivative
莱布尼茨公式　Leibniz formula
加速度　acceleration
物理解释　physical interpretation
微分的概念　concept of differential
可微函数　differentiable function
线性主部　linear principal part
自变量的微分　differential of independent variable

微分的运算法则 operational rules of differential
★ 微分形式的不变性 invariance of differential form
微分的几何意义 geometric meaning of differential
高阶微分 higher-order differential

基本概念和性质 **Basic concepts and properties**

1 和的导数等于导数的和.
The derivative of a sum is the sum of the derivatives.

2 差的导数等于导数的差.
The derivative of a difference is the difference of the derivatives.

3 积的导数等于第一个函数的导数乘以第二个函数加上第二个函数的导数乘以第一个函数.
The derivative of a product of two functions is that the second times the derivative of the first plus the first times the derivative of the second.

4 商的导数等于分子导数乘分母减去分母的导数乘分子除以分母的平方.
The derivative of a quotient of two functions is equal to the denominator times the derivative of the numerator minus the numerator times the derivative of the denominator all divided by the square of the denominator.

5 复合函数求导法：设 $u = \varphi(x)$ 在点 x_0 可导, $y = f(u)$ 在点 $u_0 = \varphi(x_0)$ 可导,则复合函数 $f \circ \varphi$ 在点 x_0 可导,且 $(f \circ \varphi)'(x_0) = f'(\varphi(x_0))\varphi'(x_0)$.
Rules for finding derivatives of composite functions: Let $u = \varphi(x)$ be derivable at x_0 and $y = f(u)$ be derivable at $u_0 = \varphi(x_0)$, then the composite function $f \circ \varphi$ is derivable at x_0 and $(f \circ \varphi)'(x_0) = f'(\varphi(x_0))\varphi'(x_0)$.

例题 Examples

例 1 设 $f(x) = \begin{cases} x^2 & x \geq 3 \\ ax+b & x < 3 \end{cases}$,试确定 a,b 的值,使 f 在 $x = 3$ 可导.

解 要使 f 在 $x = 3$ 可导, f 在 $x = 3$ 必连续,于是必左连续,因此有
$$\lim_{x \to 3^-} f(x) = \lim_{x \to 3^-}(ax+b) = 3a+b = f(3) = 9,\text{从而 } b = 9 - 3a.$$

f 在 $x = 3$ 的右导数 $f'_+(3) = \lim\limits_{x \to 3^+} \dfrac{f(x) - f(3)}{x-3} = \lim\limits_{x \to 3^+} \dfrac{x^2 - 3^2}{x-3} = 6$；

左导数为 $f'_-(3) = \lim\limits_{x \to 3^-} \dfrac{f(x) - f(3)}{x-3} = \lim\limits_{x \to 3^-} \dfrac{ax+b-3^2}{x-3} = \lim\limits_{x \to 3^-} \dfrac{ax+9-3a-9}{x-3} = a$,

即只要 $a = 6$,则 f 在 $x = 3$ 的左导数与右导数相等,从而可导. 这时 $b = -9$.

Ex. 1 Let $f(x) = \begin{cases} x^2 & x \geq 3 \\ ax+b & x < 3 \end{cases}$, try to determine the values of a and b such that f is derivable at $x = 3$.

Solution To make f is derivable at $x = 3$, it requires that f must be continuous at $x = 3$, and thus must be left-continuous, i.e.,
$$\lim_{x \to 3^-} f(x) = \lim_{x \to 3^-}(ax+b) = 3a+b = f(3) = 9, \text{ so we have } b = 9 - 3a.$$
The right-derivative and the left-derivative of f at $x = 3$ are respectively given by $f'_+(3) = \lim\limits_{x \to 3^+} \dfrac{f(x) - f(3)}{x-3} = \lim\limits_{x \to 3^+} \dfrac{x^2 - 3^2}{x-3} = 6$ and $f'_-(3) = \lim\limits_{x \to 3^-} \dfrac{f(x) - f(3)}{x-3} = \lim\limits_{x \to 3^-} \dfrac{ax+b-3^2}{x-3} = \lim\limits_{x \to 3^-} \dfrac{ax+9-3a-9}{x-3} = a$, this leads to $a = 6$. Consequently, it is required that the right-derivative and the left-derivative of f at $x = 3$ must be equal, and thus is derivable, we obtain $b = -9$.

例 2 设 $g(0) = g'(0) = 0, f(x) = \begin{cases} g(x)\sin\dfrac{1}{x}, & x \neq 0 \\ 0, & x = 0 \end{cases}$,求 $f'(0)$.

解 因为 $\lim\limits_{x\to 0}\dfrac{g(x)}{x}=\lim\limits_{x\to 0}\dfrac{g(x)-g(0)}{x}=g'(0)=0$，从而有

$$f'(0)=\lim_{x\to 0}\dfrac{f(x)-f(0)}{x-0}=\lim_{x\to 0}\dfrac{g(x)\sin\dfrac{1}{x}-0}{x}=\lim_{x\to 0}\dfrac{g(x)-g(0)}{x}\sin\dfrac{1}{x}=0.$$

Ex. 2 Let $g(0)=g'(0)=0$ and $f(x)=\begin{cases} g(x)\sin\dfrac{1}{x}, & x\neq 0 \\ 0, & x=0 \end{cases}$, compute $f'(0)$.

Solution Since $\lim\limits_{x\to 0}\dfrac{g(x)}{x}=\lim\limits_{x\to 0}\dfrac{g(x)-g(0)}{x}=g'(0)=0$, this leads to

$$f'(0)=\lim_{x\to 0}\dfrac{f(x)-f(0)}{x-0}=\lim_{x\to 0}\dfrac{g(x)\sin\dfrac{1}{x}-0}{x}=\lim_{x\to 0}\dfrac{g(x)-g(0)}{x}\sin\dfrac{1}{x}=0.$$

本章重点

1. 掌握可导与连续的关系.
2. 求哪些函数在个别点的导数需要应用导数的定义?
3. 函数 $f(x)$ 在点 x_0 的导数 $f'(x_0)$ 与函数 $f(x)$ 的微分 $\mathrm{d}y=f'(x_0)(x-x_0)$ 有什么区别?
4. 理解微分的几何意义.
5. 掌握应用导数研究函数的单调性与极值的理论、方法和步骤.

Key points of this chapter

1. Master the relationship between derivative and continuity.
2. Which kinds of derivatives of specific points of functions need using the definition of derivatives?
3. What is the difference between the derivative $f'(x_0)$ of the function $f(x)$ at the point x_0 and the differential $\mathrm{d}y=f'(x_0)(x-x_0)$ of the function $f(x)$?
4. Understand geometric meaning of a differential.
5. Master the theories, methods and steps of using derivatives to study the monotonicity and extremum of functions.

第六章 微分中值定理及其应用
Chapter 6 Mean Value Theorems of Differentials and their Applications

本章我们将给出微分学的一些基本定理和平均值定理. 这些定理能使我们用导数作为工具来进一步研究函数的特征，如函数的增减性、驻点和极值等等. 我们将介绍罗必达则，这是计算不定型极限的有力工具. 最后我们还将给出应用问题.

In this chapter, we will give some basic theorems and the mean value theorem in differential calculus, which make it possible to take derivatives as a tool to further studying some characters of a function, such as increase or decrease, critical points, and extreme, etc.. L'Hospital rule will be introduced, which is useful in evaluating limits of indeterminate forms. Finally, we will introduce some applied problems.

单词和短语 Words and expressions

★ 罗尔中值定理　Rolle mean value theorem

★ 拉格朗日中值定理　Lagrange mean value theorem

★ 柯西中值定理　Cauchy mean value theorem

★ 泰勒定理　Taylor theorem

★ 洛必达法则　L'hospital Rule

$\dfrac{0}{0}$ 型不定式极限　limit of indeterminate form of

type $\dfrac{0}{0}$

$\dfrac{\infty}{\infty}$ 型不定式　indeterminate form of type $\dfrac{\infty}{\infty}$

其他类型不定式　other indeterminate forms

带有佩亚诺型余项的泰勒公式　Taylor formula with Peano Remainder

泰勒系数　Taylor coefficient

泰勒多项式　Taylor polynomial

泰勒公式的余项　remainder of Taylor formula

带有拉格朗日型余项的泰勒公式　Taylor formula with Lagrange remainder

带有拉格朗日型余项的麦克劳林公式　Maclaurin formula with Lagrange remainder

函数的极值　extreme value of function

极值判别　test of extreme value

极值的第一充分条件　the first sufficient condition of extreme value

极值的第二充分条件　the second sufficient condition of extreme value

极值的第三充分条件　the third sufficient condition of extreme value

函数的凸性与拐点　convexity and inflection point of function

凸函数　convex function

凹函数　concave function

严格凸函数　strictly convex function

严格凹函数　strictly concave function

詹森不等式　Jenson inequality

例　题　Examples

例 1 证明:(1) 方程 $x^3-3x+c=0$(这里 c 为常数)在区间$[0,1]$内不可能有两个不同的实根;

(2) 方程 $x^n+px+q=0$(n 为正整数)当 n 为偶数时至多有两个实根;当 n 为奇数时至多有三个实根.

证明　(1) 设 $f(x)=x^3-3x+c$,由于方程 $f'(x)=3x^2-3=0$ 在 $(0,1)$ 内没有根,所以方程 $x^3-3x+c=0$ 在区间 $[0,1]$ 内不可能有两个不同的实根.

(2) 设 $f(x)=x^n+px+q$,于是 $f'(x)=nx^{n-1}+p=0$. 当 n 为偶数时,$n-1$ 为奇数,故方程 $f'(x)=nx^{n-1}+p=0$ 至多有一个实根(因为幂函数 $nx^{n-1}+p$ 严格递增),从而方程 $x^n+px+q=0$ 至多有两个实根;当 n 为奇数时,$n-1$ 为偶数,故由上述证明的关于偶数的结论有:方程 $f'(x)=nx^{n-1}+p=0$ 至多有两个实根,从而方程 $x^n+px+q=0$ 当 n 为奇数时至多有三个实根.

Ex. 1　Prove that: (1) It is impossible that the equation $x^3-3x+c=0$ has two different real roots in the interval $[0,1]$, where c is a constant;

(2) For the equation $x^n+px+q=0$, where n is a positive integer, it has at most two real roots as n is an even number and has at most three real roots as n is an odd number.

Proof　(1) Let $f(x)=x^3-3x+c$. From the equation $f'(x)=3x^2-3=0$ we know that it has no real root in $(0,1)$, and thus it is impossible that the equation $x^3-3x+c=0$ has two different real roots in the interval $[0,1]$.

(2) Let $f(x)=x^n+px+q$, we then have $f'(x)=nx^{n-1}+p=0$. As n is an even number, $n-1$ must be an odd number, and so the equation $f'(x)=nx^{n-1}+p=0$ has at most a real root since the power function $nx^{n-1}+p$ increases monotonically. Consequently, the equation $x^n+px+q=0$ has at most two real roots; while as n is an odd number, $n-1$ must be an even number, similarly, the equation $f'(x)=nx^{n-1}+p=0$ has at most two real roots, and thus the equation $x^n+px+q=0$ has at most three real roots as n is an odd number.

例 2　设 $0<\alpha<\beta<\dfrac{\pi}{2}$,证明存在 $\theta\in(\alpha,\beta)$,使得 $\dfrac{\sin\alpha-\sin\beta}{\cos\beta-\cos\alpha}=\cot\theta$.

证明　令 $f(x)=\sin x,g(x)=\cos x$,则 f,g 都在 $[\alpha,\beta]$ 连续,在 (α,β) 可导,且 f',g' 都不等于 0,$g(\alpha)\neq g(\beta)$. 由柯西中值定理,存在 $\theta\in(\alpha,\beta)$,使得 $\dfrac{\sin\beta-\sin\alpha}{\cos\beta-\cos\alpha}=\dfrac{\cos\theta}{-\sin\theta}$,即 $\dfrac{\sin\alpha-\sin\beta}{\cos\beta-\cos\alpha}=\cot\theta$.

Ex. 2 Let $0 < \alpha < \beta < \dfrac{\pi}{2}$, prove that there exists a value of $\theta \in (\alpha, \beta)$ such that $\dfrac{\sin \alpha - \sin \beta}{\cos \beta - \cos \alpha} = \cot \theta$.

Proof Let $f(x) = \sin x$ and $g(x) = \cos x$, then we know that f, g are all continuous on the interval $[\alpha, \beta]$, derivable in (α, β) and f', g' are all not equal to 0, $g(\alpha) \neq g(\beta)$. From the Cauchy mean value theorem, we know that there exists a value of $\theta \in (\alpha, \beta)$ such that $\dfrac{\sin \beta - \sin \alpha}{\cos \beta - \cos \alpha} = \dfrac{\cos \theta}{-\sin \theta}$, i. e., $\dfrac{\sin \alpha - \sin \beta}{\cos \beta - \cos \alpha} = \cot \theta$.

本章重点

1. 证明中值定理，特别是学会利用辅助函数证明问题的方法。
2. 具有应用中值定理证明问题的能力。
3. 说明中值定理的意义。
4. 掌握高阶导数并借助于泰勒公式证明问题的方法。
5. 记住几个重要函数 $e^x, \sin x, \cos x, \ln(1+x), (1+x)^\alpha$ 的麦克劳林展开式。
6. 不同类型的泰勒公式余项各有什么作用？

Key points of this chapter

1. Able to prove mean value theorems, especially the method of proving problems by using assistant functions.
2. Have the preliminary ability to prove problems by using mean value theorems.
3. State the meaning of mean value theorems.
4. Master the methods on proving problems by using higher-order derivatives and making use of remainder of Taylor formula.
5. Remember the Maclaurin expansions of several important functions, namely, e^x, $\sin x$, $\cos x$, $\ln(1+x)$ and $(1+x)^\alpha$.
6. What effects do the different types of remainder of Taylor Formulae have respectively?

第七章 实数的完备性
Chapter 7 Completeness of Real Numbers

在第一章和第二章，我们已经证明了单调有界定理、确界定理和柯西收敛准则。这三个命题反映的实数的特性通常称为完备性。本章将阐述与完备性等价的另外三个收敛数列的定理，它们是区间套定理、波尔查诺-维尔斯特拉斯定理和有限覆盖定理，然后应用这些定理证明闭区间上连续函数的性质。

In Chapter 1 and 2, we have proved the bounded monotone theorem, the principles of supremum and infimum, the Cauchy convergence criterion. The three propositions show a special property of real numbers, which is usually called completeness of real numbers. In this chapter, the completeness theorem is recast as the nested interval theorem, the Bolzano-Weierstrass theorem and the finite covering theorem for convergence sequences. Then we apply these theorems to prove the properties of continuous functions on a closed interval.

单词和短语 Words and expressions

实数集完备性的基本定理   fundamental theorems of completeness in the set of real numbers
★ 区间套定理   nested interval theorem
★ 柯西收敛准则   Cauchy convergence criterion
★ 聚点定理   accumulation theorem
★ 有限覆盖定理   finite covering theorem
实数完备性基本定理的等价性   equivalence of completeness of real numbers
上极限和下极限   upper and lower limits

例 题 Examples

例 试举例说明:在有理数集内,确界原理、单调有界定理、聚点定理和柯西收敛准则一般都不成立.

解 设 $a_n = \left(1 + \dfrac{1}{n}\right)^n, b_n = \left(1 + \dfrac{1}{n}\right)^{n+1}, n = 1, 2, \cdots$,则 $\{a_n\}$ 是单调递增的有理数列,$\{b_n\}$ 是单调递减的有理数列,且 $\lim\limits_{n \to \infty} b_n = \lim\limits_{n \to \infty} a_n = \mathrm{e}$(无理数).

(1) 点集 $\{a_n \mid n = 1, 2, \cdots\}$ 非空有上界,但在有理数集内无上确界;点集 $\{b_n \mid n = 1, 2, \cdots\}$ 非空有下界,但在有理数集内无下确界.

(2) 数列 $\{a_n\}$ 单调递增有上界,但在有理数集内无极限;$\{b_n\}$ 单调递减有下界,但在有理数集内无极限.

(3) $\{a_n \mid n = 1, 2, \cdots\}$ 是有界无限点集,但在有理数集内无聚点.

(4) 数列 $\{a_n\}$ 满足柯西收敛准则条件,但在有理数集内没有极限.

(5) $\{[a_n, b_n]\}$ 是一闭区间套,但在有理数集内不存在一点 ξ,使得 $\xi \in [a_n, b_n], n = 1, 2, \cdots$.

Ex. Try to illustrate by examples that the following propositions is not valid in rational numbers set: principles of supremum and infimum, theorem of monotonicity and boundedness, accumulation theorem and Cauchy convergence criterion.

Solution Let $a_n = \left(1 + \dfrac{1}{n}\right)^n$, $b_n = \left(1 + \dfrac{1}{n}\right)^{n+1}$, $n = 1, 2, \cdots$, then $\{a_n\}$ is an monotonically increasing rational numbers sequence and $\{b_n\}$ is an monotonically decreasing rational numbers sequence, moreover, $\lim\limits_{n \to \infty} b_n = \lim\limits_{n \to \infty} a_n = \mathrm{e}$ (an irrational number).

(1) The point set $\{a_n \mid n = 1, 2, \cdots\}$ is nonempty and has an upper bound, however, it has no supremum in rational numbers set. Similarly, $\{b_n \mid n = 1, 2, \cdots\}$ is nonempty and has a lower bound, however, it has no infimum in rational numbers set.

(2) The sequence $\{a_n\}$ increases monotonically and has an upper bound, but it has no limit in rational numbers set; while $\{b_n\}$ decreases monotonically and has a lower bound, but it has no limit in rational numbers set.

(3) $\{a_n \mid n = 1, 2, \cdots\}$ is a bounded and infinite point set, but it has no accumulation point in rational numbers set.

(4) The sequence $\{a_n\}$ satisfies the Cauchy convergence criterion, but it has no limit in rational numbers set.

(5) $\{[a_n, b_n]\}$ is a cover of closed intervals, but there is no point ξ in rational numbers set such that $\xi \in [a_n, b_n], n = 1, 2, \cdots$.

第八章 不定积分
Chapter 8　Indefinite Integrals

微分学的基本问题是求给定函数的变化率,即是求导问题.而对于给定导数函数求原函数问题是积分学的基本问题之一.本章我们将考虑该基本问题.首先我们介绍不定积分的概念和研究不定积分的基本性质;然后给出一些新的积分方法,这些新的方法是换元法、分部积分法等.为了求更多的初等函数的积分我们必须掌握这些积分方法.

The basic problem of differential calculus is to find the rate of change for a given function, that is the problem of finding derivative. While the problem to find the primitive function for a given derivative is one of the basic problems of the integral calculus. In this chapter, we shall consider such a basic problem. Firstly, we introduce the concept of indefinite integral and study its basic properties, and then present some new methods, these new methods are integration by substitution, integration by parts, and so on. We must master these integrations in order to find more integrals of elementary functions.

单词和短语 Words and expressions

★ 不定积分　indefinite integral
★ 原函数　primitive function
被积函数　integrand function
积分符号　integral sign
被积表达式　expression of integrand
积分变量　integral variable
积分常数　integral constant
★ 基本积分表　table of basic integrals
不定积分的几何意义　geometric meaning of indefinite integral
积分曲线　integral curve
初始条件　initial condition
分部积分法　integration by parts

换元积分法　integration by substitution
第一换元公式　formula of substitution of the first kind
第二换元公式　formula of substitution of the second kind
有理函数的不定积分　indefinite integral of rational function
真分式　proper fraction
假分式　improper fraction
部分分式分解　decomposition into partial fractions
待定系数法　method of undetermined coefficients

本章重点

■1 掌握原函数与不定积分的概念，以及二者之间的区别。
■2 什么样的函数存在原函数？
■3 记住分部积分公式与换元积分公式且会应用它们。
■4 掌握代数学中关于化有理函数为部分分式的方法。

Key points of this chapter

■1 Master the concepts and the difference between primitive functions and indefinite integrals.
■2 What kinds of functions have primitive functions?
■3 Remember the formula of integration by parts and the formula of integration by substitution and can apply them.
■4 Master the methods of expressing rational functions into partial fractions in algebra.

第九章　定积分
Chapter 9　Definite Integrals

本章中，为给出定积分的定义，我们首先对面积进行了简洁的讨论，并用黎曼和定义了定积分；然后我们讨论了定积分的可积条件，性质和计算；最后将证明微积分基本定理，该定理描述了微分与积分之间的关系，对于计算许多积分提供了简便的方法。定积分是微积分中的基本概念之一，是在数学、物理、机械等学科领域的强有力研究工具。

In this chapter, we briefly discuss the areas serving to motivate the definition of definite integrals and define definite integrals in term of Riemann sum. Then we discuss its conditions of integrability, properties and evaluation. Finally, the fundamental theorem of calculus is proved. It describes the relationship between integrability and differentiation and provides a simple method of computing many integrals. Definite integral is one of the basic concepts of calculus and a powerful research tool in mathematics, physics, mechanics, and other disciplines.

单词和短语 Words and expressions

★ 定积分　definite integral
曲边梯形　curvilinear trapezoid
分割　dividing / partition
模　norm / modulus
黎曼和　Riemann sum

★ 黎曼积分　Riemann integral
黎曼可积　integrability in the sense of Riemann
积分区间　interval of integration
上限和下限　upper and lower limits
定积分的几何意义　geometric meaning of defi-

nite integral
大和　upper sum
小和　lower sum
达布大和　Darboux upper sum
达布小和　Darboux lower sum
★牛顿-莱布尼茨公式　Newton-Leibniz formula
可积的必要条件　necessary condition for integrability
可积的充要条件　necessary and sufficient conditions for integrability
可积函数类　integrable function class
定积分的线性性质　linear property of definite integral
积分区间的可加性　additive with respect to integral interval / additivity over integral interval
★积分中值定理　mean value theorem of integral
积分第一中值定理　first mean value theorem of integral
平均值　average value
变上限积分　integral with variant upper limit
原函数的存在性　existence of primitive function
★原函数的存在定理　existence theorem of primitive function
换元积分法　integration by substitution
分部积分法　integration by parts
泰勒公式的积分型余项　integral form of remainder of Taylor formula

例 题　Examples

例 1　设 $f(x)=\begin{cases} xe^{-x^2}, & x \geqslant 0, \\ \dfrac{1}{1+e^x}, & x<0. \end{cases}$　计算 $I=\int_1^4 f(x-2)\mathrm{d}x$.

解　令 $t=x-2$，则有

$$I=\int_{-1}^2 f(t)\mathrm{d}t=\int_{-1}^0 f(t)\mathrm{d}t+\int_0^2 f(t)\mathrm{d}t=\int_{-1}^0 \frac{1}{1+e^t}\mathrm{d}t+\int_0^2 te^{-t^2}\mathrm{d}t$$

$$=-\int_{-1}^0 \frac{\mathrm{d}(e^{-t}+1)}{e^{-t}+1}+\frac{1}{2}\int_0^2 e^{-t^2}\mathrm{d}t^2=\ln\frac{e+1}{2}+\frac{1}{2}(1-e^{-4}).$$

Ex. 1　Let $f(x)=\begin{cases} xe^{-x^2}, & x \geqslant 0, \\ \dfrac{1}{1+e^x}, & x<0. \end{cases}$　Compute $I=\int_1^4 f(x-2)\mathrm{d}x$.

Solution　Let $t=x-2$, we then have

$$I=\int_{-1}^2 f(t)\mathrm{d}t=\int_{-1}^0 f(t)\mathrm{d}t+\int_0^2 f(t)\mathrm{d}t=\int_{-1}^0 \frac{1}{1+e^t}\mathrm{d}t+\int_0^2 te^{-t^2}\mathrm{d}t$$

$$=-\int_{-1}^0 \frac{\mathrm{d}(e^{-t}+1)}{e^{-t}+1}+\frac{1}{2}\int_0^2 e^{-t^2}\mathrm{d}t^2=\ln\frac{e+1}{2}+\frac{1}{2}(1-e^{-4}).$$

例 2　设函数 $f(x)=\dfrac{1}{2}\int_0^x (x-t)^2 g(t)\mathrm{d}t$，其中函数 $g(x)$ 在 $(-\infty,+\infty)$ 上连续，且 $g(1)=5$，$\int_0^1 g(t)\mathrm{d}t=2$. 证明 $f'(x)=x\int_0^x g(t)\mathrm{d}t-\int_0^x tg(t)\mathrm{d}t$，并计算 $f''(1)$ 和 $f'''(1)$.

解　$f(x)=\dfrac{1}{2}\int_0^x (x^2-2xt+t^2)g(t)\mathrm{d}t$

$$=\frac{1}{2}x^2\int_0^x g(t)\mathrm{d}t-x\int_0^x tg(t)\mathrm{d}t+\frac{1}{2}\int_0^x t^2 g(t)\mathrm{d}t,$$

对等式两边求导，得到

$$f'(x)=x\int_0^x g(t)\mathrm{d}t+\frac{1}{2}x^2 g(x)-\left[\int_0^x tg(t)\mathrm{d}t+x^2 g(x)\right]+\frac{1}{2}x^2 g(x)$$

$$=x\int_0^x g(t)\mathrm{d}t-\int_0^x tg(t)\mathrm{d}t.$$

再求导，得到 $f''(x)=\int_0^x g(t)\mathrm{d}t$，$f'''(x)=g(x)$，所以

$$f''(1) = 2, \quad f'''(1) = 5.$$

Ex. 2 Assume that $f(x) = \frac{1}{2}\int_0^x (x-t)^2 g(t)\mathrm{d}t$, where $g(x)$ is continuous on $(-\infty, +\infty)$ and $g(1) = 5, \int_0^1 g(t)\mathrm{d}t = 2$. Prove that $f'(x) = x\int_0^x g(t)\mathrm{d}t - \int_0^x tg(t)\mathrm{d}t$ and compute $f''(1)$ and $f'''(1)$.

Solution $f(x) = \frac{1}{2}\int_0^x (x^2 - 2xt + t^2)g(t)\mathrm{d}t$

$$= \frac{1}{2}x^2 \int_0^x g(t)\mathrm{d}t - x\int_0^x tg(t)\mathrm{d}t + \frac{1}{2}\int_0^x t^2 g(t)\mathrm{d}t.$$

Computing the derivatives of both sides of the equation, we obtain

$$f'(x) = x\int_0^x g(t)\mathrm{d}t + \frac{1}{2}x^2 g(x) - \left[\int_0^x tg(t)\mathrm{d}t + x^2 g(x)\right] + \frac{1}{2}x^2 g(x)$$

$$= x\int_0^x g(t)\mathrm{d}t - \int_0^x tg(t)\mathrm{d}t.$$

Further, we have $f''(x) = \int_0^x g(t)\mathrm{d}t, f'''(x) = g(x)$, and thus

$$f''(1) = 2, \quad f'''(1) = 5.$$

例 3 设 $(0, +\infty)$ 上的连续函数 $f(x)$ 满足 $f(x) = \ln x - \int_1^e f(x)\mathrm{d}x$, 求 $\int_1^e f(x)\mathrm{d}x$.

解 记 $\int_1^e f(x)\mathrm{d}x = a$, 则 $f(x) = \ln x - a$, 于是,

$$a = \int_1^e f(x)\mathrm{d}x = \int_1^e \ln x \mathrm{d}x - a(e-1),$$

所以 $a = \frac{1}{e}\int_1^e \ln x \mathrm{d}x = \frac{1}{e}(x\ln x - x)\Big|_1^e = \frac{1}{e}$.

Ex. 3 Assume that $f(x)$ is continuous on $(0, +\infty)$ and satisfies $f(x) = \ln x - \int_1^e f(x)\mathrm{d}x$, compute $\int_1^e f(x)\mathrm{d}x$.

Solution Let $\int_1^e f(x)\mathrm{d}x = a$, then we have $f(x) = \ln x - a$, moreover,

$$a = \int_1^e f(x)\mathrm{d}x = \int_1^e \ln x \mathrm{d}x - a(e-1).$$

Consequently, $a = \frac{1}{e}\int_1^e \ln x \mathrm{d}x = \frac{1}{e}(x\ln x - x)\Big|_1^e = \frac{1}{e}$.

本章重点

1. 知道定积分的客观背景,以及解决这些实际问题的思想方法.
2. 积分小和与积分大和对讨论函数可积性起了什么作用?
3. 函数 $f(x)$ 在 $[a,b]$ 上可积有哪些等价命题?
4. 记住定积分的性质,并清楚每个性质在解题中的作用.
5. 说明微积分基本定理的意义.
6. 函数 $f(x)$ 可积,$f(x)$ 是否存在原函数? 反之,函数 $g(x)$ 存在原函数,$g(x)$ 是否可积?

Key points of this chapter

1. Know about the background of definite integrals and the thinking method of solving these actual problems.
2. What effects do the integral lower sum and the integral upper sum have positively on the discussion of integrability of a function?
3. What equivalent propositions does the integrability have when the function $f(x)$ is defined on $[a,b]$?

4 Remember the properties of definite integrals, and show the application of each property in solving problems.

5 Show the meaning of the basic theorem of calculus.

6 If the function $f(x)$ is integrable, whether it has a primitive function? Contrarily, if the function $g(x)$ has a primitive function, whether it is integrable?

第十章 定积分的应用
Chapter 10 Applications of Definite Integrals

  本章专门研究积分的应用,定积分在数学的应用包括求体积、弧长和面积.物理的应用包括变力做功、平面流体静压力、矩和质心的计算,我们必须掌握这些积分应用公式.

  This chapter is devoted to the applications of definite integrals. The mathematical applications of definite integrals include finding volumes, arc length, and areas. The physical applications include the computations of work done by variable forces, hydrostatic force on a plane, moment and center of a mass. We must master these integral formulae for applications.

单词和短语 Words and expressions

平面图形的面积　area of plane figure
由平行截面面积求体积的方法　method of finding volume of a solid from the known area of parallel sections
平面曲线的弧长与曲率　arc length and curvature of plane curve
光滑曲线　smooth curve
弧微分　differential of arc
曲率圆　circle of curvature
曲率半径　radius of curvature

曲率中心　center of curvature
旋转曲面的面积　area of revolution surface
定积分在物理中的某些应用　some applications of definite integral in physics
引力　gravitation
功　work
定积分的近似计算　approximate computation of definite integral
梯形法　trapezoidal method
抛物线法　parabola method

本章重点

1 理解微元法的思想,说明微元法的理论基础是什么?并能应用微元法将某些几何物理等实际问题化成微积分.

2 熟练地应用本章给出的公式.

Key points of this chapter

1 Understand the thinking of infinitesimal method, and show the theoretical bases of infinitesimal method. Able to turn some actual problems of geometric physics into calculus by using infinitesimal method.

2 Apply the formulae given in this chapter skillfully.

第十一章 反常积分
Chapter 11 Improper Integrals

  之前所讨论的定积分有两个共同的特点:一是两个积分限 a,b 是常数,二是 $f(x)$ 在区间 $[a,b]$ 上有界.然而,在科学研究和实际应用中我们也经常遇到其他种类的积分,如在无穷区间上的积分或是被积函数是无界的.因此,我们需要在这两方面扩展定积分的概念,进而引入反常积分的概念.首先,我们在上两条下讨论新的概念.然后讨论无穷积分的性质与收敛判别法——比较判别法、柯西判别法、狄利克雷判别法和阿贝尔判别法.

  Definite integrals discussed in the preceding chapters have two common features. Firstly, both integral limits a and b are constants. Secondly, the integrand is bounded over the integral interval $[a,b]$.

However, we also often encounter other kinds of integrals in scientific research or actual applications. Such as a definite integral is on an infinite interval or the integrand is unbounded. We need, therefore, to extend the concept of definite integrals to the two aspects, and thus introduce the concepts of improper integrals. We firstly discuss the new concept under two items, and then discuss the properties of infinite integral and some tests of convergence, such as, comparison test, Cauchy test, Dirichlet test and Abel test.

单词和短语 Words and expressions

反常积分的概念　notion of improper integral	★ 绝对收敛　absolutely convergent
★ 无穷区间上的反常积分　improper integral on infinite interval	★ 条件收敛　conditionally convergent
	比较判别法　comparison test
★ 无界函数的反常积分　improper integral of unbounded function	柯西判别法　Cauchy test
	狄利克雷判别法　Dirichlet test
无穷积分的性质与收敛判别　property of infinite integral and test of convergence	阿贝尔判别法　Abel test

例 题　Examples

例 1 设 $\int_a^{+\infty} f(x)\mathrm{d}x$ 收敛，且 $\lim_{x\to+\infty} f(x) = A$，证明 $A = 0$.

证明 用反证法. 不妨设 $A > 0$，则对 $\varepsilon = \frac{1}{2}A > 0, \exists X > a, \forall x > X$，有 $|f(x) - A| < \frac{1}{2}A$，从而 $f(x) > \frac{1}{2}A$. 由 $\int_a^B f(x)\mathrm{d}x = \int_a^X f(x)\mathrm{d}x + \int_X^B f(x)\mathrm{d}x > \int_a^X f(x)\mathrm{d}x + \frac{1}{2}A(B-X)$ 可知，$\lim_{B\to+\infty}\int_a^B f(x)\mathrm{d}x = +\infty$，与 $\int_a^{+\infty} f(x)\mathrm{d}x$ 收敛矛盾.

同理也可证明不可能有 $A < 0$，所以 $A = 0$.

Ex. 1 Assume that $\int_a^{+\infty} f(x)\mathrm{d}x$ is convergent and $\lim_{x\to+\infty} f(x) = A$, prove that $A = 0$.

Proof Use the reduction to absurdity. Not loss of generality, let $A > 0$, then for $\varepsilon = \frac{1}{2}A > 0$, $\exists X > a, \forall x > X$, we have $|f(x) - A| < \frac{1}{2}A$, and so $f(x) > \frac{1}{2}A$. From $\int_a^B f(x)\mathrm{d}x = \int_a^X f(x)\mathrm{d}x + \int_X^B f(x)\mathrm{d}x > \int_a^X f(x)\mathrm{d}x + \frac{1}{2}A(B-X)$, we know that $\lim_{B\to+\infty}\int_a^B f(x)\mathrm{d}x = +\infty$, which contradicts with the convergence of $\int_a^{+\infty} f(x)\mathrm{d}x$.

Similarly, it is impossible that $A < 0$. Consequently, $A = 0$.

例 2 讨论下列反常积分的敛散性：

(1) $\int_1^{+\infty} \frac{1}{\sqrt{x^3 - \mathrm{e}^{-2x} + \ln x + 1}}\mathrm{d}x$；　(2) $\int_2^{\infty} \frac{\ln\ln x}{\ln x}\sin x\, \mathrm{d}x$；

(3) $\int_1^{\infty} \frac{\sin x \arctan x}{x^p}\mathrm{d}x$.

解 (1) 当 $x \to +\infty$ 时，易证 $\frac{1}{\sqrt{x^3 - \mathrm{e}^{-2x} + \ln x + 1}} \sim \frac{1}{x^{\frac{3}{2}}}$，所以积分 $\int_1^{+\infty} \frac{1}{\sqrt{x^3 - \mathrm{e}^{-2x} + \ln x + 1}}\mathrm{d}x$ 收敛.

(2) 因为 $F(A) = \int_2^A \sin x\,\mathrm{d}x$ 有界，$\frac{\ln\ln x}{\ln x}$ 在 $[2, +\infty)$ 单调，且 $\lim_{x\to+\infty}\frac{\ln\ln x}{\ln x} = 0$，由 Dirichlet 判别法，积分 $\int_2^{+\infty}\frac{\ln\ln x}{\ln x}\sin x\,\mathrm{d}x$ 收敛；

由于 $\left|\dfrac{\ln\ln x}{\ln x}\sin x\right| \geqslant \left|\dfrac{\ln\ln x}{\ln x}\right|\sin^2 x = \dfrac{1}{2}\left|\dfrac{\ln\ln x}{\ln x}\right|(1-\cos 2x)$, 而积分 $\int_2^{+\infty}\left|\dfrac{\ln\ln x}{\ln x}\right|dx$ 发散, $\int_2^{+\infty}\left|\dfrac{\ln\ln x}{\ln x}\right|\cos 2x dx$ 收敛, 所以积分 $\int_2^{+\infty}\left|\dfrac{\ln\ln x}{\ln x}\sin x\right|dx$ 发散, 即积分 $\int_2^{+\infty}\dfrac{\ln\ln x}{\ln x}\sin x dx$ 条件收敛.

(3) 当 $p>1$ 时, $\dfrac{|\sin x\arctan x|}{x^p} \leqslant \dfrac{\pi}{2x^p}$, 而 $\int_1^{+\infty}\dfrac{1}{x^p}dx$ 收敛, 所以当 $p>1$ 时积分 $\int_1^{+\infty}\dfrac{\sin x\arctan x}{x^p}dx$ 绝对收敛;

当 $0<p\leqslant 1$ 时, 因为 $F(A)=\int_1^A \sin x dx$ 有界, $\dfrac{\arctan x}{x^p}$ 在 $[1,+\infty)$ 单调, 且 $\lim\limits_{x\to +\infty}\dfrac{\arctan x}{x^p}=0$, 由 Dirichlet 判别法, 积分 $\int_1^{+\infty}\dfrac{\sin x\arctan x}{x^p}dx$ 收敛; 但因为当 $0<p\leqslant 1$ 时积分 $\int_1^{+\infty}\dfrac{\arctan x}{x^p}|\sin x|dx$ 发散, 所以当 $0<p\leqslant 1$ 时积分 $\int_1^{+\infty}\dfrac{\sin x\arctan x}{x^p}dx$ 条件收敛.

Ex. 2 Discuss the properties of convergence and divergence of the following improper integrals:

(1) $\int_1^{+\infty}\dfrac{1}{\sqrt{x^3-e^{-2x}+\ln x+1}}dx$; (2) $\int_2^{\infty}\dfrac{\ln\ln x}{\ln x}\sin x dx$;

(3) $\int_1^{\infty}\dfrac{\sin x\arctan x}{x^p}dx$.

Solution (1) As $x\to +\infty$, it is easy to show that $\dfrac{1}{\sqrt{x^3-e^{-2x}+\ln x+1}}\sim \dfrac{1}{x^{\frac{3}{2}}}$, and so the improper integral $\int_1^{+\infty}\dfrac{1}{\sqrt{x^3-e^{-2x}+\ln x+1}}dx$ is convergent.

(2) Since $F(A)=\int_2^A \sin x dx$ is bounded, moreover, $\dfrac{\ln\ln x}{\ln x}$ is monotonic on $[2,+\infty)$ and $\lim\limits_{x\to +\infty}\dfrac{\ln\ln x}{\ln x}=0$, from the Dirichlet test, we know that the integral $\int_2^{+\infty}\dfrac{\ln\ln x}{\ln x}\sin x dx$ is convergent. On the other hand, since

$$\left|\dfrac{\ln\ln x}{\ln x}\sin x\right| \geqslant \left|\dfrac{\ln\ln x}{\ln x}\right|\sin^2 x = \dfrac{1}{2}\left|\dfrac{\ln\ln x}{\ln x}\right|(1-\cos 2x),$$

and $\int_2^{+\infty}\left|\dfrac{\ln\ln x}{\ln x}\right|dx$ is divergent and $\int_2^{+\infty}\left|\dfrac{\ln\ln x}{\ln x}\right|\cos 2x dx$ is convergent, so the integral $\int_2^{+\infty}\left|\dfrac{\ln\ln x}{\ln x}\sin x\right|dx$ is divergent, i.e., $\int_2^{+\infty}\dfrac{\ln\ln x}{\ln x}\sin x dx$ is conditionally convergent.

(3) If $p>1$, we have $\dfrac{|\sin x\arctan x|}{x^p} \leqslant \dfrac{\pi}{2x^p}$, so $\int_1^{+\infty}\dfrac{1}{x^p}dx$ is convergent, consequently, $\int_1^{+\infty}\dfrac{\sin x\arctan x}{x^p}dx$ is absolutely convergent as $p>1$.

If $0<p\leqslant 1$, $F(A)=\int_1^A \sin x dx$ is bounded and $\dfrac{\arctan x}{x^p}$ is monotonic on $[1,+\infty)$, moreover, $\lim\limits_{x\to +\infty}\dfrac{\arctan x}{x^p}=0$. From the Dirichlet test, we know that the integral $\int_1^{+\infty}\dfrac{\sin x\arctan x}{x^p}dx$ is convergent. However, $\int_1^{+\infty}\dfrac{\arctan x}{x^p}|\sin x|dx$ is divergent as $0<p\leqslant 1$, and thus $\int_1^{+\infty}\dfrac{\sin x\arctan x}{x^p}dx$ is conditionally convergent as $0<p\leqslant 1$.

第十二章　数项级数
Chapter 12　Series of Number Terms

本章我们关注数项级数. 首先引入级数的一些重要理论、性质及柯西收敛准则. 关于一个级数总存在两个重要的问题：

(1) 这个级数收敛吗？

(2) 如果收敛,此级数的和是多少？

本章中我们将回答这个问题,但是我们主要以两个非常特殊类型的级数为例进行讨论:几何级数和调和级数.

对于正项(至少是非负项)级数,我们给出一些简单的且著名的收敛性检验准则.

对于任意项级数,为证明某些级数是收敛的,需要掌握收敛性检验的一些定理,如判别正项级数收敛的比较判别法、柯西判别法、达朗贝尔判别法、积分判别法、拉阿伯判别法以及判别一般级数绝对收敛和条件收敛的莱布尼茨判别法、阿贝尔判别法、狄利克雷判别法等等.

In this chapter, we are concerned with series of number terms. We first introduce some important ideas, properties and the Cauchy convergence criterion for series. For a series, there always have two important questions to ask.

(1) Does the series converge?

(2) If it converges, what is its sum?

We shall answer these questions in this chapter, but we mainly discuss two special types of series as examples, namely, geometric series and harmonic series.

For series with positive (or at least nonnegative) terms, we will present some simple and known convergence tests.

For series with terms of arbitrary signs, it is required to master some theorems of convergence test in order to show that a series is convergent, such as comparison test, Cauchy test, D'Alembert test, Integral test, Raabe test for testing the convergence of positive term series; for testing the absolute convergence and conditional convergence of common series, Leibnitz test, Abel test, Dirichlet test, and so on.

单词和短语 Words and expressions

★ 数项级数　series with number terms

无穷级数　infinite series

★ 级数的收敛性　convergence of series

部分和序列　sequence of partial sum

★ 几何级数　geometric series

★ 调和级数　harmonic series

★ 级数收敛的柯西准则　Cauchy convergence criterion for series

★ 正项级数　series with positive terms

收敛的必要条件　necessary condition for convergence

比较判别法　comparison test

根式判别法　root test

达朗贝尔判别法(比式判别法)　D'Alembert (ratio) test

比式判别法的极限形式　limit form of ratio test

柯西判别法(根式判别法)　Cauchy (root) test

根式判别法的极限形式　limit form of root test

高斯判别法　Gauss test

积分判别法　integral test

拉阿伯判别法　Raabe test

p 级数　p-series

一般项级数　series with arbitrary terms

交错级数　alternating series

莱布尼茨判别法　Leibnitz test

级数的重排　rearrangement of series

阿贝尔判别法　Abel test

狄利克雷判别法　Dirichlet test

基本概念和性质 Basic concepts and properties

正项级数收敛的当且仅当它的部分和序列有上界.

A series with positive terms converges if and only if the sequence of its partial sums has an upper bound.

例 题 Examples

例 判别下列级数的敛散性：

(1) $\sum_{n=1}^{\infty} \frac{2^n n!}{n^n}$;　　(2) $\sum_{n=1}^{\infty} \frac{n+2}{n+1} \cdot \frac{1}{\sqrt[3]{n}}$;　　(3) $\sum_{n=1}^{\infty} (-1)^n \sin \frac{1}{n}$.

解 (1) 事实上,$\lim\limits_{n\to\infty}\dfrac{2^{n+1}\cdot(n+1)!}{(n+1)^{n+1}}\cdot\dfrac{n^n}{2^n\cdot n!}=\lim\limits_{n\to\infty}\dfrac{2}{\left(1+\dfrac{1}{n}\right)^n}=\dfrac{2}{e}<1$,根据达朗贝尔判别法,该级数收敛.

(2) 该级数发散. 事实上,$\lim\limits_{n\to\infty}\dfrac{\dfrac{n+2}{n+1}\cdot\dfrac{1}{\sqrt[3]{n}}}{\dfrac{1}{\sqrt[3]{n}}}=\lim\limits_{n\to\infty}\dfrac{n+2}{n+1}=1$,已知级数 $\sum\limits_{n=1}^{\infty}\dfrac{1}{\sqrt[3]{n}}$ 发散,从而 $\sum\limits_{n=1}^{\infty}\dfrac{n+2}{n+1}\cdot\dfrac{1}{\sqrt[3]{n}}$ 发散.

(3) 该级数条件收敛. 对 $\forall n\in \mathbf{N}_+$ 有 $\sin\dfrac{1}{n}>\sin\dfrac{1}{n+1}$,$\lim\limits_{n\to\infty}\sin\dfrac{1}{n}=0$,根据莱布尼茨判别法,$\sum\limits_{n=1}^{\infty}(-1)^n\sin\dfrac{1}{n}$ 收敛;由于 $\lim\limits_{n\to\infty}\left|\dfrac{\sin\dfrac{1}{n}}{\dfrac{1}{n}}\right|=1$,已知级数 $\sum\limits_{n=1}^{\infty}\dfrac{1}{n}$ 发散,从而 $\sum\limits_{n=1}^{\infty}(-1)^n\sin\dfrac{1}{n}$ 条件收敛.

Ex Test the properties of convergence and divergence of the following series:

(1) $\sum\limits_{n=1}^{\infty}\dfrac{2^n n!}{n^n}$;　　(2) $\sum\limits_{n=1}^{\infty}\dfrac{n+2}{n+1}\cdot\dfrac{1}{\sqrt[3]{n}}$;　　(3) $\sum\limits_{n=1}^{\infty}(-1)^n\sin\dfrac{1}{n}$.

Solution (1) In fact, $\lim\limits_{n\to\infty}\dfrac{2^{n+1}\cdot(n+1)!}{(n+1)^{n+1}}\cdot\dfrac{n^n}{2^n\cdot n!}=\lim\limits_{n\to\infty}\dfrac{2}{\left(1+\dfrac{1}{n}\right)^n}=\dfrac{2}{e}<1$, from the D'Alembert test, we know that this series is convergent.

(2) This series is divergent. In fact, $\lim\limits_{n\to\infty}\dfrac{\dfrac{n+2}{n+1}\cdot\dfrac{1}{\sqrt[3]{n}}}{\dfrac{1}{\sqrt[3]{n}}}=\lim\limits_{n\to\infty}\dfrac{n+2}{n+1}=1$, it is known that $\sum\limits_{n=1}^{\infty}\dfrac{1}{\sqrt[3]{n}}$ is divergent, consequently, $\sum\limits_{n=1}^{\infty}\dfrac{n+2}{n+1}\cdot\dfrac{1}{\sqrt[3]{n}}$ is also divergent.

(3) This series is conditionally convergent. For $\forall n\in\mathbf{N}_+$ we have $\sin\dfrac{1}{n}>\sin\dfrac{1}{n+1}$ and $\lim\limits_{n\to\infty}\sin\dfrac{1}{n}=0$. From the Leibnitz test, we know that $\sum\limits_{n=1}^{\infty}(-1)^n\sin\dfrac{1}{n}$ is convergent; however, since $\lim\limits_{n\to\infty}\left|\dfrac{\sin\dfrac{1}{n}}{\dfrac{1}{n}}\right|=1$ and $\sum\limits_{n=1}^{\infty}\dfrac{1}{n}$ is divergent, and thus $\sum\limits_{n=1}^{\infty}(-1)^n\sin\dfrac{1}{n}$ is conditionally convergent.

本章重点

1. 掌握级数收敛的判别法,记住几何级数 $\sum\limits_{n=1}^{\infty}q^n$ 和广义调和级数 $\sum\limits_{n=1}^{\infty}\dfrac{1}{n^p}$ 收敛的条件.
2. 掌握级数条件收敛和绝对收敛的性质及其证明方法.
3. 具有应用级数的收敛定义和收敛级数的性质证明级数中的一些理论问题的能力.
4. 正项(同号)级数有哪些敛散性判别方法?它们的理论基础是什么?判别法之间有什么关系?
5. 判别变号级数的敛散性有哪些方法?

Key points of this chapter

1. Master the tests of convergence of series, remember the convergence conditions of the geometric se-

ries $\sum_{n=1}^{\infty} q^n$ and the generalized harmonic series $\sum_{n=1}^{\infty} \frac{1}{n^p}$.

2 Master the properties of conditional convergence and absolute convergence of series and their proving methods.

3 Have the ability of proving some theoretical problems of series by using the concept and the properties of convergent series.

4 What kinds of methods to test convergence do with positive terms series have? What are their theoretical bases? What relationships are there among these test methods?

5 What methods are there to test convergence of series with variable signs?

第十三章 函数列与函数项级数
Chapter 13 Sequences of Functions and Series of Functions

上一章中,我们一直研究所谓的常数项级数,即形为 $\sum u_n$ 的级数,其中每个 u_n 均为数. 现在我们考虑函数项级数 $\sum u_n(x)$. 当然,一旦我们取定 x,就回到了熟悉的常数项级数. 关于一个函数项级数,有三个主要问题:

1. 对于什么样的 x,级数收敛?
2. 级数对于什么函数收敛,即收敛级数的和 $S(x)$ 是什么?
3. 本章中最重要的问题是研究函数的重要性质在极限运算 $f(x) = \lim_{n \to \infty} f_n(x), x \in E$ 和 $f(x) = \sum_{n=1}^{\infty} f_n(x)$ 下是否继续保持,例如,若函数 f_n 是连续的,或者可微的,或者可积的,那么其极限函数是否也有同样的性质? f'_n 和 f' 之间的关系如何, f_n 和 f 的积分之间的关系又如何?

所有这些问题的解决都需要掌握收敛性的一个新概念—— 一致收敛.

In the previous chapter, we have studied the so-called series with constant terms, that is, series of the form $\sum u_n$, where each u_n is a number. Now we consider the series of functions $\sum u_n(x)$, of course, as soon as we substitute a value for x, we are back to the series with constant terms that we are familiar. For a series of functions, we always have three main questions to ask.

1. For what values of x does the series converge?
2. To what function does the series converge, that is, what is the sum $S(x)$ of convergent series?
3. The most important problem which arises in this chapter is to determine whether the important properties of functions are preserved under the limit operations $f(x) = \lim_{n \to \infty} f_n(x), x \in E$ and $f(x) = \sum_{n=1}^{\infty} f_n(x)$, for instance, if the functions f_n are continuous, or differentiable, or integrable, are the same true to the limit functions? What are relations between f'_n and f', or between the integrals of f_n and that of f?

To solve these questions, it is required to master a new concept of convergence, i.e., uniform convergence.

单词和短语 Words and expressions

在点 x_0 收敛　convergent at x_0
★ 收敛域　convergence domain / region of convergence
★ 和函数　sum function
极限函数　limit function
函数项级数　series of functions
★ 一致收敛性　uniform convergence
一致收敛判别法　test of uniform convergence
★ 魏尔斯特拉斯判别法　Weierstrass's test / Weierstrass uniform convergence Criterion / Weierstrass M-test for uniform convergence
一致有界　uniform boundedness
狄利克雷判别法　Dirichlet test

基本概念和性质 Basic concepts and properties

1 如果级数的各项在区间$[a,b]$上连续,且级数在$[a,b]$上一致收敛于和函数,则和函数在$[a,b]$上也是连续的。
If the general term of the series is continous on $[a,b]$, and the series is uniformly convergent with the sum function $[a,b]$. Then the sum function is also continous on $[a,b]$.

2 如果级数的各项在区间$[a,b]$上连续,且级数在$[a,b]$上一致收敛于和函数,则和函数在$[a,b]$上可逐项积分。
If the general term of the series is continoues on $[a,b]$, and the series is uniformly convergent with the sum function $[a,b]$. Then the sum function can be integrated term by term.

3 如果级数$\sum_{n=1}^{\infty}U_n(x)$在区间$[a,b]$上收敛于$S(x)$,它的各项$U_n(x)$都具有连续导数,$U'_n(x)$,且$\sum U'_n(x)$在$[a,b]$上一致收敛,则$\sum U_n(x)$在$[a,b]$上也一致收敛,且可逐项求导。
Let $\sum U_n(x)$ be a series with the function $S(x)$ on the interal $[a,b]$. If the derivative $U'_n(x)$ of the genered tems $U_n(x)$ is continous, and $\sum_{n=1}^{\infty}U'_n(x)$ is uniformly convergent on $[a,b]$. Then $\sum U_n(x)$ is also uniformly convergent on $[a,b]$, can be differentiated term by term.

例题 Examples

例 设$f(x)=\sum_{n=0}^{\infty}x^n$,证明:

(1) $f(x)$的和函数$s(x)$在区间$(-1,1)$连续;

(2) 函数级数$\sum_{n=0}^{\infty}x^n$在$(-1,1)$非一致收敛.

证明 (1) 容易求得$f(x)$的和函数为$\frac{1}{1-x}$,并且易证它在$(-1,1)$连续.

(2) $\exists \varepsilon_0=1, \forall n \in \mathbf{N}, \exists n_0 > N, \exists x_0 = 1-\frac{1}{n_0} \in (-1,1)$,有

$$|s(x_0)-s_{n_0}(x_0)| = \frac{\left(1-\frac{1}{n_0}\right)^{n_0}}{\frac{1}{n_0}} = n_0\left(1-\frac{1}{n_0}\right)^{n_0} \geq 1$$

即函数级数$\sum_{n=0}^{\infty}x^n$在$(-1,1)$非一致收敛.

Ex Let $f(x) = \sum_{n=0}^{\infty}x^n$, prove the following conclusions:

(1) The sum function $s(x)$ of $f(x)$ is continuous in $(-1,1)$;

(2) The function series $\sum_{n=0}^{\infty}x^n$ does not converge uniformly in $(-1,1)$.

Proof (1) It is easy to show that the sum function of $f(x)$ is given by $\frac{1}{1-x}$, and that it is continuous in $(-1,1)$.

(2) Obviously, $\exists \varepsilon_0 = 1, \forall n \in \mathbf{N}, \exists n_0 > N, \exists x_0 = 1-\frac{1}{n_0} \in (-1,1)$, we have

$$|s(x_0)-s_{n_0}(x_0)| = \frac{\left(1-\frac{1}{n_0}\right)^{n_0}}{\frac{1}{n_0}} = n_0\left(1-\frac{1}{n_0}\right)^{n_0} \geq 1.$$

Consequently, the function series $\sum_{n=0}^{\infty} x^n$ does not converge uniformly in $(-1, 1)$.

本章重点

1 一致收敛是本章的重要概念,和函数(或极限函数)的分析性质是本章的核心内容.
2 要求掌握判别函数列 $\{f_n(x)\}$ 在区间上一致收敛的判别方法.

Key points of this chapter

1 Uniform convergence is an important concept of this chapter, while the analytical properties of sum functions (or limit functions) are the core of the chapter.
2 Master and distinguish the test methods of uniform convergence of the function sequence $\{f_n(x)\}$ over certain intervals.

第十四章 幂级数
Chapter 14 Power Series

在这一章,我们将讨论另一类特殊的函数项级数——幂级数. 我们首先给出幂级数的一些性质. 例如,

1. 它的收敛集一定是如下种类的区间之一:
(1) 单点集 $x = a$;
(2) 区间 $(a - R, a + R)$,可能加一个或两个端点;
(3) 整个实直线.

2. 幂级数可逐项积分和逐项微分,它隐含着可微的和可积的级数的收敛半径和原级数是一样的.

这些性质一个好的应用是能从一个已知幂级数获得其他幂级数的和的公式. 在本章,我们将回答一个函数 f 能否被一个在点 $x = a$ 的幂级数所表示的问题,即泰勒定理给出了这个答案.

我们以重要的麦克劳林级数的列表来结束我们的讨论. 我们已经发现这些级数将在解决一些问题的时候将非常有用,但是更重要的是,它们的应用将贯穿于整个数学和自然科学当中.

1. $\dfrac{1}{1-x} = 1 + x + x^2 + \cdots, -1 < x < 1$;

2. $\ln(1 + x) = x - \dfrac{x^2}{2} + \dfrac{x^3}{3} - \dfrac{x^4}{4} + \cdots, -1 < x \leqslant 1$;

3. $\arctan x = x - \dfrac{x^3}{3} + \dfrac{x^5}{5} - \dfrac{x^7}{7} + \dfrac{x^9}{9} - \cdots, -1 \leqslant x \leqslant 1$;

4. $e^x = 1 + x + \dfrac{x^2}{2!} + \dfrac{x^3}{3!} + \cdots, -\infty < x < +\infty$;

5. $\sin x = x - \dfrac{x^3}{3!} + \dfrac{x^5}{5!} - \cdots, -\infty < x < +\infty$;

6. $\cos x = 1 - \dfrac{x^2}{2!} + \dfrac{x^4}{4!} - \cdots, -\infty < x < +\infty$;

7. $(1 + x)^p = 1 + \binom{p}{1} x + \binom{p}{2} x^2 + \binom{p}{3} x^3 + \cdots$.

In this chapter, we shall discuss another special function series — power series, and first present some properties of power series. For example,

1. Its convergence set is always one of the following kinds of intervals.
(1) The single point $x = a$.
(2) An interval $(a - R, a + R)$, plus possibly one or both end points.
(3) The whole real line.

2. For power series, the properties of termwise differentiation and termwise integration imply that the radius of convergence of both differentiable and integrable series are the same as the original series.

A good consequence of the properties is that we can apply it to a power series with a known sum formula to obtain sum formulae for other series. In this chapter, we answer the question on whether or not a function f can be represented by a power series at the point $x = a$, that is, the Taylor Theorem gives the answer.

We conclude our discussions of series with a list of the important Maclaurin series. We have found that these series will be useful in solving some problems. However, what is more significant, their applications will run through mathematics and science.

单词和短语 Words and expressions

★ 幂级数　　power series
　收敛区间　interval of convergence
★ 收敛半径　radius of convergence
　阿贝尔定理　Abel theorem
　幂级数的运算　operations of power series
★ 泰勒级数　Taylor series

初等函数的幂级数展开式　expansion of power series of elementary function
复变量的指数函数　exponential function of a complex variable
★ 欧拉公式　Euler Formula

例 题　Examples

例　设 $\sum\limits_{n=0}^{\infty} a_n x^n$ 与 $\sum\limits_{n=0}^{\infty} b_n x^n$ 的收敛半径分别为 R_1 和 R_2，讨论下列幂级数的收敛半径：

(1) $\sum\limits_{n=0}^{\infty} a_n x^{2n}$；　(2) $\sum\limits_{n=0}^{\infty} (a_n + b_n) x^n$；　(3) $\sum\limits_{n=0}^{\infty} a_n b_n x^n$.

解　(1) 设 $\sum\limits_{n=0}^{\infty} a_n x^{2n}$ 的收敛半径为 R，则 ① 当 $|x| < \sqrt{R_1}$ 时，$\sum\limits_{n=0}^{\infty} a_n x^{2n}$ 收敛；② 当 $|x| > \sqrt{R_1}$ 时，$\sum\limits_{n=0}^{\infty} a_n x^{2n}$ 发散，所以 $R = \sqrt{R_1}$.

(2) 设 $\sum\limits_{n=0}^{\infty} (a_n + b_n) x^n$ 的收敛半径为 R，则 ① 当 $|x| < \min(R_1, R_2)$ 时，$\sum\limits_{n=0}^{\infty} (a_n + b_n) x^n$ 收敛；② 当 $|x| > \min(R_1, R_2), R_1 \neq R_2$ 时，$\sum\limits_{n=0}^{\infty} (a_n + b_n) x^n$ 发散. 但当 $R_1 = R_2$ 时，$\sum\limits_{n=0}^{\infty} (a_n + b_n) x^n$ 的收敛半径有可能增加，例如 $\sum\limits_{n=0}^{\infty} a_n x^n = \sum\limits_{n=0}^{\infty} x^n$，收敛半径为 1，$\sum\limits_{n=0}^{\infty} b_n x^n = \sum\limits_{n=0}^{\infty} \left(\dfrac{1}{2^n} - 1\right) x^n$ 收敛半径也为 1，但 $\sum\limits_{n=0}^{\infty} (a_n + b_n) x^n$ 的收敛半径为 2. 所以 $R \geqslant \min(R_1, R_2)$.

(3) 设 $\sum\limits_{n=0}^{\infty} a_n b_n x^n$ 的收敛半径为 R，则由 $\varlimsup\limits_{n\to\infty} \sqrt[n]{|a_n b_n|} \leqslant \varlimsup\limits_{n\to\infty} \sqrt[n]{|a_n|} \cdot \varlimsup\limits_{n\to\infty} \sqrt[n]{|b_n|}$ 可知，$R \geqslant R_1 R_2$，其中不等式中的等号可能不成立，例如级数 $\sum\limits_{n=0}^{\infty} a_n x^n = \sum\limits_{n=0}^{\infty} x^{2n}$ 的收敛半径为 1，而 $\sum\limits_{n=0}^{\infty} b_n x^n = \sum\limits_{n=0}^{\infty} x^{2n+1}$ 的收敛半径也为 1，其中 n 为奇数时，$a_n = 0, b_n = 1$；n 为偶数时，$a_n = 1, b_n = 0$. 易见 $0 = \sum\limits_{n=0}^{\infty} a_n b_n x^n$ 的收敛半径为 $R = +\infty$.

Ex　Assume that the convergence radii of $\sum\limits_{n=0}^{\infty} a_n x^n$ and $\sum\limits_{n=0}^{\infty} b_n x^n$ are respectively given by R_1 and R_2, discuss the convergence radii of the following power series:

(1) $\sum\limits_{n=0}^{\infty} a_n x^{2n}$；　(2) $\sum\limits_{n=0}^{\infty} (a_n + b_n) x^n$；　(3) $\sum\limits_{n=0}^{\infty} a_n b_n x^n$.

Solution (1) Assume that the convergence radius of $\sum_{n=0}^{\infty} a_n x^{2n}$ is given by R, then ① as $|x| < \sqrt{R_1}$, we know that $\sum_{n=0}^{\infty} a_n x^{2n}$ is convergent; ② while as $|x| > \sqrt{R_1}$, $\sum_{n=0}^{\infty} a_n x^{2n}$ is divergent, and thus $R = \sqrt{R_1}$.

(2) Similarly, let R denotes the convergence radius of $\sum_{n=0}^{\infty}(a_n + b_n)x^n$, then ① as $|x| < \min(R_1, R_2)$, $\sum_{n=0}^{\infty}(a_n + b_n)x^n$ is convergent; ② as $|x| > \min(R_1, R_2)$ and $R_1 \neq R_2$, $\sum_{n=0}^{\infty}(a_n + b_n)x^n$ is divergent. However, as $R_1 = R_2$, the convergence radius of $\sum_{n=0}^{\infty}(a_n + b_n)x^n$ maybe increase, for example, the convergence radii of $\sum_{n=0}^{\infty} a_n x^n = \sum_{n=0}^{\infty} x^n$ and $\sum_{n=0}^{\infty} b_n x^n = \sum_{n=0}^{\infty}\left(\frac{1}{2^n} - 1\right)x^n$ are all 1, but that of $\sum_{n=0}^{\infty}(a_n + b_n)x^n$ is 2. Consequently, $R \geqslant \min(R_1, R_2)$.

(3) Let R denotes the convergence radius of $\sum_{n=0}^{\infty} a_n b_n x^n$, then from $\overline{\lim}_{n \to \infty} \sqrt[n]{|a_n b_n|} \leqslant \overline{\lim}_{n \to \infty} \sqrt[n]{|a_n|} \cdot \overline{\lim}_{n \to \infty} \sqrt[n]{|b_n|}$, we know that $R \geqslant R_1 R_2$, in which the equal sign of the inequality is not valid possibly, for example, the convergence radii of $\sum_{n=0}^{\infty} a_n x^n = \sum_{n=0}^{\infty} x^{2n}$ and $\sum_{n=0}^{\infty} b_n x^n = \sum_{n=0}^{\infty} x^{2n+1}$ are all 1, where $a_n = 0$, $b_n = 1$ as n is an odd number and $a_n = 1$, $b_n = 0$ as n is a even number. Obviously, the convergence radius of $0 = \sum_{n=0}^{\infty} a_n b_n x^n$ is $R = +\infty$.

本章重点

幂级数是一种特殊的函数项级数. 幂级数有许多类似于多项式的好的性质. 幂级数的重要性主要表现在两个方面:一是将函数展成幂级数,这是研究函数性质和函数值近似计算的重要方法;二是用幂级数的和函数表示新的非初等函数,这是解决某些理论问题和实际问题不可缺少的工具.

▪1 理解幂级数的意义,并知道它的一些好的性质.
▪2 会求幂级数的收敛半径,掌握和函数的分析性质.
▪3 知道怎样求缺项的幂级数的收敛半径.
▪4 弄清楚幂级数逐项求导后,收敛半径、收敛域是否发生改变的问题.
▪5 掌握把函数展成幂级数的方法.

Key points of this chapter

Power series is a special series of functions. Power series has many good properties similar to polynomials. The importance of power series is mainly presented in two aspects: one is to expand functions to power series, which is the important method to study the properties of functions and approximate computation of function values; the other is to indicate the new non-elementary functions by using sum functions of power series, which is an indispensable tool to solve some theoretical problems and actual problems.

▪1 Understand the meaning of power series, and know some of their good properties.
▪2 Be able to work out the radius of convergence of power series, and master the analytical properties of sum functions.
▪3 Know how to work out the radius of convergence of power series with missing terms.
▪4 Make clear the problem that whether the radius of convergence and the convergence domain are changed or not after the derivate of power series are operated one by one.
▪5 Master the methods of expanding functions into power series.

第十五章 傅里叶级数
Chapter 15 Fourier Series

傅里叶级数是一种特殊的函数项级数,它是研究周期函数的工具,我们主要掌握:
1. 函数的傅里叶级数展开.
2. 傅里叶级数的收敛定理.

Fourier series is a special series of functions and it is a useful tool to study periodic functions. In this chapter, we mainly master the following two items:
1. Expansion of Fourier series of functions.
2. Convergence theorem of Fourier series.

单词和短语 Words and expressions

★ 傅里叶级数　Fourier series
★ 三角级数　trigonometric series
正交函数系　system of orthogonal functions
简谐振动　simple harmonic vibration
以 2π 为周期的函数的傅里叶级数　Fourier series for functions of period 2π
角频率　angular frequency
分段光滑　piecewise smooth

傅里叶系数　Fourier coefficient
收敛定理　convergence theorem
奇函数与偶函数的傅里叶级数　Fourier series of even and odd functions
振幅　amplitude
★ 正弦级数　Sine series
★ 余弦级数　Cosine series
周期延拓　periodic extension

例题　Examples

例 求定义在任意一个长度为 2π 的区间 $[a, a+2\pi]$ 上的函数 $f(x)$ 的 Fourier 级数及其系数的计算公式.

解 设 $f(x) \sim \dfrac{a_0}{2} + \sum\limits_{n=1}^{\infty}(a_n\cos nx + b_n\sin nx)$,则

$$\int_a^{a+2\pi} f(x)\cos mx\,dx = \int_a^{a+2\pi}\left[\dfrac{a_0}{2} + \sum_{n=1}^{\infty}(a_n\cos nx + b_n\sin nx)\right]\cos mx\,dx$$

$$= \dfrac{a_0}{2}\int_a^{a+2\pi}\cos mx\,dx + \sum_{n=1}^{\infty}\left(a_n\int_a^{a+2\pi}\cos nx\cos mx\,dx + b_n\int_a^{a+2\pi}\sin nx\cos mx\,dx\right)$$

$$= a_m\pi, \quad (m = 0, 1, 2, \cdots),$$

$$\int_a^{a+2\pi} f(x)\sin mx\,dx = \int_a^{a+2\pi}\left[\dfrac{a_0}{2} + \sum_{n=1}^{\infty}(a_n\cos nx + b_n\sin nx)\right]\sin mx\,dx$$

$$= \dfrac{a_0}{2}\int_a^{a+2\pi}\sin mx\,dx + \sum_{n=1}^{\infty}\left(a_n\int_a^{a+2\pi}\cos nx\sin mx\,dx + b_n\int_a^{a+2\pi}\sin nx\sin mx\,dx\right)$$

$$= b_m\pi, \quad (m = 1, 2, \cdots).$$

所以

$$a_n = \dfrac{1}{\pi}\int_a^{a+2\pi} f(x)\cos nx\,dx \quad (n = 0, 1, 2, \cdots),$$

$$b_n = \dfrac{1}{\pi}\int_a^{a+2\pi} f(x)\sin nx\,dx \quad (n = 1, 2, \cdots).$$

Ex Present the Fourier series and the coefficients' computing formula of $f(x)$ defined on an arbitrary interval $[a, a+2\pi]$ with length 2π.

Solution Assume that $f(x) \sim \dfrac{a_0}{2} + \sum\limits_{n=1}^{\infty}(a_n\cos nx + b_n\sin nx)$, we then obtain

$$\int_a^{a+2\pi} f(x)\cos mx\,dx = \int_a^{a+2\pi}\left[\dfrac{a_0}{2} + \sum_{n=1}^{\infty}(a_n\cos nx + b_n\sin nx)\right]\cos mx$$

$$= \frac{a_0}{2}\int_a^{a+2\pi}\cos mx\,dx + \sum_{n=1}^{\infty}\left(a_n\int_a^{a+2\pi}\cos nx\cos mx\,dx + b_n\int_a^{a+2\pi}\sin nx\cos mx\,dx\right)$$
$$= a_m\pi, \quad (m=0,1,2,\cdots),$$
$$\int_a^{a+2\pi}f(x)\sin mx\,dx = \int_a^{a+2\pi}\left[\frac{a_0}{2} + \sum_{n=1}^{\infty}(a_n\cos nx + b_n\sin nx)\right]\sin mx\,dx$$
$$= \frac{a_0}{2}\int_a^{a+2\pi}\sin mx\,dx + \sum_{n=1}^{\infty}\left(a_n\int_a^{a+2\pi}\cos nx\sin mx\,dx + b_n\int_a^{a+2\pi}\sin nx\sin mx\,dx\right)$$
$$= b_m\pi, \quad (m=1,2,\cdots),$$

Consequently,
$$a_n = \frac{1}{\pi}\int_a^{a+2\pi}f(x)\cos nx\,dx \quad (n=0,1,2,\cdots),$$
$$b_n = \frac{1}{\pi}\int_a^{a+2\pi}f(x)\sin nx\,dx \quad (n=1,2,\cdots).$$

本章重点

我们知道，幂级数类似于多项式，它是研究函数很理想的工具。但是，函数在某点邻域内能展成幂级数，要求函数在该点必须存在任意阶导数。这对函数的要求太高了，因此研究不连续函数或不可导函数，幂级数这个工具就无能为力了，而傅里叶级数正是研究这类函数的有力工具，它弥补了幂级数的局限性。

1 牢记收敛定理，理解它的意义，掌握它的证明方法。
2 若函数 $f(x)$ 能展成傅里叶级数，$f(x)$ 是否必是 **R** 上的周期函数。
3 弄清傅里叶级数的优缺点。
4 会将若干简单函数展成傅里叶级数。

Key points of this chapter

As we all know, power series, similar to polynomials, are ideal tools to study functions. However, if a function can be expanded into a power series in the neighborhood of a certain point, it requires that the function must have arbitrary derivatives at the point. This is a high request for functions. Thus, to study discontinuous functions or non-derivable functions, the tool, power series, is impotent, but Fourier series is just the potent tool to study this type of functions. It has made up the limitation of power series.

1 Remember the convergence theorem, understand its meaning and master its proving method.
2 If the function $f(x)$ can be expanded into a Fourier series, whether or not $f(x)$ must be the periodical function on **R**?
3 Make clear of the advantages and disadvantages of Fourier series.
4 Be able to expand several simple functions into Fourier series.

第十六章 多元函数的极限与连续
Chapter 16 Limits and Continuity of Functions of Several Variables

本章开始研究多元函数，首先介绍 $\mathbf{R}^n$ 中的子集为开集、闭集、连通集、紧致集的概念。$\mathbf{R}^n$ 中子集的紧致性、连通性的概念完全是出于将介值定理和极值定理扩展到多元函数的目的所必需。最后将介绍多元函数极限的概念和它们的性质（运算）。

This chapter begins with the study of functions of several variables. Firstly, the concepts of sets of open, closed, connected and compact are introduced for subsets of $\mathbf{R}^n$. The notions of compactness and connectedness for a subset of $\mathbf{R}^n$ are motivated by the necessity of extending the intermediate value theorem and the extreme value theorem to functions of several variables. Finally, we introduce the concepts of limits of functions of several variables and their properties (operations).

单词和短语 Words and expressions

★ 多元函数　functions of several variables
★ 平面点集　plane point set
坐标平面　coordinate plane
内点　interior point
外点　outer point
界点　boundary point
边界　boundary
孤立点　isolated point
开集　open set
闭集　closed set
连通性　connectedness
★ 连通开集　connected open set
开域　open domain (region)
闭域　closed domain (region)
有界点集　bounded point set
无界点集　unbounded point set
闭域套定理　nested closed domain theorem
二元函数　function of two variables
n-维向量空间　n-dimensional vector space
非正常极限　improper limit
★ 二重极限　double limit
有界闭域上连续函数的性质　properties of continuous functions on bounded closed region
累次极限　repeated limits
全增量　total increment
偏增量　partial increment

例题　Examples

例1 讨论下列函数在原点的二重极限和累次极限：

(1) $f(x,y) = \dfrac{x^2 y^2}{x^2 y^2 + (x-y)^2}$；　(2) $f(x,y) = x\sin\dfrac{1}{y} + y\sin\dfrac{1}{x}$.

解　(1) 由 $\lim\limits_{\substack{x\to 0 \\ y=x+kx^2}} f(x) = \lim\limits_{x\to 0}\dfrac{x^4(1+kx)^2}{x^4(1+kx)^2 + k^2 x^4} = \dfrac{1}{1+k^2}$ 可知，其二重极限不存在.

由 $\lim\limits_{x\to 0} f(x,y) = \dfrac{0}{y^2} = 0, y\neq 0$ 可知，$\lim\limits_{y\to 0}\lim\limits_{x\to 0} f(x,y) = 0$，易得 $\lim\limits_{x\to 0}\lim\limits_{y\to 0} f(x,y) = 0$. 所以这两个累次极限存在且都等于 0.

(2) 由于 $|f(x,y)| \leqslant |x| + |y|$，所以 $\lim\limits_{\substack{x\to 0 \\ y\to 0}} f(x,y) = 0$. 由于 $\lim\limits_{x\to 0} y\sin\dfrac{1}{x} (y\neq 0)$ 和 $\lim\limits_{y\to 0} x\sin\dfrac{1}{y} (x\neq 0)$ 都不存在，所以两个累次极限都不存在.

Ex. 1 Discuss the double limit and the repeated limits of the following functions at the origin:

(1) $f(x,y) = \dfrac{x^2 y^2}{x^2 y^2 + (x-y)^2}$；　(2) $f(x,y) = x\sin\dfrac{1}{y} + y\sin\dfrac{1}{x}$.

Solution (1) From $\lim\limits_{\substack{x\to 0 \\ y=x+kx^2}} f(x) = \lim\limits_{x\to 0}\dfrac{x^4(1+kx)^2}{x^4(1+kx)^2 + k^2 x^4} = \dfrac{1}{1+k^2}$, we know that its double limit does not exist. On the other hand, from $\lim\limits_{x\to 0} f(x,y) = \dfrac{0}{y^2} = 0, y\neq 0$, we have $\lim\limits_{y\to 0}\lim\limits_{x\to 0} f(x,y) = 0$ and $\lim\limits_{x\to 0}\lim\limits_{y\to 0} f(x,y) = 0$. So the two repeated limits exist and are all equal to 0.

(2) Since $|f(x,y)| \leqslant |x| + |y|$, we have $\lim\limits_{\substack{x\to 0 \\ y\to 0}} f(x,y) = 0$. On the other hand, it is easy to show that both $\lim\limits_{x\to 0} y\sin\dfrac{1}{x} (y\neq 0)$ and $\lim\limits_{y\to 0} x\sin\dfrac{1}{y} (x\neq 0)$ do not exist, so the two repeated limits do not exist.

例2 设二元函数 $f(x,y) = \dfrac{1}{1-xy}, (x,y) \in D = [0,1) \times [0,1)$，证明：$f$ 在 D 上连续，但不一致连续.

证明　由于 f 在 D 上是初等函数，所以连续. 但因为当 $n \to +\infty$ 时，

$$\left|\left(1-\dfrac{1}{2n}, 1-\dfrac{1}{2n}\right) - \left(1-\dfrac{1}{n}, 1-\dfrac{1}{n}\right)\right| \to 0$$

而

$$f\left(1-\frac{1}{2n},1-\frac{1}{2n}\right)-f\left(1-\frac{1}{n},1-\frac{1}{n}\right)=\frac{4n^2}{4n-1}-\frac{n^2}{2n-1}=\frac{(4n-3)n^2}{(4n-1)(2n-1)}\to+\infty,$$

所以 f 在 D 上不一致连续.

Ex. 2 Assume that the function of two variables is given by $f(x,y)=\dfrac{1}{1-xy}$, $(x,y)\in D=[0,1)\times[0,1)$. Prove that f is continuous on D, but is not continuous uniformly on D.

Proof f is an elementary function on D, and thus is continuous. However, as $n\to+\infty$, we have

$$\left|\left(1-\frac{1}{2n},1-\frac{1}{2n}\right)-\left(1-\frac{1}{n},1-\frac{1}{n}\right)\right|\to 0$$

and

$$f\left(1-\frac{1}{2n},1-\frac{1}{2n}\right)-f\left(1-\frac{1}{n},1-\frac{1}{n}\right)=\frac{4n^2}{4n-1}-\frac{n^2}{2n-1}=\frac{(4n-3)n^2}{(4n-1)(2n-1)}\to+\infty,$$

Consequently, f is not continuous uniformly on D.

本章重点

我们知道研究一元函数及其极限必须掌握一维空间(实数集) **R** 的结构和连续性. 同样,研究多元函数及其极限,也必须掌握 n 维欧氏空间 $\mathbf{R}^n$ 的结构和连续性,所以我们必须掌握:

1 掌握平面点集的一些概念,如邻域、内点、界点、聚点、开集、闭集、闭区域、有界集、无界集等.

2 掌握 $\mathbf{R}^2$ 上连续函数的闭区间套定理、有限覆盖定理、聚点定理、致密性定理和柯西收敛准则及其证明方法.

3 掌握二元函数及二元函数极限的概念,弄清二重极限与累次极限的区别和联系.

4 怎样判别二元函数 $f(x,y)$ 在点 (x_0,y_0) 是否存在极限?

5 掌握二元复合函数的连续性,二元连续函数的保序性,以及在有界闭区域上连续函数的有界性、取极值性和一致连续性及证明方法,并能应用它们证明一些理论问题.

Key points of this chapter

We know that the structure and continuity of one-dimensional space (real number set) **R** must be mastered if functions of one variable and their limits are studied. Likewise, the structure and continuity of n-dimensional Euclidean space $\mathbf{R}^n$ must also be mastered if functions of several variables and their limits are studied. Thus, we must master:

1 Master some concepts of plane point set, such as neighborhood, interior point, boundary point, accumulation point, open set, closed set, closed domain, bounded set, unbounded set, etc. .

2 For continuous functions on $\mathbf{R}^2$, master the nested closed domain theorem, the finite covering theorem, the accumulation theorem, the compactness theorem, the Cauchy convergence criterion and their proving methods.

3 Master the concepts of a function of two variables and its limit, make clear of the differences and relations between double limit and repeated limits.

4 How to determine whether or not the function of two variables $f(x,y)$ has a limit at the point (x_0,y_0)?

5 Master the continuity of composite functions of two variables, the inheriting order properties of continuous functions of two variables, the properties of boundedness, taking extremum and uniform continuity of continuous functions at bounded region and their proving methods, moreover, be able to prove some theoretical problems by using them.

Chapter 17 Differential Calculus of Functions of Several Variables

本章中,我们首先给出多元函数可微性和偏导数的概念,并且研究可微的条件及其几何意义;然后我们将证明多元函数的链式法则和一阶微分形式不变性,并且给出方向导数和梯度的概念;最后将讨论泰勒公式与极值问题,包括高阶偏导数和中值定理.

In this chapter, we first present the concepts of differentiability and partial derivatives for functions of several variables, and study differentiable conditions and their geometric significance. After that, we will prove the chain rule for functions of several variables and the invariance of differential form of the first order. Moreover, we will give the concepts of directional derivatives and gradient. Finally, we will discuss the problems of Taylor formula and extreme values, including partial derivative of higher order and mean value value theorem.

单词和短语 Words and expressions

★ 全微分　total differential
★ 偏导数　partial derivative
　连续可微　continuously differentiable
　曲面的切平面　tangent plane of surface
　曲线的法平面　normal plane of curve
★ 复合函数微分法　differentiation of composite function
　多元函数的链式法则　chain rule for functions of several variables

　一阶微分形式不变性　invariance of differential form of first order
★ 方向导数与梯度　directional derivative and gradient
　可微性　differentiability
　泰勒公式　Taylor formula
　极值问题　problem of extreme value
★ 高阶偏导数　partial derivative of higher order
　混合偏导数　mixed partial derivative

例题　Examples

例1 设 $f(x,y)$ 具有连续偏导数,且 $f(x,x^2)=1, f_x(x,x^2)=x$,求 $f_y(x,x^2)$.

解 在等式 $f(x,x^2)=1$ 两边对 x 求导,有 $\frac{\partial f}{\partial x}+\frac{\partial f}{\partial y}\frac{\partial y}{\partial x}=f_x(x,x^2)+2xf_y(x,x^2)=0$,再将 $f_x(x,x^2)=x$ 代入,即可得到 $f_y(x,x^2)=-\frac{1}{2}$.

Ex. 1 Assume that the function $f(x,y)$ has continuous partial derivatives, $f(x,x^2)=1$ and $f_x(x,x^2)=x$, compute $f_y(x,x^2)$.

Solution Calculate the derivative on both sides of the equation $f(x,x^2)=1$ with respect to x, we have $\frac{\partial f}{\partial x}+\frac{\partial f}{\partial y}\frac{\partial y}{\partial x}=f_x(x,x^2)+2xf_y(x,x^2)=0$. Then substituting $f_x(x,x^2)=x$ into it, we have $f_y(x,x^2)=-\frac{1}{2}$.

例2 求方程 $f(x,x+y,x+y+z)=0$ 所确定的隐函数的偏导数 $\frac{\partial z}{\partial x},\frac{\partial z}{\partial y}$ 和 $\frac{\partial^2 z}{\partial x\partial y}$.

解 由 $f(x,x+y,x+y+z)=0$ 即可得到 $\frac{\partial z}{\partial x}=-\frac{f_1+f_2+f_3}{f_3},\frac{\partial z}{\partial y}=-\frac{f_2+f_3}{f_3}$,

$\frac{\partial^2 z}{\partial x\partial y}=\frac{\partial}{\partial x}\left(\frac{\partial z}{\partial y}\right)=-\frac{1}{f_3}\left[f_{21}+f_{22}+\left(1+\frac{\partial z}{\partial x}\right)f_{23}\right]+\frac{f_2}{f_3^2}\left[f_{31}+f_{32}+\left(1+\frac{\partial z}{\partial x}\right)f_{33}\right]=-\frac{1}{f_3^3}[f_3^2(f_{12}+f_{22})-f_2f_3f_{13}+f_2(f_1+f_2)f_{33}-f_3(f_1+2f_2)f_{23}]$.

Ex. 2 Calculate the partial derivatives of implicit function determined by the equation $f(x,x+y,x+y+z)=0$, namely, $\frac{\partial z}{\partial x},\frac{\partial z}{\partial y}$ and $\frac{\partial^2 z}{\partial x\partial y}$.

Solution From the equation $f(x,x+y,x+y+z)=0$, we can directly compute and obtain $\frac{\partial z}{\partial x}=-\frac{f_1+f_2+f_3}{f_3}, \frac{\partial z}{\partial y}=-\frac{f_2+f_3}{f_3}$ and $\frac{\partial^2 z}{\partial x\partial y}=\frac{\partial}{\partial x}\left(\frac{\partial z}{\partial y}\right)=-\frac{1}{f_3}\left[f_{21}+f_{22}+\left(1+\frac{\partial z}{\partial x}\right)f_{23}\right]+\frac{f_2}{f_3^2}\left[f_{31}+f_{32}+\left(1+\frac{\partial z}{\partial x}\right)f_{33}\right]=-\frac{1}{f_3^3}[f_3^2(f_{12}+f_{22})-f_2f_3f_{13}+f_2(f_1+f_2)f_{33}-f_3(f_1+2f_2)f_{23}]$.

例 3 求函数 $f(x,y) = \sin x + \sin y - \sin(x+y)$ 在闭区域 D 上的最大值与最小值,其中 $D = \{(x,y) \mid x \geqslant 0, y \geqslant 0, x+y \leqslant 2\pi\}$.

解 由 $\begin{cases} f_x = \cos x - \cos(x+y) = 0 \\ f_y = \cos y - \cos(x+y) = 0 \end{cases}$,得到 $\cos x = \cos y = \cos(x+y)$.

在 $D° = \{(x,y) \mid 0 < x, y < x+y < 2\pi\}$ 内,得到 $x = y = 2\pi - x - y$,即 $\left(\dfrac{2}{3}\pi, \dfrac{2}{3}\pi\right)$ 是函数在区域内部唯一的驻点.

由于在区域边界上,即当 $x = 0$ 或 $y = 0$ 或 $x+y = 2\pi$ 时,有 $f(x,y) = 0$,而在区域内部唯一的驻点上取值为 $f\left(\dfrac{2}{3}\pi, \dfrac{2}{3}\pi\right) = \dfrac{3\sqrt{3}}{2} > 0$,根据闭区域上连续函数的性质,可知函数的最大值为 $f_{\max} = \dfrac{3\sqrt{3}}{2}$,最小值为 $f_{\min} = 0$.

Ex. 3 Calculate the maximum and the minimum of $f(x,y) = \sin x + \sin y - \sin(x+y)$ on the closed domain D, where $D = \{(x,y) \mid x \geqslant 0, y \geqslant 0, x+y \leqslant 2\pi\}$.

Solution From $\begin{cases} f_x = \cos x - \cos(x+y) = 0 \\ f_y = \cos y - \cos(x+y) = 0 \end{cases}$, we obtain
$$\cos x = \cos y = \cos(x+y).$$

In the domain $D° = \{(x,y) \mid 0 < x, y < x+y < 2\pi\}$, we get $x = y = 2\pi - x - y$, namely, $\left(\dfrac{2}{3}\pi, \dfrac{2}{3}\pi\right)$ is the unique critical point.

Moreover, on the boundary, i.e., as $x = 0$ or $y = 0$ or $x+y = 2\pi$, we have $f(x,y) = 0$, on the other hand, the function value at the unique critical point is given by $f\left(\dfrac{2}{3}\pi, \dfrac{2}{3}\pi\right) = \dfrac{3\sqrt{3}}{2} > 0$. From the properties of continuous function on a closed domain, we know that the maximum and the minimum of this function are respectively given by $f_{\max} = \dfrac{3\sqrt{3}}{2}$ and $f_{\min} = 0$.

本章重点

多元函数微分学的内容与一元函数微分学的内容大体上是平行的,它们的意义和作用也是类似的. 但"多变量"与"单变量"也有某些差异,在注意"多"与"单"的共性的同时,特别要注意"多"所具有的特性. 求偏导数是多元函数微分学与积分学中的重要运算,要求:

1. 熟练掌握多元复合函数的偏导数的运算.
2. 理解全微分概念及其意义.
3. 牢记空间曲线切线方程与法平面方程;牢记曲面的切平面方程与法线方程.
4. 弄清研究多元函数有哪些基本方法.
5. 掌握二元函数的连续性、偏导数、可微之间的关系.
6. 会求二元函数的局部极值和最大值,并能计算一些简单的应用问题.

Key points of this chapter

The contents of differential calculus of functions of several variables are on the whole paralleled to those of one variable. Their meanings and effects are also similar. However, "several variables" and "one variable" are also of some differences. When we are concerned with the commonness of "several" and "one", we should pay special attention to the specialty possessed by "several". It is an important computation to work out partial derivatives and integrals in differential calculus of functions of several variables. Requests that:

1. Master the computation of partial derivate of composite functions of several variables skillfully.
2. Understand the concept and its meaning of total differential.

3. Remember the equations of tangent line and normal plane of a space curve, the equations of tangent surface and normal line of a surface.
4. Make clear how many basic methods there are to study functions of several variables.
5. Master the relations of continuity, partial derivatives, differentiability of functions of two variables.
6. Work out the local extremum and the maximum of functions of two variables, and be able to compute some simple applied problems.

第十八章 隐函数定理及其应用
Chapter 18 Implicit Function Theorems and their Applications

当 x 与 y 的关系是由一个未解出 y 的方程给出时，y 称为 x 的一个隐函数. 有时通过解方程确定隐函数，然而在许多情况下，对于一些复杂函数方程，这样确定隐函数是不可能的或太难，所以知道什么条件能求隐函数是十分重要的. 这就是我们要讨论的隐函数定理. 首先，我们将给出隐函数的概念，分析隐函数存在的条件，证明隐函数存在唯一性定理和给出例子；然后给出隐函数组的概念和定理；最后是应用：如几何应用包括平面曲线的切线和法线、空间曲线的切线和法平面. 其次，考虑条件极值问题，求极值的主要方法是拉格朗日乘数法.

If the relation between x and y is given by an equation which is not solved for y, then y is called an implicit function of x. It is sometimes possible to determine an implicit function by solving an equation. In many cases, however, for some complicated functional equations, such determination of implicit function is impossible or may be too laborious. Therefore, it is important to know what conditions we can solve. This is the implicit function theorem that we will discuss. First of all, we give the concept of implicit function, analyze the existence conditions of implicit function, prove the existence and uniqueness theorem of implicit function and consider a number of examples. After that, we will give the concepts and theorems of a system of implicit functions. The final contents are applications, such as the geometrical applications including tangent line and normal line of a plane curve, and tangent line and normal plane of a space curve. Next is to the consideration of constrained extreme problems, in which the main method of finding extreme is the Lagrange multiplier method.

单词和短语 Words and expressions

★ 隐函数存在唯一性定理　existence and uniqueness theorem of implicit functions
隐函数组　system of implicit functions
函数行列式(雅可比行列式)　functional determinant (Jacobian determinant)
反函数组　system of inverse functions
坐标变换　coordinate transformation
几何应用　geometrical application

平面曲线的切线与法线　tangent line and normal line of plane curve
空间曲线的切线与法平面　tangent line and normal plane of space curve
★ 条件极值　conditional extremum
拉格朗日乘数法　Lagrange multiplier method
拉格朗日函数　Lagrange function
拉格朗日乘数　Lagrange multiplier

本章重点

1. 理解隐函数的概念及其意义；掌握由二元方程确定的可微隐函数的充分条件及其证明，明白其几何意义.
2. 会求隐函数和隐函数组的偏导数.

Key points of this chapter

1. Understand the concept and its meaning of implicit functions; master the sufficient conditions and its proof of differentiable implicit functions determined by a duality equation, and understand its geometric meaning.
2. Be able to work out the partial derivatives of implicit functions and the system of implicit functions.

第十九章 含参量积分
Chapter 19 Integrals with Parameters

在本章我们主要研究含参量积分问题. 首先讨论含参量正常积分,包括连续性、可微性、可积性;然后讨论含参量正常积分、含参量反常积分的性质和一致收敛性及其判别法.

In this chapter, we mainly study the problems of integrals with parameters. We first discuss proper integrals with parameters including continuity, differentiability, integrability, and then discuss properties of proper integrals with parameters, uniform convergence of improper integrals with parameters and the tests.

单词和短语 Words and expressions

★ 含参量的正常积分　proper integral with parameter
累次积分　repeated integrals
★ 含参量的反常积分　improper integral with parameter
欧拉积分　Euler integral
魏尔斯特拉斯(M)判别法　Weierstrass (M) test
狄利克雷判别法　Dirichlet test

★ 无穷限的反常积分　improper integral with infinite bound
★ 无界函数的反常积分　improper integral of unbounded function
Γ- 函数　Gamma function
B- 函数　Beta function

例题 Examples

例1 用交换积分顺序的方法计算下列积分:

(1) $\int_0^1 \sin\left(\ln\frac{1}{x}\right)\frac{x^b - x^a}{\ln x}\mathrm{d}x \ (b > a > 0);$ (2) $\int_0^{\frac{\pi}{2}} \ln\frac{1 + a\sin x}{1 - a\sin x}\frac{\mathrm{d}x}{\sin x} \ (0 < a < 1).$

解 (1) $\int_0^1 \sin\left(\ln\frac{1}{x}\right)\frac{x^b - x^a}{\ln x}\mathrm{d}x = \int_0^1 \sin\left(\ln\frac{1}{x}\right)\mathrm{d}x\int_a^b x^y \mathrm{d}y$

$$= \int_a^b \mathrm{d}y \int_0^1 x^y \sin\left(\ln\frac{1}{x}\right)\mathrm{d}x,$$

由于

$$\int_0^1 x^y \sin\left(\ln\frac{1}{x}\right)\mathrm{d}x = \frac{1}{y+1}x^{y+1}\sin\left(\ln\frac{1}{x}\right)\Big|_0^1 + \frac{1}{y+1}\int_0^1 x^y \cos\left(\ln\frac{1}{x}\right)\mathrm{d}x$$

$$= \frac{1}{y+1}\int_0^1 x^y \cos\left(\ln\frac{1}{x}\right)\mathrm{d}x$$

$$= \frac{1}{(y+1)^2}x^{y+1}\cos\left(\ln\frac{1}{x}\right)\Big|_0^1 - \frac{1}{(y+1)^2}\int_0^1 x^y \sin\left(\ln\frac{1}{x}\right)\mathrm{d}x,$$

于是有 $\int_0^1 x^y \sin\left(\ln\frac{1}{x}\right)\mathrm{d}x = \dfrac{1}{1+(y+1)^2}$, 所以

$$\int_0^1 \sin\left(\ln\frac{1}{x}\right)\frac{x^b - x^a}{\ln x}\mathrm{d}x = \int_a^b \frac{\mathrm{d}y}{1+(y+1)^2} = \arctan(1+b) - \arctan(1+a).$$

(2) $\int_0^{\frac{\pi}{2}} \ln\dfrac{1+a\sin x}{1-a\sin x}\dfrac{\mathrm{d}x}{\sin x} = 2\int_0^{\frac{\pi}{2}}\mathrm{d}x \int_0^a \dfrac{\mathrm{d}y}{1-y^2\sin^2 x} = 2\int_0^a \mathrm{d}y \int_0^{\frac{\pi}{2}} \dfrac{\mathrm{d}x}{1-y^2\sin^2 x},$

由于 $\int_0^{\frac{\pi}{2}} \dfrac{\mathrm{d}x}{1-y^2\sin^2 x} = -\int_0^{\frac{\pi}{2}} \dfrac{\mathrm{d}\cot x}{\cot^2 x + 1 - y^2} = -\dfrac{1}{\sqrt{1-y^2}}\arctan\dfrac{\cot x}{\sqrt{1-y^2}}\bigg|_0^{\frac{\pi}{2}}$

$$= \frac{\pi}{2\sqrt{1-y^2}},$$

所以有 $\int_0^{\frac{\pi}{2}} \ln\dfrac{1+a\sin x}{1-a\sin x}\dfrac{\mathrm{d}x}{\sin x} = \pi\int_0^a \dfrac{\mathrm{d}y}{\sqrt{1-y^2}} = \pi\arcsin a.$

Ex. 1 Calculate the following integrals by using the method of exchanging the integral sequence:

(1) $\int_0^1 \sin\left(\ln\frac{1}{x}\right)\frac{x^b - x^a}{\ln x}dx \ (b > a > 0)$;   (2) $\int_0^{\frac{\pi}{2}} \ln\frac{1+a\sin x}{1-a\sin x}\frac{dx}{\sin x} \ (1 > a > 0)$.

Solution (1) $\int_0^1 \sin\left(\ln\frac{1}{x}\right)\frac{x^b - x^a}{\ln x}dx = \int_0^1 \sin\left(\ln\frac{1}{x}\right)dx \int_a^b x^y dy$

$$= \int_a^b dy \int_0^1 x^y \sin\left(\ln\frac{1}{x}\right)dx,$$

Since
$$\int_0^1 x^y \sin\left(\ln\frac{1}{x}\right)dx = \frac{1}{y+1}x^{y+1}\sin\left(\ln\frac{1}{x}\right)\Big|_0^1 + \frac{1}{y+1}\int_0^1 x^y \cos\left(\ln\frac{1}{x}\right)dx$$

$$= \frac{1}{y+1}\int_0^1 x^y \cos\left(\ln\frac{1}{x}\right)dx$$

$$= \frac{1}{(y+1)^2}x^{y+1}\cos\left(\ln\frac{1}{x}\right)\Big|_0^1 - \frac{1}{(y+1)^2}\int_0^1 x^y \sin\left(\ln\frac{1}{x}\right)dx,$$

we have $\int_0^1 x^y \sin\left(\ln\frac{1}{x}\right)dx = \frac{1}{1+(y+1)^2}$, and thus

$$\int_0^1 \sin\left(\ln\frac{1}{x}\right)\frac{x^b - x^a}{\ln x}dx = \int_a^b \frac{dy}{1+(y+1)^2} = \arctan(1+b) - \arctan(1+a).$$

(2) $\int_0^{\frac{\pi}{2}} \ln\frac{1+a\sin x}{1-a\sin x}\frac{dx}{\sin x} = 2\int_0^{\frac{\pi}{2}} dx \int_0^a \frac{dy}{1-y^2\sin^2 x} = 2\int_0^a dy \int_0^{\frac{\pi}{2}} \frac{dx}{1-y^2\sin^2 x}.$

Since $\int_0^{\frac{\pi}{2}} \frac{dx}{1-y^2\sin^2 x} = -\int_0^{\frac{\pi}{2}} \frac{d\cot x}{\cot^2 x + 1 - y^2} = -\frac{1}{\sqrt{1-y^2}}\arctan\frac{\cot x}{\sqrt{1-y^2}}\Big|_0^{\frac{\pi}{2}}$

$$= \frac{\pi}{2\sqrt{1-y^2}},$$

we have $\int_0^{\frac{\pi}{2}} \ln\frac{1+a\sin x}{1-a\sin x}\frac{dx}{\sin x} = \pi\int_0^a \frac{dy}{\sqrt{1-y^2}} = \pi\arcsin a.$

例 2 证明下列含参变量反常积分在指定区间上一致收敛:

(1) $\int_0^{+\infty} \frac{\cos xy}{x^2+y^2}dx, y \geqslant a > 0$;   (2) $\int_0^{+\infty} \frac{\sin 2x}{x+\alpha}e^{-\alpha x}dx, 0 \leqslant \alpha \leqslant \alpha_0$;

(3) $\int_A^{+\infty} x\sin x^4 \cos \alpha x dx, a \leqslant \alpha \leqslant b.$

解 (1) 因为 $\left|\frac{\cos xy}{x^2+y^2}\right| \leqslant \frac{1}{x^2+a^2}$, 而 $\int_0^{+\infty} \frac{1}{x^2+a^2}dx$ 收敛, 所以由 Weierstrass 判别法, $\int_0^{+\infty} \frac{\cos xy}{x^2+y^2}dx$ 在 $[a,+\infty)$ 上一致收敛.

(2) $\left|\int_0^A \sin 2x dx\right| \leqslant 1$, 即 $\int_0^A \sin 2x dx$ 关于 $\alpha \in [0,\alpha_0]$ 一致有界; $\frac{e^{-\alpha x}}{x+\alpha}$ 关于 x 单调, 且由 $\left|\frac{e^{-\alpha x}}{x+\alpha}\right| \leqslant \frac{1}{x}$, 可知当 $x \to +\infty$ 时, $\frac{e^{-\alpha x}}{x+\alpha}$ 关于 $\alpha \in [0,\alpha_0]$ 一致趋于零. 于是由 Dirichlet 判别法, 可知 $\int_0^{+\infty} \frac{\sin 2x}{x+\alpha}e^{-\alpha x}dx$ 在 $\alpha \in [0,\alpha_0]$ 上一致收敛.

(3) 由分部积分法,
$$\int_A^{+\infty} x\sin x^4 \cos \alpha x dx = -\frac{1}{4}\int_A^{+\infty} \frac{\cos \alpha x}{x^2}d\cos x^4$$

$$= -\frac{\cos \alpha x \cos x^4}{4x^2}\Big|_A^{+\infty} - \frac{1}{4}\int_A^{+\infty} \frac{\alpha \sin \alpha x \cos x^4}{x^2}dx - \frac{1}{2}\int_A^{+\infty} \frac{\cos \alpha x \cos x^4}{x^3}dx,$$

其中 $\left|\frac{\cos \alpha x \cos x^4}{x^2}\Big|_A^{+\infty}\right| \leqslant \frac{1}{A^2}$; 再由

$$\left|\frac{a\sin ax\cos x^4}{x^2}\right| \leqslant \frac{\max(|a|,|b|)}{x^2} \text{ 及 } \left|\frac{\cos ax\cos x^4}{x^3}\right| \leqslant \frac{1}{x^3},$$

可得到 $\left|\int_A^{+\infty}\frac{a\sin ax\cos x^4}{x^2}dx\right| \leqslant \max(|a|,|b|)\int_A^{+\infty}\frac{1}{x^2}dx = \frac{\max(|a|,|b|)}{A}$ 与

$\left|\int_A^{+\infty}\frac{\cos ax\cos x^4}{x^3}dx\right| \leqslant \int_A^{+\infty}\frac{1}{x^3}dx = \frac{1}{2A^2}.$

当 $A \to +\infty$ 时,上述三式关于 a 在 $[a,b]$ 上一致趋于零,所以原积分关于 a 在 $[a,b]$ 上一致收敛.

Ex. 2 Prove that the following improper integrals with parametric variables are convergent uniformly on the prescribed intervals:

(1) $\int_0^{+\infty}\frac{\cos xy}{x^2+y^2}dx, y \geqslant a > 0;$ (2) $\int_0^{+\infty}\frac{\sin 2x}{x+a}e^{-ax}dx, 0 \leqslant a \leqslant a_0;$

(3) $\int_0^{+\infty} x\sin x^4 \cos ax\, dx, a \leqslant a \leqslant b.$

Solution (1) Since $\left|\frac{\cos xy}{x^2+y^2}\right| \leqslant \frac{1}{x^2+a^2}$ and $\int_0^{+\infty}\frac{1}{x^2+a^2}dx$ is convergent, we have that $\int_0^{+\infty}\frac{\cos xy}{x^2+y^2}dx$ is convergent uniformly on $[a,+\infty)$ by using the Weierstrass test.

(2) It is easy to show that $\left|\int_0^A \sin 2x\, dx\right| \leqslant 1$, i. e., $\int_0^A \sin 2x\, dx$ is uniformly bounded with respect to $a \in [0, a_0]$ and $\frac{e^{-ax}}{x+a}$ is monotonic with x. Moreover, from $\left|\frac{e^{-ax}}{x+a}\right| \leqslant \frac{1}{x}$, we know that $\frac{e^{-ax}}{x+a}$ uniformly tends to 0 as $x \to +\infty$ for any $a \in [0, a_0]$. Thus from the Dirichlet test, we know that $\int_0^{+\infty}\frac{\sin 2x}{x+a}e^{-ax}dx$ converges uniformly for any $a \in [0, a_0]$.

(3) Using the method of integration by parts, we have

$$\int_A^{+\infty} x\sin x^4 \cos ax\, dx = -\frac{1}{4}\int_A^{+\infty}\frac{\cos ax}{x^2}d\cos x^4$$

$$= -\frac{\cos ax\cos x^4}{4x^2}\Big|_A^{+\infty} - \frac{1}{4}\int_A^{+\infty}\frac{a\sin ax\cos x^4}{x^2}dx - \frac{1}{2}\int_A^{+\infty}\frac{\cos ax\cos x^4}{x^3}dx,$$

where $\left|\frac{\cos ax\cos x^4}{x^2}\Big|_A^{+\infty}\right| \leqslant \frac{1}{A^2}.$ Furthermore, from

$$\left|\frac{a\sin ax\cos x^4}{x^2}\right| \leqslant \frac{\max(|a|,|b|)}{x^2} \text{ and } \left|\frac{\cos ax\cos x^4}{x^3}\right| \leqslant \frac{1}{x^3},$$

we obtain $\left|\int_A^{+\infty}\frac{a\sin ax\cos x^4}{x^2}dx\right| \leqslant \max(|a|,|b|)\int_A^{+\infty}\frac{1}{x^2}dx = \frac{\max(|a|,|b|)}{A}$ and

$\left|\int_A^{+\infty}\frac{\cos ax\cos x^4}{x^3}dx\right| \leqslant \int_A^{+\infty}\frac{1}{x^3}dx = \frac{1}{2A^2}.$

As $A \to +\infty$, all of the above three expressions uniformly tends to 0 on $[a,b]$ with respect to a, and thus $\int_0^{+\infty} x\sin x^4\cos ax\, dx$ converges uniformly on $[a,b]$ with a.

本章重点

含参变量积分是在高等数学中给出的一个新函数,而数学分析主要是研究函数的一些性质,如连续性、可微性、可积性等. 当然,对于这个新的函数,我们也要研究其上述性质,所以本章应该掌握:

1. 掌握含参变量有限积分和无穷积分所定义函数的分析性质及其证明方法.
2. 掌握含参变量无穷积分的一致收敛定义及其判别法,应用积分号下的可微性与可积性,会算一些定积分与反常积分.
3. 把本章无穷积分内容与无穷级数对照比较来理解,注意共性,抓住特性.

④ 本着解决问题从已知到未知的原则,掌握把无界积分化为无穷积分的方法.

Key points of this chapter

Integral with parameter is a new function given in advanced mathematics, while Mathematical Analysis is mainly to study some properties of functions: e. g. continuity, differentiability, integrability, etc.. Of course, as for the new function, we will also study the properties mentioned above. Thus, the following contents of the chapter need mastering:

① Master the analytical properties and proving methods of functions defined by finite integral and infinite integral with parameter.

② Master the concept and its testing method of uniform convergence of infinite integral with parameter, and be able to compute some definite integrals and improper integrals by using the differentiability and integrability under integral sign.

③ Understand the contents of this chapter by comparing and contrasting infinite integrals with infinite series. Pay attention to their commonness and properties.

④ Abide by the principle to solve problems from something known to something unknown, master the method of turning unbounded integrals into infinite integrals.

第二十章 重积分
Chapter 20 Multiple Integrals

在研究一元函数时,能够基本给出区间上连续函数积分存在性的全部证明.积分的研究牵涉上、下和的研究.在高维背景下,了解上、下和的概念也是重要的.它们的全部证明也会变得更复杂,因此我们将省略这些证明.然而,我们将讨论可化为累次积分来计算的积分基本定理.最后,我们将列出在极坐标系下二重积分和三重积分的各种公式,我们也给出几何理由来说明其合理性.

When studying functions of one variable, it was possible to give essentially complete proofs for the existence of an integral of a continuous function over an interval. The investigation of the integral involves lower sum and upper sum. It is important to understand the notions of upper and lower sums in higher dimensional context. Giving complete proofs for the theories becomes more complicated, and hence we shall omit the proofs. However, the basic theorem that an integral can be evaluated by repeated integrals shall be discussed in detail. Finally, we shall list various formulas giving double integral and triple integral in terms of polar coordinates, and we also give a geometric argument to make them plausible.

单词和短语 Words and expressions

内面积　inner area
外面积　outer area
曲顶柱体　cylindrical body with tip surface
细度　fineness
积分区域　integral region
★二重积分　double integral
x 型区域　x-type region
y 型区域　y-type region

★三重积分　triple integral
三重积分换元法　change of variables in triple integral
柱面坐标变换　transformation of cylindrical coordinates
球面坐标变换　transformation of spherical coordinates
曲面的面积　area of surface

例题 Examples

例1 设 $f(x)$ 在 $[0,1]$ 上连续,证明

$$\int_0^1 dy \int_y^{\sqrt{y}} e^y f(x) dx = \int_0^1 (e^x - e^{x^2}) f(x) dx.$$

证明 交换积分次序,则得到
$$\int_0^1 dy \int_y^{\sqrt{y}} e^y f(x) dx = \int_0^1 f(x) dx \int_{x^2}^{x} e^y dy = \int_0^1 (e^x - e^{x^2}) f(x) dx.$$

Ex. 1 Assume that $f(x)$ is continuous on $[0,1]$, prove that
$$\int_0^1 dy \int_y^{\sqrt{y}} e^y f(x) dx = \int_0^1 (e^x - e^{x^2}) f(x) dx.$$

Proof Exchanging the sequence of the integral, we have
$$\int_0^1 dy \int_y^{\sqrt{y}} e^y f(x) dx = \int_0^1 f(x) dx \int_{x^2}^{x} e^y dy = \int_0^1 (e^x - e^{x^2}) f(x) dx.$$

例 2 选取适当的坐标变换计算下列积分:

(1) $\iint_D y[1 + xe^{\frac{1}{2}(x^2+y^2)}] dx dy$,其中 D 为直线 $y = x, y = -1$ 和 $x = 1$ 所围的区域;

(2) $\iiint_\Omega \sqrt{1 - \frac{x^2}{a^2} - \frac{y^2}{b^2} - \frac{z^2}{c^2}} dx dy dz$,其中 Ω 为椭球 $\left\{(x,y,z) \left| \frac{x^2}{a^2} + \frac{y^2}{b^2} + \frac{z^2}{c^2} \leqslant 1 \right.\right\}$.

解 (1) $\iint_D y[1 + xe^{\frac{1}{2}(x^2+y^2)}] dx dy = \int_{-1}^{1} y dy \int_y^{1} [1 + xe^{\frac{1}{2}(x^2+y^2)}] dx$

$$= \int_{-1}^{1} \left(y - y^2 + y(e^{\frac{y^2+1}{2}} - e^{y^2}) \right) dy$$

$$= -\int_{-1}^{1} y^2 dy = -\frac{2}{3}$$

(2) 应用广义球坐标,则

$$\iiint_\Omega \sqrt{1 - \frac{x^2}{a^2} - \frac{y^2}{b^2} - \frac{z^2}{c^2}} dx dy dz = abc \int_0^{2\pi} d\theta \int_0^{\pi} \sin\varphi d\varphi \int_0^1 \sqrt{1 - r^2} \, r^2 dr$$

$$= 4\pi abc \int_0^1 \sqrt{1 - r^2} \, r^2 dr,$$

令 $r = \sin t$,则

$$\iiint_\Omega \sqrt{1 - \frac{x^2}{a^2} - \frac{y^2}{b^2} - \frac{z^2}{c^2}} dx dy dz = 4\pi abc \int_0^{\frac{\pi}{2}} \cos^2 t \sin^2 t dt = \pi abc \int_0^{\frac{\pi}{2}} \sin^2 2t dt$$

$$= \frac{1}{2} \pi abc \int_0^{\frac{\pi}{2}} (1 - \cos 4t) dt = \frac{1}{4} \pi^2 abc.$$

Ex. 2 Choose suitable transformation of coordinates to calculate the following integrals:

(1) $\iint_D y[1 + xe^{\frac{1}{2}(x^2+y^2)}] dx dy$, where D is the region enclosed by the lines $y = x$, $y = -1$ and $x = 1$;

(2) $\iiint_\Omega \sqrt{1 - \frac{x^2}{a^2} - \frac{y^2}{b^2} - \frac{z^2}{c^2}} dx dy dz$, where Ω is given by $\left\{(x,y,z) \left| \frac{x^2}{a^2} + \frac{y^2}{b^2} + \frac{z^2}{c^2} \leqslant 1 \right.\right\}$.

Solution (1) $\iint_D y[1 + xe^{\frac{1}{2}(x^2+y^2)}] dx dy = \int_{-1}^{1} y dy \int_y^{1} [1 + xe^{\frac{1}{2}(x^2+y^2)}] dx$

$$= \int_{-1}^{1} \left(y - y^2 + y(e^{\frac{y^2+1}{2}} - e^{y^2}) \right) dy$$

$$= -\int_{-1}^{1} y^2 dy = -\frac{2}{3}$$

(2) Using the transformation of generalized spherical coordinates, we have

$$\iiint_\Omega \sqrt{1-\frac{x^2}{a^2}-\frac{y^2}{b^2}-\frac{z^2}{c^2}}\,\mathrm{d}x\mathrm{d}y\mathrm{d}z = abc\int_0^{2\pi}\mathrm{d}\theta\int_0^\pi \sin\varphi\mathrm{d}\varphi\int_0^1 \sqrt{1-r^2}\,r^2\,\mathrm{d}r$$

$$= 4\pi abc\int_0^1 \sqrt{1-r^2}\,r^2\,\mathrm{d}r,$$

Let $r = \sin t$, we obtain

$$\iiint_\Omega \sqrt{1-\frac{x^2}{a^2}-\frac{y^2}{b^2}-\frac{z^2}{c^2}}\,\mathrm{d}x\mathrm{d}y\mathrm{d}z = 4\pi abc\int_0^{\frac{\pi}{2}} \cos^2 t\,\sin^2 t\,\mathrm{d}t = \pi abc\int_0^{\frac{\pi}{2}} \sin^2 2t\,\mathrm{d}t$$

$$= \frac{1}{2}\pi abc\int_0^{\frac{\pi}{2}} (1-\cos 4t)\,\mathrm{d}t = \frac{1}{4}\pi^2 abc.$$

本章重点

1. 掌握二重积分的定义、性质、可积条件.
2. 会用累次积分方法计算二重积分，能根据积分区域和被积函数的特征进行相应的变量替换，特别是极坐标替换.
3. 会应用累次积分方法计算三重积分，能根据三维空间中积分区域的特征，选取适当的累次积分的次序，简化计算.
4. 能够根据积分区域和被积函数的特征选取适当的变数替换，特别是熟练掌握柱面坐标替换和球面坐标替换.

Key points of this chapter

1. Master the concept, properties and integrable conditions of double integrals.
2. Able to compute double integrals by using the method of repeated integrals, able to substitute the corresponding variables according to the characteristics of integral region and integrand functions especially for transformation of polar coordinates.
3. Learn to compute triple integrals by using repeated integral method, able to simplify the computation by selecting the proper order of repeated integrals according to the features of integral region in three-dimension.
4. Able to select the proper variable substitutions according to the features of integral region and integrand functions, especially mastering transformation of cylindrical coordinates and transformation of spherical coordinates skillfully.

第二十一章 曲线积分
Chapter 21 Curvilinear Integrals

曲线积分是定积分的推广，它是把积分的积分区间 $[a,b]$ 用 xoy 平面上曲线 C 代替而得到的结果. 由定义计算曲线积分是笨拙的方法. 通常用曲线参数化方法较为简单，这种方法能把曲线积分转化为定积分. 最后，我们将给出曲线积分与路线无关的条件，同时把定积分中的微积分第一基本定理扩展到格林公式.

Curvilinear integral is a generalization of definite integrals, which is obtained by replacing the interval $[a,b]$ of the definite integral with a curve C in the xoy-plane. Evaluation of curvilinear integrals by means of the definition is an unwieldy process. It is usually simpler to employ a parameterization of the curve, in which the method can convert a curvilinear integral into a definite integral. Finally, we will give conditions of which the curvilinear integral is independent of path. Meanwhile, we will extend the first fundamental theorem of calculus for functions of one variable to the Green formula.

单词和短语 Words and expressions

★ 第一型曲线积分 curvilinear integrals of the first kind / line integration of 1-form

★ 第二型曲线积分 curvilinear integrals of the second kind / line integration of 2-form

两类曲线积分的联系　relation between two classes of curvilinear integrals

★ 格林公式　Green formula

★ 曲线积分与路径无关性　independence of curvilinear integrals with path

单连通区域　simple connected region

多连通区域　complex connected region

例题　Examples

例 计算下列曲线积分：

(1) $\int_{(0,1)}^{(2,3)} (x+y)dx + (x-y)dy$；　(2) $\oint_C \dfrac{xdy - ydx}{x^2+y^2}$，其中 C 是圆周 $(x-1)^2 + (y-1)^2 = 1$.

解 (1) 显然有

$$(x+y)dx + (x-y)dy = (ydx + xdy) + (xdx - ydy)$$
$$= d(xy) + d\left(\dfrac{x^2-y^2}{2}\right) = d\left(xy + \dfrac{x^2-y^2}{2}\right).$$

即上式是全微分，于是

$$\int_{(0,1)}^{(2,3)} (x+y)dx + (x-y)dy = \int_{(0,1)}^{(2,3)} d\left(xy + \dfrac{x^2-y^2}{2}\right) = \left(xy + \dfrac{x^2-y^2}{2}\right)\bigg|_{(0,1)}^{(2,3)} = 4.$$

(2) 令 $P = -\dfrac{y}{x^2+y^2}$，$Q = \dfrac{x}{x^2+y^2}$，并且注意到 $\dfrac{\partial P}{\partial y} = \dfrac{-x^2+y^2}{(x^2+y^2)^2}$，$\dfrac{\partial Q}{\partial x} = \dfrac{-x^2+y^2}{(x^2+y^2)^2}$.

进而根据格林公式，有 $\oint_C \dfrac{xdy - ydx}{x^2+y^2} = \iint_{(x-1)^2+(y-1)^2=1} 0 dxdy = 0.$

Ex Compute the following curvilinear integrals：

(1) $\int_{(0,1)}^{(2,3)} (x+y)dx + (x-y)dy$；　(2) $\oint_C \dfrac{xdy - ydx}{x^2+y^2}$, where C denotes the circle of $(x-1)^2 + (y-1)^2 = 1$.

Solution (1) Obviously,

$$(x+y)dx + (x-y)dy = (ydx + xdy) + (xdx - ydy)$$
$$= d(xy) + d\left(\dfrac{x^2-y^2}{2}\right) = d\left(xy + \dfrac{x^2-y^2}{2}\right).$$

that is to say, it is a total differential, and thus

$$\int_{(0,1)}^{(2,3)} (x+y)dx + (x-y)dy = \int_{(0,1)}^{(2,3)} d\left(xy + \dfrac{x^2-y^2}{2}\right) = \left(xy + \dfrac{x^2-y^2}{2}\right)\bigg|_{(0,1)}^{(2,3)} = 4.$$

(2) Let $P = -\dfrac{y}{x^2+y^2}$, $Q = \dfrac{x}{x^2+y^2}$, we then get $\dfrac{\partial P}{\partial y} = \dfrac{-x^2+y^2}{(x^2+y^2)^2}$, $\dfrac{\partial Q}{\partial x} = \dfrac{-x^2+y^2}{(x^2+y^2)^2}$.

From the Green formula, it yields $\oint_C \dfrac{xdy - ydx}{x^2+y^2} = \iint_{(x-1)^2+(y-1)^2=1} 0 dxdy = 0.$

本章重点

1. 掌握计算不同形式的曲线方程的第一型和第二型曲线积分，特别是用格林公式计算曲线积分.
2. 能应用曲线积分与路线无关的等价命题计算或证明某些问题.

Key points of this chapter

1. Master the computation of curvilinear integrals of the first kind and the second kind for the curve equation given by different forms, especially the computation of curvilinear integrals by the Green Formula.
2. Able to compute or prove some problems by using equivalent propositions which curvilinears are independent of paths.

第二十二章 曲面积分
Chapter 22 Surface Integrals

曲线积分是普通定积分的推广，类似地，曲面积分是二重积分的推广．首先分别讨论第一类型曲面积分和第二类型曲面积分的概念和计算．计算曲面积分，我们必须把积分的被积函数作变形，使得我们能由二重积分得到结果；然后主要讨论高斯定理和斯托克斯定理，这两个定理和格林定理都是叙述集 S 上的积分与另一个在 S 的边界上积分的关系，它们都是一元函数微积分基本定理的推广．

Curvilinear integral is a generalization of ordinary definite integrals, similarly, surface integral is a generalization of double integrals. We first discuss the concepts and the calculating methods of surface integrals of the first kind and the second kind respectively. To evaluating surface integrals, we have to make some transformations for the integrand of an integral so that we can obtain the results by double integrals. Then we mainly discuss the Gauss theorem and the Stokes theorem. The two theorems and the Green theorem all relate an integral over a set S to another integral over the boundary of S, and they are all generalizations of the fundamental theorem of calculus for functions of one variable.

单词和短语 Words and expressions

- ★ 第一型曲面积分 surface integral of the first kind
- ★ 第二型曲面积分 surface integral of the second kind
- 单侧曲面 unilateral surface
- 双侧曲面 bilateral surface / two-sided face
- 右手法则 right-hand rule
- 两类曲面积分的关系 relation between two classes of surface integrals
- 高斯公式 Gauss formula
- 斯托克斯公式 Stokes formula
- 场论初步 introduction to field
- 向量场 field of vectors
- 梯度场 gradient field
- 引力势 gravitational potential
- 散度场 divergence field
- 旋度场 rotation field
- 环流量 circulation

例题 Examples

例 1 计算下列曲面积分：

(1) $\iint\limits_{S}(x+y+z)\mathrm{d}\sigma$，其中 S 为曲面 $x^2+y^2+z^2=a^2, z\geqslant 0$；

(2) $\oiint\limits_{S} x\mathrm{d}y\mathrm{d}z + y\mathrm{d}z\mathrm{d}x + z\mathrm{d}x\mathrm{d}y$，其中 S 为球 $x^2+y^2+z^2=a^2$ 的外表面．

解 (1) 由于 $\sqrt{1+z_x^2+z_y^2}=\sqrt{1+\dfrac{x^2}{z^2}+\dfrac{y^2}{z^2}}=\dfrac{a}{\sqrt{a^2-x^2-y^2}}$，故有

$$\iint\limits_{S}(x+y+z)\mathrm{d}\sigma = \int_{-a}^{a}\mathrm{d}x\int_{-\sqrt{a^2-x^2}}^{\sqrt{a^2-x^2}}\dfrac{a}{\sqrt{a^2-x^2-y^2}}\cdot(x+y+\sqrt{a^2-x^2-y^2})\mathrm{d}y$$

$$=\int_{-a}^{a}(\pi ax+za\sqrt{a^2-x^2})\mathrm{d}x = 4a\int_{0}^{a}\sqrt{a^2-x^2}\mathrm{d}x = 4a\cdot\dfrac{\pi a^2}{4}=\pi a^3.$$

(2) 根据对称性，只要计算 $\oiint\limits_{S} z\mathrm{d}x\mathrm{d}y$，注意到上半球面 $z=\sqrt{a^2-x^2-y^2}$ 应取上侧，下半球面 $z=-\sqrt{a^2-x^2-y^2}$ 应取下侧，则有

$$\oiint\limits_{S} z\mathrm{d}x\mathrm{d}y = \iint\limits_{x^2+y^2\leqslant a^2}\sqrt{a^2-x^2-y^2}\mathrm{d}x\mathrm{d}y - \iint\limits_{x^2+y^2\leqslant a^2}(-\sqrt{a^2-x^2-y^2})\mathrm{d}x\mathrm{d}y$$

$$= 2\iint\limits_{x^2+y^2\leqslant a^2}\sqrt{a^2-x^2-y^2}\mathrm{d}x\mathrm{d}y = 2\int_{0}^{2\pi}\mathrm{d}\theta\int_{0}^{a}r\sqrt{a^2-r^2}\mathrm{d}r = \dfrac{4}{3}\pi a^3.$$

于是有 $\oiint_S x\,dydz + y\,dxdz + z\,dxdy = 3 \times \dfrac{4}{3}\pi a^3 = 4\pi a^3.$

Ex. 1 Compute the following surface integrals:

(1) $\iint_S (x+y+z)\,d\sigma$, where S is given by $x^2+y^2+z^2=a^2, z \geqslant 0$;

(2) $\oiint_S x\,dydz + y\,dxdz + z\,dxdy$, where S denotes the outer surface of the sphere $x^2+y^2+z^2=a^2$.

Solution (1) From $\sqrt{1+z_x^2+z_y^2} = \sqrt{1+\dfrac{x^2}{z^2}+\dfrac{y^2}{z^2}} = \dfrac{a}{\sqrt{a^2-x^2-y^2}}$, we have

$$\iint_S (x+y+z)\,d\sigma = \int_{-a}^{a} dx \int_{-\sqrt{a^2-x^2}}^{\sqrt{a^2-x^2}} \dfrac{a}{\sqrt{a^2-x^2-y^2}} \cdot (x+y+\sqrt{a^2-x^2-y^2})\,dy$$

$$= \int_{-a}^{a} (\pi ax + xa\sqrt{a^2-x^2})\,dx = 4a\int_0^a \sqrt{a^2-x^2}\,dx = 4a \cdot \dfrac{\pi a^2}{4} = \pi a^3.$$

(2) According to symmetry, we only need to calculate $\oiint_S z\,dxdy$. Further, the upper and the lower surfaces of the sphere are respectively denoted by $z=\sqrt{a^2-x^2-y^2}$ and $z=-\sqrt{a^2-x^2-y^2}$, so we have

$$\oiint_S z\,dxdy = \iint_{x^2+y^2 \leqslant a^2} \sqrt{a^2-x^2-y^2}\,dxdy - \iint_{x^2+y^2 \leqslant a^2} (-\sqrt{a^2-x^2-y^2})\,dxdy$$

$$= 2\iint_{x^2+y^2 \leqslant a^2} \sqrt{a^2-x^2-y^2}\,dxdy = 2\int_0^{2\pi}d\theta\int_0^a r\sqrt{a^2-r^2}\,dr = \dfrac{4}{3}\pi a^3.$$

Consequently, $\oiint_S x\,dydz + y\,dxdz + z\,dxdy = 3 \times \dfrac{4}{3}\pi a^3 = 4\pi a^3.$

例 2 计算 $\oiint_S x^3\,dydz + y^3\,dxdz + z^3\,dxdy$，其中 S 是球面 $x^2+y^2+z^2=a^2$，方向指向球面外侧.

解 由于 $P=x^3, Q=y^3, R=z^3$，从而 $\dfrac{\partial P}{\partial x}+\dfrac{\partial Q}{\partial y}+\dfrac{\partial R}{\partial z}=3(x^2+y^2+z^2)$. 于是由高斯公式，有

$$\oiint_S x^3\,dydz + y^3\,dxdz + z^3\,dxdy = 3\iiint_V (x^2+y^2+z^2)\,dxdydz$$

$$= 3\int_0^{2\pi}d\theta\int_0^{\pi}\sin\varphi\,d\varphi\int_0^a r^4\,dr$$

$$= 6\pi\int_0^{\pi}\sin\varphi\,d\varphi\int_0^a r^4\,dr = \dfrac{12}{5}\pi a^5.$$

Ex. 2 Compute $\oiint_S x^3\,dydz + y^3\,dxdz + z^3\,dxdy$, where S denotes the sphere with radius a and the direction is taken as the outer of the sphere.

Solution Since $P=x^3, Q=y^3, R=z^3$, we have $\dfrac{\partial P}{\partial x}+\dfrac{\partial Q}{\partial y}+\dfrac{\partial R}{\partial z}=3(x^2+y^2+z^2)$. Using the Gauss formula, we obtain

$$\oiint_S x^3\,dydz + y^3\,dxdz + z^3\,dxdy = 3\iiint_V (x^2+y^2+z^2)\,dxdydz$$

$$= 3\int_0^{2\pi}d\theta\int_0^{\pi}\sin\varphi\,d\varphi\int_0^a r^4\,dr$$

$$= 6\pi\int_0^{\pi}\sin\varphi\,d\varphi\int_0^a r^4\,dr = \dfrac{12}{5}\pi a^5.$$

本章重点

1. 掌握两类曲面积分的概念和性质.
2. 掌握利用对称性、参数方程、两种曲面积分的关系、高斯公式与斯托克斯公式等计算曲面积分.

Key points of this chapter

1. Master the concepts and the properties of two types of surface integrals.
2. Learn to compute surface integrals by using symmetry, parameter equation, and the relations of two types of surface integrals, moreover the Gauss formula and the Stokes formula.

练 习

一、概念

1. 能与其相反数及零表示长度的数是 _____ ;

答案 实数.

2. $\lim\limits_{x \to c^+} f(x) = L$ 表示当 x 从 _____ 边趋向于 c 时, $f(x)$ 趋向于 _____ ;

答案 右; L.

3. 如果函数 f 在点 c 有定义且 _____ $= f(c)$, 则函数 f 在点 c 连续;

答案 $\lim\limits_{x \to c} f(x)$.

4. 介值定理说明: 如果函数 f 在 $[a,b]$ 连续且 W 是位于 $f(a)$ 和 $f(b)$ 之间的数, 则存在位于 _____ 和 _____ 之间的数 c 使得 _____ ;

答案 a; b; $f(c) = W$.

5. 函数 f 在 c 点的导数定义为 $f'(c) = \lim\limits_{h \to 0}$ _____ 或者 $f'(c) = \lim\limits_{t \to c}$ _____ ;

答案 $\dfrac{f(c+h) - f(c)}{h}$; $\dfrac{f(t) - f(c)}{t - c}$.

6. 若函数 f 在点 c 可微, 则 f 在 c 点 _____ , 但反过来是错的, 例如 $f(x) =$ _____ ;

答案 连续; $|x|$.

7. 两函数积的导数等于第一个函数乘以 _____ 加第 _____ 个函数乘以第一个函数的导数, 即: $\dfrac{\mathrm{d}(f(x)g(x))}{\mathrm{d}x} =$ _____ ;

答案 第二个的导数; 二; $f(x)\dfrac{\mathrm{d}g(x)}{\mathrm{d}x} + g(x)\dfrac{\mathrm{d}f(x)}{\mathrm{d}x}$.

8. 若 f 在 $[a,b]$ 连续且若 F 是 f 的 _____ , 则 $\int_a^b f(x)\mathrm{d}x =$ _____ ;

答案 原函数; $F(b) - F(a)$.

9. 为了计算 $\int \cos^2 x \mathrm{d}x$, 我们应将它写为 _____ ;

答案 $\int \dfrac{1 + \cos 2x}{2} \mathrm{d}x$.

10. 若 $\lim\limits_{n \to \infty} a_n \neq 0$, 则级数 $\sum\limits_{n=1}^{\infty} a_n$ _____ ;

答案 发散.

二、计算下列题目:

1. 设 $x^3 y + xy^3 = 2$, 计算点 $(1,1)$ 的 y' 和 y'' ;

解 用两次隐函数微分, 得到 $x^3 y' + 3x^2 y + x(3y^2 y') + y^3 = 0$ 和 $x^3 y'' + 3x^2 y' + 3x^2 y' + 6xy + 3xy^2 y'' + y'[6xyy' + 3y^2] + 3y^2 y' = 0$, 在第一个方程中令 $x = 1, y = 1$ 得出 $y' = -1$. 在第二个方程中

令 $x=1, y=1, y'=-1$ 得出 $y''=0$。

2. 用 L'Hospital 法则计算下列极限

(1) $\lim\limits_{x\to 0}\dfrac{x+\sin 2x}{x-\sin 2x}$; (2) $\lim\limits_{x\to 0}\dfrac{e^x+e^{-x}-x^2-2}{\sin^2 x-x^2}$; (3) $\lim\limits_{x\to \pi/4}(1-\tan x)\sec 2x$;

(4) $\lim\limits_{x\to 0}\left(\dfrac{1}{x}-\dfrac{1}{e^x-1}\right)$; (5) $\lim\limits_{x\to (\pi/2)^-}(\tan x)^{\cos x}$.

解 (1) $\lim\limits_{x\to 0}\dfrac{x+\sin 2x}{x-\sin 2x}=\lim\limits_{x\to 0}\dfrac{1+2\cos 2x}{1-2\cos 2x}=\dfrac{1+2\cdot 1}{1-2\cdot 1}=-3$;

(2) $\lim\limits_{x\to 0}\dfrac{e^x+e^{-x}-x^2-2}{\sin^2 x-x^2}=\lim\limits_{x\to 0}\dfrac{e^x-e^{-x}-2x}{2\sin x\cos x-2x}=\lim\limits_{x\to 0}\dfrac{e^x-e^{-x}-2x}{\sin 2x-2x}$,

重复运用 L'Hosptital 法则,得到

$\lim\limits_{x\to 0}\dfrac{e^x+e^{-x}-2}{2\cos 2x-2}=\lim\limits_{x\to 0}\dfrac{e^x-e^{-x}}{-4\sin 2x}=\lim\limits_{x\to 0}\dfrac{e^x+e^{-x}}{-8\cos 2x}=\dfrac{1+1}{-8\cdot 1}=-\dfrac{2}{8}=-\dfrac{1}{4}$;

(3) 这是型为 $0\cdot\infty$ 的题目,$\lim\limits_{x\to \pi/4}\dfrac{1-\tan x}{\cos 2x}=\lim\limits_{x\to \pi/4}\dfrac{-\sec^2 x}{-2\sin 2x}=\dfrac{-2}{-2}=1$;

(4) 这是型为 $\infty-\infty$ 的题目,$\lim\limits_{x\to 0}\dfrac{e^x-1-x}{x(e^x-1)}=\lim\limits_{x\to 0}\dfrac{e^x-1}{xe^x+e^x-1}=\lim\limits_{x\to 0}\dfrac{e^x}{xe^x+2e^x}=\dfrac{1}{0+2}=\dfrac{1}{2}$;

(5) 这是型为 ∞^0 的题目,设 $y=(\tan x)^{\cos x}$,则 $\ln y=(\cos x)(\ln\tan x)=\dfrac{\ln\tan x}{\sec x}$,因此

$$\lim\limits_{x\to (\pi/2)^-}\ln y=\lim\limits_{x\to (\pi/2)^-}\dfrac{\ln\tan x}{\sec x}=\lim\limits_{x\to (\pi/2)^-}(\sec^2 x/\tan x)/(\sec x\tan x)$$
$$=\lim\limits_{x\to (\pi/2)^-}\dfrac{\cos x}{\sin^2 x}=\dfrac{0}{1}=0.$$

所以 $\lim\limits_{x\to (\pi/2)^-}(\tan x)^{\cos x}=1$.

3. 计算 $\int\dfrac{\cos\sqrt{x}}{\sqrt{x}}dx$;

解 设 $u=\sqrt{x}=x^{1/2}$,则 $du=\dfrac{1}{2}x^{-1/2}dx$. 因此,$2du=\dfrac{1}{\sqrt{x}}dx$,从而,

$$\int\dfrac{\cos\sqrt{x}}{\sqrt{x}}dx=2\int\cos u\,du=2\sin u+C=2\sin(\sqrt{x})+C.$$

4. 计算 $\int_1^9\sqrt{5x+4}\,dx$;

解 设 $u=5x+4$,则 $du=5dx$. 当 $x=1, u=9$,且当 $x=9, u=49$,因此,

$$\int_1^9\sqrt{5x+4}\,dx=\int_9^{49}\sqrt{u}\,\dfrac{1}{5}du=\dfrac{1}{5}\int_9^{49}u^{1/2}du=\dfrac{1}{5}\left(\dfrac{2}{3}u^{3/2}\right)\Big|_9^{49}$$
$$=\dfrac{2}{15}(49^{3/2}-9^{3/2})=\dfrac{2}{15}(7^3-3^3)=\dfrac{632}{15}.$$

5. 计算 $\lim\limits_{(x,y)\to(0,0)}\dfrac{xy}{\sqrt{x^2+y^2}}$;

解 当 $(x,y)\to(0,0)$ 时,$|x|=\sqrt{x^2}\leqslant\sqrt{x^2+y^2}$,$\left|\dfrac{xy}{\sqrt{x^2+y^2}}\right|\leqslant|y|\to 0$,所以有

$$\lim\limits_{(x,y)\to(0,0)}\dfrac{xy}{\sqrt{x^2+y^2}}=0.$$

三、证明下列题目:

1. 证明 $\lim\limits_{n\to\infty}\sqrt[n]{n}=1$;

证明 设 $\sqrt[n]{n}=1+h_n$,其中 $h_n\geqslant 0$,对于 $\forall n$(因为 $h_n\geqslant 0$)有

$$n = (1+h_n)^n = 1 + nh_n + \frac{1}{2}n(n-1)h_n^2 + \cdots + h_n^n > \frac{1}{2}n(n-1)h_n^2;$$

当 $n \geqslant 2$ 时,$h_n^2 < \frac{2}{n-1}$,或者 $n \geqslant 2$ 时,$|h_n| < \sqrt{\frac{2}{n-1}}$.

设 ε 是任意正整数,则当 $n > 1 + 2/\varepsilon^2$ 时,$|h_n| < \sqrt{\frac{2}{n-1}} < \varepsilon$.

设 m 是大于 $1 + 2/\varepsilon^2$ 的任意正整数. 因此,对 $\varepsilon > 0$,存在正整数 m 满足
$$|\sqrt[n]{n} - 1| = |h_n| < \varepsilon \quad \forall n \geqslant m.$$

因此 $\lim\limits_{n \to \infty} \sqrt[n]{n} = 1$.

2. 证明级数 $\frac{1}{(\log 2)^p} + \frac{1}{(\log 3)^p} + \cdots + \frac{1}{(\log n)^p} + \cdots$ 当 $p > 0$ 时发散;

证明 因为 $\lim\limits_{n \to \infty} \frac{(\log n)^p}{n} = 0$,所以 $\forall n > 1$,有 $(\log n)^p < n$,从而 $\frac{1}{(\log n)^p} > \frac{1}{n}$.

由于级数 $\sum\limits_{n=1}^{\infty} \frac{1}{n}$ 是发散级数,根据比较判别法,给定的级数是发散的.

3. 证明级数 $\frac{1 \cdot 2}{3^2 \cdot 4^2} + \frac{3 \cdot 4}{5^2 \cdot 6^2} + \frac{5 \cdot 6}{7^2 \cdot 8^2} + \cdots$ 收敛;

证明 将级数记为 $\sum u_n$,其中 $u_n = \frac{(2n+1)(2n+2)}{(2n+3)^2(2n+4)^2}$.

将上述级数与收敛级数 $\sum v_n$ 比较,其中 $v_n = \frac{1}{n^2}$,且

$$\lim_{n \to \infty} \frac{u_n}{v_n} = \lim_{n \to \infty} \frac{\left(2 + \frac{1}{n}\right)\left(2 + \frac{2}{n}\right)}{\left(2 + \frac{3}{n}\right)^2 \left(2 + \frac{4}{n}\right)^2} = 1.$$

因为 $\sum v_n$ 收敛,根据比较判别法,$\sum u_n$ 也收敛.

4. 证明如下定义的函数 $f(x) = \begin{cases} x \sin \frac{1}{x} & \text{当 } x \neq 0 \\ 0 & \text{当 } x = 0 \end{cases}$ 在点 $x = 0$ 连续;

证明 由于 $\lim\limits_{x \to 0} f(x) = \lim\limits_{x \to 0} \left(x \sin \frac{1}{x}\right) = 0$,即 $\lim\limits_{x \to 0} f(x) = f(0)$,所以 f 在 $x = 0$ 连续.

Exercises

Ⅰ. **Concepts**:

1. Numbers that can measure lengths (together with their negatives and zero) are called _____ numbers.

   Answer   real.

2. $\lim\limits_{x \to c^+} f(x) = L$ means that $f(x)$ gets near to _____ when x approaches c from the _____.

   Answer L; right.

3. A function f is continuous at the point c if f is defined at c and _____ $= f(c)$.

   Answer $\lim\limits_{x \to c} f(x)$.

4. The Intermediate Value Theorem says that if a function f is continuous on $[a, b]$ and W is a number between $f(a)$ and $f(b)$, then there is a number c between _____ and _____ such that _____.

   Answer a; b; $f(c) = W$.

5. The derivative of f at c is given by $f'(c) = \lim\limits_{h \to 0}$ _____; equivalently, $f'(c) = \lim\limits_{t \to c}$ _____.

Answer $\dfrac{f(c+h)-f(c)}{h}$; $\dfrac{f(t)-f(c)}{t-c}$.

6. If f is differentiable at c, then f is _____ at c. The inverse is false, as is shown by the example $f(x) =$ _____.

Answer continuous; $|x|$.

7. The derivative of a product of two functions is the first times _____ plus the _____ times the derivative of the first. In symbols, $\dfrac{d(f(x)g(x))}{dx} =$ _____.

Answer the derivative of the second; second; $f(x)\dfrac{dg(x)}{dx} + g(x)\dfrac{df(x)}{dx}$.

8. If f is continuous on $[a,b]$ and if F is a _____ of f there, then $\int_a^b f(x)dx =$ _____.

Answer primitive function; $F(b) - F(a)$.

9. To handle $\int \cos^2 x\, dx$, we first rewrite it as _____.

Answer $\int \dfrac{1+\cos 2x}{2}\, dx$.

10. If $\lim\limits_{n\to\infty} a_n \neq 0$, we can be sure that the series $\sum\limits_{n=1}^{\infty} a_n$ is _____.

Answer. divergent.

II. Evaluate the following problems:

1. Give $x^3 y + xy^3 = 2$, find y' and y'' at the point $(1,1)$;

Solution By computing the implicit differentiation twice yields
$$x^3 y' + 3x^2 y + x(3y^2 y') + y^3 = 0 \text{ and}$$
$$x^3 y'' + 3x^2 y' + 3x^2 y' + 6xy + 3xy^2 y'' + y'[6xyy' + 3y^2] + 3y^2 y' = 0.$$

Substituting $x = 1, y = 1$ in the first equation we have $y' = -1$. Then substituting $x = 1, y = 1, y' = -1$ in the second equation we have $y'' = 0$.

2. Use the L'Hospital Rule one or more times to evaluate the following limits.

(1) $\lim\limits_{x\to 0} \dfrac{x+\sin 2x}{x - \sin 2x}$;   (2) $\lim\limits_{x\to 0} \dfrac{e^x + e^{-x} - x^2 - 2}{\sin^2 x - x^2}$;   (3) $\lim\limits_{x\to \pi/4}(1-\tan x)\sec 2x$;

(4) $\lim\limits_{x\to 0}\left(\dfrac{1}{x} - \dfrac{1}{e^x - 1}\right)$;   (5) $\lim\limits_{x\to (\pi/2)^-} (\tan x)^{\cos x}$.

Solution (1) $\lim\limits_{x\to 0}\dfrac{x+\sin 2x}{x-\sin 2x} = \lim\limits_{x\to 0}\dfrac{1+2\cos 2x}{1-2\cos 2x} = \dfrac{1+2\cdot 1}{1-2\cdot 1} = -3$;

(2) It is easy to show that
$$\lim_{x\to 0}\dfrac{e^x + e^{-x} - x^2 - 2}{\sin^2 x - x^2} = \lim_{x\to 0}\dfrac{e^x - e^{-x} - 2x}{2\sin x\cos x - 2x} = \lim_{x\to 0}\dfrac{e^x - e^{-x} - 2x}{\sin 2x - 2x}.$$

Using L'Hosptital Rule repeatedly, we get
$$\lim_{x\to 0}\dfrac{e^x + e^{-x} - 2}{2\cos 2x - 2} = \lim_{x\to 0}\dfrac{e^x - e^{-x}}{-4\sin 2x} = \lim_{x\to 0}\dfrac{e^x + e^{-x}}{-8\cos 2x} = \dfrac{1+1}{-8\cdot 1} = -\dfrac{2}{8} = -\dfrac{1}{4}.$$

(3) This is of the type $0 \cdot \infty$. However, it is equal to
$$\lim_{x\to \pi/4}\dfrac{1-\tan x}{\cos 2x} = \lim_{x\to \pi/4}\dfrac{-\sec^2 x}{-2\sin 2x} = \dfrac{-2}{-2} = 1.$$

(4) This is of the type $\infty - \infty$. But it is equal to
$$\lim_{x\to 0}\dfrac{e^x - 1 - x}{x(e^x - 1)} = \lim_{x\to 0}\dfrac{e^x - 1}{xe^x + e^x - 1} = \lim_{x\to 0}\dfrac{e^x}{xe^x + 2e^x} = \dfrac{1}{0+2} = \dfrac{1}{2}.$$

(5) This is of the type ∞^0. Let $y = (\tan x)^{\cos x}$, then $\ln y = (\cos x)(\ln \tan x) = \dfrac{\ln \tan x}{\sec x}$. So we

have
$$\lim_{x \to (\pi/2)^-} \ln y = \lim_{x \to (\pi/2)^-} \frac{\ln \tan x}{\sec x} = \lim_{x \to (\pi/2)^-} (\sec^2 x/\tan x)/(\sec x \tan x)$$
$$= \lim_{x \to (\pi/2)^-} \frac{\cos x}{\sin^2 x} = \frac{0}{1} = 0.$$

so $\lim_{x \to (\pi/2)^-} (\tan x)^{\cos x} = 1$.

3. Evaluate $\int \frac{\cos \sqrt{x}}{\sqrt{x}} dx$.

Solution Let $u = \sqrt{x} = x^{1/2}$. Then $du = \frac{1}{2} x^{-1/2} dx$. i.e., $2du = \frac{1}{\sqrt{x}} dx$, we have

$$\int \frac{\cos \sqrt{x}}{\sqrt{x}} dx = 2\int \cos u \, du = 2\sin u + C = 2\sin(\sqrt{x}) + C.$$

4. Evaluate $\int_1^9 \sqrt{5x+4} \, dx$;

Solution Let $u = 5x+4$, then $du = 5dx$. We also have $u = 9$ as $x = 1$ and $u = 49$ as $x = 9$. Hence,

$$\int_1^9 \sqrt{5x+4} \, dx = \int_9^{49} \sqrt{u} \cdot \frac{1}{5} du = \frac{1}{5} \int_9^{49} u^{1/2} du = \frac{1}{5} \left(\frac{2}{3} u^{3/2} \right) \Big|_9^{49}$$
$$= \frac{2}{15}(49^{3/2} - 9^{3/2}) = \frac{2}{15}(7^3 - 3^3) = \frac{632}{15}.$$

5. Evaluate $\lim_{(x,y) \to (0,0)} \frac{xy}{\sqrt{x^2+y^2}}$;

Solution Since $|x| = \sqrt{x^2} \leqslant \sqrt{x^2+y^2}$, $\left| \frac{xy}{\sqrt{x^2+y^2}} \right| \leqslant |y| \to 0$ as $(x,y) \to (0,0)$.
So we have $\lim_{(x,y) \to (0,0)} \frac{xy}{\sqrt{x^2+y^2}} = 0$.

Ⅲ. **Prove the following problems:**

1. Show that $\lim_{n \to \infty} \sqrt[n]{n} = 1$.

Proof Let $\sqrt[n]{n} = 1 + h_n$, where $h_n \geqslant 0$, for any values of n, we obtain
$$n = (1+h_n)^n = 1 + nh_n + \frac{1}{2}n(n-1)h_n^2 + \cdots + h_n^n > \frac{1}{2}n(n-1)h_n^2,$$

Namely, $h_n^2 < \frac{2}{n-1}$ as $n \geqslant 2$ or $|h_n| < \sqrt{\frac{2}{n-1}}$ as $n \geqslant 2$.

For any positive numbers ε, we have $|h_n| < \sqrt{\frac{2}{n-1}} < \varepsilon$ as $n > 1 + 2/\varepsilon^2$.

Thus for any $\varepsilon > 0$ and for any positive integers $m > 1 + 2/\varepsilon^2$, we get
$$|\sqrt[n]{n} - 1| = |h_n| < \varepsilon \quad \forall n \geqslant m.$$

Hence $\lim_{n \to \infty} \sqrt[n]{n} = 1$.

2. Show that the series $\frac{1}{(\log 2)^p} + \frac{1}{(\log 3)^p} + \cdots + \frac{1}{(\log n)^p} + \cdots$ is divergent for $p > 0$.

Proof Let us compare the given series with the divergent series $\sum_{n=1}^{v} \frac{1}{n}$.

Since $\lim_{n \to \infty} \frac{(\log n)^p}{n} = 0$, therefore, $(\log n)^p < n$, $\forall n > 1$, i.e., $\frac{1}{(\log n)^p} > \frac{1}{n}$, $\forall n > 1$.

Since the series $\sum_{n=1}^{\infty} \frac{1}{n}$ is divergent, therefore the given series is also divergent according to the comparison test.

3. Show that the series $\frac{1\cdot 2}{3^2\cdot 4^2}+\frac{3\cdot 4}{5^2\cdot 6^2}+\frac{5\cdot 6}{7^2\cdot 8^2}+\cdots$ is convergent.

Proof Let us denote the given series by $\sum u_n$, where
$$u_n=\frac{(2n+1)(2n+2)}{(2n+3)^2(2n+4)^2}.$$

Compare it with the convergent series $\sum v_n$, where $v_n=\frac{1}{n^2}$. Now
$$\lim_{n\to\infty}\frac{u_n}{v_n}=\lim_{n\to\infty}\frac{\left(2+\frac{1}{n}\right)\left(2+\frac{2}{n}\right)}{\left(2+\frac{3}{n}\right)^2\left(2+\frac{4}{n}\right)^2}=1.$$

Since $\sum v_n$ is convergent, from the comparison test $\sum u_n$ is also convergent.

4. Show that the function defined by $f(x)=\begin{cases} x\sin\frac{1}{x} & \text{when } x\neq 0 \\ 0 & \text{when } x=0 \end{cases}$ is continuous at the point $x=0$.

Proof It is easy to show that $\lim_{x\to 0}f(x)=\lim_{x\to 0}\left(x\sin\frac{1}{x}\right)=0$, namely, $\lim_{x\to 0}f(x)=f(0)$, that is to say, f is continuous at the point $x=0$.

第二部分 高等代数
Part 2 Higher Algebra

引 言

本部分主要介绍高等代数,它是利用矩阵、向量空间和线性变换研究数学的一个分支. 这里讲的矩阵是作为求解线性方程组的一个工具. 在现代数学中,几乎都要涉及矩阵. 矩阵也同样应用在其他学科中,比如,统计学、经济学、工程学、物理学、化学、生物学和地质学等.

Introduction

The purpose of this part is to mainly provide an introduction to higher algebra, which is a branch of studying mathematics by matrices, vector spaces and linear transformations. Matrices introduced here are used as a tool for solving systems of linear equations. Moreover, matrices are also used in many areas of modern mathematics, and they also have many applications in other disciplines, such as statistics, economics, engineering, physics, chemistry, biology, geology, and so on.

第一章 多项式
Section 1 Polynomials

本章介绍实系数多项式、有理系数多项式和复系数多项式,包括多项式的根和不可约多项式.

In this chapter we introduce real coefficient polynomials, rational coefficient polynomials and complex coefficient polynomials, including roots of polynomials and irreducible polynomials.

单词和短语 Words and expressions

- ★ 实系数多项式   real coefficient polynomial
- ★ 有理系数多项式   rational coefficient polynomial
- ★ 复系数多项式   complex coefficient polynomial
- 对称多项式   symmetric polynomial
- 函数   function   ['fʌŋkʃən]
- 多项式函数   polynomial function
- 带余除法   division with remainder
- 综合除法   synthetic division
- 公因式   common factor
- 最大公因式   greatest common factor
- 单因式   unit-factor
- 重因式   repeated factor
- 单根   single root
- 重根   multiple root

本章重点

1. 掌握一元多项式的运算,特别是除法(带余除法、辗转相除、综合除法),及在除法意义下的公因式、最大公因式、互素、单因式、重因式、单根、重根等概念.
2. 掌握一元多项式在复数域、实数域,特别是有理数域上的因式分解问题.

Key points of this chapter

1. Master the computation of polynomials with one variable, especially in division (division with remainder, algorithm of division and synthetic division), and some concepts under the essence of division, namely, common factor, greatest common factor, relatively prime, unit-factor, repeated factor, single root, repeated root, and so on.
2. Master the factorization problems of polynomials with one variable in complex number field, real number field, and particularly in rational number field.

第二章 行列式
Chapter 2 Determinants

本章介绍了行列式的抽象定义和性质。因为行列式的性质在简化行列式的计算中经常用到,所以要求理解和熟练掌握行列式的性质。

In this chapter, the abstract definition and properties of determinants are introduced. Since the properties of determinants are well used in the computation of simplifying determinants, they must be understood and mastered expertly.

单词和短语 Words and expressions

排列	permutation	余子式	cofactors term
对换	transposition	★克拉默法则	Cramer rule
行列式	determinant	拉普拉斯定理	Laplace theorem

例题 Examples

例 计算行列式 $D = \begin{vmatrix} 1 & 1 & -1 & 2 \\ -1 & -1 & -4 & 1 \\ 2 & 4 & -6 & 1 \\ 1 & 2 & 4 & 2 \end{vmatrix}.$

解 $D = \begin{vmatrix} 1 & 1 & -1 & 2 \\ -1 & -1 & -4 & 1 \\ 2 & 4 & -6 & 1 \\ 1 & 2 & 4 & 2 \end{vmatrix} = \begin{vmatrix} 1 & 1 & -1 & 2 \\ 0 & 0 & -5 & 3 \\ 0 & 2 & -4 & -3 \\ 0 & 1 & 5 & 0 \end{vmatrix} = \begin{vmatrix} 0 & -5 & 3 \\ 2 & -4 & -3 \\ 1 & 5 & 0 \end{vmatrix}$

$\xrightarrow{r_2 - 2r_3} \begin{vmatrix} 0 & -5 & 3 \\ 0 & -14 & -3 \\ 1 & 5 & 0 \end{vmatrix} = \begin{vmatrix} -5 & 3 \\ -14 & -3 \end{vmatrix} = 3 \times (5+14) = 57.$

Ex. Compute $D = \begin{vmatrix} 1 & 1 & -1 & 2 \\ -1 & -1 & -4 & 1 \\ 2 & 4 & -6 & 1 \\ 1 & 2 & 4 & 2 \end{vmatrix}.$

Solution $D = \begin{vmatrix} 1 & 1 & -1 & 2 \\ -1 & -1 & -4 & 1 \\ 2 & 4 & -6 & 1 \\ 1 & 2 & 4 & 2 \end{vmatrix} = \begin{vmatrix} 1 & 1 & -1 & 2 \\ 0 & 0 & -5 & 3 \\ 0 & 2 & -4 & -3 \\ 0 & 1 & 5 & 0 \end{vmatrix} = \begin{vmatrix} 0 & -5 & 3 \\ 2 & -4 & -3 \\ 1 & 5 & 0 \end{vmatrix}$

$\xrightarrow{r_2 - 2r_3} \begin{vmatrix} 0 & -5 & 3 \\ 0 & -14 & -3 \\ 1 & 5 & 0 \end{vmatrix} = \begin{vmatrix} -5 & 3 \\ -14 & -3 \end{vmatrix} = 3 \times (5+14) = 57.$

本章重点

行列式的计算和行列式的应用,即克拉默法则.

1. 三角行列式的值是什么?
2. 具有两个相等行或列的行列式的值是什么?
3. 交换行列式的两行或列,其值怎样?
4. 对方矩阵 A 实施初等行运算 $r_i + cr_j (i \neq j)$,得到新矩阵的行列式和 $\det A$ 有什么关系?
5. 如果 $n \times n$ 矩阵 A 的每一行都乘以标量,那么给出用 $\det A$ 表示的新矩阵的行列式.

Key points of this chapter

The computation of determinants and its applications — Cramer Rule.

1 What is the value of a triangular determinant?
2 What is the value of a determinant with two identical rows or columns?
3 How does the value of a determinant change when two rows or columns are interchanged?
4 If an elementary row operation $r_i + cr_j \,(i \neq j)$ is performed on a square matrix A, what is the relation between the determinant of the new matrix and det A?
5 If each row of an $n \times n$ matrix A is multiplied by a scalar, write down the formula for the determinant of the new matrix in terms of det A.

第三章 线性方程组
Chapter 3 Systems of Linear Equations

本章的主要目的就是把一个线性方程组简化为可求解的线性方程组. 这等价于把相应的矩阵简化成为行阶梯形矩阵.

- 在一个 n 维子空间中任何 $n+1$ 个或更多个向量是线性相关的.
- 未知数的个数比方程的个数还多的齐次线性方程组一定有非平凡解.

An important aim that occurs throughout the chapter is to reduce a system of linear equations to a system that is readily solvable. This is equivalent to reducing a certain matrix into the corresponding form known as row echelon form.

- Any $n+1$ or more vectors in a subspace with n-dimension are linearly dependent.
- A system of homogeneous linear equations must have a nontrivial solution if the number of unknowns of the equations is more than the number of equations.

单词和短语 Words and expressions

向量组　vector system
向量　vector
系数矩阵　coefficient matrix
线性方程组　system of linear equations
线性组合　linear combination
★ 线性相关　linear dependence
★ 线性无关　linear independence
★ 极大线性无关　greatest linear independence
线性表示　linear expression
基（基底）　basis　['beisis]
★ 向量空间　vector space
秩　rank　[ræŋk]
齐次线性方程组　system of homogeneous linear equations
等价　equivalent

解　solution　[sə'lju:ʃən]
矩阵相等　equality of matrices
等价线性方程组　equivalent linear system
线性方程组的系数矩阵　coefficient matrix of a linear system
线性方程组的阶梯形　echelon form of a linear system
线性方程组的增广矩阵　augmented matrix of a linear system
高斯消元法　Gauss elimination method
初等行运算　elementary row operation
行阶梯形矩阵　row echelon matrix
非平凡解　nontrivial solution
简化行阶梯形矩阵　reduced row echelon matrix

例题　Examples

例 1 解线性方程组

$$\begin{cases} x_1 + x_2 + 2x_3 = 1, \\ 2x_1 - x_2 + 2x_3 = 4, \\ x_1 - 2x_2 = 3, \\ 4x_1 + x_2 + 4x_3 = 2. \end{cases}$$

解 $\bar{A} = \begin{pmatrix} 1 & 1 & 2 & 1 \\ 2 & -1 & 2 & 4 \\ 1 & -2 & 0 & 3 \\ 4 & 1 & 4 & 2 \end{pmatrix} \begin{matrix} \\ r_2-2r_1 \\ r_3-r_1 \\ r_4-4r_1 \end{matrix} \begin{pmatrix} 1 & 1 & 2 & 1 \\ 0 & -3 & -2 & 2 \\ 0 & -3 & -2 & 2 \\ 0 & -3 & -4 & -2 \end{pmatrix}$

$\begin{matrix} \\ r_3-r_2 \\ r_4-r_2 \end{matrix} \begin{pmatrix} 1 & 1 & 2 & 1 \\ 0 & -3 & -2 & 2 \\ 0 & 0 & 0 & 0 \\ 0 & 0 & -2 & -4 \end{pmatrix} \begin{matrix} \\ r_3 \leftrightarrow r_4 \\ -\frac{1}{2}r_3 \end{matrix} \begin{pmatrix} 1 & 1 & 2 & 1 \\ 0 & -3 & -2 & 2 \\ 0 & 0 & 1 & 2 \\ 0 & 0 & 0 & 0 \end{pmatrix}$

$\begin{matrix} r_2+2r_3 \\ r_1-2r_3 \end{matrix} \begin{pmatrix} 1 & 1 & 0 & -3 \\ 0 & -3 & 0 & 6 \\ 0 & 0 & 1 & 2 \\ 0 & 0 & 0 & 0 \end{pmatrix} \begin{matrix} r_1+\frac{1}{3}r_2 \\ -\frac{1}{3}r_2 \end{matrix} \begin{pmatrix} 1 & 0 & 0 & -1 \\ 0 & 1 & 0 & -2 \\ 0 & 0 & 1 & 2 \\ 0 & 0 & 0 & 0 \end{pmatrix},$

所以, $\begin{cases} x_1 = -1, \\ x_2 = -2, \\ x_3 = 2. \end{cases}$

Ex. 1 Solve the following system of linear equations:
$$\begin{cases} x_1 + x_2 + 2x_3 = 1, \\ 2x_1 - x_2 + 2x_3 = 4, \\ x_1 - 2x_2 = 3, \\ 4x_1 + x_2 + 4x_3 = 2. \end{cases}$$

Solution $\bar{A} = \begin{pmatrix} 1 & 1 & 2 & 1 \\ 2 & -1 & 2 & 4 \\ 1 & -2 & 0 & 3 \\ 4 & 1 & 4 & 2 \end{pmatrix} \begin{matrix} \\ r_2-2r_1 \\ r_3-r_1 \\ r_4-4r_1 \end{matrix} \begin{pmatrix} 1 & 1 & 2 & 1 \\ 0 & -3 & -2 & 2 \\ 0 & -3 & -2 & 2 \\ 0 & -3 & -4 & -2 \end{pmatrix}$

$\begin{matrix} r_3-r_2 \\ r_4-r_2 \end{matrix} \begin{pmatrix} 1 & 1 & 2 & 1 \\ 0 & -3 & -2 & 2 \\ 0 & 0 & 0 & 0 \\ 0 & 0 & -2 & -4 \end{pmatrix} \begin{matrix} r_3 \leftrightarrow r_4 \\ -\frac{1}{2}r_3 \end{matrix} \begin{pmatrix} 1 & 1 & 2 & 1 \\ 0 & -3 & -2 & 2 \\ 0 & 0 & 1 & 2 \\ 0 & 0 & 0 & 0 \end{pmatrix}$

$\begin{matrix} r_2+2r_3 \\ r_1-2r_3 \end{matrix} \begin{pmatrix} 1 & 1 & 0 & -3 \\ 0 & -3 & 0 & 6 \\ 0 & 0 & 1 & 2 \\ 0 & 0 & 0 & 0 \end{pmatrix} \begin{matrix} r_1+\frac{1}{3}r_2 \\ -\frac{1}{3}r_2 \end{matrix} \begin{pmatrix} 1 & 0 & 0 & -1 \\ 0 & 1 & 0 & -2 \\ 0 & 0 & 1 & 2 \\ 0 & 0 & 0 & 0 \end{pmatrix},$

so we have $\begin{cases} x_1 = -1, \\ x_2 = -2, \\ x_3 = 2. \end{cases}$

例2 求向量组的秩:
$\boldsymbol{\alpha}_1 = (1, 1, 3, 1), \boldsymbol{\alpha}_2 = (-1, 1, -1, 3), \boldsymbol{\alpha}_3 = (5, -2, 8, -9), \boldsymbol{\alpha}_4 = (-1, 3, 1, 7).$

解 $A = \begin{pmatrix} 1 & -1 & 5 & -1 \\ 1 & 1 & -2 & 3 \\ 3 & -1 & 8 & 1 \\ 1 & 3 & -9 & 7 \end{pmatrix} \begin{matrix} \\ r_2-r_1 \\ r_3-3r_1 \\ r_4-r_1 \end{matrix} \begin{pmatrix} 1 & -1 & 5 & -1 \\ 0 & 2 & -7 & 4 \\ 0 & 2 & -7 & 4 \\ 0 & 4 & -14 & 8 \end{pmatrix}$

$\begin{matrix} r_3-r_2 \\ r_4-2r_2 \end{matrix} \begin{pmatrix} 1 & -1 & 5 & -1 \\ 0 & 2 & -7 & 4 \\ 0 & 0 & 0 & 0 \\ 0 & 0 & 0 & 0 \end{pmatrix}.$

所以，$R(A) = 2$.

Ex. 2 Find the rank of the vectors:
$$\alpha_1 = (1,1,3,1), \alpha_2 = (-1,1,-1,3), \alpha_3 = (5,-2,8,-9), \alpha_4 = (-1,3,1,7).$$

Solution $A = \begin{pmatrix} 1 & -1 & 5 & -1 \\ 1 & 1 & -2 & 3 \\ 3 & -1 & 8 & 1 \\ 1 & 3 & -9 & 7 \end{pmatrix} \begin{matrix} r_2 - r_1 \\ r_3 - 3r_1 \\ r_4 - r_1 \end{matrix} \begin{pmatrix} 1 & -1 & 5 & -1 \\ 0 & 2 & -7 & 4 \\ 0 & 2 & -7 & 4 \\ 0 & 4 & -14 & 8 \end{pmatrix}$

$\begin{matrix} r_3 - r_2 \\ r_4 - 2r_2 \end{matrix} \begin{pmatrix} 1 & -1 & 5 & -1 \\ 0 & 2 & -7 & 4 \\ 0 & 0 & 0 & 0 \\ 0 & 0 & 0 & 0 \end{pmatrix}.$

So $R(A) = 2$.

本章重点

1 掌握向量组的线性相关性（线性组合、线性表示、线性相关、线性无关、极大线性无关、向量组的秩、矩阵的秩）。

2 掌握线性方程组解的判定（唯一解、无穷多解、没有解）、解的结构和解的求法。

3 每个矩阵都能通过适当的初等行运算化为行阶梯形矩阵和简化行阶梯形矩阵。

4 未知数的个数多于方程的个数的齐次线性方程组总有非平凡解。

5 什么是相容和不相容方程组？举出例子。

6 写出关于线性方程组（矩阵）的3种初等（行）变换。

7 什么是方程组的系数矩阵和增广矩阵？举例说明。

8 详细定义矩阵的行阶梯形和简化行阶梯形。

9 用你自己的语言写出把一个矩阵化为(i)行阶梯形和(ii)简化行阶梯形的步骤。

10 齐次线性方程组有非平凡解是什么意思？

11 你能给出 m 个方程 n 个未知数的齐次线性方程组有非平凡解的充分条件吗？

Key points of this chapter

1 Master the linear relations in vector systems (linear combination, linear presentation, linear dependence, linear independence, greatest linear independence, rank of vector system, and rank of matrices).

2 Master the judgment of solutions to linear equation system (unique solution, infinitely solutions and no solution), the structure of solutions and methods for finding the solutions.

3 Every matrix can be reduced to row echelon form as well as to reduced row echelon form by performing appropriate elementary row operations.

4 A homogeneous linear system in which the number of unknowns is more than the number of equations always has a nontrivial solution.

5 What is consistent and inconsistent linear systems? Give an example.

6 Write out three types of elementary (row) operations on a system of linear equations (matrix).

7 What are the coefficient matrix and the augmented matrix of a linear system? Explain by means of an example.

8 Carefully define row echelon form and reduced row echelon form of a matrix.

9 Write out steps in your own words to reduce a matrix into (i) row echelon form and (ii) reduced row echelon form.

10 What does it mean that a homogeneous linear system has a nontrivial solution?

11 Can you present the sufficient conditions for a homogeneous linear system of m equations in n unknowns to possess a nontrivial solution?

第四章 矩阵
Chapter 4 Matrices

矩阵的加法、减法、标量乘法的定义都非常自然,而且这些运算也都很容易.尽管矩阵的乘法是求解线性方程组的核心工具,但是它的定义不那么自然,并且你会发现使用矩阵的乘法的公式有一定难度.

- 方阵乘积的行列式等于行列式的乘积.
- 一个方阵可逆当且仅当它的行列式不为零(它的秩等于它的阶数).
- 对任何一个矩阵,列秩等于行秩,称之为秩.
- 矩阵加法满足交换律和结合律.
- 矩阵乘法满足结合律,但不满足交换律.
- 非零矩阵的乘积可以是零矩阵.
- 矩阵乘积的转置等于倒序的矩阵转置的乘积.
- 矩阵是可逆的当且仅当它的行阶梯形中没有零行.
- 对矩阵 A 实施初等行运算的矩阵恰好是矩阵 EA,其中 E 是对同阶单位矩阵实施同样的初等行运算得到的矩阵.对初等列运算也有类似的叙述,不同的是 E 乘在 A 的右边.

Addition, subtraction and scalar multiplication of matrices are defined naturally, and these operations are easy to apply. But multiplication of matrices, although motivated by a system of linear equations for finding its solutions, does not appear naturally, and you might find it is difficult to appreciate the formula for multiplication of matrices.

- The determinant of the product of square matrices is the product of their determinants.
- A square matrix is invertible if and only if its determinant is nonzero(if and only if its rank is equal to its size).
- For any matrices, the row rank equals to the column rank, and the value is called the rank.
- Matrix addition is commutative and associative.
- Matrix multiplication is associative but not commutative.
- The product of nonzero matrices can be a zero matrix.
- The transpose of the product of matrices is the product of the transposes of the matrices in reverse order.
- The inverse of a matrix exists if and only if its row echelon form has no zero rows.
- Performing an elementary row operation on a matrix A results in a matrix is equal to EA, where E is the matrix obtained from the identity matrix of suitable size by performing the same elementary row operation. Similar statement holds when an elementary column operation is performed, except for the difference that E is multiplied on the right of A.

单词和短语 Words and expressions

★ 单位矩阵　identity matrix
方阵　square matrix
对角矩阵　diagonal matrix
初等矩阵　elementary matrix
★ 伴随矩阵　adjoint matrix
满秩矩阵　full-rank matrix
零矩阵　zero matrix
行　row
行矩阵　row matrix
列　column
列矩阵　column matrix

上三角矩阵　upper triangular matrix
下三角矩阵　lower triangular matrix
★ 矩阵的逆　inverse of matrix
矩阵的加法　addition of matrix
负矩阵　negative matrix
矩阵的乘法　multiplication of matrix
矩阵的标量乘法　scalar multiplication of matrix
矩阵的转置　transpose of matrix
★ 矩阵的秩　rank of matrix
矩阵的列秩　column rank of matrix
矩阵的行秩　row rank of matrix

左分配律　left distributive law　　　结合律　associative law
右分配律　right distributive law

例 题　Examples

例 1 设 $A = \begin{pmatrix} -1 & 4 & -5 \\ 2 & 0 & 1 \end{pmatrix}, B = \begin{pmatrix} 3 & 0 & -7 \\ -1 & 1 & 2 \end{pmatrix}$,求 $2A - 3B$.

解 $2A - 3B = 2\begin{pmatrix} -1 & 4 & -5 \\ 2 & 0 & 1 \end{pmatrix} - 3\begin{pmatrix} 3 & 0 & -7 \\ -1 & 1 & 2 \end{pmatrix}$

$= \begin{pmatrix} -2 & 8 & -10 \\ 4 & 0 & 2 \end{pmatrix} - \begin{pmatrix} 9 & 0 & -21 \\ -3 & 3 & 6 \end{pmatrix} = \begin{pmatrix} -11 & 8 & 11 \\ 7 & -3 & -4 \end{pmatrix}$.

Ex. 1 Let $A = \begin{pmatrix} -1 & 4 & -5 \\ 2 & 0 & 1 \end{pmatrix}, B = \begin{pmatrix} 3 & 0 & -7 \\ -1 & 1 & 2 \end{pmatrix}$, compute $2A - 3B$.

Solution $2A - 3B = 2\begin{pmatrix} -1 & 4 & -5 \\ 2 & 0 & 1 \end{pmatrix} - 3\begin{pmatrix} 3 & 0 & -7 \\ -1 & 1 & 2 \end{pmatrix}$

$= \begin{pmatrix} -2 & 8 & -10 \\ 4 & 0 & 2 \end{pmatrix} - \begin{pmatrix} 9 & 0 & -21 \\ -3 & 3 & 6 \end{pmatrix} = \begin{pmatrix} -11 & 8 & 11 \\ 7 & -3 & -4 \end{pmatrix}$.

例 2 设 $A = \begin{pmatrix} 2 & 2 & 3 \\ 1 & -1 & 0 \\ -1 & 2 & 1 \end{pmatrix}$,求 A^{-1}.

解 $(A \mid E) = \begin{pmatrix} 2 & 2 & 3 & 1 & 0 & 0 \\ 1 & -1 & 0 & 0 & 1 & 0 \\ -1 & 2 & 1 & 0 & 0 & 1 \end{pmatrix} \xrightarrow[r_2 \leftrightarrow r_3]{r_1 \leftrightarrow r_2} \begin{pmatrix} 1 & -1 & 0 & 0 & 1 & 0 \\ -1 & 2 & 1 & 0 & 0 & 1 \\ 2 & 2 & 3 & 1 & 0 & 0 \end{pmatrix}$

$\xrightarrow[r_3 - 2r_1]{r_2 + r_1} \begin{pmatrix} 1 & -1 & 0 & 0 & 1 & 0 \\ 0 & 1 & 1 & 0 & 1 & 1 \\ 0 & 4 & 3 & 1 & -2 & 0 \end{pmatrix} \xrightarrow{r_3 - 4r_2} \begin{pmatrix} 1 & -1 & 0 & 0 & 1 & 0 \\ 0 & 1 & 1 & 0 & 1 & 1 \\ 0 & 0 & -1 & 1 & -6 & -4 \end{pmatrix}$

$\xrightarrow{r_2 + r_3} \begin{pmatrix} 1 & -1 & 0 & 0 & 1 & 0 \\ 0 & 1 & 0 & 1 & -5 & -3 \\ 0 & 0 & -1 & 1 & -6 & -4 \end{pmatrix} \xrightarrow[(-1)r_3]{r_1 + r_2} \begin{pmatrix} 1 & 0 & 0 & 1 & -4 & -3 \\ 0 & 1 & 0 & 1 & -5 & -3 \\ 0 & 0 & 1 & -1 & 6 & 4 \end{pmatrix}$,

所以,　　　　　　　　$A^{-1} = \begin{pmatrix} 1 & -4 & -3 \\ 1 & -5 & -3 \\ -1 & 6 & 4 \end{pmatrix}$.

Ex. 2 Let $A = \begin{pmatrix} 2 & 2 & 3 \\ 1 & -1 & 0 \\ -1 & 2 & 1 \end{pmatrix}$, compute A^{-1}.

Solution $(A \mid E) = \begin{pmatrix} 2 & 2 & 3 & 1 & 0 & 0 \\ 1 & -1 & 0 & 0 & 1 & 0 \\ -1 & 2 & 1 & 0 & 0 & 1 \end{pmatrix} \xrightarrow[r_2 \leftrightarrow r_3]{r_1 \leftrightarrow r_2} \begin{pmatrix} 1 & -1 & 0 & 0 & 1 & 0 \\ -1 & 2 & 1 & 0 & 0 & 1 \\ 2 & 2 & 3 & 1 & 0 & 0 \end{pmatrix}$

$\xrightarrow[r_3 - 2r_1]{r_2 + r_1} \begin{pmatrix} 1 & -1 & 0 & 0 & 1 & 0 \\ 0 & 1 & 1 & 0 & 1 & 1 \\ 0 & 4 & 3 & 1 & -2 & 0 \end{pmatrix} \xrightarrow{r_3 - 4r_2} \begin{pmatrix} 1 & -1 & 0 & 0 & 1 & 0 \\ 0 & 1 & 1 & 0 & 1 & 1 \\ 0 & 0 & -1 & 1 & -6 & -4 \end{pmatrix}$

$\xrightarrow{r_2 + r_3} \begin{pmatrix} 1 & -1 & 0 & 0 & 1 & 0 \\ 0 & 1 & 0 & 1 & -5 & -3 \\ 0 & 0 & -1 & 1 & -6 & -4 \end{pmatrix} \xrightarrow[(-1)r_3]{r_1 + r_2} \begin{pmatrix} 1 & 0 & 0 & 1 & -4 & -3 \\ 0 & 1 & 0 & 1 & -5 & -3 \\ 0 & 0 & 1 & -1 & 6 & 4 \end{pmatrix}$,

So　　　　　　　　$A^{-1} = \begin{pmatrix} 1 & -4 & -3 \\ 1 & -5 & -3 \\ -1 & 6 & 4 \end{pmatrix}$.

本章重点

1. 掌握矩阵的运算,特别是矩阵的逆运算的一些结论.
2. 掌握初等矩阵的定义、性质和功能.
3. 在什么条件下我们可以定义矩阵的加法和减法,写出定义矩阵加法的法则.
4. 如果 A 是一个 $m \times n$ 矩阵,B 是一个 $k \times l$ 矩阵,写出下面运算的条件:(1)AB 有意义;(2)BA 有意义;(3)AB 有意义,但 BA 没意义.
5. 给出对称矩阵的例子.设 A 和 B 是两个 $n \times n$ 对称矩阵,说明 A^2,$A+B$,AB 哪些是对称的.
6. 叙述对已知矩阵 A 实施初等行运算得到的矩阵与对单位矩阵 E 实施同样的初等行运算得到的矩阵之间的关系.
7. 每个矩阵都有逆矩阵吗? 解释你的回答.
8. 构造两个可逆矩阵 A 和 B,使得 $A+B$ 是不可逆的.
9. 构造两个不可逆矩 A 和 B,使得 $A+B$ 是可逆的.

Key points of this chapter

1. Master the computation of matrices, especially some conclusions in computation of the inverse of a matrix
2. Master the definition, properties and functions of an elementary matrix.
3. How can we define addition and subtraction of two matrices? Write out the rule for defining addition of matrices.
4. If A is a $m \times n$ matrix and B is a $k \times l$ matrix, write out the conditions under which (i) AB is well defined, (ii) BA is well defined, (iii) AB is well defined but BA is not well defined.
5. Give examples of symmetric matrices. For two $n \times n$ symmetric matrices A and B, which of the matrices A^2, $A+B$, AB are symmetric.
6. State a general fact that gives the relation between the matrix obtained by performing elementary row operations on a given matrix A and the matrix obtained by performing the same elementary row operations on identity matrix E.
7. Does every matrix possess an inverse? Give a reason to justify your answer.
8. Construct two invertible matrices A and B such that $A+B$ is not invertible.
9. Construct two noninvertible matrices A and B such that $A+B$ is invertible.

第五章 二次型
Chapter 5 Quadratic Forms

本章介绍了二次型、正定二次型和非奇异线性变换的定义和性质.同样讲述了如何把一个二次型化为标准型.

In this chapter we introduce the definitions and properties of quadratic forms, positive definite quadratic forms and non-singular linear transformation. This chapter also presents how to reduce quadratic forms to normal forms.

单词和短语 Words and expressions

★ 二次型　quadratic form	★ 正定二次型　positive definite quadratic form
★ 对称矩阵　symmetric matrix	正定矩阵　positive definite matrix
线性替换　linear substitution	非奇异线性变换　nonsingular linear transformation
★ 标准型　normal form	

本章重点

1. 把二次型用非奇异线性变换化为标准型和规范型.
2. 掌握正定二次型的定义、性质及判定.

Key points of this chapter

1. Apply a non-singular linear transformation to change a quadratic form into a normal form or standard form.
2. Master the definition, properties and identification methods of a positive quadratic form.

第六章 线性空间
Chapter 6　Linear Spaces

本章将介绍线性空间、子空间、基、维数和坐标的定义和性质.

In this chapter we will introduce the definitions and various properties of linear space, subspace, basis, dimension and coordinate.

单词和短语 Words and expressions

- ★ 线性空间　linear space
- ★ 子空间　subspace
- ★ 子空间的基　basis of subspace
- 子空间的生成元　generators of subspace
- ★ 子空间的维数　dimension of subspace
- 同构空间　isomorphism space
- 一一映射　one-to-one mapping
- 线性映射的逆　inverse of linear mapping
- 同构　isomorphism
- 零映射（函数）　zero mapping (function)
- 域　field
- 最大线性无关组　maximal set of linear independence
- 满射　surjection

本章重点

1. 掌握线性空间的定义和性质.
2. 会叙述维数、基（底）、坐标的概念.
3. 给出同一个向量在不同基（底）下坐标之间的关系，即基变换与坐标变换.
4. 掌握子空间的定义、性质、运算（交、和、直和）及子空间的维数公式.
5. 你知道子空间可以有无穷基吗？至少给出 R^2 和 R^3 的两个基. 两个不同基共同的特点是什么？

Key points of this chapter

1. Master the definition and properties of linear spaces.
2. Be able to recite the concepts of dimension, basis and coordinates.
3. Present the relations between the coordinates of the same vector on different bases, namely, the change of bases and change of coordinates.
4. Master the definition, properties, computation (intersection, sum and direct sum) and dimension formulas of subspaces.
5. Do you know that a subspace can have infinite bases? Give an example of at least two bases of R^2 and R^3. What is a common feature of two different bases?

第七章 线性变换
Chapter 7　Linear Transformations

本章介绍了线性变换、特征值和特征向量的定义和性质. 一旦你知道它们的定义，就可以直接计算线性变换的特征值和特征向量.

- F^n 的任何一个线性变换都可以看成一个适当的 $n\times n$ 矩阵.
- 任何一个 $n\times n$ 矩阵可以对角化当且仅当它有 n 个线性无关的特征向量.
- 每个矩阵都可以三角化.

In this chapter we present the definitions and various properties of linear transformation, eigenvalue and eigenvector. Computations of eigenvalues and eigenvectors can be performed directly once you know their definitions.

- Every linear transformation of the vector space F^n can be identified with a suitable $n \times n$ matrix.
- A $n \times n$ matrix can be diagonalizable if and only if it has n linearly independent eigenvectors.
- Every matrix can be changed into a triangular matrix.

单词和短语 Words and expressions

★ 线性变换　linear transformation
★ 特征值　eigenvalue
★ 特征向量　eigenvector
★ 特征多项式　characteristic polynomial
哈密顿-凯莱定理　Hamilton-Cayley theorem
方阵的特征值　eigenvalue of square matrix
方阵的特征向量　eigenvector of square matrix
矩阵的迹　trace of matrix

核　kernel ['kə:nl]
矩阵的特征多项式　characteristic polynomial of matrix
特征值的重数　multiplicity of eigenvalue
迹　trace ['treis]
相似矩阵　similar ['similə] matrix
可对角化矩阵　diagonalizable matrix

本章重点

1. 掌握线性变换的定义和性质.
2. 知道线性变换与线性变换在一个固定基底下的矩阵之间的对应关系及同一个线性变换在不同基底下矩阵之间的相似关系.
3. 掌握线性变换的特征值和特征向量的定义和性质.
4. 掌握(线性变换在一个固定基底下的)矩阵可以对角化的条件及方法.
5. 掌握线性变换的值域、核、不变子空间的定义和性质.

Key points of this chapter

1. Master the definition and properties of a linear transformation.
2. Know the correspondence of a linear transformation and its matrix under a given basis, the similarities between the matrices of the same linear transformation under different bases.
3. Master the definitions and properties of eigenvalue and eigenvector of a linear transformation.
4. Master the conditions and methods of a diagonalizable matrix (of a linear transformation on a given bases).
5. Master the definitions and properties of range, kernel and invariant subspace of a linear transformation.

第八章　λ-矩阵
Chapter 8　λ-Matrices

本章介绍了 λ-矩阵和若尔当典范形的定义和性质.

This chapter contains the abstract definitions and various properties of λ-matrix and Jordan canonical form.

单词和短语 Words and expressions

若尔当典范形　Jordan canonical form
★ 行列式因子　determinant factor
★ 不变因子　invariant factor

初等因子　elementary factor
初等变换　elementary transformation

本章重点

1. 掌握行列式因子、不变因子、初等因子的概念.
2. 知道若尔当典范形的推导.

Key points of this chapter

1. Master the concepts of determinant factor, invariant factor and elementary factor.

2. Know the deduction of Jordan canonical form.

第九章 欧几里得空间
Chapter 9 Euclidean Space

本章与几何有最密切的联系.会使用施密特正交化方法来求一个正交集合.

The contents in this chapter has the closest connection with geometry. Be able to use the Gram-Schmidt method to find an orthogonal set.

单词和短语 Words and expressions

★ 度量矩阵　metric matrix
正交基　orthogonal basis
★ 标准正交基　normal orthogonal basis
正交矩阵　orthogonal matrix
正交向量　orthogonal vector

★ 实对称矩阵　real symmetric matrix
内积　dot product　['prɔdəkt]
向量的长度　length [leŋθ] of vector
施密特正交化过程　Schmidt orthogonalization process

例 题 Examples

例 判断方阵 $A=\begin{pmatrix} 2 & -2 & 0 \\ -2 & 1 & -2 \\ 0 & -2 & 0 \end{pmatrix}$ 是否可对角化？若可对角化，求出可逆矩阵 U，使 $U^{-1}AU$ 为对角矩阵．

解 由 $|A-\lambda E|=\begin{vmatrix} 2-\lambda & -2 & 0 \\ -2 & 1-\lambda & -2 \\ 0 & -2 & -\lambda \end{vmatrix}=(\lambda+2)(\lambda-1)(\lambda-4)$ 可知，A 的特征值为 $\lambda_1=-2, \lambda_2=1, \lambda_3=4$．

当 $\lambda_1=-2$ 时，$A+2E=\begin{pmatrix} 4 & -2 & 0 \\ -2 & 3 & -2 \\ 0 & -2 & 3 \end{pmatrix} \sim \begin{pmatrix} 0 & 1 & -1 \\ -2 & 1 & 0 \\ 0 & 0 & 0 \end{pmatrix}$，$p_1=(1,2,2)^T$．

当 $\lambda_2=1$ 时，$A-E=\begin{pmatrix} 1 & -2 & 0 \\ -2 & 0 & -2 \\ 0 & -2 & -1 \end{pmatrix} \sim \begin{pmatrix} 1 & 0 & 1 \\ 0 & 0 & 0 \\ 0 & -2 & -1 \end{pmatrix}$，$p_2=(-2,-1,2)^T$．

当 $\lambda_3=4$ 时，$A-4E=\begin{pmatrix} -2 & -2 & 0 \\ -2 & -3 & -2 \\ 0 & -2 & -4 \end{pmatrix} \sim \begin{pmatrix} 0 & 1 & 2 \\ -1 & -1 & 0 \\ 0 & 0 & 0 \end{pmatrix}$，$p_3=(2,-2,1)^T$．

因为 A 有三个线性无关的特征向量，所以 A 能对角化．

取 $U=\begin{pmatrix} 1 & -2 & 2 \\ 2 & -1 & -2 \\ 2 & 2 & 1 \end{pmatrix}$，$U^{-1}AU=\begin{pmatrix} -2 & 0 & 0 \\ 0 & 1 & 0 \\ 0 & 0 & 4 \end{pmatrix}$．

Ex Is the square matrix A diagonalizable? If it is, find an invertible matrix U such that $U^{-1}AU$ is a diagonal matrix.

Solution $|A-\lambda E|=\begin{vmatrix} 2-\lambda & -2 & 0 \\ -2 & 1-\lambda & -2 \\ 0 & -2 & -\lambda \end{vmatrix}=(\lambda+2)(\lambda-1)(\lambda-4)$, so the eigenvalues of the matrix A are given by $\lambda_1=-2, \lambda_2=1, \lambda_3=4$.

When $\lambda_1=-2$, $A+2E=\begin{pmatrix} 4 & -2 & 0 \\ -2 & 3 & -2 \\ 0 & -2 & 3 \end{pmatrix} \sim \begin{pmatrix} 0 & 1 & -1 \\ -2 & 1 & 0 \\ 0 & 0 & 0 \end{pmatrix}$, $p_1=(1,2,2)^T$.

When $\lambda_2=1$, $A-E=\begin{pmatrix} 1 & -2 & 0 \\ -2 & 0 & -2 \\ 0 & -2 & -1 \end{pmatrix} \sim \begin{pmatrix} 1 & 0 & 1 \\ 0 & 0 & 0 \\ 0 & -2 & -1 \end{pmatrix}$, $p_2=(-2,-1,2)^T$.

When $\lambda_3=4$, $A-4E=\begin{pmatrix} -2 & -2 & 0 \\ -2 & -3 & -2 \\ 0 & -2 & -4 \end{pmatrix} \sim \begin{pmatrix} 0 & 1 & 2 \\ -1 & -1 & 0 \\ 0 & 0 & 0 \end{pmatrix}$, $p_3=(2,-2,1)^T$.

For matrix A has 3 linearly independent eigenvectors, so matrix A is diagonalizable. Let $U=\begin{pmatrix} 1 & -2 & 2 \\ 2 & -1 & -2 \\ 2 & 2 & 1 \end{pmatrix}$, then $U^{-1}AU=\begin{pmatrix} -2 & 0 & 0 \\ 0 & 1 & 0 \\ 0 & 0 & 4 \end{pmatrix}$.

本章重点

1. 掌握欧几里得空间、一个向量的长度、两个向量的夹角的概念.
2. 知道度量矩阵的定义、意义.
3. 知道标准正交基的求法和意义.
4. 掌握正交变换、对称变换的定义和性质.
5. 通过施密特方法可以构造任何子空间的正交基.

Key points of this chapter

1. Master the definitions of an Euclidean space, the length of a vector and the angle between two vectors.
2. Know the definition and its significance of metric matrix.
3. Know the methods for finding the normal orthogonal bases and the significance of doing so.
4. Master the definition and the properties of orthogonal transformations and symmetric transformations.
5. An orthogonal basis can be constructed of any subspace by Schmidt method.

第十章 双线性函数
Chapter 10 Bilinear Functions

本章介绍了双线性函数的定义和性质.

This chapter contains the definition and various properties of the bilinear function.

单词和短语 Words and expressions

★ 双线性函数　bilinear function
对偶基　dual basis
对偶空间　dual spaces
反对称双线性函数　antisymmetric bilinear function
对称双线性函数　symmetric bilinear function

本章重点

掌握线性函数和双线性函数的定义和性质.

Key points of this chapter

Master the definitions and properties of linear functions and bilinear functions.

练　习

1. 求 $\det(A)$, $\det(B)$ 和 $\det(AB)$ 的值，其中

$$A=\begin{pmatrix} 1 & -2 & 3 \\ 0 & 3 & 5 \\ 3 & 4 & -2 \end{pmatrix}, B=\begin{pmatrix} 5 & -1 & 0 \\ -3 & 2 & 4 \\ 2 & 5 & 0 \end{pmatrix}$$

证明 $\det(AB) = \det(A)\det(B)$ 对这种特殊情况是正确的.

2. 假设 A 是可逆的矩阵,证明 $\det(A^{-1}) = \dfrac{1}{\det(A)}$.

3. 假设 A 是方矩阵且 $A^2 = O$,证明 $\det(A) = 0$.

4. 假设矩阵 A 满足 $A^2 = A$,证明 $\det(A) = 0$ 或 1.

5. 求解 x 使 $\det\begin{pmatrix} 1-x & -2 \\ -2 & 1-x \end{pmatrix} = 0$.

6. 求 2×2 矩阵 A, B,使得 $AB = O$,但 $A \neq O$ 且 $B \neq O$.

7. 对下面矩阵 A,求一个矩阵 P 且把它表示成初等矩阵的乘积,使得 PA 为行阶梯形.

(1) $A = \begin{pmatrix} 1 & 1 & 1 \\ 2 & 3 & 1 & 2 \\ 1 & -1 & 3 & 2 \end{pmatrix}$ (2) $A = \begin{pmatrix} 1 & 3 & -1 & 1 \\ 2 & 5 & 1 & 5 \\ 1 & 1 & 1 & 3 \end{pmatrix}$

8. 设 A 是一个 $n \times n$ 矩阵且满足 $A^3 - 2A^2 - I = O$,其中 O 代表 $n \times n$ 零矩阵,证明矩阵 A 可逆.

9. (1) 证明矩阵 $A = \begin{pmatrix} a & 1 & 1 & 1 \\ 1 & b & 2 & -1 \\ 2 & 2 & 3 & -1 \\ -1 & 1 & 1 & 1 \end{pmatrix}$ 可逆当且仅当 $a \neq -1$ 且 $b \neq \dfrac{1}{2}$.

(2) 利用(1)的结果求线性方程组 $AZ = C$ 的唯一解,其中 $C = [1, 2, 3, 4]^T$.

10. 设 $A = \begin{pmatrix} 1 & 2 \\ -1 & 0 \end{pmatrix}, B = \begin{pmatrix} 2 & 1 \\ -1 & 3 \\ 5 & 2 \end{pmatrix}, C = \begin{pmatrix} 3 & 0 \\ 1 & 1 \end{pmatrix}$.

下面哪些表达式有意义,并计算有意义表达式的结果.
(1) $A - 3C$,   (2) $A(B+C)$,   (3) $B(A+C)$,   (4) $(A+C)B$.

Exercises

1. Evaluate $\det(A)$, $\det(B)$ and $\det(AB)$, where
$$A = \begin{pmatrix} 1 & -2 & 3 \\ 0 & 3 & 5 \\ 3 & 4 & -2 \end{pmatrix}, \quad B = \begin{pmatrix} 5 & -1 & 0 \\ -3 & 2 & 4 \\ 2 & 5 & 0 \end{pmatrix}.$$
Verify that the general result $\det(AB) = \det(A)\det(B)$ is true for this particular case.

2. Suppose A is an invertible matrix. Show that $\det(A^{-1}) = \dfrac{1}{\det(A)}$.

3. Suppose A is a square matrix such that $A^2 = O$. Show that $\det(A) = 0$.

4. Suppose A is a matrix with $A^2 = A$. Show that $\det(A) = 0$ or 1.

5. Let $\det\begin{pmatrix} 1-x & -2 \\ -2 & 1-x \end{pmatrix} = 0$, solve for x.

6. Find 2×2 matrices A, B, such that $AB = O$, but $A \neq O$ and $B \neq O$.

7. For each of the following matrices A, find a matrix P, which is expressed as a product of elementary matrices, such that PA is a row echelon matrix:

(1) $A = \begin{pmatrix} 1 & 1 & 1 \\ 2 & 3 & 1 & 2 \\ 1 & -1 & 3 & 2 \end{pmatrix}$ (2) $A = \begin{pmatrix} 1 & 3 & -1 & 1 \\ 2 & 5 & 1 & 5 \\ 1 & 1 & 1 & 3 \end{pmatrix}$

8. Let A be a $n \times n$ matrix such that $A^3 - 2A^2 - I = O$, where O denotes the $n \times n$ zero matrix. Show that A has an inverse matrix.

9. (1) Show that the matrix $A = \begin{pmatrix} a & 1 & 1 & 1 \\ 1 & b & 2 & -1 \\ 2 & 2 & 3 & -1 \\ -1 & 1 & 1 & 1 \end{pmatrix}$ is invertible if and only if a and b are any values except $a = -1$ and $b = \frac{1}{2}$.

(2) Use part (1) to obtain a unique solution of the linear system $AZ = C$, where $C = [1, 2, 3, 4]^T$.

10. Give the matrices $A = \begin{pmatrix} 1 & 2 \\ -1 & 0 \end{pmatrix}$, $B = \begin{pmatrix} 2 & 1 \\ -1 & 3 \\ 5 & 2 \end{pmatrix}$, $C = \begin{pmatrix} 3 & 0 \\ 1 & 1 \end{pmatrix}$, which of the following expressions can be well defined? Calculate those expressions which are well defined.

(1) $A - 3C$, (2) $A(B+C)$, (3) $B(A+C)$, (4) $(A+C)B$.

第三部分 解析几何
Part 3　Analytic Geometry

引 言

16世纪以前,代数和几何被认为是不同的学科。笛卡尔(Descartes,1596~1650)第一个发现这两个学科是相互关联的,并且每个学科都能推动另外一个学科的发展。我们把这两个学科的结合称为解析几何,这种结合取得的成果已经远远超过笛卡尔的想象。

解析几何架起了代数和几何之间的一座桥梁,这使得一些几何问题可用代数的方法(或解析的方法)得到解决。另外,我们也可用几何的方法解决代数问题,特别是当把数引入本质几何概念中时,前者显得更加重要。代数和几何之间的联系就是把数与点对应起来。

Introduction

Before the 16th century, Algebra and Geometry were regarded as separate subjects. It was Descartes (1596~1650) who first noticed that these two subjects could be united, and that each subject could contribute to the development of the other. This union, which we now call Analytic Geometry, has been fruitful far beyond the widest dreams of Descartes.

Analytic Geometry provides a bridge between Algebra and Geometry that makes it possible for geometric problems to be solved algebraically (or analytically). Moreover, it also allows us to solve algebraic problems geometrically, but the former is far more important, especially when numbers are assigned to essentially geometric concepts. The relation between Algebra and Geometry is made by assigning numbers to points.

第一章 矢量与坐标
Chapter 1　Vectors and Coordinates

本章系统介绍了矢量代数的基本知识,它实质上是一个使空间几何结构代数化的过程,换句话说,通过矢量把代数运算引入几何中来。

In this chapter, we introduce systematically the basic knowledge of vector algebra, which is a process essentially making the geometric structure in space algebraization. In other words, we can apply algebraic calculation to geometry by vector.

单词和短语 Words and expressions

标量	scalar	['skeilə]
★矢量	vector	['vektə]
★坐标	coordinate	[kəu'ɔdinit]
空间	space	[speis]
系统	system	['sistəm]
代数	algebra	['ældʒibrə]
量	quantity	['kwɔntiti]
计算	calculation	[kælkju'leiʃən]
力学	mechanics	[mi'kæniks]
物理学	physics	['fiziks]
解	solution	[sə'ljuːʃən]
概念	concept	['kɔnsept]
长度	length	[leŋθ]
面积	area	['eəriə]

体积	volume	['vɔljum]
位移	displacement	[dis'pleismənt]
大小	magnitude	['mægnitjuːd]
方向	direction	[di'rekʃən]
线段	line segment	
有向线段	directed line segment	
始点	initial point	
终点	terminal point	
模	module	['mɔdjuːl]
等于	equal to	
零	zero	['ziərəu]
平行四边形	parallelogram	
邻边	adjacent side	
折线	broken line	

法则	rule [ruːl]
共线(的)	collinear
共面(的)	coplanar
加法运算	additive operation
三角形	triangle ['traiæŋgl]
对角线	diagonal
平移	translation [træns'leiʃən]
交换律	law of commutation
结合律	law of association
分配律	law of distribution
减	minus ['mainəs]
★线性运算	linear operation
乘法	multiplication [mʌltipli'keiʃən]
分解	decomposition
系数	coefficient [kəui'fiʃənt]
★基底	base [beis]
等式	equality [i(ː)'kwɔliti]
三棱形	triangular prism
六面体	hexahedron
平分线	bisector
连线	connecting line
线性相关	linear dependence
线性无关	linear independence
线性组合	linear combination
标架	frame [freim]
直角	right angle
仿射	affine
左旋/手标架	left-handed frame
右旋/手标架	right-handed frame
空间矢量	space vector
★笛卡尔坐标系	Cartesian coordinate system
第一象限	first quadrant
分母	denominator
分子 [化]	numerator [məˈlekjulə] [数]
仿射坐标系	affine coordinate system
坐标平面	coordinate plane
坐标轴	coordinate axis
顶点	vertex
重心	barycenter

中点	median point
中线	median line
射影	projection
射影矢量	projective vector
射线	ray [rei]
垂直	perpendicularity
夹角	included angle
余弦	cosine
方向余弦	direction cosine
正弦	sine
反交换	anti-commutative
反矢量	inverse vector
垂足	pedal ['pedl]
两倍	two-fold
★混合积	mixed product
★数量积/内积	inner product
绝对值	absolute value
★单位矢量	unit vector
零矢量	zero vector
平方	square [skweə]
平方和	sum of squares
高	height [hait]
高线	altitude ['æltitjuːd]
距离	distance ['distəns]
分量	component [kəm'pəunənt]
左手系	left-handed system
★右手系	right-handed system
棱/边	edge [edʒ]
正数	positive number
复数	complex number
符号	symbol ['simbəl]
初等几何	elementary geometry
四面体	tetrahedron
★矢量积	cross product
三重向量积	triple vector product
实数域	real number field
★克拉默法则	Cramer rule
拉格朗日恒等式	Lagrange identity

基本概念和性质 Basic concepts and properties

① 既有大小又有方向的量叫做矢量。
　A quantity with magnitude and direction is said to be a vector.
② 矢量的大小叫做矢量的模，也称矢量的长度。矢量 a 的模记做 $|a|$。
　The magnitude of a vector is said to be module or the length of the vector, denoted by $|a|$。
③ 当矢量 b 与矢量 c 的和等于矢量 a，即 $b+c=a$ 时，我们把矢量 c 叫做矢量 a 与 b 的差，并记做 $c=a-b$。

A vector c is said to be the difference of a and b if the sum of b and c is equal to a, namely, $b+c=a$, denoted by $c=a-b$.

4 实数 λ 与矢量 a 的乘积是一个矢量，记做 λa，它的模 $|\lambda a|=|\lambda||a|$；λa 的方向，当 $\lambda>0$ 时与 a 相同，当 $\lambda<0$ 时与 a 相反。我们把这种运算叫做数量与矢量的乘法，简称为数乘。

The product of a real number λ and a vector a is a vector, denoted by λa. Its module is $|\lambda a|=|\lambda||a|$ and its direction is the same as a while $\lambda>0$, and converse to a while $\lambda<0$. This kind of operation is said to be multiplication between the scalar and the vector, which is also briefly called scalar multiplication.

5 两个矢量 a 和 b 的模和它们夹角的余弦的乘积叫做矢量 a 和 b 的数性积(也称内积)，记做 $a \cdot b$，即 $a \cdot b = |a||b| \cos \angle(a,b)$.

The product of module of two vectors a and b with the cosine of their included angle is said to be an inner product of a and b, denoted by $a \cdot b$, i.e., $a \cdot b = |a||b| \cos \angle(a,b)$.

例 题 Examples

例1 判断 $A=(1,7), B=(0,3), C=(-2,-5)$ 三点是否共线？

解
$$|AB| = \sqrt{(0-1)^2+(3-7)^2} = \sqrt{17}$$
$$|BC| = \sqrt{(-2-0)^2+(-5-3)^2} = \sqrt{68} = 2\sqrt{17}$$
$$|AC| = \sqrt{(-2-1)^2+(-5-7)^2} = \sqrt{153} = 3\sqrt{17}$$

因而 $|AC|=|AB|+|BC|$，则三点必共线。若不共线，三点将构成三角形，且任意一边都小于其他两边之和。

Ex. 1 Determine whether $A=(1,7)$, $B=(0,3)$, and $C=(-2,-5)$ are collinear.

Solution
$$|AB| = \sqrt{(0-1)^2+(3-7)^2} = \sqrt{17}$$
$$|BC| = \sqrt{(-2-0)^2+(-5-3)^2} = \sqrt{68} = 2\sqrt{17}$$
$$|AC| = \sqrt{(-2-1)^2+(-5-7)^2} = \sqrt{153} = 3\sqrt{17}$$

So $|AC|=|AB|+|BC|$, the three points must be collinear. If they were not, they would form a triangle and the length of any one side would be less than the sum of the other two.

本章重点

本章主要利用矢量代数的基本知识，使空间几何结构代数化，从而利用矢量把代数运算引入到几何中来，把代数与几何结合起来。

主要掌握如下几点：

1 矢量的概念。

2 矢量的运算。

3 坐标表示。

Key points of this chapter

In this chapter, we mainly apply the basic knowledge of vector algebra to the process of making geometric structures in space algebraization, and hence to the introduction of algebraic calculation into geometry through vectors, thus we combine Algebra and Geometry. The following points must be mastered:

1 Concept of vectors.

2 Computation of vectors.

3 Expression of coordinates.

第二章 轨迹与方程
Chapter 2　Loci and Equations

本章主要建立了轨迹与其方程的对应,这样就把几何问题归结为代数问题,从而用代数的方法来解决几何的问题.

In this chapter, we mainly establish the correspondence between locus and its equation, which converts a geometric problem into an algebraic problem. Therefore, it allows us to solve geometric problems algebraically.

单词和短语 Words and expressions

★ 轨迹　locus
方程　equation　[iˈkweiʃn]
★ 平面　plane　[plein]
实数　real number
对应　correspondence　[kɔrisˈpɔndəns]
★ 曲线　curve　[kəːv]
★ 曲面　surface　[ˈsəːfis]
★ 直线　straight line
充要条件(当且仅当)　if and only if
图/图表　graph　[grɑːf]
圆　circle　[ˈsəːkl]
半径　radius　[ˈreidjəs]
公式　formula　[ˈfɔːmjulə]
原点　origin　[ˈɔridʒin]
圆心　center of a circle
反比　proportion by inversion
参数　parameter
区间　interval　[ˈintəvəl]
对称的　symmetric　[siˈmetrik]
定点　fixed point
准方程　quasi-equation
一次方程　linear equation
法矢量　normal vector
重合　coincidence
交角　intersection angle
交点　intersection point
坐标系　coordinate system
轴　axis　[ˈæksis]
有向角　directed angle
旋轮线/摆线　cycloid
拱形　arch　[ɑːtʃ]
尖点　cusp
切线　tangent line
切点/触点　tangent point

渐近切线　asymptotic tangent
渐伸线/切展线　involvent
星形线　astroid
初等函数　elementary function
椭圆　ellipse　[iˈlips]
斜率　slope　[sləup]
等价　equivalence
抛物线　parabola
心形线　cardioid
抛物柱面　parabolic cylinder
虚曲面　imaginary surface
二次方程　quadratic equation
一元二次方程　quadratic equation with one variable
配方　complete square
球/球面　sphere　[sfiə]
圆柱面　cylinder　[ˈsilində]
椭圆柱面　elliptic cylinder
双曲柱面　hyperbolic cylinder
母线　generators / generatrix
★ 准线　directrix
质点　particle　[ˈpɑːtikl]
角速度　angular velocity
维安尼曲线　Viviani curve
螺旋线　spiral
螺旋面　spiral surface
射影柱面　projective cylinder
右手坐标系　right-handed coordinate system
圆锥　cone
圆锥面　cone surface
因式/因子　factor　[ˈfæktə]
因式分解　factoring
联立方程　simultaneous equations

基本概念和性质　Basic concepts and properties

1. 在平面上,如果一个方程与一条曲线有着关系:① 满足方程的 (x,y) 必是曲线某一点的坐标;② 曲线上任何一点的坐标 (x,y) 满足这个方程,那么这个方程就叫做这条曲线的方程,而这条曲线叫做这个方程的图形.

In a plane, if an equation and a curve have the following relations: ① The point (x,y) satisfying the equation must be one point's coordinates of this curve; ② An arbitrary point's coordinates (x,y) on this curve satisfies this equation. Then this equation is said to be the equation of this curve and this curve is said to be the figure of this equation.

2. 如果一个方程 $F(x,y,z)=0$ 与一个曲面 Σ 有着关系:① 满足方程的点 (x,y,z) 是曲面 Σ 上的点的坐标;② 曲面 Σ 上的任何一点的坐标 (x,y,z) 满足这个方程,那么这个方程就叫做曲面 Σ 的方程,而曲面 Σ 叫做这个方程的图形.

If an equation $F(x,y,z)=0$ and a surface Σ have the following relations: ① The point (x,y,z) satisfying the equation must be one point's coordinates of the surface Σ. ② An arbitrary point's coordinates (x,y,z) of the surface Σ satisfies this equation. Then this equation is said to be the equation of the surface Σ and the surface Σ is said to be the figure of this equation.

3. 空间曲线,可以看成两个曲面的交线.

A curve in space can be seen as an intersection line of two surfaces.

例题　Examples

例 1 求在 xoy 平面内所有与点 $(1,3)$ 和 $(-2,5)$ 距离相等的点集的方程.

解 设满足条件的点为 (x,y),则
$$\sqrt{(x-1)^2+(y-3)^2} = \sqrt{(x+2)^2+(y-5)^2},$$
$$(x-1)^2+(y-3)^2 = (x+2)^2+(y-5)^2,$$
$$x^2-2x+1+y^2-6y+9 = x^2+4x+4+y^2-10y+25.$$

因此有 $6x-4y+19=0$.

Ex. 1 Find an equation for the set of all points in the xoy plane which are equidistant from $(1,3)$ and $(-2,5)$.

Solution Let (x,y) be one such point satisfying the condition. Then
$$\sqrt{(x-1)^2+(y-3)^2} = \sqrt{(x+2)^2+(y-5)^2},$$
$$(x-1)^2+(y-3)^2 = (x+2)^2+(y-5)^2,$$
$$x^2-2x+1+y^2-6y+9 = x^2+4x+4+y^2-10y+25.$$

So $6x-4y+19=0$.

例 2 把椭圆的普通方程 $\dfrac{x^2}{a^2}+\dfrac{y^2}{b^2}=1$ 改写为参数方程.

解 设 $x=a\cos\theta$,代入原方程得 $y=\pm b\sin\theta$. 如果取 $y=-b\sin\theta$,令 $\theta=-t$,那么 $x=a\cos\theta$, $y=-b\sin\theta$ 可以变形为 $x=a\cos t$, $y=b\sin t$,所以取 θ 为参数,且 $-\pi\leqslant\theta<\pi$,那么椭圆的参数方程为
$$\begin{cases} x=a\cos\theta \\ y=b\sin\theta \end{cases}, \quad (-\pi\leqslant\theta<\pi).$$

Ex. 2 Find the parametric equation of ellipse with the general equation $\dfrac{x^2}{a^2}+\dfrac{y^2}{b^2}=1$.

Solution Let $x=a\cos\theta$, according to the original equation, we have $y=\pm b\sin\theta$. Take $y=-b\sin\theta$ and let $\theta=-t$, then $x=a\cos\theta$, $y=-b\sin\theta$ can be expressed by $x=a\cos t$, $y=b\sin t$, take θ as a pa-

rameter satisfying $-\pi \leqslant \theta < \pi$, so the parametric equation of ellipse is
$$\begin{cases} x = a\cos\theta \\ y = b\sin\theta \end{cases}, \quad (-\pi \leqslant \theta < \pi).$$

本章重点

  这一章主要介绍了曲线轨迹,并将其与方程对应起来,这样进一步建立了代数与几何的结合,从而用代数的方法来解决几何的问题。主要掌握如下几点:

1 曲面与空间曲线方程的概念.

2 母线平行于坐标轴的柱面方程的特征.

Key points of this chapter

  This chapter mainly introduces the trajectories of curves and corresponds it with equations, thus the combination of Algebra and Geometry is further established to provide algebraic methods for solving geometric problems. The following points must be mastered:

1 The concepts of equations of a surface and a curve in space.

2 The characteristics of a cylindrical equation whose generators runs parallel to axis of coordinates.

第三章 平面与空间直线
Chapter 3 Planes and Straight Lines in Space

  在这一章中,我们用代数的方法定量地研究了空间中最简单而又最基本的图形——平面与空间直线,建立了它们的各种形式的方程,导出了它们之间位置关系的解析表达式,以及距离、交角等计算公式.

  In this chapter, using the algebraic method, we study quantitatively the simplest but the most basic figures in space — planes and straight lines in space, establish their equations of various forms and induce the analytic expressions of the relations between their positions, calculating formulae of distance and intersectional angle.

单词和短语 Words and expressions

方位	azimuth	相关	correlate ['kɔrileit]
截距式	intercept form	标准方程	standard equation
常数	constant ['kɔnstənt]	方向数	direction number
垂线	vertical line	射影平面	projective plane
径矢	radius of vector	公垂线	common perpendicular
异号	opposite sign	★ 直线束	pencil of lines
对边	opposite side	★ 平面束	plane pencil /pencil of planes
方向矢量	direction vector	平行平面束	pencil of parallel planes
离差	dispersion	平行平面	parallel planes
相交	intersection	平行切线	parallel tangents
不等式	inequality	平行移动	parallel translation
二面角	interfacial angle		

基本概念和性质 Basic concepts and properties

1 空间中任意平面的方程都可以表示成一个关于变量 x,y,z 的一次方程;反过来,每一个关于变量 x,y,z 的一次方程都表示一个平面,即平面的一般方程可表示为 $Ax + By + Cz + D = 0$.

  In a space, the equation of an arbitrary plane can be expressed by a linear equation about variables x, y and z; conversely, each linear equation about variables x, y and z represents a plane, i. e., the general equation of a plane is $Ax + By + Cz + D = 0$.

2 点 $M_0(x_0, y_0, z_0)$ 与平面 $Ax + By + Cz + D = 0$ 间的距离为

$$d = \frac{|Ax_0 + By_0 + Cz_0 + D|}{\sqrt{A^2 + B^2 + C^2}}.$$

The distance between a point $M_0(x_0, y_0, z_0)$ and a plane $Ax + By + Cz + D = 0$ is

$$d = \frac{|Ax_0 + By_0 + Cz_0 + D|}{\sqrt{A^2 + B^2 + C^2}}.$$

3 由直线上的一点 $M_0(x_0, y_0, z_0)$ 与直线的方向矢量 $v = \{X, Y, Z\}$ 能确定一条直线 $l: \frac{x - x_0}{X} = \frac{y - y_0}{Y} = \frac{z - z_0}{Z}$,此方程称为直线 l 的标准方程.

A point $M_0(x_0, y_0, z_0)$ and a direction vector $v = \{X, Y, Z\}$ can determine a line $l: \frac{x - x_0}{X} = \frac{y - y_0}{Y} = \frac{z - z_0}{Z}$, this equation is said to be a standard equation of the line l.

4 空间的直线,可以看成两个平面的交线.

A line in a space can be seen as an intersection line of two planes.

例 题　Examples

例 1　求过点 $(1, 4, 3)$ 且垂直于 $\frac{x-1}{2} = \frac{y+3}{1} = \frac{z-2}{4}$ 与 $\frac{x+2}{3} = \frac{y-4}{2} = \frac{z+1}{-2}$ 的直线方程.

解　沿这两条给定直线的向量为 $u = 2i + j + 4k$ 与 $v = 3i + 2j - 2k$,且 $u \times v = -10i + 16j + k$ 均与两直线垂直.因此所要求的直线为

$$\frac{x-1}{-10} = \frac{y-4}{16} = \frac{z-3}{1}.$$

Ex. 1　Find equations for the line passing through $(1, 4, 3)$ and being perpendicular to

$$\frac{x-1}{2} = \frac{y+3}{1} = \frac{z-2}{4} \quad \text{and} \quad \frac{x+2}{3} = \frac{y-4}{2} = \frac{z+1}{-2}.$$

Solution　Vectors along the two given lines are $u = 2i + j + 4k$ and $v = 3i + 2j - 2k$; and thus $u \times v = -10i + 16j + k$ is perpendicular to both of them. therefore, The desired line is,

$$\frac{x-1}{-10} = \frac{y-4}{16} = \frac{z-3}{1}.$$

例 2　求通过点 $P(2, -1, -1), Q(1, 2, 3)$ 且垂直于平面 $2x + 3y - 5z + 6 = 0$ 的平面方程.

解　因为 $QP = \{1, -3, -4\}$,平面的法矢量 $n_1 = \{2, 3, -5\}$,则 $QP \times n_1 = \begin{vmatrix} i & j & k \\ 1 & -3 & -4 \\ 2 & 3 & -5 \end{vmatrix} = 27i - 3j + 9k$,取 $n = \{9, -1, 3\}$.所以有

$$9(x - 2) - (y + 1) + 3(z + 1) = 0, \text{ 即}: 9x - y + 3z - 16 = 0.$$

Ex. 2　Find the equation of the plane which goes through the points $P(2, -1, -1)$ and $Q(1, 2, 3)$ and is perpendicular to the plane $2x + 3y - 5z + 6 = 0$.

Solution　Since $QP = \{1, -3, -4\}$, the normal vector of the plane is $n_1 = \{2, 3, -5\}$, then $QP \times n_1 = \begin{vmatrix} i & j & k \\ 1 & -3 & -4 \\ 2 & 3 & -5 \end{vmatrix} = 27i - 3j + 9k$, take $n = \{9, -1, 3\}$, the desired plane is

$$9(x - 2) - (y + 1) + 3(z + 1) = 0, \text{ i.e., } 9x - y + 3z - 16 = 0.$$

本章重点

本章主要定量地研究了平面与空间直线，运用代数方法给出了它们各种形式的方程，以及它们之间位置关系的解析表达式等计算公式．主要掌握如下几点：

1 平面的方程．
2 直线的方程．
3 平面与直线相互间的关系．

Key points of this chapter

In this chapter, we study quantitatively planes and straight lines in space and induce the calculation formulae by using algebraic methods to provide their various equations, including the analytic expressions of the relations between the positions of spatial lines and planes. The following points must be mastered:

1 The equation of a plane.
2 The equation of a line.
3 Relations between planes and lines.

第四章 柱面、锥面、旋转曲面与二次曲面
Chapter 4 Cylinders, Cone Surfaces, Rotation Surfaces and Quadratic Surfaces

本章主要介绍柱面、锥面、旋转曲面与二次曲面．在这些曲面中，有的具有较为突出的几何特征，有的在方程上却表现出特殊的简单形式．

In this chapter, we mainly introduce cylinders, cone surfaces, rotation surfaces and quadratic surfaces. Among these surfaces, some have conspicuous geometric features, while others display special simple forms in equations.

单词和短语 Words and expressions

★ 二次曲面　quadratic surface
★ 旋转曲面　rotation surface
齐次方程　homogeneous equation
直线族　family of straight lines
★ 二次锥面　quadratic cone
虚锥面　imaginary cone
实轴　real axis
虚轴　imaginary axis
环面　torus
椭球面　ellipsoid
旋转椭球面　ellipsoid of revolution
旋转抛物面　paraboloid of revolution
单叶　univalent
旋转轴　rotation axis
定直线　fixed line
主平面　principal plane
主轴　principal axis
对称平面　symmetry plane
长半轴　semi-major axis
中半轴　semi-mean axis

短半轴　semi-minor axis
虚半轴　imaginary semi-axis
长方体　cuboid
长轴　major axis
短轴　minor axis
截口/线/面　section ['sekʃən]
主截线/面　principal section
分割/切断　segmentation
截割　cutting
曲面图形　surface chart
双曲面　hyperboloid
单叶双曲面　hyperboloid of one sheet
双曲线　hyperbola
单叶旋转双曲面　hyperboloid of one sheet of revolution
双叶双曲面　hyperboloid of two sheets
旋转双叶双曲面　hyperboloid of two sheets of revolution
双曲抛物面　hyperbolic paraboloid
马鞍曲面　anticlastic surface

无心二次曲面　quadratic surface without center
中心　center
对称中心　symmetric center
对称轴　symmetric axis

焦点　focus ['fəukəs]
矩形　rectangle ['rektæŋgl]
弧　arc [ɑːk]
直纹曲面　ruled surface

基本概念和性质　Basic concepts and properties

1. 在空间,一条曲线 Γ 绕定直线 l 旋转一周所产生的曲面叫做旋转曲面. 曲线 Γ 叫做旋转曲面的母线,定直线 l 叫做旋转曲面的旋转轴,简称为轴.

In a space, wheeling around the fixed line l, a curve Γ generates a surface which is said to be rotation surface. The curve Γ is said to be generators of this rotation surface and the fixed line l is said to be its rotation axis.

2. 椭球面的标准方程为 $\frac{x^2}{a^2}+\frac{y^2}{b^2}+\frac{z^2}{c^2}=1$,其中 a,b,c 为任意的正常数.

The standard equation of ellipsoid is $\frac{x^2}{a^2}+\frac{y^2}{b^2}+\frac{z^2}{c^2}=1$, where a, b and c are all arbitrary positive constants.

3. 单叶双曲面的标准方程为 $\frac{x^2}{a^2}+\frac{y^2}{b^2}-\frac{z^2}{c^2}=1$,其中 a,b,c 为任意的正常数.

The standard equation of hyperboloid of one sheet is $\frac{x^2}{a^2}+\frac{y^2}{b^2}-\frac{z^2}{c^2}=1$, where a, b and c are all arbitrary positive constants.

4. 双叶双曲面的标准方程为 $\frac{x^2}{a^2}+\frac{y^2}{b^2}-\frac{z^2}{c^2}=-1$,其中 a,b,c 为任意的正常数.

The standard equation of hyperboloid of two sheets is $\frac{x^2}{a^2}+\frac{y^2}{b^2}-\frac{z^2}{c^2}=-1$, where a, b and c are all arbitrary positive constants.

例 题　Examples

例1　锥面的顶点在原点,且准线为 $\begin{cases}\frac{x^2}{a^2}+\frac{y^2}{b^2}=1\\ z=c\end{cases}$,求锥面的方程.

解　设 $M_1(x_1,y_1,z_1)$ 为准线上的任意点,那么过 M_1 的母线为

$$\frac{x}{x_1}=\frac{y}{y_1}=\frac{z}{z_1}, \tag{1}$$

且有

$$\frac{x_1^2}{a^2}+\frac{y_1^2}{b^2}=1, \tag{2}$$

$$z_1=c. \tag{3}$$

由(1)和(3)得

$$x_1=c\frac{x}{z},\ y_1=c\frac{y}{z}. \tag{4}$$

(4)代入(2)得所求的锥面方程为

$$\frac{x^2}{a^2}+\frac{y^2}{b^2}-\frac{z^2}{c^2}=0$$

这个锥面叫做二次锥面.

Ex. 1　Find the equation of the cone such that the vertex of the cone is on the origin and its directrix is $\begin{cases}\frac{x^2}{a^2}+\frac{y^2}{b^2}=1\\ z=c\end{cases}$.

Solution Let $M_1(x_1, y_1, z_1)$ be an arbitrary point on the directrix, then the generators across M_1 is

$$\frac{x}{x_1} = \frac{y}{y_1} = \frac{z}{z_1}, \tag{1}$$

and

$$\frac{x_1^2}{a^2} + \frac{y_1^2}{b^2} = 1, \tag{2}$$

$$z_1 = c. \tag{3}$$

From (1) and (3), we have

$$x_1 = c\frac{x}{z}, \quad y_1 = c\frac{y}{z}. \tag{4}$$

On substitution (4) into (2), then the desired equation of cone is given by

$$\frac{x^2}{a^2} + \frac{y^2}{b^2} - \frac{z^2}{c^2} = 0$$

This cone is said to be quadratic cone.

例 2 已知椭球面的轴与三坐标轴重合, 且通过椭圆 $\frac{x^2}{9} + \frac{y^2}{16} = 1, z = 0$ 与点 $M(1, 2, \sqrt{23})$, 求这个椭球面的方程.

解 设椭球面方程为 $\frac{x^2}{a^2} + \frac{y^2}{b^2} + \frac{z^2}{c^2} = 1$, 它与 xoy 面的交线为 $\begin{cases} \frac{x^2}{a^2} + \frac{y^2}{b^2} = 1 \\ z = 0 \end{cases}$, 与已知椭圆 $\begin{cases} \frac{x^2}{9} + \frac{y^2}{16} = 1 \\ z = 0 \end{cases}$ 比较知 $a^2 = 9, b^2 = 16$. 又因为椭圆通过点 $M(1, 2, \sqrt{23})$, 所以又有 $\frac{1}{9} + \frac{4}{16} + \frac{23}{c^2} = 1$, 得 $c^2 = 36$.

因此所求椭球面的方程为 $\frac{x^2}{9} + \frac{y^2}{16} + \frac{z^2}{36} = 1$.

Ex. 2 Find the equation of the ellipsoid that its axis coincides with three coordinate axes and goes through the ellipse $\frac{x^2}{9} + \frac{y^2}{16} = 1, z = 0$ and the point $M(1, 2, \sqrt{23})$.

Solution Let the equation of the ellipsoid be $\frac{x^2}{a^2} + \frac{y^2}{b^2} + \frac{z^2}{c^2} = 1$. It intersects with the xoy plane. The line of intersection of them is an ellipse $\begin{cases} \frac{x^2}{a^2} + \frac{y^2}{b^2} = 1 \\ z = 0 \end{cases}$.

Compare with $\begin{cases} \frac{x^2}{9} + \frac{y^2}{16} = 1 \\ z = 0 \end{cases}$, we obtain $a^2 = 9, b^2 = 16$. Since it crosses the point $M(1, 2, \sqrt{23})$, then $\frac{1}{9} + \frac{4}{16} + \frac{23}{c^2} = 1$, and thus $c^2 = 36$.

So the desired equation of ellipsoid is $\frac{x^2}{9} + \frac{y^2}{16} + \frac{z^2}{36} = 1$.

本章重点

本章主要介绍了柱面、锥面、旋转曲面与二次曲面的基本概念、基本性质和一些较为突出的几何特征.

主要掌握如下几点:

1. 旋转曲面的定义.
2. 旋转曲面的方程.

Key points of this chapter

This chapter aims at explaining the definitions, properties and certain conspicuous geometric features of cylinders, cone surfaces, rotation surfaces and quadratic surfaces. The following points must be mastered:

1 Definition of the rotation surface.
2 Equation of the rotation surface.

第五章 二次曲线的一般理论
Chapter 5　General Theory of Quadratic Curves

本章主要研究了一般二次曲线的几何理论，讨论了一般二次曲线的渐近方向、中心以及在直角坐标变换下的"不变量"等重要概念.

In this chapter, we mainly study the geometric theory of general quadratic curves and discuss some important concepts such as asymptotic direction, center and "invariant" under the rectangular coordinate transformation.

单词和短语 Words and expressions

★ 二次曲线　　quadratic curve
虚数　　imaginary number
虚元素　　imaginary element
虚点　　imaginary point
共轭虚点　　conjugate imaginary points
共轭复数　　conjugate complex number
虚矢量　　imaginary vector
虚直线　　imaginary line
虚系数　　imaginary coefficient
偏导数　　partial derivative
复点　　complex point
复矢量　　complex vector
对角元素　　diagonal element
行列式　　determinant
行　　row [rəu]
列　　column ['kɔləm]
弦　　chord [kɔːd]
椭圆型　　elliptic type
抛物型　　parabolic type
双曲型　　hyperbolic type
渐近方向　　asymptotic direction
虚方向　　imaginary direction
中心直线　　central line
★ 中心二次曲线　　quadratic curve with center
★ 无心二次曲线　　quadratic curve without center
中心曲线　　central curve
无心曲线　　noncentral curve
正常点　　proper point
切线方程　　tangential equation

恒等式　　identity
奇异点/奇点　　singular point
中点轨迹　　midpoint locus
平行弦　　parallel chords
主方向　　principal direction
主直径　　principal diameter
共轭方向　　conjugate direction
共轭直径　　conjugate diameter
共轭弦　　conjugate chord
特征方程　　characteristic equation
特征根　　characteristic value
坐标变换　　transformation of coordinates
判别式　　discriminant
旋转轴　　rotation axis
虚圆　　imaginary circle
虚椭圆　　imaginary ellipse
二次项　　quadratic element
旋转角　　angle of rotation
可逆变换　　reversible transformation
逆变换　　inverse transformation
余切　　cotangent
成比例　　proportional [prə'pɔːʃənl]
增广矩阵　　augmented matrix
共轭虚直线　　conjugate imaginary line
不变量　　invariant
多项式　　polynomial
三角函数　　trigonometric function
★ 柯西-施瓦茨不等式　　Cauchy-Schwarz inequality

基本概念和性质　Basic concepts and properties

1. 在平面上，由二元二次方程
$$a_{11}x^2 + 2a_{12}xy + a_{22}y^2 + 2a_{13}x + 2a_{23}y + a_{33} = 0$$
所表示的曲线，叫做二次曲线.

In a plane, a curve shown by the following quadratic equation with two unknowns is said to be a quadratic curve, namely,
$$a_{11}x^2 + 2a_{12}xy + a_{22}y^2 + 2a_{13}x + 2a_{23}y + a_{33} = 0.$$

2. 如果点 C 是二次曲线的通过它的所有弦的中点（因而 C 是二次曲线的对称中心），那么点 C 叫做二次曲线的中心.

If the point C is the midpoint of a quadratic curve through all its chords, so C is the symmetric center of the quadratic curve, then the point C is said to be the center of the quadratic curve.

3. 通过二次曲线的中心，而且以渐近方向为方向的直线叫做二次曲线的渐近线.

A line across the center of a quadratic curve with asymptotic direction as its direction is said to be an asymptotic line of this quadratic curve.

4. 如果直线与二次曲线相交于相互重合的两个点，那么这条直线就叫做二次曲线的切线，这个重合的交点叫做切点.

If a line intersects with a quadratic curve at two coincident points, the line is said to be the tangent line of the quadratic curve and the coincident point is said to be its tangent point.

5. 二次曲线的一族平行弦的中点轨迹是一条直线.

The midpoint's locus of a family of parallel chords on the quadratic curve is a line.

例 题　Examples

例　求抛物线 $y^2 = 2px$ 的直径.

解　令 $F(x,y) \equiv 2px - y^2 = 0$，$F_1(x,y) = p$，$F_2(x,y) = -y$，所以共轭于非渐近方向 $X:Y$ 的直径为 $Xp - Yy = 0$，即 $y = \dfrac{X}{Y}p$.

因此抛物线 $y^2 = 2px$ 的直径平行于它的渐近方向 $1:0$.

Ex　Find the diameter of the parabola $y^2 = 2px$.

Solution　Let $F(x,y) \equiv 2px - y^2 = 0$, $F_1(x,y) = p$, $F_2(x,y) = -y$, then the diameter conjugating to the non-asymptotic direction $X:Y$ is $Xp - Yy = 0$, i. e., $y = \dfrac{X}{Y}p$.

So the diameter of the parabola $y^2 = 2px$ runs parallel to its asymptotic direction $1:0$.

本章重点

掌握二次曲线的化简与分类.

Key points of this chapter

Master the simplification and classification of a quadratic curve.

第六章　二次曲面的一般理论
Chapter 6　General Theory of Quadratic Surfaces

这一章介绍如何从二维空间（即平面）关于一般二次曲线方程的讨论推广到三维空间的一般二次曲面方程的情形.

This chapter introduces how to extend the equations of general quadratic curves in two-dimensional space (i. e., plane) to the equations of general quadratic surfaces in three-dimensional space.

单词和短语 Words and expressions

二重根	double root	切平面	tangent plane
主子式	principal minor determinant	主径面	principal radial plane
虚根	imaginary root	齐次线性变换	homogeneous linear transform
实根	real root	正交条件	orthogonality condition
★ 径面	radial plane	交叉	cross [krɔs]
矩阵	matrix		

基本概念和性质 Basic concepts and properties

1 在空间，由三元二次方程 $a_{11}x^2 + a_{22}y^2 + a_{33}z^2 + 2a_{12}xy + 2a_{13}xz + 2a_{23}yz + 2a_{14}x + 2a_{24}y + 2a_{34}z + a_{44} = 0$ 所表示的曲面叫做二次曲面.

In a space, a surface shown by the quadratic equation with three unknowns, i. e., $a_{11}x^2 + a_{22}y^2 + a_{33}z^2 + 2a_{12}xy + 2a_{13}xz + 2a_{23}yz + 2a_{14}x + 2a_{24}y + 2a_{34}z + a_{44} = 0$, is said to be a quadratic surface.

2 如果点 C 是二次曲面的通过它的所有弦的中点（因而 C 是二次曲面的对称中心），那么点 C 叫做二次曲面的中心.

If the point C is the midpoint of a quadratic surface through all its chords, so C is the symmetric center of the quadratic surface, then the point C is said to be the center of the quadratic surface.

3 如果直线与二次曲面相交于两个相互重合的点，那么这条直线叫做二次曲面的切线，那个重合的交点叫做切点. 如果直线全部在二次曲面上，这条直线也叫做二次曲面的切线，直线上的每一点都是切点.

If a line intersects with a quadratic surface at two coincident points, the line is said to be the tangent line of the quadratic surface and the coincident point is said to be its tangent point. If the line is on the quadratic surface, it is also said to be a tangent line of the quadratic surface and all points on the line are all tangent points.

4 二次曲面一族平行弦的中点轨迹是一个平面.

The midpoint's locus of a family of parallel chords on a quadratic surface is a plane.

例 题 Examples

例 求 $x^2 - xy + y^2 - 2 = 0$ 旋转 $45°$ 后的表达式.

解 因为 $\sin 45° = \cos 45° = 1/\sqrt{2}$，旋转变换 $x = \dfrac{x' - y'}{\sqrt{2}}$ 与 $y = \dfrac{x' + y'}{\sqrt{2}}$，代入原方程可得

$$\frac{(x'-y')^2}{2} - \frac{x'-y'}{\sqrt{2}} \cdot \frac{x'+y'}{\sqrt{2}} + \frac{(x'+y')^2}{2} - 2 = 0,$$

$$\frac{x'^2 - 2x'y' + y'^2 - x'^2 + y'^2 + x'^2 + 2x'y' + y'^2}{2} = 2,$$

因此 $x'^2 + 3y'^2 = 4$.

Ex. Find the new representation of $x^2 - xy + y^2 - 2 = 0$ after rotating through an angle of $45°$.

Solution Since $\sin 45° = \cos 45° = 1/\sqrt{2}$, the rotation transformations are

$$x = \frac{x'-y'}{\sqrt{2}} \text{ and } y = \frac{x'+y'}{\sqrt{2}}.$$

Substituting them into the original equation, we have

$$\frac{(x'-y')^2}{2} - \frac{x'-y'}{\sqrt{2}} \cdot \frac{x'+y'}{\sqrt{2}} + \frac{(x'+y')^2}{2} - 2 = 0,$$

$$\frac{x'^2 - 2x'y' + y'^2 - x'^2 + y'^2 + x'^2 + 2x'y' + y'^2}{2} = 2,$$

and thus $x'^2 + 3y'^2 = 4$.

练 习

1. 设 $a+b+c=0$,$|a|=3$,$|b|=1$,$|c|=4$,求 $a \cdot b+b \cdot c+c \cdot a$.
2. 已知 $a+3b$ 与 $7a-5b$ 垂直,且 $a-4b$ 与 $7a-2b$ 垂直,求 a 与 b 的夹角.
3. 在直角坐标系中,已知 $a=(2,3,-1)$,$b=(1,-2,3)$,求与 a 和 b 都垂直,且满足如下条件之一的向量 c:
   (1) c 为单位向量;(2) $c \cdot d=10$,其中 $d=(2,1,-7)$.
4. 在直角坐标系中,已知三点 $A(1,2,3)$,$B(2,0,4)$,$C(2,-1,3)$,求 $\triangle ABC$ 的面积.
5. 求过点 $(3,-5,1)$ 和 $(4,1,2)$,垂直于平面 $x-8y+3z-1=0$ 的平面方程.
6. 求过点 $(2,-3,-5)$ 且与平面 $6x-3y-5z+2=0$ 垂直的直线方程.
7. 求通过直线 $\frac{x-2}{1}=\frac{y+3}{-5}=\frac{z+1}{-1}$ 且与直线 $\begin{cases} 2x-y+z-3=0 \\ x+2y-z-5=0 \end{cases}$ 平行的平面方程.
8. 求直线 $\frac{x}{2}=y=\frac{z-1}{1}$ 绕直线 $x=-y=\frac{z-1}{2}$ 旋转所得旋转曲面的方程.
9. 指出下列曲面的名称:
   (1) $y^2-z^2=2x$; (2) $y^2-x^2=2x$; (3) $3x^2-4y^2=5z^2+k$.
10. 求曲线 $5x^2+6xy+5y^2=8$ 经过点 $(0,2\sqrt{2})$ 的切线方程.

Exercises

1. Find $a \cdot b+b \cdot c+c \cdot a$ such that $a+b+c=0$, where $|a|=3$, $|b|=1$, $|c|=4$.
2. Find the included angle of a and b such that $a+3b$ is perpendicular to $7a-5b$ and $a-4b$ is perpendicular to $7a-2b$.
3. In rectangular coordinates system, let $a=(2,3,-1)$ and $b=(1,-2,3)$, find the vector c which is perpendicular to a and b and satisfies one of the following conditions:
   (1) c is a vector which its module is equal to 1; (2) $c \cdot d=10$ while $d=(2,1,-7)$.
4. In rectangular coordinates system, for the three given points $A(1,2,3)$, $B(2,0,4)$ and $C(2,-1,3)$, find the area of the triangle $\triangle ABC$.
5. Find the equation of the plane which goes through the point $(3,-5,1)$, $(4,1,2)$ and is perpendicular to the plane $x-8y+3z-1=0$.
6. Find the equation of the line which goes through the point $(2,-3,-5)$ and is perpendicular to the plane $6x-3y-5z+2=0$.
7. Find the equation of the plane which goes through the line $\frac{x-2}{1}=\frac{y+3}{-5}=\frac{z+1}{-1}$ and runs parallel to the line $\begin{cases} 2x-y+z-3=0 \\ x+2y-z-5=0 \end{cases}$.
8. The line $\frac{x}{2}=y=\frac{z-1}{1}$ circles around the line $x=-y=\frac{z-1}{2}$ which generates a rotation surface, find the equation of the rotation surface.
9. Point out the names of the following surfaces:
   (1) $y^2-z^2=2x$; (2) $y^2-x^2=2x$; (3) $3x^2-4y^2=5z^2+k$.
10. Find the equation of the tangent line of the curve $5x^2+6xy+5y^2=8$ going through the point $(0,2\sqrt{2})$.

数学与应用数学专业课程

第一部分 数学模型
Part 1 Mathematical Model

引言

数学是一门重要的学科,是人们解决实际问题的基本工具.人们总是从实际问题出发,对其进行分析、简化、转化为一个数学问题,然后用适当的数学方法去解决,这就是数学建模.数学建模是用数学方法解决实际问题的第一步,而且是很关键的一步,两千多年前欧几里德几何的创立、17世纪万有引力定律的发现,都是数学建模的成功范例.随着科学与技术的发展,新技术、新工艺不断出现,提出了许多新的数学问题,需要对传统的数学模型进行改进或者建立新的数学模型,同时,数学与其他学科的交叉,学科,也为数学建模的发展提供了广阔的新天地.

数学建模可以看做是用数学公式对各种现象的描述过程.我们把现实世界的问题用数学模型来表达,应用现代数学和计算机技术来解决这些问题,并表达出自己的结论,使得一般人即使不经过计算也能明白.

数学建模是一门艺术,它将实际问题转化为易处理的数学公式,其理论和数值分析提供了对原始应用非常有用的内在本质、答案和指导.

数学建模
- 对很多应用来讲是必不可少的.
- 对许多进一步的应用非常成功.
- 提供了精确的、直接的问题的解.
- 使我们对模型系统有了彻底的了解.
- 提供了更好的设计和控制一个系统的途径.
- 允许我们有效地应用现有的计算能力.

Introduction

Mathematics is an important subject and is a fundamental tool to solve some practical problems. People always analyze, simplify practical problems, change them into mathematical problems, and sowe them by using suitable mathematical method. this is the procedure of mathematical modeling. To solve practical problems by using mathematical methods, mathematical modeling is the first step and is also the key step. Foundation of Euclid geometry 2000 years ago and discovery of gravity law in the 17th century are all successful examples of mathematical modeling. With the development of science and technology, new technique and new craftwork come forth continuously, this causes that many new mathematical problems must be proposed, so we need to improve classical mathematical model or construct new model. Also, the interdiscipline est of mathematics and other subjects brings more and more applications of mathematical modeling.

Mathematical modeling can be viewed as the process of describing all kinds of phenomena in terms of mathematical formulas. We formulate mathematical models for real-world problems and solve them by using modern mathematical and computer techniques, and then express our conclusions in the way that normal intelligent people can understand, even if they never took calculations.

Mathematical modeling is a class of art, which translates practicl problems into tractable mathematical formulations whose theoretical and numerical analysis provides useful inherent essence, answers, and guidance for the original application.

Mathematical modeling
- is indispensable in many applications.
- is successful in many further applications.
- gives the precise and direct solutions for problems.
- enables a thorough understanding of the model system.
- prepares the way for better design or control of a system.
- allows the efficient use of modern computing capabilities.

第一章 数学模型概论
Chapter 1　Overview of Mathematical Model

本章首先通过几个实际案例强调了建立数学模型的必要性,并且对各种模型进行了分类,最后介绍了数学建模的基本方法和步骤.

This chapter first emphasizes the necessity of mathematical modeling by some practical examples, and classifies all kinds of models. Finally, we introduce the basic methods and steps of mathematical modeling.

数学建模应用举例　Lists of Mathematical Modeling Applications

单词和短语 Words and expressions

头骨的建模、分类和重建　modeling, classifying and reconstructing of skulls
虚拟现实　virtual reality
人工智能　artificial intelligence
计算机视觉　computer vision
图像分析　image analysis
机器人技术　robotics
语音识别　speech recognition
光学符号识别　optical character recognition
计算机动画　computer animation
行星系统发现　detection of planetary systems
哈勃望远镜的修正　correcting the Hubble telescope
恒星演化　evolution of stars
人类基因计划　human genome project
人口动态　population dynamics
血统进化　evolutionary pedigrees
疾病传播　spreading of infectious diseases
动植物繁殖　animal and plant breeding
化学反应动力学　chemical reaction dynamics
分子模型　molecular model

电子结构计算　electronic structure calculations
图像处理　image processing
指纹识别　finger print recognition
人脸识别　face recognition
风险分析　risk analysis
地震预报　earth quake prediction
自动翻译　automatic translation
结构优化　structural optimization
爆破仿真　crash simulation
计算机辅助X射线断层　computer-aided tomography
血液循环模型　blood circulation model
天气预报　weather prediction
声音的分析与合成　analysis and synthesis of sounds
神经网络　neural network
弹道设计　trajectory planning
飞行模拟　flight simulation
航天飞机返航　shuttle reentry
空中交通安排　air traffic scheduling
自动驾驶仪　automatic pilot

数学模型的分类　Classification of Mathematical Model

单词和短语 Words and expressions

初等模型　elementary model

运筹学模型　operational research model

微分方程模型	differential equation model	白箱模型	white-box model
概率统计模型	probabilistic and statistical model	灰箱模型	gray-box model
人口模型	population model	黑箱模型	black-box model
交通模型	traffic model	确定性模型	deterministic model
经济预测模型	economic prediction model	随机性模型	stochastic model
金融模型	finance model	动态模型	dynamic model
环境模型	environment model	静态模型	static model
生态模型	ecological model	离散模型	discrete model
企业管理模型	enterprise management model	连续模型	continuous model

数学建模的方法　Methods of Mathematical Modeling

数学建模的方法主要包括：机理分析法、统计分析法、系统分析法.

Methods of mathematical modeling mainly contain: analysis method of mechanism, analysis method of statistic, analysis method of system.

数学建模的步骤　Steps for Mathematical Modeling

准备→假设→分析、简化问题→建立模型→模型求解→评价、改进模型→模型应用.

Preparation→Hypothesis→Analysis and Simplification→Modeling→Solving→Estimating and Improving→Application.

第二章　最优化模型
Chapter 2　Optimization Models

最优化模型（运筹学模型）广泛应用于工程和管理中. 它提供了将实际问题转化为在一定约束之下, 如何使得目标函数达到最优的数学问题的方法.

Optimization model (Operational research model) is widely used in engineering and management. It provides the method on how to convert the object function of a practical problem to an optimal mathematical problem under certain constraints.

单词和短语　Words and expressions

线性规划模型	linear programming model	决策模型	decision model
整数规划模型	integer programming model	投入产出分析模型	input-output analysis model
网络模型	network model	评价模型	evaluating model
非线性规划模型	nonlinear programming model	层次分析法模型（AHP 模型）	analytic hierarchy process model
目标规划模型	goal programming model		
库存模型	storage model	数据包络分析模型（DEA 模型）	data envelopment analysis model
对策论模型	game theory model		
随机规划模型	stochastic programming model		

■ 最优化模型的数学描述

将一个最优化问题用数学语言来描述, 即为求函数 $u = f(x), x = (x_1, x_2, \cdots, x_n), x \in \Omega$ 在约束条件 $h_i(x) = 0, i = 1, 2, \cdots, m$ 和 $g_i(x) \geqslant 0 (g_i(x) \leqslant 0), i = 1, 2, \cdots, p$ 下的最大值或最小值, 其中 $f(x)$ 为目标函数, x 为决策变量, Ω 为可行域.

Mathematical Description of Optimization Model

In mathematical language, an optimization problem is to find the maximum or the minimum of the function $u = f(x), x = (x_1, x_2, \cdots, x_n), x \in \Omega$ under the constraint conditions $h_i(x) = 0, i = 1, 2, \cdots, m$ and $g_i(x) \geqslant 0 (g_i(x) \leqslant 0), i = 1, 2, \cdots, p$, where $f(x)$ is the objective function, x is the decision variable and Ω is the feasible field.

2 建立最优化模型的一般步骤

(1) 确定决策变量和目标变量;
(2) 确定目标函数的表达式;
(3) 寻找约束条件.

General Steps of Optimization Modeling

(1) Determine the decision variable and the objective variable;
(2) Determine the expression of the objective function;
(3) Search the constraint conditions.

例题 Examples

例 随机贮存模型 (s,S) 策略.

由于顾客对一种商品的需求是随机的,因此在实际生活中,有一种进货策略 (s,S) 策略被广为采用:商店老板每隔一定时间要对商品的存货进行清点,只有当存货数量不足 s 时才决定进货,且一次进货的订货量取 S 与当前存货数量的差值.

一、模型假设

1. 假设商店经营的商品单一,商店采用周期进货策略:每隔一定时间,如一周,商店老板要对商品的存货进行清点,以决定是否进货.只有当存货数量 q 不足 s 时才决定进货,且一次进货的订货量取 S 与当前存货数量的差值,x 表示进货量.

2. 顾客在一周时间内对该物品的需求量 r 是一随机变量,$\rho(r)$ 表示随机变量 r 的概率密度函数.

3. 商店在一周可能支付的费用有:每次的订货费 c_0,其取值与进货数量无关;每件商品在一周的贮存费 c_1. a,b 分别表示一件商品的购进价格和售出价格.

4. 商店在一周的销售活动全部集中在一周的周初,因此商店须为剩余商品支付一周的贮存费用.

二、模型建立

首先考虑 S 的确定,设当前存货数量为 q,且决定进货,这时进货数量 x 成为决策变量.因此,x 的取值应当在期望值的意义上使得利润最大化.

$$f(x,r) = \begin{cases} (b-a) \cdot r - [c_0 + c_1 \cdot (x+q-r)] & r \leqslant x+q \\ (b-a) \cdot (x+q) - c_0 & r > x+q \end{cases} \tag{1}$$

为进货数量取 x,而需求量为 r 时商店在下周的利润. 取其数学期望,得:

$$\begin{aligned}\bar{f}(x) &= \int_0^{+\infty} f(x,r) \cdot \rho(r) \cdot \mathrm{d}r \\ &= \int_0^{x+q}\{(b-a)r - [c_0 + c_1(x+q-r)]\}\rho(r)\mathrm{d}r + \int_{x+q}^{+\infty}[(b-a)(x+q) - c_0]\rho(r)\mathrm{d}r\end{aligned} \tag{2}$$

若记 $L(u) = \int_0^u [(b-a)r - c_1(u-r)]\rho(r)\mathrm{d}r + \int_u^{+\infty}(b-a)u \cdot \rho(r)\mathrm{d}r$,则

$$\bar{f}(x) = L(x+q) - c_0. \tag{3}$$

三、模型求解

令 $\dfrac{\mathrm{d}\bar{f}}{\mathrm{d}x} = 0$,得最优性条件: $c_1 \cdot \int_0^{x+q} \rho(r)\mathrm{d}r = (b-a) \cdot \int_{x+q}^{+\infty}\rho(r)\mathrm{d}r$.

我们也直接从最优性条件获得,不论当前存货数量 q 取何值,只要决定进货,那么最优的订货量总是使得下期起初的货物量 $x+q$ 达到确定的值:

$$\int_0^{x+q}\rho(r)\mathrm{d}r = \frac{b-a}{c_1+(b-a)}, \text{ 即 } S \text{ 应满足} \int_0^S \rho(r)\mathrm{d}r = \frac{b-a}{c_1+(b-a)}.$$

按照前面的进货策略,根据当前存货数量 q,要么选择进货 $x = S-q$,这时下周销售利润的期望 $\bar{f} = L(S) - c_0$;要么选择不进货,这时下周销售利润的期望 $\bar{f} = L(q)$. 显然,若 $L(S) - c_0 < L(q)$ 时,应当选择不进货.如下图所示,函数 $L(q)$ 在 $[0,+\infty)$ 上通常为一单峰曲线,可得 $s = \min\{q > 0 \mid L(q) \geqslant L(S) - c_0\}$,也即关于变量 q 的方程 $L(S) - c_0 = L(q)$ 在 $(0,S)$ 内的解.

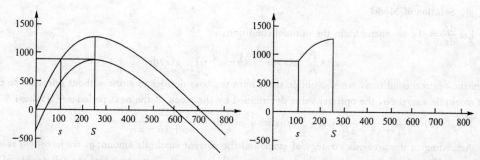

四、点评

本例属于贮存模型，它被归结为最优化问题或利润最大化问题，这并非偶然，因为人类所从事的一切生产或社会活动均是有目的的，其行为总是在特定的价值观念或审美取向的支配下进行的。因此，当可行方案不唯一的前提下，总是在某种评价指标下选择最优的方案。可以说，最优化思想和方法是数学建模的灵魂。

Ex Model of Random Storage (s, S) Strategy.

Since the request of a commodity for a patron is random, in practical lives, and thus a stock strategy is extensively adopted — (s, S) strategy: The boss of a store maybe check the amount of merchandises of commodity every other certain time, and decides to stock only when the amount of merchandise is less than s and each stock amount is taken as the difference between stock S and current merchandise.

I. Hypothesis of Model

1. Assume that the commodity of a store is unique, and adopts the strategy of periodic stock: such as the boss will check the amount of merchandise after a week, and then decides whether or not to stock, and decides to stock only when the amount of merchandise q is less than s, and each stock amount is taken as the difference between stock S and current merchandise, x means the stock amount.

2. The requirement, denoted by r, of patrons in a week is a random variable and $\rho(r)$ denotes the probability density function with respect to r.

3. The fees of possibly paying of a store in a week includes: fees of each stock is denoted by c_0, independent of the stock amount; the stockpile fee of each commodity in a week is taken as c_1. The prices of purchase and sale are respectively denoted by a and b.

4. The salerooms of a store in a week are all focus on the prime of a week, and thus the store must offer the stockpile fees of the surplus commodity of the next week.

II. Formulation of Model

First of all, consider the determination of S. Assume that the current stockpile amount is given by q and that the boss decides to stock, in this case, the stock amount x becomes the decision variable. In the meaning of expectation value, the value of x must be made the profits maximum, and so, the profits of the next week is given by (1).

$$f(x,r) = \begin{cases} (b-a) \cdot r - [c_0 + c_1 \cdot (x+q-r)] & \text{if } r \leqslant x+q \\ (b-a) \cdot (x+q) - c_0 & \text{if } r > x+q \end{cases} \quad (1)$$

The mathematical expectation is given by (2).

$$\begin{aligned}\overline{f}(x) &= \int_0^{+\infty} f(x,r) \cdot \rho(r) \cdot dr \\ &= \int_0^{x+q} \{(b-a)r - [c_0 + c_1(x+q-r)]\}\rho(r)dr + \int_{x+q}^{+\infty} [(b-a)(x+q) - c_0]\rho(r)dr \end{aligned} \quad (2)$$

Let $L(u) = \int_0^u [(b-a)r - c_1(u-r)]\rho(r)dr + \int_u^{+\infty}(b-a)u \cdot \rho(r)dr$, then (3) is valid.

$$\overline{f}(x) = L(x+q) - c_0. \quad (3)$$

III. Solution of Model

Let $\dfrac{d\bar{f}}{dx} = 0$, we then obtain the optimal condition:

$$c_1 \cdot \int_0^{x+q} \rho(r)dr = (b-a) \cdot \int_{x+q}^{+\infty} \rho(r)dr$$

From the optimal condition, we also obtain that, once the boss decides to stock without reference to current stockpile amount q, the optimal value determined by the stock of the next period $x+q$ takes:

$$\int_0^{x+q} \rho(r)dr = \frac{b-a}{c_1 + (b-a)}, \text{ i.e., } S \text{ satisfies } \int_0^S \rho(r)dr = \frac{b-a}{c_1+(b-a)}.$$

According to the previous strategy of stock and the current stockpile amount q, we have two results to choose: If the boss chooses to stock $x = S - q$, in this case, the expectation of the sell profits of the next week is given by $\bar{f} = L(S) - c_0$; otherwise, $\bar{f} = L(q)$. Obviously, if $L(S) - c_0 < L(q)$, it should not choose to stock. As shown in the following figures, the function $L(q)$ is always a curve with single apex on $[0, +\infty)$, this leads to $s = \min\{q > 0 \mid L(q) \geqslant L(S) - c_0\}$, namely, the solution of the equation $L(S) - c_0 = L(q)$ with respect to q in $(0, S)$.

IV. Remarks

This example, which belongs to a storage model, is summarized as the optimization problem or the maximum profit problem. However, this is necessary, because all the manufacture and the campaign of society made by human being are purposeful and their behaviors are always dominated by the specifically valuable notion or taste direction. In the preconditions of several feasible projects, we always choose the optimal project under certain estimation indices. In sum, the idea and the method of optimization is the spirit of mathematical modeling.

第三章 微分方程模型
Chapter 3 Models of Differential Equations

在很多实际应用中，经常要涉及各变量的变化率的问题，这一类问题的解决通常要建立相应的微分方程模型. 微分方程模型描述的是变量之间的间接关系，因此，要得到直接关系，需要求解微分方程. 求解微分方程通常有三种方法：求精确解、求数值解（近似解）、定性理论方法.

The change rate of variables is often involved in many applications. In order to solve these problems, the relevant differential equation model is always used. A differential equation model describes the indirect relationship of variables, and so, to obtain the direct relation, it requires solving the differential equation. In general, there are three methods to solve the differential equation, namely, exactly solving method, numerical solving method (approximation method), qualitative theory method.

单词和短语 Words and expressions

常微分模型	ordinary differential model	战争模型	battle model
偏微分模型	partial differential model	人口动态模型	population dynamics model
差分方程模型	difference equation model	传染病模型	epidemiology model
经济增长模型	economic growth model	莱斯利模型	Leslie model

建立微分方程模型的方法

1. 根据规律列方程

利用数学、力学、物理、化学等学科中的定理或经过实验检验的规律等来建立微分方程模型.

2. 微元分析法

与第一种方法不同的是，微元分析法利用已知的定理与规律寻找微元之间的关系式.

3. 模拟近似法

建模过程中，在不同的假设下去模拟实际的现象，建立能近似反映问题的微分方程，然后从数学上求解或分析所建方程及其解的性质，再去同实际情况对比，检验此模型能否刻画、模拟某些实际现象.

Methods of Differential Equation Modeling

1. Formulate equations based on rules

Formulate equations by using the theorems in Mathematics, Mechanics, Physics, Chemistry, etc, or the rules of experiments.

2. Method of micro-element analysis

Method of micro-element analysis is to look for the expression of micro-elements by using the known theorems and rules, which is different from the first method.

3. Method of simulative approximation

During the course of modeling, we always simulate the practical phenomena under different assumption and formulate the differential equations to approximately describe problems. Then we solve the founded equations or analyze the properties of solutions. Finally, we contrast them with the practical cases and verify whether the model can describe and simulate some actual phenomena.

例 题　Examples

例　种群竞争.

问题：在自然环境中，生物种群丰富多彩，它们之间通常存在着或是相互竞争，或是相互依存，或是弱肉强食等这样的三种基本关系. 从稳定状态的角度来看，需要对具有如上提及的某种关系的两个种群数量的发展进行讨论.

设想有两个种群为了争夺有限的同一食物来源和生活空间时，从长远的眼光来审视，其最终结局是它们中的竞争力弱的一方首先被淘汰，然后另一方独占全部资源而以单种群模式发展；还是存在某种稳定的平衡状态，两个物种按照某种规模构成双方长期共存？这里不妨将讨论的对象想象为生活在同一草原上的羚羊和老鼠.

一、模型假设

以 $x_1(t)$ 和 $x_2(t)$ 分别表示处于相互竞争关系中甲、乙二种群在时刻 t 的数量.

1. 资源有限，设为 1, N_1 和 N_2 分别表示甲、乙二种群在单种群情况下自然资源所能承受的最大种群数量.

2. 种群数量的增长率 $\dot{x}_i(t)(i=1,2)$ 与该种群数量 $x_i(t)(i=1,2)$ 成正比，同时也与有限资源 $s_i(t)(i=1,2)$ 成正比.

3. 各种群在对所占据资源的利用上是不充分的，σ_1 和 σ_2 分别表示甲、乙二种群对对方已占用资源的相对挑剔程度. 例如，若 $\sigma_1 \in (0,1)$，表示在甲种群看来，乙种群是"奢侈的"；若 $\sigma_1 > 1$，说明甲种群在食物选择上是"过分"挑剔的，或者可理解为，对于甲种群，乙种群在资源利用上对资源有破坏性；换一个说法，σ_1 和 σ_2 反映了甲、乙二种群的适应能力，σ_1 越小、σ_2 越大，则甲种群的相对适应能力越强.

4. r_1 和 r_2 分别表示甲、乙二种群的固有增长率.

二、模型建立

根据模型假设，可得如下简化的数学模型：

$$\begin{cases} \dot{x}_1 = r_1 x_1 (1 - x_1/N_1 - \sigma_1 x_2/N_2) \\ \dot{x}_2 = r_2 x_2 (1 - \sigma_2 x_1/N_1 - x_2/N_2) \end{cases} \tag{1}$$

三、模型求解

令 $\begin{cases} r_1 x_1 (1 - x_1/N_1 - \sigma_1 x_2/N_2) = 0 \\ r_2 x_2 (1 - \sigma_2 x_1/N_1 - x_2/N_2) = 0 \end{cases}$，可得该模型的四个平衡点：

$$P_1(0,0), P_2(N_1, 0), P_3(0, N_2), P_4\left(\frac{1-\sigma_1}{1-\sigma_1\sigma_2}N_1, \frac{1-\sigma_2}{1-\sigma_1\sigma_2}N_2\right). \tag{2}$$

先讨论平衡点 $P_1(0,0)$ 的稳定性. 微分方程(1)关于 $P_1(0,0)$ 的线性化方程为 $\begin{cases} \dot{x}_1 = r_1 x_1 \\ \dot{x}_2 = r_2 x_2 \end{cases}$，此时系数矩阵 $A = \begin{pmatrix} r_1 & 0 \\ 0 & r_2 \end{pmatrix}$，它的二个特征值分别为 $\lambda_1 = r_1 > 0$ 和 $\lambda_2 = r_2 > 0$，故 $P_1(0,0)$ 是不稳定的结点.

对于平衡点 $P_2(N_1,0)$，微分方程（1）关于 $P_2(N_1,0)$ 的线性化方程为
$\begin{cases} \dot{x}_1 = -r_1(x_1-N_1) - r_1\sigma_1 x_2 N_1/N_2 \\ \dot{x}_2 = r_2 x_2(1-\sigma_2) \end{cases}$，它的系数矩阵为 $A = \begin{pmatrix} -r_1 & -\sigma_1 r_1 N_1/N_2 \\ 0 & (1-\sigma_2)r_2 \end{pmatrix}$，特征值为 $\lambda_1 = -r_1 < 0$ 和 $\lambda_2 = (1-\sigma_2)r_2$. 因此，当且仅当 $\sigma_2 > 1$ 时，平衡点 $P_2(N_1,0)$ 是（局部）稳定的.

类似可以得平衡点 $P_3(0,N_2)$ 是（局部）稳定的充要条件为 $\sigma_1 > 1$.

平衡点 $P_4\left(\dfrac{1-\sigma_1}{1-\sigma_1\sigma_2}N_1, \dfrac{1-\sigma_2}{1-\sigma_1\sigma_2}N_2\right)$ 只有在第一象限内方有实际意义，为此应有 σ_1 和 σ_2 同时大于"1"或同时小于"1"，采用类似的分析，可以得到当 σ_1 和 σ_2 同时大于"1"时，平衡点 P_4 为一鞍点，是不稳定的；当 σ_1 和 σ_2 同时小于"1"时，平衡点 P_4 为一稳定的结点.

四、点评

本例通过一些基本假设建立了种群竞争问题的数学模型，将问题归结为一个非线性常微分方程组，然后利用微分方程的定性理论讨论了各平衡点的稳定性，给出了平衡点局部稳定的充要条件.

Ex Competition of Species.

In natural environments, biology species are rich and colorful, their relations can be come down to three basic relations, namely, maybe competition each other, maybe dependence each other and maybe law of the jungle. In the sight of stable state, it requires discussing the developing amount of two species with certain relations mentioned above.

As two species compete for the same and finite food source and life space, in the sight of future, there are two possibly ultimate results, namely, one possibility is that the species with weak competition will be first washed out and that the other will monopolize the whole resources and develop in the mode of single species, another possibility is that there exists a certain stable equilibrium state such that the two species coexists under some dimensions. In this case, the objects are assumed to be antelope and mice living in the same grassland.

Ⅰ. Hypothesis of Model

Let $x_1(t)$ and $x_2(t)$ respectively denote the amount of the first and the second species in the relation of competition at time t.

1. Assume that the finite sources is equal to 1. N_1 and N_2 respectively denote the maximal amount of two species that natural source can be endured in single species.

2. Assume that the increasing rate, $\dot{x}_i(t)(i=1,2)$, of the species amount have direct ratio with the species amount $x_i(t)(i=1,2)$, and have direct ratio with the surplus source $s_i(t)(i=1,2)$.

3. Since the utilization of sources occupied by each species is inadequate, σ_1 and σ_2 respectively denote the fastidious degree of source occupied by the other side. For example, if $\sigma_1 \in (0,1)$, it means that the second species is luxurious in the sight of the first species, while if $\sigma_1 > 1$, this means that the choice on food of the first species is too fastidious, it can also be understood as, for the first species, the utilization of sources occupied by the second species has devastating. In other word, σ_1 and σ_2 reflect the suitable ability of two species. As the value of σ_1 is small and small, σ_2 is larger and larger, this also means that the suitable ability of the first species is better and better.

4. r_1 and r_2 respectively denote the inherently increasing rate of two species.

Ⅱ. Formulation of Model

From the hypothesis of model, we have the following mathematical model which is simplified: to be (1)

$$\begin{cases} \dot{x}_1 = r_1 x_1(1-x_1/N_1 - \sigma_1 x_2/N_2) \\ \dot{x}_2 = r_2 x_2(1-\sigma_2 x_1/N_1 - x_2/N_2) \end{cases} \quad (1)$$

Ⅲ. Solution of Model

Let $\begin{cases} r_1 x_1(1-x_1/N_1 - \sigma_1 x_2/N_2) = 0 \\ r_2 x_2(1-\sigma_2 x_1/N_1 - x_2/N_2) = 0 \end{cases}$, we then obtain four equilibrium points of the model, given by (2).

Firstly, we discuss stability of the equilibrium point $P_1(0,0)$. The linearized equation of Eq. (1) corresponding to $P_1(0,0)$ is that $\begin{cases} \dot{x}_1 = r_1 x_1 \\ \dot{x}_2 = r_2 x_2 \end{cases}$. The coefficient matrix is that $A = \begin{pmatrix} r_1 & 0 \\ 0 & r_2 \end{pmatrix}$, and its eigenvalues are respectively given by $\lambda_1 = r_1 > 0$ and $\lambda_2 = r_2 > 0$, and thus $P_1(0,0)$ is an unstable node.

For the equilibrium point $P_2(N_1,0)$, the linearized equation of Eq. (1) corresponding to $P_2(N_1,0)$ is that $\begin{cases} \dot{x}_1 = -r_1(x_1 - N_1) - r_1\sigma_1 x_2 N_1/N_2 \\ \dot{x}_2 = r_2 x_2 (1-\sigma_2) \end{cases}$. The coefficient matrix is that $A = \begin{pmatrix} -r_1 & -\sigma_1 r_1 N_1/N_2 \\ 0 & (1-\sigma_2)r_2 \end{pmatrix}$, and its eigenvalues are respectively given by $\lambda_1 = -r_1 < 0$ and $\lambda_2 = (1-\sigma_2)r_2$, and thus $P_2(N_1,0)$ is (local) stable if and only if $\sigma_2 > 1$.

Similarly, the equilibrium point $P_3(0,N_2)$ is (local) stable if and only if $\sigma_1 > 1$.

Since the equilibrium point $P_4\left(\dfrac{1-\sigma_1}{1-\sigma_1\sigma_2}N_1, \dfrac{1-\sigma_2}{1-\sigma_1\sigma_2}N_2\right)$ only has actual meaning in the first quadrant, this leads to σ_1 and σ_2 are greater than "1" simultaneously or less than "1" simultaneously. With the same analysis, we have that P_4 is a saddle as σ_1 and σ_2 are greater than "1" simultaneously, and thus it is unstable; that P_4 is a stable node as σ_1 and σ_2 are less than "1" simultaneously.

IV. Remarks

In this example, we found the mathematical model of competition between species under some basic hypotheses and formulate this problem into a nonlinear ordinary differential equations. Moreover, we discuss stability of each equilibrium point by using the qualitative theory of differential equation and present the sufficient and necessary conditions such that the equilibrium points are stable.

第四章 概率统计模型
Chapter 4 Models of Probability and Statistics

现实世界的变化受着众多因素的影响,包括确定的和随机的. 解决这类问题的工具就是概率统计的知识. 如果从建模的背景、目的和手段看,主要因素是确定的,随机因素可以忽略,或者随机因素的影响可以简单地以平均值的作用出现,那么就能够建立确定性模型. 如果随机因素对研究对象的影响必须考虑,就应建立随机模型. 本章首先讨论如何用随机变量和概率分布描述随机因素的影响,建立随机模型——概率模型. 另一方面,如果由于客观事物内部规律的复杂性及人们认识程度的限制,无法分析实际对象内在的因果关系,建立合乎机理规律的模型,那么通常要搜集大量的数据,基于对数据的统计分析建立模型,这就是本章将要讨论的用途非常广泛的另一类随机模型——统计回归模型.

Variation of practical world is influenced by many factors, which includes determinate and stochastic. Probability and statistics are often used for solving these problems. In the sight of background, intention and artifice of modeling, the main factors are determinate and the stochastic factors can be ignored, or the effects of stochastic factors can simply appear as average action, that is to say, we can found determinate models. If the effects of stochastic factors on studying object must be taken into account, we need to found stochastic models. This chapter first discusses how to describe the effects of stochastic factors by using stochastic variables and probability distribution, and then found the stochastic model — Probability Model. On the other hand, if we cannot analyze the inner consequence of practical object owing to the complexity of inner discipline of impersonal things and constraint of recognizant degree of people, to found models corresponding to mechanism rule, we must search abundant data and statistic these data. Therefore, this chapter is to discuss another stochastic model that is well used — Statistical Regression Model.

单词和短语 Words and expressions

预测模型　　prediction model
时间序列分析模型　time series analysis model
线性回归模型　linear regression model
正交实验设计模型　orthogonal experiment design model
马尔可夫链模型　Markov chain model
经济计量模型　economical metrology model

例　题　Examples

例　报童的诀窍.

问题:报童每天清晨从报社购进报纸零售,晚上将没有卖掉的报纸退回.报童每天购进报纸太多,卖不完会赔钱;购进太少,不够卖会少挣钱.试为报童筹划一下每天购进报纸的数量,以获得最大收入.购进量由需求量确定,需求量是随机的.

一、模型假设

设报纸每份的购进价为 b,零售价为 a,退回价为 c,假设 $a>b>c$,即报童售出一份报纸赚 $a-b$,退回一份报纸赔 $b-c$.假定报童已通过自己的经验或其他渠道掌握了需求量的随机规律,即在他的销售范围内每天报纸的需求量为 r 份的概率是 $f(r),r=0,1,2,\cdots$. 有了 $f(r)$ 和 a,b,c 就可以建立关于购进量的优化模型.

二、模型建立

假设每天购进量是 n 份,需求量 r 是随机的,r 是可以小于、等于或大于 n 的,所以报童每天的收入也是随机的. 那么,作为优化模型的目标函数,不能取每天的收入,而应取长期卖报(月,年)的日平均收入. 从概率论大数定律的观点看,这相当于报童每天收入的期望值,简称平均收入.

记报童每天购进 n 份报纸的平均收入为 $G(n)$,如果这天的需求量 $r\leqslant n$,则售出 r 份,退回 $n-r$ 份;如果需求量 $r>n$,则将 n 份全部售出. 需求量 r 的概率是 $f(r)$,则

$$G(n) = \sum_{r=0}^{n}[(a-b)r-(b-c)(n-r)]f(r) + \sum_{r=n+1}^{\infty}(a-b)nf(r). \tag{1}$$

因此,问题归结为:当 $f(r)$ 和 a,b,c 为已知时,求 n 使 $G(n)$ 最大.

三、模型求解

通常需求量 r 和购进量 n 都相当大,将 r 视为连续变量便于分析和计算,这时概率 $f(r)$ 转化为概率密度函数 $p(r)$,则

$$G(n) = \int_{0}^{n}[(a-b)r-(b-c)(n-r)]p(r)\mathrm{d}r + \int_{n}^{\infty}(a-b)np(r)\mathrm{d}r \tag{2}$$

计算

$$\begin{aligned}\frac{\mathrm{d}G}{\mathrm{d}n} &= (a-b)np(n) - \int_{0}^{n}(b-c)p(r)\mathrm{d}r - (a-b)np(n) + \int_{n}^{\infty}(a-b)p(r)\mathrm{d}r \\ &= -(b-c)\int_{0}^{n}p(r)\mathrm{d}r + (a-b)\int_{n}^{\infty}p(r)\mathrm{d}r.\end{aligned} \tag{3}$$

令 $\dfrac{\mathrm{d}G}{\mathrm{d}n}=0$,得到

$$\frac{\int_{0}^{n}p(r)\mathrm{d}r}{\int_{n}^{\infty}p(r)\mathrm{d}r} = \frac{a-b}{b-c} \tag{4}$$

使报童日平均收入达到最大的购进量 n 应满足上式. 因为 $\int_{0}^{\infty}p(r)\mathrm{d}r=1$,所以有

$$\int_{0}^{n}p(r)\mathrm{d}r = \frac{a-b}{a-c}. \tag{5}$$

根据需求量的概率密度 $p(r)$ 的图形可以确定购进量 n. 在图中,P_1,P_2 分别表示曲线 $p(r)$ 下的两块面

积, 则

$$\frac{P_1}{P_2} = \frac{a-b}{b-c}. \tag{6}$$

当购进 n 份报纸时, $P_1 = \int_0^n p(r)\,dr$ 是需求量 r 不超过 n 的概率, 即卖不完的概率; $P_2 = \int_n^\infty p(r)dr$ 是需求量 r 超过 n 的概率, 即卖完的概率. 所以式(1)表明, 购进的份数 n 应该是卖不完与卖完的概率之比, 恰好等于卖出一份赚的钱 $a-b$ 与退回一份赔的钱 $b-c$ 之比.

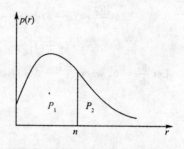

四、评价

当报童与报社签订的合同使报童每份赚钱与赔钱之比越大时, 报童购进的份数就应该越多.

Ex Newsboy's Knack.

Everyday, a newsboy retails newspapers from the newspaper office in the morning and withdraws surplus in the evening. Further, if the newsboy purchases too many newspapers, he will be out of pocket for the return newspapers, otherwise, he will gain little for the inadequate newspapers to sale if he purchases too few. Try to design the purchase amount of newspaper for the newsboy such that he gains the maximal income. Purchase amount is determined by demand, and demand is stochastic.

I. Hypothesis of Model

Assume that b, a and c respectively denote the prices of purchase, retail and return of each newspaper, where $a > b > c$, namely, the newsboy gains $a-b$ if he sails a newspaper and pays for $b-c$ if he returns a newspaper. Furthermore, assume that the newsboy has master the stochastic rule of demand through the experience of himself or other ways, namely, the probability of demand r in his range of sale is given by $f(r)$, $r = 0, 1, 2, \cdots$. With the known $f(r)$ and a, b, c, we can found the optimization model concerning the purchase amount.

II. Formulation of Model

Assume that n denotes the purchase amount of everyday and the demand r is stochastic. r maybe less than, equal to or greater than n, and thus the income of the newsboy is also stochastic. As an object function of the optimization model, it does not take the income of everyday but the average income of a long time sale (month, year). In the sight of Law of Large Numbers in Probability, it corresponds to the expectation value of newsboy of everyday income, i.e., average income.

Denote $G(n)$ by the average income of n purchase by the newsboy everyday. If the demand of that day satisfies $r \leq n$, he will sale r newspaper and return $n-r$; if $r > n$, he will sale n. The probability of demand r is denoted by $f(r)$, then (1) is valid

$$G(n) = \sum_{r=0}^{n}[(a-b)r - (b-c)(n-r)]f(r) + \sum_{r=n+1}^{\infty}(a-b)nf(r). \tag{1}$$

Therefore, the problem is come down to: with the known $f(r)$ and a, b, c, find a value of n such that $G(n)$ takes the maximum.

III. Solution of Model

In general, the demand r and the purchase amount n are all quite large, and r is assumed to be a continuous variable to convenient for calculation. In this case, $f(r)$ is translated into the probability density function $p(r)$, and (2) is valid

$$G(n) = \int_0^n [(a-b)r - (b-c)(n-r)]p(r)dr + \int_n^\infty (a-b)np(r)dr \tag{2}$$

Compute

$$\frac{dG}{dn} = (a-b)np(n) - \int_0^n (b-c)p(r)dr - (a-b)np(n) + \int_n^\infty (a-b)p(r)dr$$
$$= -(b-c)\int_0^n p(r)dr + (a-b)\int_n^\infty p(r)dr. \tag{3}$$

Let $\frac{dG}{dn} = 0$, where $\frac{dG}{dn} = 0$ is given by (3). we obtain that (4) is valid.

$$\frac{\int_0^n p(r)dr}{\int_n^\infty p(r)dr} = \frac{a-b}{b-c}. \tag{4}$$

To make the average income of the newsboy maximal, the purchase amount n may satisfy the above expression. Since $\int_0^\infty p(r)dr = 1$, (5) is valid.

$$\int_0^n p(r)dr = \frac{a-b}{a-c}. \tag{5}$$

From the figure of the probability density function $p(r)$ of demand, we can determine the purchase amount n. In the following figure, P_1, P_2 respectively denote two pieces of areas under the curve $p(r)$, this leads to (6) is valid.

$$\frac{P_1}{P_2} = \frac{a-b}{b-c}. \tag{6}$$

As the newsboy purchases n newspaper, $P_1 = \int_0^n p(r)dr$ is the probability of $r \leqslant n$, i.e., the probability of surplus; $P_2 = \int_n^\infty p(r)dr$ is the probability of $r > n$, i.e., the probability of sale. The above expression (1) means that the amount of purchase n should be the ratio between surplus and sale and is equal to ratio between $a-b$ and $b-c$.

Ⅳ. Remarks

As ratio between in and out of pocket of each newspaper of the contract signed by the newsboy and the newspaper office is larger and larger, the newsboy should purchase more and more newspapers.

第五章 基本算法工具包
Chapter 5 Basic Algorithmic Toolkits

单词和短语 Words and expressions

数值线性代数 numerical linear algebra	非线性规划 nonlinear programming
线性方程组 system of linear equations	数值统计 numerical statistics
特征值问题 eigenvalue problems	可视化 Visualization
线性规划 linear programming	参数估计 parameter estimation
线性最优化 linear optimization	最小二乘估计 least squares estimation
函数求值 function evaluation	最大似然率估计 maximum likelihood estimation
自动微分法和数值微分法 automatic and numerical differentiation	预测 prediction
	分类 classification
插值 interpolation	时间序列分析 time series analysis
近似(Padé)方法 approximation (Padé) method	信号处理 signal processing
最小二乘法 least squares method	频谱分析 spectral analysis
积分 integration	隐马尔可夫模型 hidden Markov model
傅里叶变换 Fourier transform	蒙特-卡洛方法 Monte Carlo method
非线性方程组 system of nonlinear equations	常微分方程 ordinary differential equation
最优化 optimization	初值问题 initial value problem

边值问题　boundary value problem
稳定性　stability
偏微分方程　partial differential equation
有限差分法　finite difference method
有限元法　finite element method
边界元法　boundary element method
网格生成　mesh generation
自适应网格　adaptive mesh

随机微分方程　stochastic differential equation
积分方程和正则化　integral equations and regularization
符号方法（计算代数）　symbolic method（computer algebra）
排序　sorting
压缩　compression
密码学　cryptography

第六章　最新算法
Chapter 6　Newest Algorithms

单词和短语 Words and expressions

模拟退火算法　simulated annealing algorithm（SA）
蚁群算法　ant colony algorithm
人工神经网络算法　neural network algorithm
贪婪算法　greedy method

遗传算法　genetic algorithm（GA）
模糊数学方法　fuzzy mathematical method
数据挖掘　data mining

第七章　数学建模竞赛
Chapter 7　Mathematical Contest in Modeling（MCM）

规则　rules（from http://www.comap.com/undergraduate/contests/mcm/）

1 每一个队至多有 4 名成员.
Student teams can contain at most 4 individuals.

2 可以有指导教师但不是必须的.
A team advisor is recommended but not necessary.

3 论文用 HTML 格式上传，包括所有的图和数学语法.
Papers are to be uploaded in HTML format, along with all graphics and mathematical syntax.

4 参赛者可以查看任何已有的资料.
Students are free to look at existing papers if they choose.

5 参赛队可以在任何时间进入论文阶段.
A team is free to enter a paper whenever it chooses.

6 必须按照模型解提交模式提交论文.如果合适的话,指导教师按照评论提交模式提交对学生论文的评论.
Submissions are uploaded through the modeling solution submission form. Advisors, if applicable, are to upload commentary on the students through the commentary submission form.

7 论文必须遵守要求的格式.
Papers must follow the required format guidelines.

网上资源 Internet Resource

1. http://www.comap.com/undergraduate/contests/mcm/
2. http://www.mathmodels.org/contests/
3. http://www.shumo.com/
4. http://mcm.edu.cn
5. http://mcm.ustc.edu.cn

第二部分 复变函数与积分变换
Part 2 Functions of Complex Variable and Integral Transforms

引 言

复变函数论又称为复分析,是最具活力和最有用的数学分支之一.尽管它是在神秘、猜疑和不信任的背景下被开创的,但"虚数"和"复数"却不断出现在演讲及著作中. 最终,在19世纪,通过Cauchy、Riemann、Weierstrass、Gauss 等其他数学家的努力为这门学科奠定了坚实的理论基础. 积分变换是通过积分运算,把一个函数变成另一个函数的变换,这里所说的积分变换指的是傅立叶变换和拉普拉斯变换,它与复变函数有着密切的联系.

今天,工程学家、物理学家、数学家和其他的科学家认为,这门学科是数学背景下最本质的部分. 从理论的观点来看,是由于用复变函数的理论观点可以阐明和统一许多数学概念,从应用观点来看,复变函数和积分变换广泛地应用于自然科学的众多领域,如理论物理、空气动力学、电磁学、流体力学、弹性力学等其他科学和工程领域.

这门课需要考虑到数学、物理、空气动力学、弹性力学和其他领域在复变函数方面的应用.

本部分在每一章的开始给出了有关定义、原理和带有例证性的定理及其他描述性的材料. 然后按等级给出了几套附加题和答案,这些题用于详细解释和说明本章的理论,对学生难以掌握的重点和要点提供重复性而有效的练习,在题中还包含了理论证明和公式推导. 这些附加题和答案其实也是学生对每一章内容进行全面复习的材料.

本部分的主要内容有:复数的代数和几何表示、复微分和积分,包括泰勒级数和罗朗级数在内的无穷级数、留数定理及其应用、共形映射及其应用、傅立叶变换和拉普拉斯变换.

Introduction

The theory of function of complex variable, also called complex analysis, it is one of the most beautiful as well as useful branches of mathematics. Although originating in an atmosphere of mystery, suspicion and distrust, as evidence by the terms "imaginary" and "complex" present in the literature, it was finally placed on a sound foundation in the 19th century through the efforts of Cauchy, Riemann, Weierstrass, Gauss and other great mathematicians. Integral transform is to transform a function into another function through integrating, where the integral transforms include Fourier transform and Laplace transform, and link up function of complex variable.

Today the subject is recognized as an essential part of the mathematical background by engineers, physicists, mathematicians and other scientists. From the theoretical viewpoint this is because many mathematical concepts become clarified and unified when examined in the light of complex variable theory. From the applied viewpoint the theory is of tremendous value in the solutions of problems of theoretical physics, aerodynamics, electromagnetic theory, fluid mechanics, elasticity and many other fields of science and engineering.

This course should also be of considerable value to those taking courses in mathematics, physics, aerodynamics, elasticity or any of the numerous other fields in which complex variable methods are employed.

Each chapter begins with a clear statement of pertinent definitions, principles and theorems together with illustrative and other descriptive materials. This is followed by graded sets of solved and supplementary problems. The solved problems serve to illustrate and amplify the theory, bring into sharp focus those fine points without which the student continually feels himself on unsafe ground, and provide the repletion of basic principles so vital to effective learning. Numerous proofs of theorems and derivations of formulas are included among the solved problems. The large numbers of supplementary problems with answers serve as a complete review of the material in each chapter.

Topics in this part include representations of algebra and geometry of complex numbers, complex differential and integral, infinite series including Taylor and Laurent series, theory of residues with applications, conformal mapping with applications, Fourier transforms and Laplace transforms.

第一章　复数与复变函数
Chapter 1　Complex Numbers and Functions of Complex Variable

本章学习复数概念、复数运算及其表示；复变函数概念及其极限、连续等内容，这些内容是全书的基础。

In this chapter we introduce the concept of complex number, operation of complex numbers and its expression; the concept of functions of complex variable and its limit and continuity. These contents are the foundation of the whole book.

单词和短语 Words and expressions

★ 复数　complex number
虚数单位　imaginary unit
实部　real part
虚部　imaginary part
纯虚数　pure imaginary number
共轭复数　complex conjugate number
运算　operation
减法　subtraction
乘法　multiplication
除法　division
复平面　complex plane
分配律　distribute rule
交换律　exchange rule
复合函数　combining function
★ 复数的三角形式　trigonometrical form of complex number
模　modulus
辐角　argument

乘方　power
开方　extraction
开集　open set
闭集　closed set
邻域　neighborhood
充分必要条件　sufficient and necessary condition
边界点　boundary point
有界集　bounded set
区域　domain
简单闭曲线　simple closed curve
★ 连通区域　connected region
分段光滑　piecewise smooth
无穷远点　point at infinity
★ 复变函数　function of complex variable
单值函数　single-valued function
多值函数　multi-valued function
连续　continuity
不等式　inequality

基本概念和性质　Basic concepts and properties

1 复数的代数形式为 $z = \alpha + \mathrm{i}\beta$.

The algebraic form of a complex number is given by $z = \alpha + \mathrm{i}\beta$.

2 复数的指数形式为 $z = r\mathrm{e}^{\mathrm{i}\theta}$.

The exponential form of a complex number is given by $z = r\mathrm{e}^{\mathrm{i}\theta}$.

3 复数的三角形式为 $z = r(\cos\theta + \mathrm{i}\sin\theta)$.

The trigonometrical form of a complex number is given by $z = r(\cos\theta + \mathrm{i}\sin\theta)$.

4 复数乘积的模等于模的乘积，乘积的幅角等于幅角的和.

The modulus of a product of two numbers is equal to the products of modulus, and the argument of a product of two numbers is equal to the sum of arguments.

5 欧拉公式为 $\mathrm{e}^{\mathrm{i}\theta} = \cos\theta + \mathrm{i}\sin\theta$.

The Euler formula is given by $\mathrm{e}^{\mathrm{i}\theta} = \cos\theta + \mathrm{i}\sin\theta$.

6 棣莫弗公式为 $(\cos\theta + \mathrm{i}\sin\theta)^n = \cos n\theta + \mathrm{i}\sin n\theta$.

The De Moivre formula is given by $(\cos\theta + \mathrm{i}\sin\theta)^n = \cos n\theta + \mathrm{i}\sin n\theta$.

7 定义：设函数 $f(z)$ 在 z_0 的某去心邻域 $U(z_0,\rho)$ 内有定义，如果有一确定的数 A 存在，对于任意给定的 $\varepsilon > 0$，存在正数 $\delta(\varepsilon)$ $(0 < \delta \leqslant \rho)$，使得当 $0 < |z-z_0| < \delta$ 时，有 $|f(z)-A| < \varepsilon$，则称 A 为当 z 趋向于 z_0 时的极限，记做

$$\lim_{z \to z_0} f(z) = A \text{ 或 } f(z) \to A(z \to z_0).$$

Definition: Let $f(z)$ be a function defined in a deleted neighborhood $U(z_0,\rho)$ of z_0, and A be a constant. If for any given $\varepsilon > 0$, there exists a number $\delta(\varepsilon)$ $(0 < \delta \leqslant \rho)$, such that $|f(z)-A| < \varepsilon$ for all z satisfying $0 < |z-z_0| < \delta$, then the constant A is called the limit of the function $f(z)$ as $z \to z_0$, and is denoted by

$$\lim_{z \to z_0} f(z) = A \text{ or } f(z) \to A(z \to z_0).$$

8 函数 $f(z)$ 在 z_0 连续当且仅当 $\lim_{z \to z_0} f(z) = f(z_0)$.

The function $f(z)$ is said to be continuous at z_0 if and only if $\lim_{z \to z_0} f(z) = f(z_0)$.

例 题　Examples

例 求 n 次单位根.

解 把 1 写成　　$1 = 1\exp[i(0+2k\pi)]$　$(k=0,\pm 1,\pm 2,\cdots)$，并且

$$1^{\frac{1}{n}} = \sqrt[n]{1}\exp\left[i\left(\frac{0}{n}+\frac{2k\pi}{n}\right)\right] = \exp\left[i\frac{2k\pi}{n}\right] \quad (k=0,\pm 1,\pm 2,\cdots).$$

当 $n=2$ 时，1 的根是 ± 1。$n \geqslant 3$ 时，它的根是顶点在单位圆 $|z|=1$ 上的规则多边形的顶点，一个顶点对应着主根 $z=1(k=0)$。

若令 $\omega_n = \exp\left[i\frac{2\pi}{n}\right]$，则 $\omega_n^k = \exp\left[i\frac{2k\pi}{n}\right]$，则 1 的 n 个不同的单位根为

$$1, \omega_n, \omega_n^2, \cdots, \omega_n^{n-1}.$$

Ex. Find the nth roots of unity.

Solution We write $1 = 1\exp[i(0+2k\pi)]$ $(k=0,\pm 1,\pm 2,\cdots)$, and find that

$$1^{\frac{1}{n}} = \sqrt[n]{1}\exp\left[i\left(\frac{0}{n}+\frac{2k\pi}{n}\right)\right] = \exp\left[i\frac{2k\pi}{n}\right] \quad (k=0,\pm 1,\pm 2,\cdots).$$

When $n=2$, these roots are, of course, ± 1. When $n \geqslant 3$, the regular polygon at whose vertices the roots lie is inscribed in the unit circle $|z|=1$, with one vertex corresponding to the principal root $z=1(k=0)$. If we write $\omega_n = \exp\left[i\frac{2\pi}{n}\right]$, it follows from $\omega_n^k = \exp\left[i\frac{2k\pi}{n}\right]$. Hence the distinct n roots of unity just found are simply

$$1, \omega_n, \omega_n^2, \cdots, \omega_n^{n-1}.$$

本章重点

1 注意掌握用复数形式的方程（或不等式）表示平面图形来解决有关几何问题的方法.
2 正确理解复变函数及与之有关的概念.
3 正确理解区域、单连通区域、多连通区域、简单曲线等概念.

Key points of this chapter

1 Master the methods of applying equations or inequalities in complex number form to express plane graphics and to solve geometric problems.
2 Correctly understand functions of complex variable and relevant concepts.
3 Correctly understand the concepts of domain, simply connected domains, multiple connected domains and simple curves.

第二章 解析函数
Chapter 2 Analytic Functions

本章主要讨论复变函数的导数与解析函数等基本概念，以及判断函数解析的方法。

In this chapter, the basic concepts of derivative of functions of complex variable and analytic functions are mainly discussed, as well as the methods of judging analytic functions.

单词和短语 Words and expressions

微分	differential	双曲函数	hyperbolic function
奇点	singularity	幂函数	power function
★解析函数	analytic function	高阶导数	higher order derivative
导数	derivative	求导法则	derivation rule
★柯西-黎曼方程	Cauchy-Riemann equation	链式法则	chain rule
★调和函数	harmonic function	定义域	domain
指数函数	exponential function	导函数	derivative function
对数函数	logarithm function	反函数	inverse function
三角函数	trigonometric function		

基本概念和性质 Basic concepts and properties

1 设 $f(z) = u(x,y) + iv(x,y)$，$f'(z)$ 在 $z_0 = x_0 + iy_0$ 存在。则 $u(x,y)$ 和 $v(x,y)$ 在 (x_0, y_0) 的一阶偏导数一定存在，且满足柯西-黎曼方程 $u_x = v_y, u_y = -v_x$，同时有 $f'(z_0) = u_x(x_0, y_0) + iv_x(x_0, y_0)$。

Assume that $f(z) = u(x,y) + iv(x,y)$ and that $f'(z)$ exists at the point $z_0 = x_0 + iy_0$. Then the first-order partial derivatives of $u(x,y)$ and $v(x,y)$ at (x_0, y_0) must exist, and they must satisfy the Cauchy-Riemann equations $u_x = v_y, u_y = -v_x$, there also, $f'(z)$ can be written as $f'(z_0) = u_x(x_0, y_0) + iv_x(x_0, y_0)$.

2 如果一个复变函数在某一开集的每一点可导，则函数在这个开集上解析。

A function of complex variable is analytic in an open set if it has a derivative at each point in that set.

3 如果二元实函数 $H(x,y)$ 在区域 D 内具有二阶的连续偏导数，并且满足拉普拉斯方程 $H_{xx}(x,y) + H_{yy}(x,y) = 0$，则称 $H(x,y)$ 为区域 D 内的调和函数。

The real-valued function $H(x,y)$ of two real variables is said to be harmonic in the domain D if it has continuous partial derivatives of the second order and satisfies the Laplace equation $H_{xx}(x,y) + H_{yy}(x,y) = 0$.

4 函数 $f(z) = u(x,y) + iv(x,y)$ 在区域 D 内解析当且仅当虚部 $v(x,y)$ 是实部 $u(x,y)$ 的共轭调和函数。

The function $f(z) = u(x,y) + iv(x,y)$ is analytic in the domain D if and only if the imaginary part $v(x,y)$ is a conjugate harmonic function of the real part $u(x,y)$.

例题 Examples

例 1 求 $u(x,y) = y^3 - 3x^2 y$ 的共轭调和函数。

解 $u(x,y) = y^3 - 3x^2 y$ 在整个 xy 平面上看做是调和函数。因为 $u(x,y)$ 与它的共轭调和函数 $v(x,y)$ 满足 Cauchy-Riemann 方程 $u_x = v_y, u_y = -v_x$，由此可得

$$v_y(x,y) = -6xy$$

对上面的式子关于 y 积分得 $v(x,y) = -3xy^2 + \varphi(x)$，这里 φ 是 x 的函数，由 Cauchy-Riemann 方程的第二个方程得

$$3y^2 - 3x^2 = 3y^2 - \varphi'(x),$$

即 $\varphi'(x) = 3x^2$。从而 $\varphi(x) = x^3 + C$，C 是任意实数，可得 $u(x,y)$ 的共轭调和函数 $v(x,y) = -3x^2 y +$

$x^3 + C$. 相应的解析函数为 $f(z) = i(z^3 + C)$.

Ex. 1 Find the harmonic conjugate function of $u(x, y) = y^3 - 3x^2 y$.

Solution $u(x, y) = y^3 - 3x^2 y$ is readily seen to be harmonic throughout the xy-plane. Since a harmonic conjugate $v(x, y)$ is related to $u(x, y)$ by means of the Cauchy-Riemann equations

$$u_x = v_y, u_y = -v_x.$$

This means that

$$v_y(x, y) = -6xy.$$

Holding x fixed and integrating each side here with respect to y, we find that

$$v(x, y) = -3xy^2 + \varphi(x),$$

where φ is, at present, an arbitrary function of x. Using the second of the Cauchy-Riemann equations we have $3y^2 - 3x^2 = 3y^2 - \varphi'(x)$ or $\varphi'(x) = 3x^2$. Thus $\varphi(x) = x^3 + C$, where C is an arbitrary real number. Then the function $v(x, y) = -3x^2 y + x^3 + C$ is a harmonic conjugate of $u(x, y)$. The corresponding analytic function is $f(z) = i(z^3 + C)$

例 2 求 $(2z^2 + i)^5$ 的导数.

解 令 $w = 2z^2 + i$ 且 $W = w^5$, 则 $\dfrac{d}{dz}(2z^2 + i)^5 = 5w^4 \cdot 4z = 20z(2z^2 + i)^4$.

Ex. 2 Find the derivative of $(2z^2 + i)^5$.

Solution Let $w = 2z^2 + i$ and $W = w^5$, then $\dfrac{d}{dz}(2z^2 + i)^5 = 5w^4 \cdot 4z = 20z(2z^2 + i)^4$.

本章重点

1 解析函数具有很好的性质. C-R 条件是判断函数可微和解析的主要条件, 函数 $f(z)$ 在区域 D 内可微, 等价于函数 $f(z)$ 在 D 内解析; 但 $f(z)$ 在一点 z_0 可微, 却不等价于 $f(z)$ 在 z_0 解析.

2 要求掌握从已知的调和函数求共轭的调和函数以组成解析函数的方法. 已知两个共轭调和函数, 便可构成解析函数; 反之, 已知解析函数, 则它的实部与虚部均为调和函数, 且虚部是实部的共轭调和函数; 随便给两个调和函数并不一定组成解析函数. 对这些关系要有清晰的了解和深刻的认识.

3 要清楚认识复变量初等函数其实是相应的实变量初等函数在复平面上的推广, 其关键所在是推广后的函数所必须具备的解析性, 如幂函数、指数函数、正弦函数在复平面上解析; 根式函数、对数函数在单值分支内连续且解析等.

4 要注意每一个函数的基本特性及一些运算法则.

Key points of this chapter

1 Analytic functions have fine properties. C-R premise is the main condition in the judgment of whether a function is differentiable and analytic. If the function $f(z)$ is differentiable in D, the situation is equivalent to $f(z)$ is analytic in D; however, if $f(z)$ is differentiable at the point z_0, it is not to say that $f(z)$ is analytic at the point z_0.

2 It is required that one should grasp the methods of finding the conjugate harmonic function of a given harmonic function to form an analytic function. If two conjugate harmonic functions are given, they can form an analytic function; whereas if an analytic function is given, it can be inferred that both the real part and the imaginary part should be harmonic functions, with the imaginary part as the conjugate harmonic function of the real part; but two randomly given harmonic functions are not necessarily to form an analytic function. To learn this chapter well requires a clear and thorough comprehension of these relations.

3 If one is to recognize that a complex variable elementary function is essentially an extension of its corresponding real variable elementary function on the complex plane, the crucial point is that the extension of a function must be analytic. For instance, power functions, exponential functions and sine functions are analytic on the complex plane; surd functions, logarithm functions are continuous analytic on the single-valued sub branch and so on.

4 Attention should be paid both on every fundamental feature of functions and on rules in operation of functions.

第三章 复变函数的积分
Chapter 3 Integrals of functions of complex variable

本章研究解析函数的积分理论。在引入复变函数积分概念与积分性质的基础上，对解析函数积分及运算性质等一系列特性进行了讨论。给出了柯西积分定理，从而揭示了区域与沿其内任一闭曲线积分的联系，进而得到柯西积分公式，使得闭区域上一点的函数值与其边界上的积分相联系，从而揭示了解析函数的一些内在联系。

Integral theory of analytic function will be studied in this chapter. Based on the concepts of integral of functions of complex variable and properties of integral, integral and calculation properties of analytic functions are discussed. The Cauchy integral theorem is presented, thereby the relation between domain and integral along any closed curve is discovered, and among these theorems is the Cauchy integral formula. Thus the value of a function at a certain point is related with its integration on any closed curve. Consequently the inherent affiliation within analytic functions is discovered.

单词和短语 Words and expressions

★ 柯西积分公式　Cauchy integral formula ｜ ★ 柯西不等式　Cauchy inequality

基本概念和性质　Basic concepts and properties

1 柯西-古萨定理：

如果函数 $f(z)$ 在单连通区域 D 内处处解析，则函数 $f(z)$ 沿 D 内的任何一条封闭曲线 C 的积分为零，即 $\oint_C f(z) = 0$.

Cauchy-Goursat theorem：

If the function $f(z)$ is analytic at all points in the simply connected domain D, C be an any simply closed curve in the D, then $\oint_C f(z) = 0$.

2 柯西积分公式：

如果 $f(z)$ 在区域 D 内处处解析，C 为 D 内任何一条正向简单闭曲线，z_0 为 C 内任一点，那么 $f(z_0) = \frac{1}{2\pi i} \oint_C \frac{f(z)}{z - z_0} dz$.

Cauchy integral formula：

Let $f(z)$ be analytic everywhere inside D and let C be a simply closed curve taking in the positive sense. If z_0 is any point in C, then $f(z_0) = \frac{1}{2\pi i} \oint_C \frac{f(z)}{z - z_0} dz$.

3 解析函数的高阶导数：

设函数 $f(z)$ 在简单曲线 C 所围成的区域 D 内解析，在 $\overline{D} = D \bigcup C$ 上连续，则 $f(z)$ 的各阶导数均在 D 内解析，对 D 内任一点 z，有

$$f^{(n)}(z) = \frac{n!}{2\pi i} \oint_C \frac{f(s)}{(s-z)^{n+1}} ds \quad (n = 1, 2, \cdots)$$

Derivatives of higher order of analytic function：

Suppose that the function $f(z)$ is analytic in the domain D which by the simple closed curve C, and $f(z)$ is contiuous in $\overline{D} = D \bigcup C$, then the n-th derivatives of the analytic functions $f(z)$ is analytic in D, for any z belongs to D

$$f^{(n)}(z) = \frac{n!}{2\pi i} \oint_C \frac{f(s)}{(s-z)^{n+1}} ds \quad (n = 1, 2, \cdots)$$

例题 Examples

例 求 $I = \int_C \bar{z} dz$ 其中 C 是 $z = 2e^{i\theta} \left(-\dfrac{\pi}{2} \leqslant \theta \leqslant \dfrac{\pi}{2}\right)$ 从 $z = -2i$ 到 $z = 2i$ 的 $|z| = 2$ 的右半圆周.

解 根据积分的定义 $I = \int_{-\pi/2}^{\pi/2} \overline{2e^{i\theta}} (2e^{i\theta})' d\theta$. 因为 $\overline{e^{i\theta}} = e^{-i\theta}$ 和 $(e^{i\theta})' = ie^{i\theta}$,

$$I = \int_{-\pi/2}^{\pi/2} 2e^{-i\theta} \cdot 2ie^{i\theta} d\theta = 4i \int_{-\pi/2}^{\pi/2} d\theta = 4\pi i.$$

Ex. Find the value of the integral $I = \int_C \bar{z} dz$ where C is the right-hand half $z = 2e^{i\theta}$ $\left(-\dfrac{\pi}{2} \leqslant \theta \leqslant \dfrac{\pi}{2}\right)$ of the circle $|z| = 2$, from $z = -2i$ to $z = 2i$.

Solution According to the definition of integration

$$I = \int_{-\pi/2}^{\pi/2} \overline{2e^{i\theta}} (2e^{i\theta})' d\theta.$$

Since $\overline{e^{i\theta}} = e^{-i\theta}$ and $(e^{i\theta})' = ie^{i\theta}$, this means that

$$I = \int_{-\pi/2}^{\pi/2} 2e^{-i\theta} \cdot 2ie^{i\theta} d\theta = 4i \int_{-\pi/2}^{\pi/2} d\theta = 4\pi i.$$

本章重点

1. 复变函数的积分和实平面曲线积分有什么不同？有什么联系？
2. 柯西积分定理的条件和结论是什么？
3. 柯西积分定理和柯西积分公式有什么联系？

Key points of this chapter

1. What are the differences and relations between integrals of complex variable function and real plane curve?
2. What are the conditions and conclusions of Cauchy integral theorem?
3. What relations between Cauchy integral theorem and Cauchy integral formula?

第四章 解析函数的级数表示
Chapter 4 Series Expressions of Analytic Functions

本章研究复变函数的幂级数与洛朗级数，表述了幂级数与解析函数的密切联系，一方面幂级数在一定区域内收敛于一个解析函数；另一方面，一个解析函数在其解析点的邻域内，能展开成幂级数.

洛朗级数是幂级数的进一步发展. 它是由一个普通幂级数同一个只含负次幂的级数组合而成的. 洛朗级数的性质可由幂级数的性质推导出来.

Power series and Laurent series of functions of complex variable are discussed in this chapter, as well as the relationship between power series and analytic functions. On the one hand, a power series converges to an analytic function in some domains; on the other hand, an analytic function can be represented as the sum of a power series in some neighborhoods of its analytic point.

Laurent series is a further evolution of power series. It is made up of commonly power series and series only containing negative power. Properties of Laurent series can be deduced from the properties of power series.

单词和短语 Words and expressions

复函数序列	sequences of complex function	函数项级数	series of functions
级数	series	收敛性	convergence
幂级数	power series	收敛半径	radius of convergence

★ 泰勒级数 Taylor series
★ 洛朗级数 Laurent series
发散 divergence
麦克劳林级数 Maclaurin series

泰勒级数展开 Taylor series expansion
绝对收敛 absolutely convergent
一致收敛 uniform convergence
部分和 partial sum

基本概念和性质   Basic concepts and properties

1 级数 $\sum_{n=1}^{\infty} z^n$ 收敛的必要条件为 $\lim_{n \to \infty} z_n = 0$.

A necessary condition for the convergence of series $\sum_{n=1}^{\infty} z^n$ is that $\lim_{n \to \infty} z_n = 0$.

2 绝对收敛的复数项级数一定收敛.

Absolute convergence of a series of complex numbers implies the convergence of that series.

3 泰勒定理：设 $f(z)$ 在 $|z - z_0| < R_0$ 内解析，则 $f(z)$ 有幂级数展开式

$$f(z) = \sum_{n=0}^{\infty} a_n (z - z_0)^n, \quad (|z - z_0| < R_0), \text{ 其中 } a_n = \frac{f^{(n)}(z_0)}{n!}.$$

Taylor theorem: Suppose that the function $f(z)$ is analytic throughout $|z - z_0| < R_0$. Then $f(z)$ has the power series expansion

$$f(z) = \sum_{n=0}^{\infty} a_n (z - z_0)^n, \quad (|z - z_0| < R_0), \text{ where } a_n = \frac{f^{(n)}(z_0)}{n!}.$$

4 $f(z) = \sum_{n=0}^{\infty} a_n (z - z_0)^n$ 称为 $f(z)$ 在 z_0 点的泰勒级数；当 $z_0 = 0$ 时，$f(z) = \sum_{n=0}^{\infty} \frac{f^{(n)}(0)}{n!} z^n$ 称为 $f(z)$ 的麦克劳林级数.

$f(z) = \sum_{n=0}^{\infty} a_n (z - z_0)^n$ is the expansion of Taylor series of $f(z)$ at the point z_0; When $z_0 = 0$, $f(z) = \sum_{n=0}^{\infty} \frac{f^{(n)}(0)}{n!} z^n$ is called a Maclaurin series.

例题   Examples

例 1 把 $f(z) = \dfrac{1 + 2z^2}{z^3 + z^5}$ 展开成 z 的幂级数.

解 $f(z) = \dfrac{1 + 2z^2}{z^3 + z^5} = \dfrac{1}{z^3}\left(2 - \dfrac{1}{1 + z^2}\right)$，我们不能直接求 $f(z)$ 的麦克劳林级数，因为 $f(z)$ 在 $z = 0$ 处不解析，但有 $\dfrac{1}{1 + z^2} = 1 - z^2 + z^4 - z^6 + \cdots (|z| < 1)$. 因此，当 $0 < |z| < 1$ 时，有

$$f(z) = \frac{1}{z^3}(2 - 1 + z^2 - z^4 + z^6 - \cdots) = \frac{1}{z^3} + \frac{1}{z} - z + z^3 - \cdots$$

Ex. 1 Expand the function $f(z) = \dfrac{1 + 2z^2}{z^3 + z^5}$ into a series involving the power of z.

Solution $f(z) = \dfrac{1 + 2z^2}{z^3 + z^5} = \dfrac{1}{z^3}\left(2 - \dfrac{1}{1 + z^2}\right)$.

We cannot find a Maclaurin series for $f(z)$ since it is not analytic at $z = 0$. But we know the expansion $\dfrac{1}{1 + z^2} = 1 - z^2 + z^4 - z^6 + \cdots (|z| < 1)$.

Hence, when $0 < |z| < 1$, we have

$$f(z) = \frac{1}{z^3}(2 - 1 + z^2 - z^4 + z^6 - \cdots) = \frac{1}{z^3} + \frac{1}{z} - z + z^3 - \cdots$$

例2 求函数 $\dfrac{-1}{(z-1)(z-2)} = \dfrac{1}{z-1} - \dfrac{1}{z-2}$ 在下面区域的洛朗级数：
$$D_1: |z|<1, \quad D_2: 1<|z|<2, \quad D_3: |z|>2$$

解 (1) $\dfrac{1}{1-z} = \sum_{n=0}^{\infty} z^n$，$|z|<1$ 是麦克劳林级数. 由

$$f(z) = \dfrac{1}{z-1} + \dfrac{1}{2} \cdot \dfrac{1}{1-\dfrac{z}{2}}$$

得 $|z|<1$ 和 $\left|\dfrac{z}{2}\right|<1$. 在 D_1 内，

$$f(z) = -\sum_{n=0}^{\infty} z^n + \sum_{n=0}^{\infty} \dfrac{z^n}{2^{n+1}} = \sum_{n=0}^{\infty} (2^{-n-1} - 1) z^n \quad |z|<1.$$

(2) 当 $1<|z|<2$ 时，$\left|\dfrac{1}{z}\right|<1$，$\left|\dfrac{z}{2}\right|<1$，则

$$\dfrac{1}{1-z} = -\dfrac{1}{z} \cdot \dfrac{1}{1-\dfrac{1}{z}} = -\dfrac{1}{z}\left(1 + \dfrac{1}{z} + \dfrac{1}{z^2} + \cdots\right)$$

$$f(z) = \dfrac{1}{z-1} + \dfrac{1}{2} \cdot \dfrac{1}{1-\dfrac{z}{2}}$$

$$= \dfrac{1}{z}\left(1 + \dfrac{1}{z} + \dfrac{1}{z^2} + \cdots + \dfrac{1}{z^n} + \cdots\right) + \dfrac{1}{2}\left(1 + \dfrac{z}{2} + \dfrac{z^2}{2^2} + \cdots\right)$$

$$= \dfrac{1}{z} + \dfrac{1}{z^2} + \dfrac{1}{z^3} + \cdots + \dfrac{1}{z^n} + \cdots + \dfrac{1}{2} + \dfrac{z}{2^2} + \dfrac{z^2}{2^3} + \cdots$$

(3) 当 $2<|z|<\infty$ 时，$|z|>2$，$\left|\dfrac{2}{z}\right|<1$，则

$$\dfrac{1}{2-z} = -\dfrac{1}{z} \cdot \dfrac{1}{1-\dfrac{2}{z}} = -\dfrac{1}{z}\left(1 + \dfrac{2}{z} + \dfrac{4}{z^2} + \cdots\right)$$

$$f(z) = \dfrac{1}{z-1} - \dfrac{1}{z} \cdot \dfrac{1}{1-\dfrac{2}{z}}$$

$$= \dfrac{1}{z}\left(1 + \dfrac{1}{z} + \dfrac{1}{z^2} + \cdots + \dfrac{1}{z^n} + \cdots\right) - \dfrac{1}{z}\left(1 + \dfrac{2}{z} + \dfrac{4}{z^2} - \cdots\right)$$

$$= -\dfrac{1}{z^2} - \dfrac{3}{z^3} - \dfrac{7}{z^4} - \cdots - \dfrac{2^{n-1}-1}{z^n} - \cdots$$

Ex. 2 The function $\dfrac{-1}{(z-1)(z-2)} = \dfrac{1}{z-1} - \dfrac{1}{z-2}$ is analytic in the following domain
$$D_1: |z|<1, \quad D_2: 1<|z|<2, \quad D_3: |z|>2$$
Find the Laurent series of the function in these domains.

Solution (1) $\dfrac{1}{1-z} = \sum_{n=0}^{\infty} z^n$, $|z|<1$.

The representation in D_1 is a Maclaurin series. To find it, we write

$$f(z) = \dfrac{1}{z-1} + \dfrac{1}{2} \cdot \dfrac{1}{1-\dfrac{z}{2}}$$

and observe that, since $|z|<1$ and $\left|\dfrac{z}{2}\right|<1$ in D_1,

$$f(z) = -\sum_{n=0}^{\infty} z^n + \sum_{n=0}^{\infty} \dfrac{z^n}{2^{n+1}} = \sum_{n=0}^{\infty} (2^{-n-1} - 1) z^n \quad |z|<1.$$

(2) when $1 < |z| < 2$, $\left|\frac{1}{z}\right| < 1$, $\left|\frac{z}{2}\right| < 1$, then

$$\frac{1}{1-z} = -\frac{1}{z} \cdot \frac{1}{1-\frac{1}{z}} = -\frac{1}{z}\left(1 + \frac{1}{z} + \frac{1}{z^2} + \cdots\right)$$

$$f(z) = \frac{1}{z-1} + \frac{1}{2} \cdot \frac{1}{1-\frac{z}{2}}$$

$$= \frac{1}{z}\left(1 + \frac{1}{z} + \frac{1}{z^2} + \cdots + \frac{1}{z^n} + \cdots\right) + \frac{1}{2}\left(1 + \frac{z}{2} + \frac{z^2}{2^2} + \cdots\right)$$

$$= \frac{1}{z} + \frac{1}{z^2} + \frac{1}{z^3} + \cdots + \frac{1}{z^n} + \cdots + \frac{1}{2} + \frac{z}{2^2} + \frac{z^2}{2^3} + \cdots$$

(3) when $2 < |z| < \infty$, $|z| > 2$, $\left|\frac{2}{z}\right| < 1$, then

$$\frac{1}{2-z} = -\frac{1}{z} \cdot \frac{1}{1-\frac{2}{z}} = -\frac{1}{z}\left(1 + \frac{2}{z} + \frac{4}{z^2} + \cdots\right)$$

$$f(z) = \frac{1}{z-1} - \frac{1}{z} \cdot \frac{1}{1-\frac{2}{z}}$$

$$= \frac{1}{z}\left(1 + \frac{1}{z} + \frac{1}{z^2} + \cdots + \frac{1}{z^n} + \cdots\right) - \frac{1}{z}\left(1 + \frac{2}{z} + \frac{4}{z^2} - \cdots\right)$$

$$= -\frac{1}{z^2} - \frac{3}{z^3} - \frac{7}{z^4} - \cdots - \frac{2^{n-1}-1}{z^n} - \cdots$$

本章重点

1 应将函数展开成什么级数? 幂级数还是洛朗级数?
2 函数能不能展开? 怎样展开? 在哪些区域里展开?

Key points of this chapter

1 What kind of series should functions be extended into? Power series or Laurent series?
2 Is it possible to expand a function and how to do so? In which domain should they be expanded?

第五章　留数及其应用
Chapter 5　Residues and their Applications

本章主要介绍孤立奇点的分类、单值解析函数在孤立奇点的留数的概念.

We introduce the concept and classification of isolated singularities, the concept of residue of a single-valued analytic function at an isolated singularity.

单词和短语 Words and expressions

★ 留数　residue
孤立奇点　isolated singularity
可去奇点　removable singularity
本性奇点　essential singularity

★ 极点　pole
m 阶极点　pole of order m
当且仅当　if and only if
★ 亚纯函数　meromorphic function

基本概念和性质　Basic concepts and properties

1 z_0 称为函数 $f(z)$ 的奇点指 $f(z)$ 在 z_0 不解析但在 z_0 的任何一个邻域的某些点解析. 如果函数 $f(z)$ 在 z_0 不解析, 但在 z_0 的某一去心邻域 $0 < |z - z_0| < \varepsilon$ 内处处解析, 则称 z_0 为 $f(z)$ 的孤立奇点.

The point z_0 is called a singular point of the function $f(z)$ if $f(z)$ fails to be analytic at z_0 but is analytic at some point in every neighborhood of z_0. The singular point z_0 is said to be isolated if, in addition, there is a deleted neighborhood of $0 < |z - z_0| < \varepsilon$ of z_0 such that $f(z)$ is analytic.

2 Cauchy 留数定理：设 C 是一条正向简单闭曲线，函数 $f(z)$ 在曲线 C 包围的区域内除有限个点 $z_k(k = 1, 2, \cdots, n)$ 处处解析，则 $\int_C f(z)\mathrm{d}z = 2\pi \mathrm{i} \sum_{k=1}^{n} \operatorname*{Res}_{z_k} f(z)$.

Cauchy residue theorem: Let C be a simple closed curve taking the positive sense. If the function $f(z)$ is analytic in the domain enclosed by C except for finite number of singular points $z_k (k = 1, 2, \cdots, n)$ inside C, then $\int_C f(z)\mathrm{d}z = 2\pi \mathrm{i} \sum_{k=1}^{n} \operatorname*{Res}_{z_k} f(z)$.

3 如果一个函数 $f(z)$ 在一个正向简单闭曲线 C 所围区域内除有限个点外处处解析，则 $\int_C f(z)\mathrm{d}z = 2\pi \mathrm{i} \operatorname*{Res}_{z=0} \left[\frac{1}{z^2} f\left(\frac{1}{z}\right)\right]$.

If the function $f(z)$ is analytic everywhere in the finite plane except for finite number of singular points interior to a positively oriented simple closed curve C, then $\int_C f(z)\mathrm{d}z = 2\pi \mathrm{i} \operatorname*{Res}_{z=0} \left[\frac{1}{z^2} f\left(\frac{1}{z}\right)\right]$.

例 题 Examples

例 利用积分公式计算 $\int_C \frac{5z-2}{z(z-1)}\mathrm{d}z$，其中 C 是圆周 $|z| = 2$ 的逆时针方向.

解 被积函数在 C 内有两个孤立奇点 $z = 0$ 和 $z = 1$，我们可借助于麦克劳林级数计算函数在这两点的留数 $B_1: z = 0$ 和 $B_2: z = 1$，即

$$\frac{1}{1-z} = 1 + z + z^2 + z^3 + \cdots \quad (|z| < 1),$$

$$\frac{5z-2}{z(z-1)} = \frac{5z-2}{z} \cdot \frac{-1}{1-z} = \left(5 - \frac{2}{z}\right)(-1 - z - z^2 - \cdots),$$

$$\frac{5z-2}{z(z-1)} = \frac{5(z-1)+3}{z-1} \cdot \frac{1}{1-(1-z)} = \left(5 + \frac{3}{z-1}\right)[1 + (1-z) + (1-z)^2 + \cdots],$$

则 $B_1 = 2, B_2 = 3$,

$$\int_C \frac{5z-2}{z(z-1)}\mathrm{d}z = 2\pi \mathrm{i}(B_1 + B_2) = 10\pi \mathrm{i}.$$

Ex Use the integral formula to evaluate $\int_C \frac{5z-2}{z(z-1)}\mathrm{d}z$, where C is the circle $|z| = 2$ taking the counterclockwise.

Solution The integrand has two isolated singularities $z = 0$ and $z = 1$, both of which are interior to C. We find the residues $B_1: z = 0$ and $B_2: z = 1$ with the aid of the Maclaurin series

$$\frac{1}{1-z} = 1 + z + z^2 + z^3 + \cdots \quad (|z| < 1),$$

$$\frac{5z-2}{z(z-1)} = \frac{5z-2}{z} \cdot \frac{-1}{1-z} = \left(5 - \frac{2}{z}\right)(-1 - z - z^2 - \cdots),$$

$$\frac{5z-2}{z(z-1)} = \frac{5(z-1)+3}{z-1} \cdot \frac{1}{1-(1-z)} = \left(5 + \frac{3}{z-1}\right)[1 + (1-z) + (1-z)^2 + \cdots],$$

then $B_1 = 2, B_2 = 3$, and

$$\int_C \frac{5z-2}{z(z-1)}\mathrm{d}z = 2\pi \mathrm{i}(B_1 + B_2) = 10\pi \mathrm{i}.$$

本章重点

1 理解并熟练运用孤立奇点的定义及分类.

❷ 理解留数定义及留数定理.
❸ 利用留数定理熟练计算积分.

Key points of this chapter

❶ Understand and use proficiently the definition and classification of isolated singularity.
❷ Understand the definition of residue and the residue theorem.
❸ Apply residue theorem in evaluating integrals proficiently.

第六章　共形映射
Chapter 6　Conformal Mappings

本章介绍共形映射的概念,检验解析与共形之间的关系,讨论基本初等函数的共形映射,确定了由半平面变换到一个多边形区域的共形映射的结构.

In this chapter we introduce the concept of conformal mapping, and examine the relationship between analyticity and conformality. After this we discuss conformal mappings of fundamental elementary functions and determine the structure of a conformal mapping from a half-plane onto a general polygonal region.

单词和短语 Words and expressions

从 A 到 B 的转角　oriented angle from A to B　　不动点　fixed point
★ 保角映射　angle-preserving mapping　　分式线性变换　linear fractional transformation
自映射　self-mapping　　多边形　polygon

本章重点

本章通过对导函数的几何特性的分析,引入共形映射的概念,并借助这一概念来进一步认识解析函数的映射特征与应用.
(1)保形映射是解析函数特有的性质.
(2)分式线性函数具有保形性、保对称性以及保圆性等一些性质.

Key points of this chapter

Through an analysis of the geometric properties of derived function, this chapter introduces the concept of conformal mapping, which further helps understand the mapping features and applications of analytic functions.
(1)Shape-preserving mapping is a specific feature of analytic function.
(2)Linear fractional function has the properties of shape-preserving, symmetry-preserving and circle-preserving.

第七章　傅里叶变换
Chapter 7　Fourier Transforms

本章介绍了傅里叶变换、傅里叶积分,以及傅里叶变换的性质及应用.

Fourier transform, Fourier integral and their properties and applications are introduced in this chapter.

单词和短语 Words and expressions

★ 傅里叶变换　Fourier transform　　延迟性　time shifting
★ 傅里叶积分　Fourier integral　　积分变换　integral transform
★ 卷积　convolution　　反演公式　inversion formula
线性性　linearity　　共轭傅里叶积分　conjugate Fourier integral
对称性　symmetry　　广义傅里叶积分　generalized Fourier integral

傅里叶逆变换　inverse Fourier transform
傅里叶反演公式　Fourier inversion formula
傅里叶正弦变换　Fourier sine transform
傅里叶余弦变换　Fourier cosine transform

基本概念和性质　Basic concepts and properties

1 傅里叶级数的三角形式（trigonometrical form of Fourier series）

$$f(t) = \frac{a_0}{2} + \sum_{n=1}^{+\infty}(a_n \cos n\omega t + b_n \sin n\omega t), \omega = \frac{2\pi}{T},$$

$$a_n = \frac{2}{T}\int_{-T/2}^{T/2} f(t) \cos n\omega t \, dt, (n = 0, 1, 2, \cdots)$$

$$b_n = \frac{2}{T}\int_{-T/2}^{T/2} f(t) \sin n\omega t \, dt. (n = 1, 2, \cdots)$$

2 傅里叶级数的复指数形式（exponential form of Fourier series）

$$f(t) = \sum_{n=-\infty}^{+\infty} c_n e^{in\omega t}, \quad c_n = \frac{1}{T}\int_{-T/2}^{T/2} f(t) e^{-in\omega t} \, dt.$$

3 傅里叶积分公式（Fourier integral formula）

$$f(t) = \frac{1}{2\pi}\int_{-\infty}^{+\infty}\left[\int_{-\infty}^{+\infty} f(t) e^{-i\omega t} \, dt\right] e^{i\omega t} \, d\omega.$$

本章重点

1 理解傅里叶变换，利用定义求函数的傅里叶变换．
2 掌握傅里叶变换的性质并利用性质求函数的傅里叶变换．
3 应用傅里叶变换解决一些计算和应用问题．

Key points of this chapter

1 Understand Fourier transform and apply its definition to find the Fourier transform of a function.
2 Grasp the properties of Fourier transform and apply the properties to find the Fourier transform of a function.
3 Apply the Fourier transform to solve the problems in calculation and application.

第八章　拉普拉斯变换
Chapter 8　Laplace Transforms

　　本章从傅氏变换引出拉氏变换的概念，讨论拉氏变换的基本性质以及拉氏逆变换的求解方法，并介绍了它在求解微分方程方面的应用．

　　In this chapter, we deduce the concept of Laplace transform from Fourier transform, discuss the basic properties of Laplace transform and the solving methods for finding the inverse Laplace transform, and introduce its application on differential equations.

单词和短语　Words and expressions

拉普拉斯变换　Laplace transform
像　image
逆变换　inverse transform

本章重点

1 理解拉普拉斯变换和逆变换的定义，会求函数的拉普拉斯变换．
2 熟练掌握拉普拉斯变换的性质，并利用它们求解函数的拉普拉斯变换和逆变换．
3 应用拉普拉斯变换解微分方程．

Key points of this chapter

1 Understand the definitions of Laplace transform and inverse Laplace transform, and know how to find the Laplace transform of a function.

2 Grasp proficiently the properties of Laplace transform and learn to use them in finding the Laplace transform and inverse transform of a function.

3 Apply Laplace transform to solve differential equations.

练 习

1. 计算 $h'(z)$，其中 $h(z) = [(z^2-1)/(z^2+1)]^{10}$。

解 注意到 $h = g \circ f$，这里 $f(z) = (z^2-1)/(z^2+1)$，$g(z) = z^{10}$，计算 $f'(z) = 4z/(z^2+1)^2$，$g'(z) = 10z^9$，应用链式规则有：

$$h'(z) = g'(f(z))f'(z) = 10\left(\frac{z^2-1}{z^2+1}\right)^9 \frac{4z}{(z^2+1)^2} = \frac{40z(z^2-1)^9}{(z^2+1)^{11}}.$$

2. 计算 $\int_0^{2\pi} (a+\cos\theta)^{-1} d\theta$ $(a>1)$。

解 这类积分的计算可由书中定理而得到，这里令 $R(x,y) = (a+x)^{-1}$，考虑函数：

$$f(z) = \frac{1}{z} R\left(\frac{z+z^{-1}}{2}, \frac{z-z^{-1}}{2i}\right) = \frac{1}{z} \frac{1}{a+[(z+z^{-1})/2]} = \frac{2}{z^2+2az+1}$$

是一个只有一个极点的有理函数，它的极点本来为 $-a+\sqrt{a^2-1}$ 和 $-a-\sqrt{a^2-1}$。因为 $a>1$，所以只有 $z_1 = -a+\sqrt{a^2-1}$ 在 $\Delta(0,1)$ 内，由公式得

$$\text{Res}(z_1,f) = \lim_{z\to z_1}(z-z_1)f(z) = \lim_{z\to z_1}\frac{2(z-z_1)}{z^2+2az+1} = \lim_{z\to z_1}\frac{2}{2z+2a} = \frac{1}{\sqrt{a^2-1}},$$

根据留数定理得

$$\int_0^{2\pi} \frac{d\theta}{a+\cos\theta} = \frac{2\pi}{\sqrt{a^2-1}} \quad (a>1).$$

3. 找出 $f(z) = (e^z-1)^{-3}$ 在 $z=0$ 处的奇异部分，并利用它计算 $\text{Res}(0,f)$。

解 本例中 $z=0$ 是这个函数的 3 阶极点，首先计算：

$$(e^z-1)^3 = \left(z+\frac{z^2}{2!}+\frac{z^3}{3!}+\frac{z^4}{4!}+\cdots\right)^3 = z^3\left[1+\frac{z}{2}+\frac{z^2}{6}+O(z^3)\right]^3$$

$$= z^3\left[1+\frac{z}{2}+\frac{z^2}{6}+O(z^3)\right]^2\left[1+\frac{z}{2}+\frac{z^2}{6}+O(z^3)\right]$$

$$= z^3\left[1+z+\frac{7z^2}{12}+O(z^3)\right]\left[1+\frac{z}{2}+\frac{z^2}{6}+O(z^3)\right]$$

$$= z^3\left[1+\frac{3z}{2}+\frac{5z^2}{4}+O(z^3)\right]$$

得几何级数：

$$\left[1+\frac{3z}{2}+\frac{5z^2}{4}+O(z^3)\right]^{-1} = 1-\left[\frac{3z}{2}+\frac{5z^2}{4}+O(z^3)\right]+\left[\frac{3z}{2}+\frac{5z^2}{4}+O(z^3)\right]^2+O(z^3)$$

$$= 1-\frac{3z}{2}-\frac{5z^2}{4}+\frac{9z^2}{4}+O(z^3)$$

$$= 1-\frac{3z}{2}+z^2+O(z^3).$$

因此得到以下展开式：

$$f(z) = (e^z-1)^{-3} = \frac{1}{z^3}-\frac{3}{2z^2}+\frac{1}{z}+O(1).$$

所求奇异部分为：

$$S(z) = \frac{1}{z^3}-\frac{3}{2z^2}+\frac{1}{z},$$

这表明：$\text{Res}(0,f) = 1$.

4. 讨论 $\sum_{n=-\infty}^{+\infty} 2^{-|n|} z^n$ 的收敛性.

解 把级数分成 $n \geq 0$ 和 $n < 0$ 两部分，讨论非负指数部分：

$$\sum_{n=0}^{+\infty} 2^{-|n|} z^n = \sum_{n=0}^{+\infty} \left(\frac{z}{2}\right)^n = \frac{1}{1-(z/2)} = \frac{2}{2-z}, \ (|z/2| < 1 \text{ 即 } |z| < 2)$$

对其余的 z，级数是发散的. 对于负指数部分可得：

$$\sum_{n=-\infty}^{-1} 2^{-|n|} z^n = \sum_{n=1}^{+\infty} 2^{-|n|} z^{-n} = \sum_{n=1}^{+\infty} \left(\frac{1}{2z}\right)^n = \frac{1}{2z} \sum_{n=1}^{+\infty} \left(\frac{1}{2z}\right)^{n-1}$$

$$= \frac{1}{2z} \sum_{n=0}^{+\infty} \left(\frac{1}{2z}\right)^n = \frac{1}{2z} \cdot \frac{1}{1-(1/2z)} = \frac{1}{2z-1},$$

$(|1/(2z)| < 1 \text{ 即 } |z| > 1/2)$

对其余的 z，级数是发散的.

因此有：$\sum_{n=-\infty}^{+\infty} 2^{-|n|} z^n = \frac{2}{2-z} + \frac{1}{2z-1} = \frac{3z}{(2-z)(2z-1)}$, $(1/2 < |z| < 2)$.

5. 把 $f(z) = 2z(z^2-1)^{-1}$ 在 $z_0 = i$ 处展开成幂级数.

解 把 $f(z)$ 分成两部分，分别展开成级数：

$$\frac{2z}{z^2-1} = \frac{1}{z-1} + \frac{1}{z+1} = \frac{1}{(i-1)+(z-i)} + \frac{1}{(i+1)+(z-i)}$$

$$= \frac{1}{i-1} \cdot \frac{1}{1-\left(-\frac{z-i}{i-1}\right)} + \frac{1}{i+1} \cdot \frac{1}{1-\left(-\frac{z-i}{i+1}\right)}$$

$$= \frac{1}{i-1} \sum_{n=0}^{+\infty} (-1)^n \left(\frac{z-i}{i-1}\right)^n + \frac{1}{i+1} \sum_{n=0}^{+\infty} (-1)^n \left(\frac{z-i}{i+1}\right)^n$$

$$= \sum_{n=0}^{+\infty} (-1)^n [(i-1)^{-n-1} + (i+1)^{-n-1}](z-i)^n.$$

我们已经得到了 f 在 i 处的泰勒展开，可以用下面更简捷的方法表示展开式的系数：

$$(i-1)^{-n-1} + (i+1)^{-n-1} = \frac{(-1)^{n+1}[(i+1)^{n+1} + (i-1)^{n+1}]}{2^{n+1}}.$$

对于任何正整数 k 有：

$$(i-1)^k + (i+1)^k = 2^{\frac{k}{2}} e^{\frac{k\pi i}{4}} + 2^{\frac{k}{2}} e^{\frac{3k\pi i}{4}} = 2^{\frac{k}{2}} e^{\frac{k\pi i}{2}} \left(e^{\frac{k\pi i}{4}} + e^{-\frac{k\pi i}{4}}\right) = 2^{\frac{k+2}{2}} e^{\frac{k\pi i}{2}} \cos\frac{k\pi}{4}.$$

因此得到修改后的展开式：

$$\frac{2z}{z^2-1} = \sum_{n=0}^{\infty} (-1) 2^{-\frac{n}{2}} e^{\frac{n\pi i}{2}} \text{ 的 } \frac{n\pi}{4} (z-i)^n, \quad (z \in \Delta(i,\sqrt{2})).$$

6. 计算 $f(t) = \delta(t)$.

解 $f(t) = \delta(t)$ 的傅里叶变换：$F(\omega) = \int_{-\infty}^{+\infty} e^{-i\omega t} \delta(t) dt = 1$.

因此得傅里叶对：$\delta(t) \leftrightarrow 1$. 函数 $k\delta(t)$ 代表从原点出发，长度为 k 的有向线段. 平移后脉冲函数 $\delta(t-t_0)$ 的傅立叶展开为 $e^{-i\omega t_0}$，因此有 $\delta(t-t_0) \leftrightarrow e^{-i\omega t_0}$.

上式的幅值为 1，且具有线性相位，用傅里叶逆变换得到：

$$\delta(t) = \frac{1}{2\pi} \int_{-\infty}^{+\infty} e^{i\omega t} d\omega = \frac{1}{2\pi} \int_{-\infty}^{+\infty} \cos\omega t \, d\omega.$$

7. 假设 $f(t)$ 的拉氏变换为 $F(p) = \frac{p}{(p+1)(p+2)}$，计算 $f(t)$.

解 $F(p) = \dfrac{p}{(p+1)(p+2)}$，极点为 $p_1 = -1, p_2 = -2$.

因此，$F(p_1) = \dfrac{p}{p+2}\bigg|_{p=-1} = -1$, $F(p_2) = \dfrac{p}{p+1}\bigg|_{p=-2} = 2$.

所以 $f(t) = -e^{-t} + 2e^{-t}, t > 0$.

Exercises

1. Calculate $h'(z)$ for $h(z) = [(z^2-1)/(z^2+1)]^{10}$.

   Solution Noting that $h = g \circ f$, where $f(z) = (z^2-1)/(z^2+1)$ and $g(z) = z^{10}$, we compute $f'(z) = 4z/(z^2+1)^2$, $g'(z) = 10z^9$, and apply the chain rule to obtain
   $$h'(z) = g'(f(z))f'(z) = 10\left(\dfrac{z^2-1}{z^2+1}\right)^9 \dfrac{4z}{(z^2+1)^2} = \dfrac{40z(z^2-1)^9}{(z^2+1)^{11}}.$$

2. Evaluate $\int_0^{2\pi}(a + \cos\theta)^{-1}d\theta \ (a > 1)$.

   Solution This integral is of the type covered by theorem in this book. Here $R(x, y) = (a+x)^{-1}$. We consider the function f given by
   $$f(z) = \dfrac{1}{z}R\left(\dfrac{z+z^{-1}}{2}, \dfrac{z-z^{-1}}{2i}\right) = \dfrac{1}{z}\dfrac{1}{a + [(z+z^{-1})/2]} = \dfrac{2}{z^2 + 2az + 1}$$
   A rational function whose only poles are simple ones located at the points $-a + \sqrt{a^2-1}$ and $-a - \sqrt{a^2-1}$. Since $a > 1$, just the first of these, $z_1 = -a + \sqrt{a^2-1}$, finds itself in $\Delta(0,1)$ Formula gives
   $$\text{Res}(z_1, f) = \lim_{z \to z_1}(z-z_1)f(z) = \lim_{z \to z_1}\dfrac{2(z-z_1)}{z^2+2az+1} = \lim_{z \to z_1}\dfrac{2}{2z+2a} = \dfrac{1}{\sqrt{a^2-1}},$$
   Referring to the residue theorem we conclude that
   $$\int_0^{2\pi}\dfrac{d\theta}{a+\cos\theta} = \dfrac{2\pi}{\sqrt{a^2-1}} \quad \text{for} \quad a > 1.$$

3. Find the singular part of $f(z) = (e^z - 1)^{-3}$ at the origin, and use it to determine $\text{Res}(0, f)$.

   Solution The function we are dealing with in this example has a pole of order 3 at the origin. We first calculate
   $$\begin{aligned}(e^z-1)^3 &= \left(z + \dfrac{z^2}{2!} + \dfrac{z^3}{3!} + \dfrac{z^4}{4!} + \cdots\right)^3 = z^3\left[1 + \dfrac{z}{2} + \dfrac{z^2}{6} + O(z^3)\right]^3\\ &= z^3\left[1 + \dfrac{z}{2} + \dfrac{z^2}{6} + O(z^3)\right]^2\left[1 + \dfrac{z}{2} + \dfrac{z^2}{6} + O(z^3)\right]\\ &= z^3\left[1 + z + \dfrac{7z^2}{12} + O(z^3)\right]\left[1 + \dfrac{z}{2} + \dfrac{z^2}{6} + O(z^3)\right]\\ &= z^3\left[1 + \dfrac{3z}{2} + \dfrac{5z^2}{4} + O(z^3)\right]\end{aligned}$$
   An appeal to the geometric series gives
   $$\begin{aligned}\left[1 + \dfrac{3z}{2} + \dfrac{5z^2}{4} + O(z^3)\right]^{-1} &= 1 - \left[\dfrac{3z}{2} + \dfrac{5z^2}{4} + O(z^3)\right] + \left[\dfrac{3z}{2} + \dfrac{5z^2}{4} + O(z^3)\right]^2 + O(z^3)\\ &= 1 - \dfrac{3z}{2} - \dfrac{5z^2}{4} + \dfrac{9z^2}{4} + O(z^3)\\ &= 1 - \dfrac{3z}{2} + z^2 + O(z^3)\end{aligned}$$
   Consequently we obtain the expansion
   $$f(z) = (e^z-1)^{-3} = \dfrac{1}{z^3} - \dfrac{3}{2z^2} + \dfrac{1}{z} + O(1).$$

The requested singular part is now seen to be
$$S(z) = \frac{1}{z^3} - \frac{3}{2z^2} + \frac{1}{z},$$
which reveals that Res $(0, f) = 1$.

4. Discuss the convergence of $\sum_{n=-\infty}^{+\infty} 2^{-|n|} z^n$.

Solution We split the given series into the parts corresponding to $n \geq 0$ and $n < 0$, respectively, and consider each of these subsidies on its own. For the portion whose terms go with non-negative indices
$$\sum_{n=-\infty}^{+\infty} 2^{-|n|} z^n = \sum_{n=0}^{+\infty} \left(\frac{z}{2}\right)^n = \frac{1}{1-(z/2)} = \frac{2}{2-z},$$
when $|z/2| < 1$, i.e., $|z| < 2$ and that this series diverges for all remaining z. For the negatively indexed half of the series we obtain
$$\sum_{n=-\infty}^{-1} 2^{-|n|} z^n = \sum_{n=1}^{+\infty} 2^{-|-n|} z^{-n} = \sum_{n=1}^{+\infty} \left(\frac{1}{2z}\right)^n = \frac{1}{2z} \sum_{n=1}^{+\infty} \left(\frac{1}{2z}\right)^{n-1}$$
$$= \frac{1}{2z} \sum_{n=0}^{+\infty} \left(\frac{1}{2z}\right)^n = \frac{1}{2z} \frac{1}{1-(1/2z)} = \frac{1}{2z-1},$$
when $|1/(2z)| < 1$, i.e., $|z| > 1/2$, but we observe divergence otherwise. Thus we can say that
$$\sum_{n=-\infty}^{+\infty} 2^{-|n|} z^n = \frac{2}{2-z} + \frac{1}{2z-1} = \frac{3z}{(2-z)(2z-1)}, \text{ when } 1/2 < |z| < 2,$$
whereas this series is divergent for every other z.

5. Expand $f(z) = 2z(z^2 - 1)^{-1}$ into a power series at $z_0 = i$.

Solution Divide $f(x)$ into two parts and then expand them into series:
$$\frac{2z}{z^2-1} = \frac{1}{z-1} + \frac{1}{z+1} = \frac{1}{(i-1)+(z-i)} + \frac{1}{(i+1)+(z-i)}$$
$$= \frac{1}{i-1} \frac{1}{1-\left(-\frac{z-i}{i-1}\right)} + \frac{1}{i+1} \frac{1}{1-\left(-\frac{z-i}{i+1}\right)}$$
$$= \frac{1}{i-1} \sum_{n=0}^{\infty} (-1)^n \left(\frac{z-i}{i-1}\right)^n + \frac{1}{i+1} \sum_{n=0}^{+\infty} (-1)^n \left(\frac{z-i}{i+1}\right)^n$$
$$= \sum_{n=0}^{\infty} (-1)^n [(i-1)^{-n-1} + (i+1)^{-n-1}](z-i)^n.$$

We have determined the Taylor expansion of f about i. With a small amount of extra effort, however, we can express its coefficients in a more readily computed form. First, by rationalizing denominators we see that
$$(i-1)^{-n-1} + (i+1)^{-n-1} = \frac{(-1)^{n+1}[(i+1)^{n+1} + (i-1)^{n+1}]}{2^{n+1}}.$$

Next for any positive integer k we have
$$(i-1)^k + (i+1)^k = 2^{\frac{k}{2}} e^{\frac{k\pi i}{4}} + 2^{\frac{k}{2}} e^{\frac{3k\pi i}{4}} = 2^{\frac{k}{2}} e^{\frac{k\pi i}{2}} (e^{\frac{k\pi i}{4}} + e^{-\frac{k\pi i}{4}}) = 2^{\frac{k+2}{2}} e^{\frac{k\pi i}{2}} \cos\frac{k\pi}{4}.$$

These observations lead to the revised expansion
$$\frac{2z}{z^2-1} = \sum_{n=0}^{\infty} 2^{-\frac{n-1}{2}} \frac{1}{2ni+3} \cos\frac{(n+1)\pi}{4} (z-i)^n.$$

It is valid for z in $\Delta(i, \sqrt{2})$, the largest open disk with center at i that lies in the set $U = C \sim \{\pm 1\}$ where f is analytic.

6. Evaluate $f(t) = \delta(t)$.

Solution The Fourier transform of the delta function $f(t) = \delta(t)$ is given by
$$F(\omega) = \int_{-\infty}^{+\infty} e^{-i\omega t}\delta(t)dt = 1.$$
we thus have the pair $\delta(t) \leftrightarrow 1$.

The function $k\delta(t)$ is represented graphically by a directed line with the letter k next to it.

The transform of the shifted impulse $\delta(t-t_0)$ is given by $e^{-i\omega t_0}$
$$\delta(t-t_0) \leftrightarrow e^{-i\omega t_0}$$

It has a constant amplitude equal to one and linear phase, we now examine the validity of the inversion formula; We clearly have
$$\delta(t) = \frac{1}{2\pi}\int_{-\infty}^{+\infty} e^{i\omega t}dt = \frac{1}{2\pi}\int_{-\infty}^{+\infty}\cos \omega t\, d\omega.$$

7. Assume that the Laplace transform of $f(t)$ is $F(p) = \dfrac{p}{(p+1)(p+2)}$, evaluate $f(t)$.

Solution $F(p) = \dfrac{p}{(p+1)(p+2)}$, its poles are given by $p_1 = -1, p_2 = -2$.

Hence, $F(p_1) = \dfrac{p}{p+2}\bigg|_{p=-1} = -1$, $F(p_2) = \dfrac{p}{p+1}\bigg|_{p=-2} = 2$.

Therefore, $f(t) = -e^{-t} + 2e^{-t}$, $t > 0$.

第三部分 运筹与优化
Part 3 Operations Research and Optimization

引 言

这门学科、运筹学或管理科学(尽管可能存在的哲学差异,我们使用两个术语交替),已经被许多该领域的研究人员所认可。定义的范围从"科学的方法去做决定,"到"为源自现实生活的系统使用定量工具"、"科学的决策"及其他。在 1970 年代中期,美国的运营研究协会(当时的两大领域的专业协会)定义这门学科如下:"运筹学是涉及科学的决策:通常在稀缺资源需要分配情况下,如何最好的设计和操作人机系统。"

运筹学的根源可以追溯到几十年以前,当早期,曾试图在机构的管理方面采用科学的方法。然而,这项研究活动一开始就服务于第二次世界大战早期的美国军队。由于战争,迫切需要以有效方式分配在各种军事行动和活动中的稀缺资源。因此,英国和美国的管理者要求大量的科学家应用科学的方法来处理这个问题和其他战略和战术问题。实际上,要求他们对(军事)运作做研究。这些科学家们组成的团队是第一个 OR 团队。通过开发这些新雷达工具的有效使用方法,在赢得空中的英国战役中这些团队起了重要的作用。通过他们在如何更好地管理车队和反潜作战方面的研究,在赢得北大西洋的战争中也发挥了重大作用;类似的努力协助出现在太平洋岛战役。

运筹学主要研究经济活动和军事活动中能用数量来表达的有关策划、管理方面的问题。当然,随着客观实际的发展,运筹学的许多内容不但研究经济和军事活动,有些已经深入到日常生活当中去了。运筹学可以根据问题的要求,通过数学上的分析、运算,得出各种各样的结果,最后提出综合性的合理安排,以达到最好的效果。

运筹学作为一门用来解决实际问题的学科,在处理千差万别的各种问题时,一般有以下几个步骤:确定目标、制定方案、建立模型、制定解法。

随着科学技术和生产的发展,运筹学已渗入很多领域里,发挥了越来越重要的作用。运筹学本身也在不断发展,现在已经是一个包括好几个分支的数学部门了。比如:数学规划(又包含线性规划;非线性规划;整数规划;组合规划等)、图论、网络流、决策分析、排队论、可靠性数学理论、库存论、博弈论、搜索论、模拟等等。

事实上,运筹学已被广泛应用在不同领域。如制造,交通,建筑,电信,金融,医疗,军事,和公共服务。

Introduction

The subject matter, operations research or management science (even though there may be philosophical differences, we use the two terms interchangeably), has been defined by many researchers in the field. Definitions range from "a scientific approach to decision making," to "the use of quantitative tools for systems that originate from real life," "scientific decision making," and others. In the mid-1970s, the Operations Research Society of America (then one of the two large professional societies in the field) defined the subject matter as follows: "Operations Research is concerned with scientifically deciding how to best design and operate man-machine systems usually under conditions requiring the allocation of scarce resources."

The roots of Operations Research can be traced back many decades, when early attempts were made to use a scientific approach in the management of organizations. However, the beginning of the activity called operations research has generally been attributed to the military services early in World War II. Because of the war effort, there was an urgent need to allocate scarce resources to the various military operations and to the activities within each operation in an effective manner. Therefore, the British and the U.S. military management called upon a large number of scientists to apply a scientific approach to dealing with this and other strategic and tactical problems. In effect, they were asked to do research on (military) operations. These teams of scientists were the first OR teams. By developing effective methods of using the new tool of radar, these teams were instrumental in winning the Air Battle of Britain. Through their research on how to better manage convoy and antisubmarine operations, they also played a major role in winning the Battle of the North Atlantic. Similar efforts assisted the Island Campaign in the Pacific.

Operations Research studies mainly about planning, management problems in economic activities and military activities that can use the quantity relation express. Of course, with the objective reality of the development of operations research and management science, many content not only in the study of economic and military activities, some already penetrated into daily life. Operations Research can be based on the requirement of the problem, through mathematical analysis, operation, obtains various results. Finally, put forward the comprehensive and reasonable arrangement in order to achieve the best results.

Operations Research as a subject used to solve practical problems in the processing course, differ in thousands of ways of various problems, generally has the following stages: determine a goal, make plan, establish model, formulating method.

With the development of science and technology and the development of production, operations research has infiltrated into many fields, plays a more and more important role. Operations research itself is also in constant development, is now a consists of several branches of mathematics department. For example: mathematical programming (including linear programming; nonlinear programming; integer programming; combinatorial planning), graph theory, network flow, decision analysis, queuing theory, reliability theory, inventory theory, game theory, search theory, simulation and so on.

In fact, Operations Research has been applied extensively in such diverse areas as manufacturing, transportation, construction, telecommunications, financial planning, health care, the military, and public services, to name just a few. Therefore, the breadth of application is unusually wide.

第一章 线性规划模型
Chapter 1 THE LINEAR PROGRAMMING MODEL

线性规划的发展一直名列20世纪进展的最重要的科学,我们必须同意这一评估。自1950以来它的影响一直是巨大的。今天它是一个标准的工具,挽救了各工业化国家的许多数千或数百万美元的,甚至中等大小的公司或企业;它的应用在社会其他行业已迅速蔓延。大部分计算机上所有的科学计算专门是利用线性规划。许多教科书都写入线性规划,而发表文章描述其重要应用的有数百篇。事实上,从配置的生产设施,产品的分配到国内需求的国家资源分配,从投资组合到选择航运模式,从农业规划的设计到放射治疗等等,都离不开线性规划模型。

The development of linear programming has been ranked among the most important scientific advances of the mid-20th century, and we must agree with this assessment. Its impact since just 1950 has been extraordinary. Today it is a standard tool that has saved many thousands or millions of dollars for most companies or businesses of even moderate size in the various industrialized countries of the world; and its use in other sectors of society has been spreading rapidly. A major proportion of all scientific computation on computers is devoted to the use of linear programming. Dozens of textbooks have been written about linear programming, and published articles describing important applications now number in the hundreds. Indeed, ranging from the allocation of production facilities to products to the allocation of national resources to domestic needs, from portfolio selection to the selection of shipping patterns, from agricultural planning to the design of radiation therapy, and so on, are inseparable from the linear programming model.

单词和短语 Words and expressions

模型 model	线性规划 Linear programming

可行解	feasible solution	约束条件	constraint condition
可行域	feasible region	计划	schedule
最优解	optimal solution	原材料	raw material
目标函数	objective function	可获利的	profitable

基本概念和性质　Basic concepts and properties

the mathematical model for this general problem

$$\max \ (\min) \quad z = c_1 x_1 + c_2 x_2 + \cdots\cdots + c_n x_n$$

$$a_{11} x_1 + a_{12} x_2 + \cdots\cdots + a_{1n} x_n \leqslant (= \cdot \geqslant) b_1$$
$$\vdots \quad \vdots \quad \vdots \quad \vdots \quad \vdots \quad \vdots$$
$$a_{m1} x_1 + a_{m2} x_2 + \cdots\cdots + a_{mn} x_n \leqslant (= \cdot \geqslant) b_m$$
$$x_1 \geqslant 0, \cdots, \cdots, x_n \geqslant 0$$

A feasible solution is a solution for which all the constraints are satisfied.

An infeasible solution is a solution for which at least one constraint is violated.

The feasible region is the collection of all feasible solutions.

An optimal solution is a feasible solution that has the most favorable value of the objective function.

例　题　Examples

例 1 某工厂在计划期内要安排甲、乙两种产品的生产,已知生产单位产品所需的设备台时及 A、B 两种原材料的消耗以及资源的限制,如下表：

	甲	乙	资源限制
设备	1	1	300 台时
原料 A	2	1	400 千克
原料 B	0	1	250 千克
单位产品获利	50 元	100 元	

问题：工厂应分别生产多少单位甲、乙产品才能使工厂获利最多？

线性规划模型：

目标函数：Max　　　　　　$z = 50 x_1 + 100 x_2$

约束条件：s.t.　　　　　　$x_1 + x_2 \leqslant 300$

　　　　　　　　　　　　$2 x_1 + x_2 \leqslant 400$

　　　　　　　　　　　　$x_2 \leqslant 250$

　　　　　　　　　　　　$x_1, x_2 \geqslant 0$

Ex. 1 A factory in the planning period to schedule A, B two production, known manufacturing unit product required equipment platform and A, B two raw material consumption and resource constraints, the following table:

	A	B	Resource constraints
Equipment	1	1	300equipment number multtplied by time
raw material A	2	1	400kilogram
raw material B	0	1	250kilogram
Unit product profitability	50yuan	100yuan	

Question: how many units production A, B are factory products can make factory the most profitable?

Linear programming model:

Objective function: Max $z = 50x_1 + 100x_2$

Constraint condition: s.t. $x_1 + x_2 \leq 300$

$2x_1 + x_2 \leq 400$

$x_2 \leq 250$

$x_1, x_2 \geq 0$

例2 运输问题：甲、乙、丙3个煤矿供应3个城市用煤，各煤矿产量及各城市煤炭需求量见下表1，各煤矿到各城市之间的运输距离见表2，问如何安排调运计划，使总运输(周转)量最小。

煤矿	日产量(t)	城市	日需量(t)
甲	700	A	1200
乙	400	B	500
丙	900	C	300

城市煤矿	A	B	C
甲	30	110	100
乙	10	90	80
丙	70	40	50

线性规划模型：设 X_{ij} 为从第 i 个煤矿运往第 j 个城市的数量。

Ex. 2 Transportation problem: A, B, C three coal mine supply three city with coal, the coal mine production and the city's coal demand in Table 1 below, the distance between the coal mine to the city the transport see table 2, ask how to arrange the transportation plan, making the total transportation volume (turnover) minimum.

coal mine	Daily output(t)	city	On demand(t)
A	700	A	1200
B	400	B	500
C	900	C	300

城市煤矿	A1	B1	C1
A	30	110	100
B	10	90	80
C	70	40	50

Linear programming model: let X_{ij} is the decision variable from the i coal bound for the j city

$\min z = 30x_{11} + 110x_{12} + 100x_{13} + 10x_{21} + 90x_{22} + 80x_{23} + 70x_{31} + 40x_{32} + 50x_{33}$

$$s.t. \begin{cases} x_{11} + x_{12} + x_{13} = 700 \\ x_{21} + x_{22} + x_{23} = 400 \\ x_{31} + x_{32} + x_{33} = 900 \\ x_{11} + x_{21} + x_{31} = 1200 \\ x_{12} + x_{22} + x_{32} = 500 \\ x_{13} + x_{23} + x_{33} = 300 \\ x_{ij} \geq 0 (i,j = 1,2,3) \end{cases}$$

本章重点

1. 了解线性规划模型解决哪些实际问题；
2. 熟练建立线性规划模型的步骤。

Key points of this chapter

1. Understanding the linear programming model to how solve the practical problem;
2. Skilled steps of establishing a linear programming model.

第二章　线性规划的解法
Chapter 2　The Solving Methods of Linear Programming

我们现在已经准备好开始研究单纯形方法，它是一个解决线性规划问题的通用程序。它是由数学家乔治·丹茨在1947年开发的，已经被证明是一种非常有效的方法，用在当今的计算机上，经常解决今天的巨大问题。除了它对较少变量的问题的使用外，这种方法一直在计算机上实现，并加入到复杂的软件包内被广泛使用。单纯形方法的扩展和变化也可以用于执行职位最优分析（包括灵敏度分析）的模型。

单纯形法是一个代数过程。然而，其基本概念是几何。理解这些几何的概念，对如何运作单纯形方法以及如何使它更有效提供了一种强有力的直观感觉。因此，在深入钻研代数的细节之前，我们需要用几何的视点来关注大局。

We now are ready to begin studying the simplex method, a general procedure for solving linear programming problems. Developed by George Dantzig in 1947, it has proved to be a remarkably efficient method that is used routinely to solve huge problems on today's computers. Except for its use on tiny problems, this method is always executed on a computer, and sophisticated software packages are widely available. Extensions and variations of the simplex method also are used to perform post optimality analysis (including sensitivity analysis) on the model.

The simplex method is an algebraic procedure. However, its underlying concepts are geometric. Understanding these geometric concepts provides a strong intuitive feeling for how the simplex method operates and what makes it so efficient. Therefore, before delving into algebraic details, we focus on the big picture from a geometric viewpoint.

单词和短语　Words and expressions

单纯形法　simplex method
迭代　iteration
角点可行点　CPF solutions
初始化　initialization
增广的　augmented

松弛变量　slack variables
解释　interpretation
基本可行解　basic feasible (BF) solution
非基变量　nonbasic variables
调整　adjust

基本概念和性质　Basic concepts and properties

1 对于具有 n 个变量的线性规划问题，其每一个角点解都是 n 个约束边界的交点。
For a linear programming problem with n decision variables, each of its corner-point solutions lies at the intersection of n constraint boundaries.

2 一个增广解是松弛变量的对应值增广的原始变量（决策变量）解。
An augmented solution is a solution for the original variables (the decision variables) that has been augmented by the corresponding values of the slack variables.

3 一个基本解是一个增广的角点解。
A basic solution is an augmented corner-point solution.

4 一个基本可行解是一个增广的 CPF 解。
A basic feasible (BF) solution is an augmented CPF solution.

例题　Examples

例1　用图解法求解线性规划问题
Ex. 1　A graphical method to solve linear Programming problems

$$\max\ z = 2x_1 + x_2$$
$$\text{s.t.} \begin{cases} x_1 + 1.9x_2 \geq 3.8 \\ x_1 - 1.9x_2 \leq 3.8 \\ x_1 + 1.9x_2 \leq 10.2 \\ x_1 - 1.9x_2 \geq -3.8 \\ x_1, x_2 \geq 0 \end{cases}$$

解 (Solusion)：

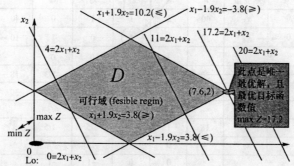

例 2 用单纯形法求下列线性规划的最优解

Ex. 2 solving the following linear programming by Simplex Method

$$\max Z = 3x_1 + 4x_2$$
$$\begin{cases} 2x_1 + x_2 \leq 40 \\ x_1 + 3x_2 \leq 30 \\ x_1, x_2 \geq 0 \end{cases}$$

解 (1)将问题化为标准型，加入松弛变量 x_3、x_4 则标准型为：

$$\max Z = 3x_1 + 4x_2$$
$$\begin{cases} 2x_1 + x_2 + x_3 = 40 \\ x_1 + 3x_2 + x_4 = 30 \\ x_1, x_2, x_3, x_4 \geq 0 \end{cases}$$

Solution: (1) Convert the problem into a standard, by adding slack variables x_3, x_4, then the standard for:

(2)求出线性规划的初始可行解，列出初始单纯形表。

(2)Find the initial basic feasible solution of the linear programming, listed in the initial simplex table.

c_j			3	4	0	0	
cB	基	b	x_1	x_2	x_3	x_4	θ_i
0	x_3	40	2	1	1	0	
0	x_4	30	1	3	0	1	
σ_j			3	4			

检验数

$\lambda_1 = c_1 - (c_3 a_{11} + c_4 a_{21}) = 3 - (0 \times 2 + 0 \times 1) = 3$

(3)进行最优性检验

(3)make an optimal test

（4）从一个基可行解转换到另一个目标值更大的基可行解，列出新的单纯形表。

(4) From a basic feasible solution converted to another target greater basic feasible solution, list the new simplex table.

	c_j		3	4	0	0	θ_i
c_B	基变量	b	x_1	x_2	x_3	x_4	
0	x_3	40	2	1	1	0	40
0	x_4	30	1	3	0	1	10
	σ_j		3	4	0	0	
0	x_3	30	5/3	0	1	−1/3	18
4	x_2	10	1/3	1	0	1/3	30
	σ_j		5/3	0	0	4/3	
3	x_1	18	1	0	3/5	−1/5	
4	x_2	4	0	1	−1/5	−2/5	
	σ_j		0	0	−1	−1	

（换入列：b_i/a_{i2}，$a_{i2}>0$；换出行；将3化为1；乘以1/3后得到）

（5）重复（3）、（4）步直到计算结束为止。

(5) Repeat (3)、(4) step until the end of the calculation.

本章重点

1 线性规划解的概念以及3个基本定理；
2 熟练掌握单纯形法的解题思路及求解步骤。

Key points of this chapter

1 The concept of a linear programming solutions as well as the three fundamental theorem;
2 Master problem-solving ideas and solving steps of the simplex method.

第三章　对偶理论与灵敏度分析
Chapter 3　Duality theory and sensitivity analysis

在线性规划早期的发展中，一个最重要的发现是对偶概念和它的许多重要成果。这个发现显示，每一个线性规划问题都同它的另一个称之为对偶线性规划问题相关联。对偶问题和原始问题（称为初始）之间的关系被证明在许多方面是非常有用的。

对偶理论的一个关键应用之一在于解释和执行灵敏度分析。灵敏度分析在几乎每一个线性规划研究中都是一个非常重要的部分。因为使用在原始模型中的大多数参数值仅仅是未来条件的估算，如果其他条件相反，则最优解的效果需要进一步研究。此外，某些参数的值（如资源数量）可能代表管理决策，在这种情况下，选择参数值可能是研究问题时主要的考虑，这可以通过灵敏度分析。

One of the most important discoveries in the early development of linear programming was the concept of duality and its many important ramifications. This discovery revealed that every linear programming problem has associated with it another linear programming problem called the dual. The relationships between the dual problem and the original problem (called the primal) prove to be extremely useful in a variety of ways.

One of the key uses of duality theory lies in the interpretation and implementation of sensitivity analysis. Sensitivity analysis is a very important part of almost every linear programming study. Because most of the parameter values used in the original model is just estimates of future conditions, the effect on the optimal solution if other conditions prevail instead needs to be investigated. Furthermore, certain parameter values (such as resource amounts) may represent managerial decisions, in which case the choice of the parameter values may be the main issue to be studied, which can be done through sensitivity analysis.

单词和短语 Words and expressions

对偶　　Duality
灵敏度　　sensitivity
参数　　parameter
标准型　　standard form
原始—对偶表　　primal-dual table

最优性条件　　Condition for Optimality
完备对偶　　complete dual
弱对偶　　Weak duality
互补解　　Complementary solutions
保持非负　　remain nonnegative

基本概念和性质 Basic concepts and properties

1 已知左边的原始问题（也许从另一种形式的转换后）的标准形式，其对偶问题为右侧所示的形式。

Given our standard form for the primal problem at the left (perhaps after conversion from another form), its dual problem has the form shown to the right.

$P: \max Z = C^T X$
$\begin{cases} AX \leqslant b \\ X \geqslant 0 \end{cases}$

$D: \min W = Y^T b$
$\begin{cases} A^T Y \geqslant C^T \\ Y \geqslant 0 \end{cases}$

2 弱对偶性质：如果 x 是原问题的一个可行解决，y 是对偶问题的一个可行解，则 $c^T x \leqslant y^T b$

Weak duality property: If x is a feasible solution for the primal problem and y is a feasible solution for the dual problem, then

$$c^T x \leqslant y^T b$$

3 强对偶性质：如果 $x*$ 是原问题的一个最优解，$y*$ 是对偶问题的一个最优解，则 $c^T x^* = y^{*T} b$

Strong duality property: If $x*$ is an optimal solution for the primal problem and $y*$ is an optimal solution for the dual problem, then

$$c^T x^* = y^{*T} b$$

例题 Examples

例 1 写出下述线性规划的对偶问题

Ex.1 Write the duality of the following linear programming

$\max z = 2x_1 + x_2$
$\begin{cases} 5x_2 \leqslant 15 \\ 6x_1 + 2x_2 \leqslant 24 \\ x_1 + x_2 \leqslant 5 \\ x_1, x_2 \geqslant 0 \end{cases}$

$\min w = 15y_1 + 24y_2 + 5y_3$
$\begin{cases} 6y_2 + y_3 \geqslant 2 \\ 5y_1 + 2y_2 + y_3 \geqslant 1 \\ y_1, y_2, y_3 \geqslant 0 \end{cases}$

例 2 已知线性规划问题（Known linear programming problem）：

$\max z = 3x_1 + 4x_2 + x_3$
$\begin{cases} x_1 + 2x_2 + x_3 \leqslant 10 \\ 2x_1 + 2x_2 + x_3 \leqslant 16 \\ x_j \geqslant 0, j = 1, 2, 3 \end{cases}$

它的最优解是 $X^* = (6, 2, 0)^T$，求其对偶问题的最优解 Y^*。

Its optimal solution is $X^* = (6, 2, 0)^T$, to find the optimal solution Y^* of the dual problem.

解 写出原问题的对偶问题，即

Solution: to write the dual problem of the original problem, namely,

$$\min w = 10y_1 + 16y_2$$
$$\begin{cases} y_1 + 2y_2 - y_3 = 3 \\ 2y_1 + 2y_2 - y_4 = 4 \\ y_1 + y_2 - y_5 = 1 \\ y_1, y_2, y_3, y_4, y_5 \geq 0 \end{cases} \xrightarrow{\text{标准化}} \begin{cases} \min w = 10y_1 + 16y_2 \\ y_1 + 2y_2 \geq 3 \\ 2y_1 + 2y_2 \geq 4 \\ y_1 + y_2 \geq 1 \\ y_1, y_2 \geq 0 \end{cases}$$

设对偶问题最优解为 $Y^* = (y_1, y_2)$，由互补松弛性定理可知，X^* 和 Y^* 满足：

Set up the optimal solution of the dual problem is $Y^* = (y_1, y_2)^T$, by the complementary slackness theorem shows that X^* and Y^* to meet：

$$(y_3, y_4, y_5)(x_1, x_2, x_3)^T = 0$$
$$(y_1, y_2)(x_4, x_5)^T = 0$$

因为 $X_1 \neq 0, X_2 \neq 0$，所以对偶问题的第一、二个约束的松弛变量等于零，即 $y_3 = 0, y_4 = 0$，带入方程中：

Since $X_1 \neq 0, X_2 \neq 0$, so, First, two constraint slack variables of the dual problem equal to zero, i.e., $y_3 = 0, y_4 = 0$, into the equation：

$$\begin{cases} y_1 + 2y_2 = 3 \\ 2y_1 + 2y_2 = 4 \end{cases}$$

解此线性方程组得 $y_1 = 1, y_2 = 1$，从而对偶问题的最优解为：$Y^* = (1,1)^T$，最优值 $w = 26$。

The solution of linear equations $y_1 = 1$, $y_2 = 1$, the optimal solution of dual problem is：$Y^* = (1, 1)^T$, the optimal value is $w = 26$.

本章重点

1. 熟悉对偶问题与原问题的对应关系及对偶问题的经济意义；
2. 理解影子价格的含义，了解影子价格的应用；
3. 熟悉灵敏度分析的概念和内容，掌握灵敏度分析的基本方法。

Key points of this chapter

1. Familiar with the corresponding relation of the dual problem and the original problem, and the economic meaning of dual problem；
2. Understand the meaning of shadow price and the application of shadow price；
3. Familiar with the concept and contents of sensitivity analysis, and master the basic method of sensitivity analysis.

第四章 运输问题及其解法
Chapter 4 Transportation problem and its solution

第 2 章强调了线性规划广泛的适用性。在本章，我们将继续扩大我们的视野，讨论两个特别重要的类型（及相关）的线性规划问题。一个类型，即所谓的交通问题，得到这个名字是因为许多应用涉及如何以最佳方式运送货物。进一步，它的一些重要的应用（例如，生产调度）与运输无关。

第二类称为指派问题，涉及到分配任务到人这样的一些应用。虽然它的应用看起来不同于交通问题，但我们也可以把指派问题作为特殊类型的交通问题。

运输和分配问题的应用往往需要非常多的约束和变量，所以单纯形法在计算机上的一个直接应用可能需要高昂的计算代价。幸运的是，这些问题的主要特点是：在约束中的大部分 a_{ij} 系数是零和相当少的非零系数以一个独特的模式出现。于是，这就有可能发展为几种特殊简化算法，该算法利用这个问题的特殊结构，实现技巧性的计算储蓄。因此，重要的是要充分熟悉这些特殊类型的问题，当他们出现和运用正确的计算过程时，你能够识别出他们来。

Chapter 2 emphasized the wide applicability of linear programming. We continue to broaden our horizons in this chapter by discussing two particularly important (and related) types of linear programming problems. One type, called the transportation problem, received this name because many of its applications involve determining how to optimally transport goods. However, some of its important applications (e. g. , production scheduling) actually have nothing to do with transportation.

The second type, called the assignment problem, involves such applications as assigning people to tasks. Although its applications appear to be quite different from those for the transportation problem, we shall see that the assignment problem can be viewed as a special type of transportation problem.

Applications of the transportation and assignment problems tend to require a very large number of constraints and variables, so a straightforward computer application of the simplex method may require an exorbitant computational effort. Fortunately, a key characteristic of these problems is that most of the a_{ij} coefficients in the constraints are zeros, and the relatively few nonzero coefficients appear in a distinctive pattern. As a result, it has been possible to develop special streamlined algorithms that achieve dramatic computational savings by exploiting this special structure of the problem. Therefore, it is important to become sufficiently familiar with these special types of problems that you can recognize them when they arise and apply the proper computational procedure.

单词和短语 Words and expressions

运输	transportation	供应	supply
指派	assignment	需求	demand
简洁的	streamlined	参数表	a parameter table
网络单纯性型	network simplex method	虚拟的	Dummy
资源	sources	操作	manipulations

基本概念和性质 Basic conception and Properties

1 需求的假设：每个资源都有一个固定的供应单位，这整个供应必须分发到目的地。（我们让 s_i 表示由资源 i 提供的单位数目，$i=1,2,\ldots,m$）。同样地，每个目的地都有一个固定单位的需求，而整个的需求必须从这些资源上得到。（我们让 d_j 表示由目的地 j 收到的单位数量，$j=1,2,\ldots,n$。）

The requirements assumption: Each source has a fixed supply of units, where this entire supply must be distributed to the destinations. (We let s_i denote the number of units being supplied by source i, for i =1, 2... m.) Similarly, each destination has a fixed demand for units, where this entire demand must be received from the sources. (We let d_j denote the number of units being received by destination j, for j = 1, 2, ..., n.)

2 可行解属性：交通问题将有可行的解决方案，当且仅当

$$\sum_{i=1}^{m} s_i = \sum_{j=1}^{n} d_j$$

The feasible solutions property: a transportation problem will have feasible solutions if and only if

$$\sum_{i=1}^{m} s_i = \sum_{j=1}^{n} d_j$$

3 成本假设：从任何特定资源分配单位成本是和单位分布的数量成正比。因此，这种成本仅仅是分布的单位成本乘以单位分布的数目。（我们让 c_{ij} 表示资源 i 与分布 j 的这个单位成本。）

The cost assumption: the cost of distributing units from any particular source to any particular destination is directly proportional to the number of units distributed. Therefore, this cost is just the unit cost of distribution times the number of units distributed. (We let c_{ij} denote this unit cost for source i and destination j.)

4 整数解的属性：对于交通问题的每个 s_i 及 d_j 都有一个整数的值，在每一个基本可行（BF）解（包括一个最优解）中的所有的基本变量也具有整数值。

Integer solutions property: for transportation problems where every s_i and d_j have an integer value, all the basic variables (allocations) in every basic feasible (BF) solution (including an optimal one) also have integer values.

例 题 Examples

例 1 某公司从两个产地 A_1、A_2 将物品运往三个销地 B_1，B_2，B_3，各产地的产量、各销地的销量和各产地运往各销地每件物品的运费如下表所示，问：应如何调运可使总运输费用最小？

Ex. 1 A company transport the three sales to B_1, B_2, B3 from two origin A_1, A_2, the production of various origin, the sales of various marketing and the freight of each item from various origin destined to each sale in the following table as shown, in Q: how should transporting enable minimum transportation costs?

	B_1	B_2	B_3	产量 production
A_1	6	4	6	200
A_2	6	5	5	300
销量 sales	150	150	200	

解 产销平衡问题：总产量＝总销量＝500

设 x_{ij} 为从产地 A_i 运往销地 B_j 的运输量，得到下列运输量表：

Solution: the balance between production and sales: total production = total sales = 500

Let x_{ij} is the transport amount of Origin Ai destined for sales to Bj, Given the following transport Scale:

	B_1	B_2	B_3	产量 production
A_1	x_{11}	x_{12}	x_{13}	200
A_2	x_{21}	x_{22}	x_{23}	300
销量 sales	150	150	200	

运输问题的一般形式：产销平衡

The general form of the transport problem: balance production and sales.

$$\text{Min } C = 6x_{11} + 4x_{12} + 6x_{13} + 6x_{21} + 5x_{22} + 5x_{23}$$

$$\text{s. t.} \quad \begin{aligned} x_{11} + x_{12} + x_{13} &= 200 \\ x_{21} + x_{22} + x_{23} &= 300 \\ x_{11} + x_{21} &= 150 \\ x_{12} + x_{22} &= 150 \\ x_{13} + x_{23} &= 200 \\ x_{ij} \geq 0 \quad (i = 1,2; &\ j = 1,2,3) \end{aligned}$$

一般地，设 $A_1, A_2, \cdots, A_m$ 表示某物资的 m 个产地；$B_1, B_2, \cdots, B_n$ 表示某物质的 n 个销地；a_i 表示产地 A_i 的产量；b_j 表示销地 B_j 的销量；c_{ij} 表示把物资从产地 A_i 运往销地 B_j 的单位运价。设 x_{ij} 为从产地 A_i 运往销地 B_j 的运输量，得到下列一般运输量问题的模型：

In general, assume that $A_1, A_2, \ldots, A_m$, denote the m material origins; $B_1, B_2, \ldots, B_n$, denote the n sales to a substance; a_i said that the yield of the origin A_i; b_j said the sales volume of sales to Bj; c_{ij} denotes unit freight from the origin A_i shipped off to B_j. X_{ij} is the amount of sales to be shipped from the origin A_i transport B_j, the following general traffic model:

$$\text{s. t} \begin{cases} \sum_{j=1}^{n} x_{ij} = a_i & i = 1, \cdots, m \\ \sum_{i=1}^{m} x_{ij} = b_j & j = 1, \cdots, n \\ x_{ij} \geq 0, & i = 1, \cdots, m; j = 1, \cdots, n \end{cases}$$

变化：

(1)有时目标函数求最大。如求利润最大或营业额最大等；

(2)当某些运输线路上的能力有限制时，在模型中直接加入约束条件(等式或不等式约束)；

(3)产销不平衡时，可加入假想的产地(销大于产时)或销地(产大于销时)。

Changes：

(1) Sometimes the objective functions for the largest. Such as to maximize profit or turnover, etc.;

(2) When there are the capacity limitations on certain transportation routes, the constraints (equality or inequality constraints) are directly added in the model;

(3) Production and marketing imbalance, can be added to the supposed place of origin (sales greater than production) or sales to (capacity greater than sales).

例 2 在例 1 的表中，如产量大于销量，各产地的产量、各销地的销量和各产地运往各销地每件物品的运费如下表所示，问：应如何调运可使总运输费用最小？

Ex. 2 In the table of Example 1, such as yield greater than the sales, how should transporting enable minimum transportation costs？

	B_1	B_2	B_3	B_4	产量 production
A_1	6	4	6	0	300
A_2	6	5	5	0	300
销量 sales	150	150	200	100	

解 增加一个虚设的销地运输费用为 0，其他同前。

Add a dummy pins, the cost of transportation for 0, Other with the former.

本章重点

1. 熟悉产销平衡与产销不平衡模型的建立；
2. 掌握用表上作业法求解运输问题的方法；
3. 掌握求检验数的两种方法—闭回路法和位势法

Key points of this chapter

1. Familiar with the models for the balance of produce and sale and unbalanced production-marketing;
2. Grasp the table-manipulation method for solving transportation problem;
3. Master two kinds of methods for testing closed loop method and potential method.

第五章 多目标规划
Chapter 5　5 Multi-objective Programming

作为已经在前面章节里提出的不同问题，他们共享一个共同的特点：它们都具有一个单一的目标函数，其结果是最佳的解决方案(或多个最优，在双退化的情况下)。然而，最优的概念只适用于一个目标的情况下。如果我们声明某个东西是"最好"或最优，我们始终有一个心中的目标：最快的车，最舒适的车辆，汽车是最便宜的操作，等等。每当第二个甚至更多的目标包括在一个问题里，最优的概念不再适用。例如，如果车辆的最高速度和它的耗油量是要同时关注的，那么用一辆车做比较，其速度可能会高达每小时 110 英里和每 20 英里一加仑(公路等级)和车辆可能会高达每小时 90 英里和每 25 英里一加仑不再是一个简单的问题：前车是在牺牲燃油效率快。它现在将取决于决策者的两个标准，这被认为是更重要的。换句话说，决策者将迟早有指定的标准之间的权衡。这是在这一章中考虑的问题类型。

而在这一领域的术语不太规范，我们通常指的问题是作为多准则决策形成的多评价标准，或者多目标决策。这些问题主要有两个子类，一个叫多属性决策（MADM 的），另一个是多目标（线性）规划（MOLP）。

As diverse as the problems in the previous chapters have been, they share one common feature: they all have one single objective function and the result is an optimal solution (or multiple optima, in case of dual degeneracy). However, the concept of optimality applies only in case of a single objective. If we state that something is "the best" or optimal, we always have an objective in mind: the fastest car, the most comfortable vehicle, the automobile that is cheapest to operate, and so forth. Whenever a second or even more objectives are included in a problem, the concept of optimality no longer applies. For instance, if the top speed of a vehicle and its gas mileage are relevant concerns, then the comparison between a car, whose speed may go up to 110 miles per hour and which gives 20 miles to the gallon (highway rating) and a vehicle that can go up to 90 miles per hour and which gives 25 miles to the gallon is no longer a simple one: the former car is faster at the expense of fuel efficiency. It will now depend on the decision maker which of the two criteria is considered more important. In other words, the decision maker will sooner or later have to specify a tradeoff between the criteria. This is the type of problems considered in this chapter.

While the terminology in this field is not quite standardized, we typically refer to problems with multiple objectives or evaluation criteria as multi-criteria decision making problems or MCDM. There are two major subclasses of these problems, one called multi-attribute decision making (MADM), and the other being multiobjective (linear) programming (MOLP).

单词和短语 Words and expressions

多目标的　　Multiobjective　　　　　　加权法　　weighting method
多准则的　　multicriteria　　　　　　　目标规划　　Goal Programming
改善锥　　　improvement cone　　　　　公度（成比例，相称）　commensurability
平凡的　　　nondominated　　　　　　　图示的　　graphical
帕累托最优　pareto-optimal　　　　　　　分离图　　separate graph

基本概念和性质　Basic concepts and properties

1 如果将 X 向其改进锥移动总是会导致丧失可行性，则点 X 被称为非支配。

If moving out of X into its improvement cone will always result in the loss of feasibility, then the point X is called nondominated .

2 所有非支配点的集合称为有效前沿。

The collection of all nondominated points is called the efficient frontier.

3 d^+ —— 超出目标的偏差，称正偏差变量。

Deviation above the target, said the positive deviation variable.

4 d^- —— 未达到目标的偏差，称负偏差变量。

Does not reach the target deviation, known as negative deviation variable.

5 在一个目标规划的模型中，为达到某一目标可牺牲其他一些目标，称这些目标是属于不同层次的优先级。优先级层次的高低可分别通过优先因子 $P1, P2, \cdots$ 表示。对于同一层次优先级的不同目标，按其重要程度可分别乘上不同的权系数。权系数是一个个具体数字，乘上的权系数越大，表明该目标越重要。

In a goal programming model, in order to achieve some goal may sacrifice some other targets, say the goal is to belong to different levels of priority. Priority level can be expressed respectively by priority factors P1, P2 ... For the same level of priority of the different target, according to the important degree; can be respectively by different weight coefficient. Weight coefficient is a specific digital, multiplied by the weight coefficient is bigger, show that the target is more important.

例题　Examples

例 1 用图解法求解下列目标规划问题
Using the graphic method to solve the following goal programming problem

$$\min z = P_1 d_1^+ + P_2(d_2^- + d_2^+) + P_3 d_3^-$$

$$\begin{cases} 2x_1 + x_2 \leq 11 & (1) \\ x_1 - x_2 + d_1^- - d_1^+ = 0 & (2) \\ x_1 + 2x_2 + d_2^- - d_2^+ = 10 & (3) \\ 8x_1 + 10x_2 + d_3^- - d_3^+ = 56 & (4) \\ x_1, x_2, d_i^-, d_i^+ \geq 0 \, (i=1,2,3) \end{cases}$$

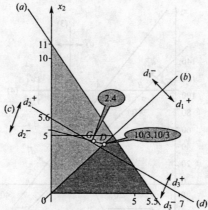

满意解是线段 GD 上任意点，其中 G 点 $X=(2,4)^T$，D 点 $X=(10/3,10/3)^T$。
Satisfactory solution is arbitrary point on GD segment which G point, $X=(2,4)^T$, D point $X=(10/3,10/3)^T$.

例 2 已知一个生产计划的线性规划模型如下，其中目标函数为总利润，x_1, x_2 为产品 A、B 产量。
Given a production planning model is as follows, in which the objective function is the total profit, x_1, x_2 are yields for A, B products.

$$\max Z = 30x_1 + 12x_2$$

$$\begin{cases} 2x_1 + x_2 \leq 140 & (甲资源) \\ x_1 \leq 60 & (乙资源) \\ x_2 \leq 100 & (丙资源) \\ x_{1-2} \geq 0 \end{cases}$$

现有下列目标：
1. 要求总利润必须超过 2500 元；
2. 考虑产品受市场影响，为避免积压，A、B 的生产量不超过 60 件和 100 件；
3. 由于 A 资源供应比较紧张，不要超过现有量 140。
试建立目标规划模型，并用图解法求解。

The following objectives:
1. Requirements total profit must be more than 2500 Yuan;
2. Consider the product subject to market, in order to avoid the backlog, A, B production of no more than 60 pieces and 100 pieces;

3. Due to the A resource is in short supply, do not exceed the available volume 140.
Try to establish the model of goal programming, and using the graphic method to solve.

解 以产品 A,B 的单件利润比 2.5 :1 为权系数,模型如下:

Solution: the profit ratio of 2.5:1 of a single piece for product A, B is the weight coefficient, the model is as follows:

$$\min Z = P_1 d_1^- + 2.5 P_2 d_3^+ + P_2 d_4^+ + P_3 d_2^+$$

$$\begin{cases} 30 x_1 + 12 x_2 + d_1^- - d_1^+ = 2500 \\ 2 x_1 + x_2 + d_2^- - d_2^+ = 140 \\ x_1 + d_3^- - d_3^+ = 60 \\ x_2 + d_4^- - d_4^+ = 100 \\ x_1 \leqslant 60 \\ x_2 \leqslant 100 \\ x_{1-2} \geqslant 0, d_l^+, d_l^- \geqslant 0 (l = 1,2,3,4) \end{cases}$$

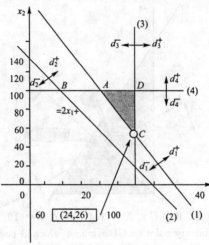

Solution C $(60, 58.3)^T$ is the demand satisfaction.

本章重点

① 熟悉企业经营中多目标模型的问题;
② 学会用偏差变量处理目标和约束;
③ 学会正确设置目标的优先级与权系数;
④ 掌握用图解法法求解多目标规划。

Key points of this chapter

① Familiar with the business object model problem;
② Learn to use deviation variable treatment to goals and constraints;
③ Learn how to correctly set the target priority and weight coefficient;
④ Master of graphic method for solving multiple objective programming.

第六章 整数规划
Chapter 6 Integer Programming

在第 2 章里,你已看到几个例子中的各种不同线性规划的应用。然而,阻碍更多应用的一个关键性限制是可分性的假设,这需要决策变量中允许非整数值的存在。在许多实际问题中,只有决策变量为整数值时，才有意义的。例如，在指派人，机器，和车辆的活动中，整数数量常常是必要的。如果需要整数的值是唯一的方式，则使该问题从线性规划的公式中偏离出来，那么它是一个整数规划问题（IP）。（更完整的名字是整数线性规划，但线性的这个形容词通常是指下降的，除非这个问题是对比更深奥的整数非线性规划问题，这就超出了本书的范围。）

整数线性规划问题的种类：

纯整数线性规划：指全部决策变量都必须取整数值的整数线性规划。混合整数线性规划：决策变量中有一部分必须取整数值，另一部分可以不取整数值的整数线性规划。0—1型整数线性规划：决策变量只能取值0或1的整数线性规划。

In Chap. 2 you saw several examples of the numerous and diverse applications of linear programming. However, one key limitation that prevents many more applications is the assumption of divisibility, which requires that no integer values be permissible for decision variables. In many practical problems, the decision variables actually make sense only if they have integer values. For example, it is often necessary to assign people, machines, and vehicles to activities in integer quantities. If requiring integer values is the only way in which a problem deviates from a linear programming formulation, then it is an integer programming (IP) problem. (The more complete name is integer linear programming, but the adjective linear normally is dropped except when this problem is contrasted with the more esoteric integer nonlinear programming problem, which is beyond the scope of this book.)

Integer linear programming question types：

Pure integer linear programming：refers to all of the decision variables must be integer values of linear integer programming；Mixed integer linear programming：decision variables are in part must take integer values, the other part can not take integer values in linear programming；0-1Integer linear programming：decision variables can only be a value of 0 or 1 in linear programming.

单词和短语 Words and expressions

线性规划	Integer Programming	应急决策	contingent decisions
分离	divisibility	二元决策	BIP
偏离	deviates from	互补的	auxiliary
混合线性规划	MIP	割平面法	cutting plane techniques
二元变量	binary variables	二进制表示	binary representation
分支定界法	branch and bound method	是否类型	yes-or-no type

基本概念和性质　Basic concepts and properties

1 纯整数线性规划：指全部决策变量都必须取整数值的整数线性规划。

Pure integer linear programming：refers to all of the decision variables must be integer values of linear integer programming.

2 混合整数线性规划：决策变量中有一部分必须取整数值，另一部分可以不取整数值的整数线性规划。

Mixed integer linear programming：decision variables are in part must take integer values, the other part can not take integer values in linear programming.

3 0-1型整数线性规划：决策变量只能取值0或1的整数线性规划。

0-1Integer linear programming：decision variables can only be a value of 0 or 1 in linear programming.

4 整数规划问题的可行解集合是它松弛问题可行解集合的一个子集，任意两个可行解的凸组合不一定满足整数约束条件，因而不一定仍为可行解。

Feasible solution set of the integer programming problem is a subset of the feasible solution set for its relaxed problem, convex combination of any two feasible solutions does not necessarily meet the integer constraints, and thus not necessarily still feasible solution.

5 整数规划问题的可行解一定是它的松弛问题的可行解（反之不一定），但其最优解的目标函数值不会优于后者最优解的目标函数值。

A feasible solution of the integer programming problem must be the feasible solutions of the relaxation (not necessarily vice versa), but the goal function value of the optimal solution is not superior to the objective function value of the latter optimal solution.

例 题 Examples

例 1 用分支定界法求解下面整数规划

Ex. 1 A branch and bound method to solve the following integer programming

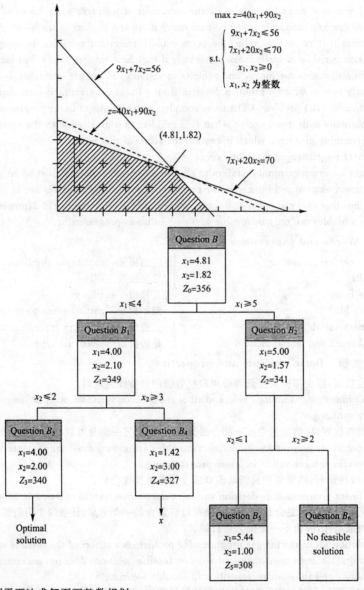

例 2 用割平面法求解下面整数规划

Ex. 2 Using the cutting plane techniques for solving the integer programming

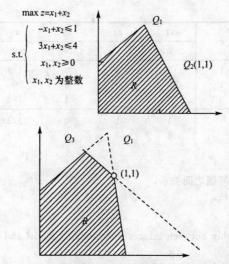

Simplex table calculation

行	x	c	x_1	x_2	x_3	x_4	b
			1	1	0	0	
1	x_1	1	1	0	$-1/4$	$1/4$	$3/4$
2	x_2	1	0	1	$3/4$	$1/4$	$7/4$
4		z_j	1	1	$1/2$	$1/2$	$5/2$
5		$z_j - c_j$	0	0	$1/2$	$1/2$	

$$x_1 - \frac{1}{4}x_3 + \frac{1}{4}x_4 = \frac{3}{4} \quad \cdot \quad x_1 = 3/4, x_2 = 7/4$$

$$x_2 + \frac{3}{4}x_3 + \frac{1}{4}x_4 = \frac{7}{4} \quad \cdot \quad \max z = 5/2$$

将系数和常数项都分解成整数和非负真分数之和,移项上两式变为:
Divided the coefficients and constant terms into the sum of integer and non-negative proper fraction, transposed into:

$$x_1 - x_3 = \frac{3}{4} - \left(\frac{3}{4}x_3 + \frac{1}{4}x_4\right)$$

$$x_2 - 1 = \frac{3}{4} - \left(\frac{3}{4}x_3 + \frac{1}{4}x_4\right)$$

再考虑整数约束,由模型知所有变量都是非负整数,所以等式左边是整数;等式右边是一个分数与一个正数的差值,差值应为整数,所以等式两端只能取小于或等于 0 的整数值。
Consider the integer constraints, known by the model all variables are non-negative integer, so the left of the equation is an integer; the right hand side is the difference of a fraction with a positive, the difference should be an integer, so the equation at both ends can only take the integer value with a small than or equal to 0.

$$\frac{3}{4} - \left(\frac{3}{4}x_3 + \frac{1}{4}x_4\right) \leqslant 0$$

即
$$-3x_3 - x_4 \leqslant -3$$

称为切割方程,作为增加的约束条件,增加松弛变量 x_5,单纯形表计算:
Called cutting equation as additional constraints to increase the slack variable x_5, simple tableau cal-

culation:

行	x	c	x_1	x_2	x_3	x_4	x_5	b
			1	1	0	0	0	
1	x_1	1	1	0	0	1/3	−1/12	1
2	x_2	1	0	0	0	0	1/4	1
3	x_3	0	0	1	1	1/3	−1/3	1
4	z_j		1	1	0	1/3	1/3	2
5	$z_j - c_j$		0	0	0	1/3	1/3	

本章重点

1. 整数规划模型与相应松弛问题之间关系;
2. 掌握分支定界法和割平面法。

Key points of this chapter

1. Familiar with the relationship between integer programming model and the corresponding relaxation problem;
2. Master the branch and bound method and cut plane techniques.

第七章 动态规划
Chapter 7 Dynamic Programming

动态规划是一个非常有用的数学方法,它形成一系列相互关联的决策。它为确定最佳组合决策提供了一个系统化的过程。

同线性规划相比,动态规划问题不存在一个"标准"的数学公式。相反,动态规划是解决问题的一般类型方法,并且使用过的特定方程必须改进以适应每一种情况。因此,对动态规划问题一般结构形成的一定程度的独创性与洞察力是何时以及如何通过动态规划程序可以解决问题必须认识到的。这些能力最好通过接触各种各样的动态编程的应用来发展,同时研究所有这些情况公共的特点。有大量事例说明了这一点。

Dynamic programming is a useful mathematical technique for making a sequence of interrelated decisions. It provides a systematic procedure for determining the optimal combination of decisions.

In contrast to linear programming, there does not exist a "standard" mathematical formulation of the dynamic programming problem. Rather, dynamic programming is a general type of approach to problem solving, and the particular equations used must be developed to fit each situation. Therefore, a certain degree of ingenuity and insight into the general structure of dynamic programming problems is required to recognize when and how a problem can be solved by dynamic programming procedures. These abilities can best be developed by an exposure to a wide variety of dynamic programming applications and a study of the characteristics that are common to all these situations. A large number of illustrative examples are presented for this purpose.

单词和短语 Words and expressions

动态规划　Dynamic Programming
驿站马车问题　Stagecoach Problem
直接成本　immediate cost
阶段　stages
最优策略　optimal policy
马尔科夫性质　Markovian property

递归关系　recursive relationship
逐步　stage by stage
概率的　probabilistic
用图解　diagrammatically
对策　prescription
均衡(性)　proportionality

基本概念和性质 Basic concepts and properties

1 动态规划可以分为阶段,每个阶段都需要策略决策。
Dynamic Programming can be divided into stages, with a policy decision required at each stage.

2 从一开始,每个阶段都和许多阶段相关联。
Each stage has a number of states associated with the beginning of that stage.

3 在每个阶段,策略决策的效果是把目前的状态转移到与下一状态开始相关的某个状态(可能要根据一个概率分布)。
The effect of the policy decision at each stage is to transform the current state to a state associated with the beginning of the next stage (possibly according to a probability distribution).

4 设计求解过程是为了找到一个整体问题的最优策略,即在每个阶段的对策为每一个可能状态的最优策略。
The solution procedure is designed to find an optimal policy for the overall problem, i.e., a prescription of the optimal policy decision at each stage for each of the possible states.

5 已知目前的状态,其余阶段的最优策略是独立于前阶段采纳的策略。因此,最佳的直接决策只取决于当前状态而不去管你如何到达那里。这是动态规划的最优性原则。
Given the current state, an optimal policy for the remaining stages is independent of the policy decisions adopted in previous stages. Therefore, the optimal immediate decision depends on only the current state and not on how you got there. This is the principle of optimality for dynamic programming.

6 求解过程是由最后一个阶段找到最优策略开始。
The solution procedure begins by finding the optimal policy for the last stage.

例题 Examples

例 一个线路网络图,从 A 到 E 要修建一条石油管道,必须 在 B、C、D 处设立加压站。各边上的数为长度,现需要找一条路使总长度最短。

A line network diagram, from A to E to the construction of an oil pipeline, the pressure station must be set up in B, C, and D. The number of sides is the length, now need to find a way to reach the total length of the shortest.

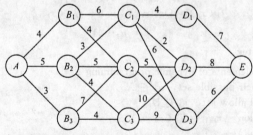

解 (solution):可分成 4 个阶段:A 到 B、B 到 C、C 到 D、D 到 E;
每个阶段 k 的起点称为状态 S_k;
从 k 阶段的起点出发可以做一选择,即决定到下一阶段的哪个节点,称为决策 X_k;
可见,S_k+1 是由 S_k 和 X_k 所决定的。
那麼,从 A 出发经过 4 个阶段:A 到 B、B 到 C、C 到 D、D 到 E,逐次作出决策,构成从 A 到 E 的一条路线,记为 u。即,
$u = S_1 \ X_1 \ S_2 \ X_2 \ S_3 \ X_3 \ S_4 \ X_4 \ S_5$ 其中 $S_1 = A, S_5 = E$
记 d 为两个相邻节点之间的长度,如 $d(A, B3) = 3$。
① 记 $f_k(S_k)$ 为从 S_k 到 E 的最短长度,称为从 S_k 到 E 的距离。
那么,$f_1(A)$ 是从 A 到 E 的最短距离,即最优策略的值。
② 最短路问题的特点:如果从 A 到 E 的最优策略经过某节点,那么这个策略的从该节点到 E 的一

段,必定是该节点到 E 的所有线路中 S_k 最短的一条,即这一段的长度为 $f_k(S_k)$。
　　(1)逆序法:从 E 到 A
　　(2)顺序法:对节点 S_k,从 A 到 S_k 所有线路中,最短的一条的长度记为 $\varphi_k(S_k)$,例如 $\varphi_1(A)=0$,称为问题的边界条件。

The process can be divided into 4 stages: A to B, B to C, and C to D, D to E;
The starting point of each stage k called state S_k;
From the starting point of k stage can make a choice, which decided to which node of the next stage, called the decision in X_k; Visible, S_k+1 is decided by S_k and X_k. Then, starting from the A through 4 stages: A to B, B to C, C to D, D to E, successive decisions, constituted a route from A to E, denoted by U. That is,
$$U = S_1 X_1 S_2 X_2 S_3 X_3 S_4 X_4 S_5 \text{ where } S_1 = A, S_5 = E$$
Denote that D is the length between two adjacent nodes, such as D$(A, B_3) = 3$.
① Denote that $f_k(S_k)$ is the shortest length from S_k to E, called the distance from S_k to E. So, F1(A) is the shortest distance from A to E, i. e. optimal strategic value y.
② The shortest path characteristics: if the optimal strategy from A to E passes through a node, then this segment of the strategy from the node to a E must be one of the shortest S_k from the node to E all line, i. e. The length is of $f_k(S_k)$.
(1) The method of reverse order: from E to A
(2) sequential method: on node S_k, through all lines from A to S_k, one of the shortest length recorded for $\varphi_k(S_k)$, for example a $\varphi_1(A) = 0$, known as f the boundary condition of the problem.

本章重点

1 熟悉动态规划的四大要素:
　　① 状态变量及其可能集合　　$x_k \in X_k$
　　② 决策变量及其允许集合　　$u_k \in U_k$
　　③ 状态转移方程　　$x_{k+1} = T_k(x_k, u_k)$
　　④ 阶段效应　　$r_k(x_k, u_k)$
2 掌握动态规划基本方程:
$$f_{n+1}(x_{n+1}) = 0 (边界条件)$$
$$f_k(x_k) = \underset{u}{\mathrm{Opt}}\{r_k(x_k, u_k) + f_{k+1}(x_{k+1})\}$$
$$k = n, \cdots, 1$$

Key points of this chapter

1 Familiar with the four elements of dynamic programming;
　　① state variables and their possible set　　$x_k \in X_k$
　　② decision variables and allow set　　$u_k \in U_k$
　　③ state transition equation　　$x_{k+1} = T_k(x_k, u_k)$
　　④ stage effect　　$r_k(x_k, u_k)$
2 Master the basic equation of dynamic programming.
$$f_{n+1}(x_{n+1}) = 0$$
$$f_k(x_k) = \underset{u}{\mathrm{Opt}}\{r_k(x_k, u_k) + f_{k+1}(x_{k+1})\}$$
$$k = n, \cdots, 1$$

第八章　图与网络分析
Chapter 8　Graph Theory and Network Analysis

　　图论的历史可以追溯到 1736 年,当时瑞士数学家伦纳德·欧拉研究了著名的"哥尼斯堡 7 桥"难题。当时有七座桥梁横跨在穿过哥尼斯堡城的普雷格尔河上,欧拉猜想是否有可能开始从城市的某个地方开始,穿越每个桥梁只走一次,并返回到起始点。在经过一年的研究之后,29 岁的欧拉提交了《哥尼斯堡七桥》的论文,圆满解决了这一问题,同时开创了数学新一分支——图论。

二百年后的1936,匈牙利数学家 Denès König 写出了开创性的论著"论有限和无限图",奠定了现代图论的基础。这个题目第一次被运筹学采纳是在1950年代,代表人物是最杰出的L.R.福特和D.R.富尔克森。

Graph theory, the subject at the root of this chapter, dates back to 1736, when the Swiss mathematician Leonard Euler considered the now famed "Köngsborg bridge problem." At that time, there were seven bridges across the River Pregel that ran through the city of Köngsborg on the Baltic Sea, and Euler wondered whether or not it would be possible to start somewhere in the city, walk across each of the bridges exactly once, and return to where he came from. After a year of study, 29 years of Euler submitted "Seven Bridges" thesis, solved this problem, while creating a new branch of mathematics, graph theory.

Two hundred years later in 1936, the Hungarian mathematician Denès Köngsborg wrote the seminal book "The Theory of Finite and Infinite Graphs," that laid the foundations of modern graph theory. The subject was first used by operations researchers in the 1950s, most prominently by L. R. Ford and D. R. Fulkerson.

单词和短语 Words and expressions

网络分析	Network Analysis	瓶颈	bottlenecks
节点	vertices	最短路径	Shortest Path
无向图	undirected graph,	离散	discretized
可达的	reachable	用符号表示	symbolize
连通图	connected graph	克鲁斯克尔方法	Kruskal technique
基尔霍夫结点	Kirchhoff node	增广链	Augmented chain

基本概念和性质 Basic concepts and properties

1 若用点表示研究的对象,用边表示这些对象之间的联系,则图 G 可以定义为点和边的集合,记作:
$$G = \{V, E\}$$
With the point said the object of study, with a side said the link between these objects, and then Graph G can be defined as a collection of vertices and edges, Notes:
$$G = \{V, E\}$$

2 图中的点用 v 表示,边用 e 表示。对每条边可用它所连接的点表示,记作: $e_k = [v_i, v_j]$
With v said point in the graph, said the edge by e. Each edge can be expressed by the points with connected to it, denoted by:
$$e_k = [v_i, v_j]$$

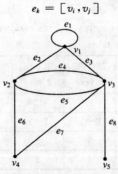

3 与某一个点 v_i 相关联的边的数目称为点 v_i 的次(也叫做度),记作 $d(v_i)$。
The number of edges associated with a point v_i is called as times of the point (also called degree),

denoted by $d(v_i)$.

4 链：由两两相邻的点及其相关联的边构成的点边序列。
The point edge sequence constituted by each group of adjacent points and their associated side called chain.

5 起点与终点重合的链称作圈。如果每一对顶点之间至少存在一条链，称这样的图为连通图，否则称图不连通。
The chain of starting point and terminal point of coincidence is called as ring. If exists at least a chain between every pair of vertices, called such graph as a connected graph, otherwise known graph as connectivity.

6 设图 $G=(V,E)$，对 G 的每一条边 (v_i,v_j) 相应赋予数量指标 w_{ij}，w_{ij} 称为边 (v_i,v_j) 的权，赋予权的图 G 称为网络（或赋权图）。端点无序的赋权图称为无向网络，端点有序的赋权图称为有向网络。
Assume graph G = (V, E), each edge on G is given a number index w_{ij} of corresponding, w_{ij} known as the weights of the edges, the graph G assigned to weight is called the network (or a weighted graph). The endpoint of disordered weighted graph called undirected network, the weighted graph with endpoint ordered is called a directed network.

7 一个图的次等于各点的次之和。
The time of a graph is equal to the sum of the time of all points.

8 有向图中，所有顶点的入次之和等于所有顶点的出次之和。
In directed graph, the sum of into-time equal to that of out-time to all the vertices.

9 任何图中，顶点次数之和等于所有边数的 2 倍。
To any graph, the sum of vertex time equal to two times of all edges.

例 题 Examples

例 1 由 $Dijkstra$ 算法求设备更新问题
To solve the equipment update problem by $Dijkstra$ algorithm

某工厂使用一台设备，每年年初工厂要作出决定：继续使用，购买新的？如果继续使用旧的，要负维修费；若要购买一套新的，要负购买费。试确定一个 5 年计划，使总支出最小。

若已知设备在各年的购买费，及不同机器役龄时的残值与维修费，如表所示：

Some factory use one device, the factory needs to make a decision at the beginning of each year: to continue to use, or buy new? If continue to use the old, must to bear the repair costs; if you want to buy a new, take the purchase expenses. Determine a plan for 5 years, bringing the total expenditure minimum.

Known equipment purchase costs in each year, the residual value of useful age to different machines and maintenance costs, as shown in the table:

项目 Project	第1年 1 year	第2年 2 year	第3年 3 year	第4年 4 year	第5年 5 year
购买费 purchase costs	11	12	13	14	14
机器役龄 Useful age	0—1	1—2	2—3	3—4	4—5
维修费 maintenance costs	5	6	8	11	18
残值 residual value	4	3	2	1	0

解 把这个问题化为最短路问题。
Translate this problem into a shortest path problem.

用点 v_i 表示第 i 年初购进一台新设备,虚设一个点 v_6 ,表示第 5 年底。边(v_i , v_j)表示第 i 年购进的设备一直使用到第 j 年初(即第 j-1 年底)。边(v_i , v_j)上的数字表示第 i 年初购进设备,一直使用到第 j 年初所需支付的购买费、维修的全部费用(可由表 8-2 计算得到)。见下图:

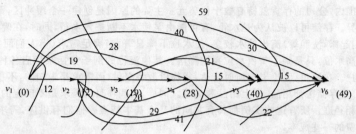

Using v_i expresses a new piece of equipment purchased at the beginning of the i year, hypothetical one point V_6, said at the end of the fifth year. The edge(v_i, v_j)said the time using the equipment purchased at the beginning of the i year to the beginning of the j year (namely, the j-1 year end). The figures on the edge (v_i, v_j) expresses the total cost of purchase of equipment required to pay the purchase cost, maintenance costs from at the beginning of the i year to one of the j year, (by table calculated). See below:

这样设备更新问题就变为:求从 v_1 到 v_6 的最短路问题。

So updating equipment question becomes the shortest path problem from V_1 to V_6.

1) v_1 (0)。

2) min{(k_{12}), k_{13}, k_{14}, k_{15}, k_{16}}=12, given v_2 by the label (12),(v_1 , v_2)adds a bold line.

3) min{(k_{13}), k_{14}, k_{15}, k_{16}, k_{23}, k_{24}, k_{25}, k_{26}}={19, ⋯ 13+12, ⋯}=19,
given v_3 by the label,(19),(v_1 , v_3)adds a bold line.

4) min{(k_{14}), k_{15}, k_{16}, k_{24}, k_{25}, k_{26}, k_{34}, k_{35}, k_{36}}={59,⋯19+30,⋯12+20,⋯}=28,
given v_4 by the label(28),(v_1 , v_4)adds a bold line.

5) min{(k_{15}), k_{16}, k_{25}, k_{26}, (k_{35}), k_{36}, k_{45}, k_{46}}={40,⋯41,40,⋯43, ⋯}=40, Corresponding two sides:
given v_5 by the label(40),(v_1 , v_5),(v_3 , v_5)add a bold line.

6) min{k_{16}, k_{26},(k_{36}), k_{46}, k_{56}}= min{59,53,49,50,55}=49
given v_6 by the label(49),(v_3 , v_6)adds a bold line.

Computing result: $v_1 \to v_3 \to v_6$ as shortest path, the cut long 49。

即:在第一年、第三年初各购买一台新设备为最优决策。这时 5 年的总费用为 49。

Namely:, The purchase of a new piece of equipment at the beginning of the first and the third year, should be optimal decision. Now the total cost of 5 years is 49.

本章重点

① 熟练 Kruskal 方法求最小支撑树问题;
② 熟练 Dijkstra 算法求最短路径的应用问题;
③ 熟悉并掌握求最大流问题的方法。

Key points of this chapter

① Skilled Kruskal method to calculate the minimum spanning tree problem;
② Master Dijkstra algorithms for the shortest path problem and its application;
③ Familiar with and master the method for the maximum flow problem.

第九章 存储论
Chapter 9 Inventory Models

在世界范围内,公司的存货就持有数十亿美元。主要的原因是创建一个缓冲区,平衡商品的流入和流出之间的差异。存储可以被认为是水箱:有可能由泵使水不断进入水箱,同时,在晚上流出的水较少,在早晨(当人们起床,洗澡等)流出的水较多,使水位下降显著,当满足需求(当人们回家,洗衣服等)时,再次在晚上增高水位,只是到了午夜水位再次降落。其他流行的例子包括杂货店各种食品的库存等待出售给他的客户。在这里,无论卡车何时抵达,货物以散装交付,而需求是未知的,不稳定的。举个医院的例子,他们都要存储医疗用品、床上用品和血浆。同样,这些项目的需求是不确定的,从一天到接下来的一天可能大相径庭。所有这些实例都具有共同的几个基本特点。他们有供应,需求,和为获取,保持和处置存货花费的一些成本。

与储存("携带")库存相关的成本也是非常大,也许是一季度存货价值。因此,在美国每年用于储存库存的成本费将运行到数千亿美元。降低存储成本,避免不必要的库存量可以大大加强公司的竞争力。

一些日本公司率先引入零库存系统,该系统强调规划和调度,用料达到"即时"使用。因此,靠把库存水平降到很低从而实现了庞大的节约。

在世界其他地区的许多公司也已按照这种方式管理自己的库存。在这一领域中的应用运筹学技术(有时也被称为科学的库存管理)为获得竞争优势提供了有力的工具。

Worldwide, companies hold billions of dollars in inventories. The main reason is to create a buffer that balances the differences between the inflow and outflow of goods. Inventories can be thought of as water tanks: there may be a constant inflow of water that is pumped into the tank by a pump, while the outflow is low at night, high in the morning (when people get up, take a shower, etc), it then decreases significantly until the demand again increases in the evening (when people come home, do laundry, etc), just to fall off again for the night. Other, popular, examples include grocery stores whose inventories consist of various foodstuffs awaiting sale to its customers. Here, the delivery of the goods is in bulk whenever a delivery truck arrives, while the demand is unknown and erratic. In the case of hospitals, they have in stock medical supplies, bed linen and blood plasma. Again, the demand for these items is uncertain and may differ widely from one day to the next. All these instances have a few basic features in common. They have a supply, a demand, and some costs to obtain, keep, and dispose of inventories.

The costs associated with storing ("carrying") inventory are also very large, perhaps a quarter of the value of the inventory. Therefore, the costs being incurred for the storage of inventory in the United States run into the hundreds of billions of dollars annually. Reducing storage costs by avoiding unnecessarily large inventories can enhance any firm's competitiveness.

Some Japanese companies were pioneers in introducing the just-in-time inventory system—a system that emphasizes planning and scheduling so that the needed materials arrive "just-in-time" for their use. Huge savings are thereby achieved by reducing inventory levels to a bare minimum.

Many companies in other parts of the world also have been revamping the way in which they manage their inventories. The application of operations research techniques in this area (sometimes called scientific inventory management) is providing a powerful tool for gaining a competitive edge.

单词和短语 Words and expressions

存储策略　inventory policy	持有费用　holding cost
装配　assemble	订购费用　cost of ordering
单位生产费用　unit production cost,	行政费用　administrative cost
设置成本　setup cost	拖后　backlogging
短缺费用　shortage cost	收入　revenue

回收价值　salvage value
贴现率　discount rate

经济订货批量模式　Economic order quantity mode

基本概念和性质　Basic concepts and properties

1 库存费：货物进入储存到使用或销售出去这段时间内发生的成本。它的大小取决于库存量的大小和存放时间的长短。一般给出单位时间单位货物的存贮费，

Inventory costs：the cost of storage cargo eduring this time entering the store to use or sold. Its size depends on the length of storage time and the size of the inventory. Generally, given the storage fee for the per unit of goods in unit time.

2 缺货费：供不应求时引起的损失费用。它的大小与缺货时间和缺货量有关。一般给出单位时间单位货物的缺货费。

Shortage cost：the amount of the commodity required (demand) exceeds the available stock. Its size is related to the lack of time and shortage amount. Generally, given shortage cost for per unit of goods in the unit time.

3 订购费：可通过（通过购买或生产这一数额）由函数 $c(z)$ 来表达的成本。这个函数最简单的形式是 z 正比于订购的数量，也就是，cz，其中 c 代表单位支付的价格。另一种常见的假设是，当 $z>0$，$c(z)$ 两部分组成：第一项是正比于订购的数量，另一项是常数项 K。当 $z=0$，这个函数等于零。即，

$$c(z) = \begin{cases} 0 & z=0 \\ K+cz & z>0 \end{cases}$$

Ordering cost：The cost of ordering an amount z (either through purchasing or producing this amount) can be represented by a function $c(z)$. The simplest form of this function is one that is directly proportional to the amount ordered, that is, cz, where c represents the unit price paid. Another common assumption is that $c(z)$ is composed of two parts: a term that is directly proportional to the amount ordered and a term that is a constant K for z positive and is 0 for $z=0$. For this case,

$$c(z) = \begin{cases} 0 & z=0 \\ K+cz & z>0 \end{cases}$$

4 生产费用包括两项费用：生产准备费用（它与组织生产的次数有关，与产品数量无关）和生产本身的成本（对应于订货成本，它与产品数量有关）。

The production costs include two costs: the production preparation costs (It is related with the number of the organization of production, and nothing to do with the number of products) and production itself costs (Corresponding to the ordering cost, and to do with the number of products).

例题　Examples

例1 某商品单位成本为5元，每天保管费为成本的 0.1%，每次订购费为10元。已知对该商品的需求是 100 件/天，不允许缺货。假设该商品的进货可以随时实现。问应怎样组织进货，才能最经济。

Ex. 1 The unit cost of some commodity is 5 Yuan, storage fee is 0.1% of the cost every day, cost of ordering is 10 per order. Known demand for this product is 100 / day, not allowing shortage. Assume that the purchase of good can be achieved at any time. Ask how should organize the stock, to the economy.

解 已知 $K=5$ 元/件，$C_1 = 5 \times 0.1\% = 0.005$ 元/件·天，$C_3 = 10$ 元，$R = 100$ 件/天。

由经济订购批量（Economic ordering quantity）公式

$$t^* = \sqrt{\frac{2C_3}{C_1 R}} = \sqrt{\frac{2 \times 10}{0.005 \times 100}} = 6.32(\text{day})$$

$$Q^* = Rt^* = 100 \times 6.32 = 632$$

$$C(t^*) = \sqrt{2C_1C_3R} = \sqrt{2 \times 0.005 \times 10 \times 100} = 3.16(\text{Yuan/day})$$

Known number $K = 5$ Yuan / piece, $C_1 = 5 \times 0.1\% = 0.005$ Yuan / piece. Days, $C_3 = 10$ Yuan, $R = 100$ piece / day.

By Economic ordering quantity

$$t^* = \sqrt{\frac{2C_3}{C_1R}} = \sqrt{\frac{2 \times 10}{0.005 \times 100}} = 6.32(\text{day})$$

$$Q^* = Rt^* = 100 \times 6.32 = 632$$

$$C(t^*) = \sqrt{2C_1C_3R} = \sqrt{2 \times 0.005 \times 10 \times 100} = 3.16(\text{Yuan/day})$$

本章重点

1. 熟悉库存问题的建模思想；
2. 抓住三个主要环节：存贮状态图、费用函数和经济批量公式；

Key points of this chapter

1. Familiar with inventory problem modeling;
2. Hold three key links: storage state diagram, cost function and Economic ordering quantity formula.

第十章 排队论
Chapter 10　Queuing Theory

队列（排队）是一个日常生活的一部分。我们都要排队买电影票、去银行存款、购买食品、邮包裹，在一家自助餐厅获取食物，在一个娱乐公园开始乘骑等等，我们已经习惯了数量可观的等待，但是仍然会讨厌异常长时间的等待。

然而，要等待不仅仅是一个狭隘的个人烦恼。一个国家的民众通过排队等候浪费的时间量是有关生活质量和国家经济效率的一个主要因素。例如，苏联解体之前，它的公民常常不得不忍受只是购买基本生活必需品去排巨大的长队，这使它成为世界笑话。即使在如今的美国，据估计，美国人每年要花费 37000000000 小时用于排队等候。相反，如果这个时间可以有效使用，相当于 2000 万人一年的有用工作量！

即使这个惊人的数字并没有告诉你造成过多等待影响的全部。那么，当其他种类的等待相比于排队的人时，会产生巨大的低效。例如，使机器等待被修复，可能会导致生产损失；需要等待卸载的车辆（包括船舶和卡车）可能延迟后续的出货量；等待起飞或降落的飞机，可能会破坏以后的旅行日程；由于饱和线在电信传输的延迟，可能会导致数据的毛刺；导致制造业工作岗位等执行可能会扰乱后续生产；超出了他们的截止日期，推迟服务工作，可能会导致未来的业务损失；等等。

排队理论是研究所有这些各种形式的等待问题。它使用排队模型来表示在实践中出现的各种类型的排队系统（系统涉及队列某些种类）。每个模型公式表明应该执行相应的排队系统，包括在各种不同的情形下将发生的等待平均量。因此，这些排队模型是非常有助于确定如何以最有效的方法操作排队系统。提供太多的服务能力来操作系统涉及成本过高。但是没有提供足够的服务能力会导致过多的等待和它所有的不幸后果。这些模型使我们在服务的成本和等待量之间找到合适的平衡。

Queues (waiting lines) are a part of everyday life. We all wait in queues to buy a movie ticket, make a bank deposit, pay for groceries, mail a package, obtain food in a cafeteria, start a ride in an amusement park, etc. We have become accustomed to considerable amounts of waiting, but still get annoyed by unusually long waits.

However, having to wait is not just a petty personal annoyance. The amount of time that a nation's populace wastes by waiting in queues is a major factor in both the quality of life there and the efficiency of the nation's economy. For example, before its dissolution, the U. S. S. R. was notorious for the tremendously long queues that its citizens frequently had to endure just to purchase basic necessities. Even in the

United States today, it has been estimated that Americans spend 37000000000 hours per year waiting in queues. If this time could be spent productively instead, it would amount to nearly 20 million person-years of useful work each year!

Even this staggering figure does not tell the whole story of the impact of causing excessive waiting. Great inefficiencies also occur because of other kinds of waiting than people standing in line. For example, making machines wait to be repaired may result in lost production. Vehicles (including ships and trucks) that need to wait to be unloaded may delay subsequent shipments. Airplanes waiting to take off or land may disrupt later travel schedules. Delays in telecommunication transmissions due to saturated lines may cause data glitches. Causing manufacturing jobs to wait to be performed may disrupt subsequent production. Delaying service jobs beyond their due dates may result in lost future business, etc.

Queuing theory is the study of waiting in all these various guises. It uses queuing models to represent the various types of queuing systems (systems that involve queues of some kind) that arise in practice. Formulas for each model indicate how the corresponding queuing system should perform, including the average amount of waiting that will occur, under a variety of circumstances. Therefore, these queuing models are very helpful for determining how to operate a queuing system in the most effective way. Providing too much service capacity to operate the system involves excessive costs. But not providing enough service capacity results in excessive waiting and all its unfortunate consequences. The models enable finding an appropriate balance between the cost of service and the amount of waiting.

单词和短语 Words and expressions

排队　Queues
输入源　input source
服务机制　service mechanism
间隔时间　interarrival time
假设　assumption
先到先得　First-come-first-served
服务渠道　service channels

开始　Commencement
指数的　exponential
退化分布　degenerate distribution
爱尔朗分布　Erlang distribution
利用系数　utilization factor
稳态　steady-state
瞬态工况　transient condition

基本概念和性质　Basic concepts and properties

1 排队系统有三个组成部分
输入过程：顾客总体数（来源无限或有限），顾客到来方式（单个或成批），顾客流的概率分布（泊松流、定长、爱尔朗分布等）；
服务规则：损失制（服务台满时顾客立即离去），等待制（先到先服务，后到先服务，随机服务，优先权），混合制（队长有限制，排队时间有限制）；
服务机构：服务台数量及布置形式（单/多服务台，串、并列或结合），某一时刻接受服务的顾客数（每服务台每次服务顾客数），服务时间分布（负指数、定长、爱尔朗分布等）.
The queuing system has three components
Input process: the number of customers overall (source of infinite or finite), the customer arrival (single or batch), the probability distribution of customer flow (Poisson, fixed-length, Erlang distribution, etc.);
Service rule: loss systeml When the desk is full, the customer immediately leave), waiting for the system (first come first served, and then to first-served, random, priority), mixed-mode (captain limited, queuing time limited);
Services: help desk number and arrangement (the single / multi-server, serial, parallel or a combination), the number of customer to accept service at a time (the customer service number at per desk every time), the service time distribution (negative exponential, fixed long, the Erlang distri-

bution，etc.）．

2 研究目的：通过对排队系统中概率规律的研究，使系统达到最优设计和最优控制，以最小费用实现系统的最大效益。

Research Objective：to study the probability law of the queuing system，optimize the system design and optimal control so that achieve the maximum effectiveness of the system by minimizing the costs.

3 顾客到达过程是一个参数为 λ 的泊松过程的充分必要条件为：相应的顾客到达间隔 $\{T_n\}$ 是一族相互独立的随机变量，且每个随机变量都服从下面的负指数分布：

$$P(T_n \leqslant t) = \begin{cases} 1-e, t \geqslant 0 \\ 0, t \geqslant 0 \end{cases}$$

The necessary and sufficient condition that the customer arrival process is a Poisson process with a parameter λ for：the corresponding customer arrival interval $\{T_n\}$ is a family of independent random variables，and each random variable obeys the following negative exponential distribution：

$$P(T_n \leqslant t) = \begin{cases} 1-e, t \geqslant 0 \\ 0, t \geqslant 0 \end{cases}$$

4 如果 $\xi_1, \xi_2, \cdots, \xi_k$ 是 k 个相互独立具有相同的负指数分布（参数为 μ）的随机变量，则随机变量 $\tau = \xi_1 + \xi_2 + \cdots + \xi_k$ 服从 k 阶爱尔朗分布、

If $\xi_1, \xi_2, \cdots, \xi_k$ are mutually independent random variables with the same negative exponential distribution (Parameters for μ)，then the random variables obey Erlang distribution with order k.

例 题 Examples

例 1 某火车站售票处有三个窗口，同时售各车次的车票。顾客到达服从泊松分布，平均每分钟到达 $\lambda = 0.9$（人），服务时间服从负指数分布，平均服务率 $\mu = 24$（人/h），分两种情况：
1. 顾客排成一队，依次购票；2. 顾客在每个窗口排一队，不准串队。
求：(1) 售票处空闲的概率；(2) 平均等待时间和逗留时间。(3) 队长和队列长。

Ex.1 The ticket office of some train station has three windows, while sale tickets of all trips. Customer arrives to obey Poisson distribution, the arrivals on average every minute：$\lambda = 0.9$ (person), , service time to obey negative exponential distribution , the average service rate $\mu = 24$ (person / h), in two situations：

(1). A customer lined up, followed by tickets；(2). Customers at each window row a team are not allowed to string team.

To solve：(1) the free probability of ticket Office；(2) the average waiting time and length of stay. (3). the captain and the queue length

解 1. M/M/3/∞/∞   单位应相同(Unit should be the same)：
$\mu = 0.4$(person/ minute)
set $\rho = \lambda/(3\mu)$

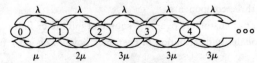

- 稳态概率(steady-state probability)：
  - if $n < 3$， $P_n = [\lambda/(n\mu)]P_{n-1} = (\lambda^n/n!)\rho P_0 = 3^n/n! \cdot \rho^n P_0$
  - If $n \geqslant 3$， $P_n = [\lambda/(c\mu)]P_{n-1} = (3^3/3!)\rho^n P_0 = 4.5\rho^n P_0$
  $\rho = 0.9/(0.4 * 3) = 0.75$

By $\sum_{n=0}^{\infty} P_n$，

$$P_0 + 3*0.75P_0 + 4.5*0.75^2 P_0 + \sum_{n=3}^{\infty} 4.5\rho^n P_0 = 1$$

$$P_0 = 1/(1 + 2.25 + 2.53125 + 4.5\rho 3/(1-\rho)) = 1/13.375 = 0.0748$$

$$P_1 = 0.1683 \quad P_2 = 0.1893$$

Hence: $L_q = 4.5\rho^4 P_0/(1-\rho)^2 = 1.704$,

$$\lambda_e = \lambda, \quad W_q = L_q/\lambda_e = 1.704/0.9 = 1.893$$

$$W_s = W_q + 1/\mu = 1.893 + 2.5 = 4.393(minute), \quad L_s = \lambda_e W_s = 3.954$$

Therefore: ticket free probability is 0.0748; the average waiting time for $W_q = 1.893$ minutes, the average length of stay for $W_s = 4.393$ minutes, captain $L_s = 3.954$ (person) $L_q = 1.704$ (person).

2. M/M/1/∞/∞

三个系统并联(Three parallel system): $\lambda = 0.3, \mu = 0.4, \rho = \lambda/\mu = 0.75$

$P_0 = 1 - \rho = 0.25$, Three service stations have free time, $P_0^3 = 0.0156$

$L_s = \rho/(1-\rho) = 3, \lambda_e = \lambda = 0.3, L_q = L_s - \lambda/\mu = 2.25$

$W_s = L_s/\lambda = 10, W_q = W_s - 1/\mu = 7.5$

Therefore: ticket free probability is 0.0748; the average waiting time for $Wq = 0.0156$.

The average waiting time is $Wq = 7.5$ minutes, the average length of stay is $W_s = 10$ minutes, Captain $L_s = 3$, three team $3 + 3 + 3 = 9$, queue length 6.75 $Lq = 2.25$ (man), by contrast, a team shared three service station efficiency.

本章重点

1. 掌握排队论的基础理论:包括各类模型的分布理论;
2. 运用排队论的方法对与性能分析有关的问题进行建模和分析。

Key points of this chapter

1. Master the basic theory of queuing theory: including various models of distribution theory;
2. Utilize the method of queuing theory to modeling and analysis for the problems with performance analysis.

第十一章 对策论
Chapter 11 Game Theory

生活充满了冲突和竞争。参与冲突的对手包括室内游戏,军事斗争,政治运动,通过竞争性业务公司的广告和营销活动等等,例子不胜枚举。在许多情况下反应的一个基本特征是最后的结局主要取决于由对手选择的组合战略。博弈论是以一种正式的,抽象的方式来处理一般特征的竞争情况的数学理论。它特别强调对手的决策过程。

对策论的零星研究可上溯到18世纪初甚至更早,现代对策理论中的一些经典的博弈模型,如关于产量决策的古诺模型和关于价格决策的伯特兰德模型则是古诺和伯特兰德分别于1838年和1883年提出的,但对策论的真正发展还是在100年前。早在1912年,E. Zermelo用集合论的方法研究过下棋,他著有《关于集合论在象棋对策中的应用》。法国数学家Borel在1921年,也用数学方法研究过下棋时的一些个别现象,并且引入了"最优策略"的概念。20世纪40年代以来,由于生产与战争的需要,运筹学的各学科纷纷出现。特别是战争中兵力的调配、军队的部署、监视对方、侦察对方兵器等活动,迫切要求指挥者拿出最好方案,用已有的条件去取得较大的胜利,于是方案对策论的数学模型很快就形成了。当时,各参战国组织了大批的科学家参加了这方面的研究工作。

伴随着对策论的研究,经济科学的研究与对策论很快结合到一起,1944年诺依曼和摩根斯坦合著《博弈论和经济行为》一书的出版标志着系统的对策理论的初步形成。从此,对策论的研究才系统化与公理化。五六十年代是对策论研究、发展最重要的阶段,一些重要的对策论的概念就是在这个阶段发展起来的,如"纳什均衡"等。不过对策论真正得到重视并被看作重要的经济理论还是近十多年的事。1994年三位长期致力于对策论的理论和应用研究、实践的学者纳什、海萨尼和塞尔顿共同获得诺贝尔经济学奖,则更使对策论作为重要的经济学科或运筹学分支的地位和作用得到了最具权威性的肯定。

博弈论继续深入的研究便成为比较复杂类型的竞争情况。然而,在本章关注的是在最简单的情况下,所谓两个人的"零和游戏"。顾名思义,这些游戏涉及只有两个对手或球员(他们可能是军队,团队,公司,等等)。之所以被称为零和游戏是因为一名玩家赢得了的同时另一个玩家输了,所以,他们的净奖金的总和是零。

Life is full of conflict and competition. Numerous examples involving adversaries in conflict include parlor games, military battles, political campaigns, advertising and marketing campaigns by competing business firms, and so forth. A basic feature in many of these situations is that the final outcome depends primarily upon the combination of strategies selected by the adversaries. Game theory is a mathematical theory that deals with the general features of competitive situations like these in a formal, abstract way. It places particular emphasis on the decision-making processes of the adversaries.

The sporadic research on game theory can be traced back to the early eighteenth century or even earlier, some classic game of model modern game theory, such as Cournot model on the output decision and Bert Rand model on price decision is presented by the Cournot and Bertrand respectively in 1838 and 1883, but the real development of the game theory in 100 years ago. Early in 1912, E. Zermelo studied chess with the method of set theory, he wrote "the application of set theory in chess countermeasures". The French mathematician Borel in 1921, also use mathematics method to study playing some of the individual phenomenon, and introduce the concept of "optimal strategy". Due to the production and the needs of the war since the forties of this century, Operations researches in the various disciplines have been appearing. Especially the deployment of troops in the war, military deployment, monitoring each other, detect each other weapons, and other activities, urgently requires the commander to come up with the best program to use the existing conditions to obtain a great victory, so the scheme theory mathematical model quickly formed. At that time, the veterans organized a large number of scientists participated in the research work of this respect.

With the game theory research, economic science research and game theory soon jointed together. 1944 John Von Neumann and Morgenstern potato "theory of games and economic behavior", the publication of the book marked the initial formation of the theory of system countermeasure. Since then, the research of game theory begins to systematic and axiomatic. Fifty or sixty time is the most important stages of the game theory research and development, some important concept on countermeasures was developed at this stage, such as the "Nash equilibrium". But that Game theory really gets attention and is regarded as the important economic theory is thing of nearly more than 10 years. In 1994, three scholar of long-term commitment to the theory and applied research, practice on countermeasures, Nash, Harsanyi and Selten shred the Nobel Prize for economics, which more make the status and role of game theory obtain the most authoritative affirmation as an important economic discipline or a branch of operations research.

The research on game theory continues to delve into rather complicated types of competitive situations. However, the focus in this chapter is on the simplest case, called two-person, zero-sum games. As the name implies, these games involve only two adversaries or players (who may be armies, teams, firms, and so on). They are called zero-sum games because one player wins whatever the other one loses, so that the sum of their net winnings is zero.

单词和短语 Words and expressions

对策论　Game theory　　　　　　　　收益表　payoff table
伯特兰德模型　BertRand model　　　　合理标准　rational criteria
纳什均衡　Nash equilibrium　　　　　　对策　countermeasure

零和游戏　zero-sum games
平衡点　equilibrium point
无关紧要的　inconsequential
消除策略　eliminating strategy
鞍点　saddle point
担保　guarantee
极大极小准则　minimax criterion
合作博弈　cooperative game

基本概念和性质　Basic concepts and properties

1. 对策行为:具有竞争或对抗性质的行为。对策模型:具有对策行为的模型。
   Countermeasures behavior: the behavior of competitive or adversarial nature.
   Game Models: a model of response behavior.

2. 局中人:一个对策行为(或一局对策)中有权决定自己行动方案的参加者。
   Players: the participant that has the right to decide their own action programs to in a the countermeasures behavior (or a bureau of countermeasures).

3. 策略:在一局对策中,可供局中人选择的一个实际可行的行动方案。
   Strategies: a practical plan of action to choose for players in a Bureau of countermeasures.

4. 得失:一局博弈结局时的结果。
   Payoffs: The results of the outcome in a board game

5. 在一局对策中,每个局中人都选定一个策略形成一个策略组,称为一个局势。如果第 i 个局中人的一个策略为 $s_i \in S_i$,则 n 个局中人的策略组为 $s = (s_1, s_2, \cdots, s_n)$,是一个局势。
   In a Board Game, each player has selected a strategy to form a strategy group, called a situation. If the strategy of first i player is $s_i \in S_i$, then the strategy group of n player's $s = (s_1, s_2, \cdots, s_n)$ is a situation.

6. 对任意局势,每个局中人都可以得到一个赢得,记 $H_i(s)$,则称为第 i 个局中人的赢得函数。
   For any situation, each player can get one payoff, denote as $H_i(s)$, called a Payoff function of the first i player.

7. 纳什均衡(Nash Equilibrium):在一策略组合中,所有的参与者面临这样一种情况,当其他人不改变策略时,他此时的策略是最好的。
   Nash equilibrium: in a strategy combination, all participants are facing such a situation, when other people do not change their tactics, his strategy is the best.

8. 如果在一局对策中包含两个局中人,二局中人都只有有限个策略可供选择,在任一个局势下,两个局中人的赢得之和总是等于零,则称此对策为矩阵对策。
   In a board of two players, two players can choose from only a finite number of strategies, in any one situation, the sum to winning for the two players is always equal to zero, then called this countermeasure as the matrix countermeasure.

9. 假定 $G = \{S_1, S_2; A\}$ 是矩阵对策,
   $$S_1 = \{\alpha_1, \alpha_2, \cdots, \alpha_m\}, \quad S_2 = \{\beta_1, \beta_2, \cdots, \beta_n\}, \quad A = (a_{ij})_{m \times n}$$
   如果等式 $\max_i \min_j a_{ij} = \min_j \max_i a_{ij} = a_{i^* j^*}$ 是成立的,记 $V_G = a_{i^* j^*}$,则称 $V_G = a_{i^* j^*}$ 为矩阵对策 G 的值,对策的局势 $(\alpha_{i^*}, \beta_{j^*})$ 是对策 G 的解,$\alpha_{i^*}, \beta_{j^*}$ 分别称为局中人 Ⅰ、Ⅱ 的最优策略。
   Assume that $G = \{S_1, S_2; A\}$ is matrix countermeasure,
   $$S_1 = \{\alpha_1, \alpha_2, \cdots, \alpha_m\}, \quad S_2 = \{\beta_1, \beta_2, \cdots, \beta_n\}, \quad A = (a_{ij})_{m \times n}.$$
   If equality $\max_i \min_j a_{ij} = \min_j \max_i a_{ij} = a_{i^* j^*}$ is hold, denote that $V_G = a_{i^* j^*}$, then said $V_G = a_{i^* j^*}$ as the value of the matrix countermeasure G, The corresponding situation $(\alpha_{i^*}, \beta_{j^*})$ is the solution for Countermeasures G, $\alpha_{i^*}, \beta_{j^*}$ are respectively called Ⅰ, Ⅱ the optimal pure strategy.

10. 矩阵对策 $G = \{S_1, S_2; A\}$ 在纯策略下有解的充要条件是:存在局势 $(\alpha_{i^*}, \beta_{j^*})$,使得对一切 $i = 1, 2, \cdots, m; j = 1, 2, \cdots, n$ 均有 $a_{ij^*} \leq a_{i^* j^*} \leq a_{i^* j}$。

In pure strategies, the necessary and sufficient condition for matrix game $G = \{S_1, S_2; A\}$, having solution: there is the situation $(\alpha_{i*}, \beta_{j*})$, for all $i = 1, 2, \cdots, m; j = 1, 2, \cdots, n$,
$$a_{ij*} \leqslant a_{i*j*} \leqslant a_{i*j}.$$

11 设对策矩阵为 A,如果存在 $(i*,j*)$ 对任意 i,j 有 $a_{ij*} \leqslant a_{i*j*} \leqslant a_{i*j}$,则称为矩阵 A 的一个鞍点。鞍点的实际意义:$(i*,j*)$ 为一个鞍点,相应的 $(\alpha_{i*}, \beta_{j*})$ 是对策的一个解,是一个平衡局势,是双方理智的选择。

Denote that countermeasure matrix is A, if there exists $(i*, j*)$, for any i, j, $a_{ij*} \leqslant a_{i*j*} \leqslant a_{i*j}$, then $(i*, j*)$ is called is a saddle point for matrix A. The practical significance of saddle point: for a saddle point $(i*, j*)$, the corresponding $(\alpha_{i*}, \beta_{j*})$ is a solution of countermeasure, is a balanced situation, and is both a rational choice.

例 题 Examples

例 1 给定矩阵对策 $G = \{S_1, S_2; A\}$,其中
$$A = \begin{pmatrix} -7 & 1 \\ 3 & 2 \\ 16 & -1 \\ -3 & 0 \end{pmatrix},$$

则 $\max\limits_{i}\min\limits_{j} a_{ij} = \min\limits_{j}\max\limits_{i} a_{ij} = a_{22} = 2V_G = 2$,$\alpha_2$ 和 β_2 是 I 和 II 的最优纯策略。

Ex. 1 Given matrix game $G = \{S_1, S_2; A\}$, where
$$A = \begin{pmatrix} -7 & 1 \\ 3 & 2 \\ 16 & -1 \\ -3 & 0 \end{pmatrix},$$

We have $\max\limits_{i}\min\limits_{j} a_{ij} = \min\limits_{j}\max\limits_{i} a_{ij} = a_{22} = 2V_G = 2$, α_2 and β_2 are the optimal pure strategy for players I and II.

本章重点

1 理解对策论的基本概念,掌握矩阵对策的求解方法;
2 理解纳什均衡的概念及相应的求解方法;
3 理解二人的无限零和对策及有限的二人非零和对策问题。

Key points of this chapter

1 Understand the basic concept of game theory; grasp the method of calculating the matrix game;
2 Understand the concept of Nash equilibrium and the corresponding solving methods;
3 Understand two infinite zero-sum countermeasures and finite two nonzero-sum countermeasures problems.

第十二章 决策分析
Chapter 12　Decision Analysis

在世界各地,在每一个时刻,数以百万计的人都在做出自己的分散决策:什么时候在早晨起床,穿什么衣服,午餐或晚餐吃什么,晚上做什么(去剧院或看电视),到哪里度假,甚至更多。同样,当运送公司的产品给客户时,将决定使用哪个运输方式路,他们的产品在哪里可以找到区域配送中心,制定什么新的产品线路,等等。

古今中外的许多政治家、军事家、外交家、企业家都曾做出过许许多多出色的决策,至今被人们所称颂。决策的正确与否会给国家、企业、个人带来重大的经济损失或丰厚的利益。在国际市场的竞争中,

一个错误的决策可能会造成几亿、几十亿甚至更多的损失。真可谓一着不慎,满盘皆输。

关于决策的重要性,著名的诺贝尔经济学获奖者西蒙(H. A. Simon)有一句名言:"管理就是决策,管理的核心就是决策"。决策是一种选择行为的全部过程,其中最关键的部分是回答"是"与"否"。决策分析在经济及管理领域具有非常广泛的应用,在投资、产品开发、市场营销、项目可行性研究等方面的应用都取得过辉煌的成就。决策科学本身内容也非常广泛,包括决策数量化方法、决策心理学、决策支持系统、决策自动化等。

Everywhere in the world, at each moment, millions of people make their own decentralized decisions: when to get up in the morning, what tie to wear, what to eat for lunch or dinner, what to do in the evening (go to the theater or watch television), where to vacation, and many more. Similarly, firms will decide which mode of transportation to use when routing their products to customers, where to locate regional distribution centers, what new product lines to develop, etc.

Many politicians, military strategists, diplomats, entrepreneurs have to make many good decisions, which has been lauded at all times. That decision-making is correct or not would be bring major economic loss or substantial interest to the state, enterprises and individuals. In the international market competition, a wrong decision may cause the loss of hundreds of millions, billions or even more. Really be described as a careless loser.

About the importance of the decision-making, the famous Nobel winners H. A. Simon have a famous saying: "Management is decision-making; the core of management is decision-making". A decision is a choice behavior of the whole process, which one of the most critical parts is to answer "yes" or "No". Decision analysis has a very wide range of applications in the field of economic and management, the applications in terms of investment, product development, marketing, project feasibility, and so study have made brilliant achievements. Decision - making science itself is also very extensive, including quantitative decision-making methods, decision - making psychology, decision support systems, and decision-making automation.

单词和短语 Words and expressions

决策分析　decision analysis
分散　decentralized
回报　payoff
可视化　visualization
优势　dominances
瓦尔德规则　Wald's rule
预期收益　anticipated payoffs

极小极大遗憾准则　minimax regret criterion
鲁棒优化　robust optimization
遗憾矩阵　regret matrix
货币价值　monetary values
灵敏度分析　sensitivity analyses
先验概率　prior probability

基本概念和性质　Basic concepts and properties

1 决策:两个以上可供选择的行动方案,记为 d_j。
　Making Decision: more than two alternative courses of action, denoted by d_j.

2 状态(事件):决策实施后可能遇到的自然状况,记为 θ_i。
　Status (event): the natural condition may be encountered after decision implementation.

3 状态概率:对各状态发生可能性大小的主观估计,记为 $P(\theta_i)$。
　State probability: the subjective estimate of the likelihood of the occurrence of each state denoted by $P(\theta_i)$.

4 结局(损益):当决策 d_j 实施后遇到状态 θ_i 时所产生的效益(利润)或损失(成本),记为 u_{ij}。
　Outcome (profit or loss): the benefit (profit) or loss (cost) produced after the implementation of the decision-making d_j when encountered state θ_i. denoted by u_{ij}.

5 确定型:状态只有一种。

Certainty: there is only one state.

6 不确定型：状态不只一种；又可分为完全不确定型（状态概率未知）和风险型（状态概率可知）。
Uncertainty: the state more than one; can be divided into completely uncertain type (state probabilities unknown) and risk-based type(state probability shows).

7 称采用最优期望益损值作为决策准则的决策方法为期望值法。
Said the decision - making method to use the optimal expected gain or loss values as decision-making criteria as the expectation method.

8 悲观准则又称华尔德准则或保守准则。按悲观准则决策时，决策者是非常谨慎保守的，为了"保险"，从每个方案中选择最坏的结果，在从各个方案的最坏结果中选择一个最好的结果，该结果所在的方案就是最优决策方案。
Pessimistic criteria, also known as Walter standards or conservative standards, when make decision according to the pessimistic criteria, policy-makers is very conservative, for "insurance", Select the worst results from each program, and then from the worst results of the various options. Select one best result, the program where the results is in is the best decision program.

9 当决策者对客观状态的估计持乐观态度时，可采用乐观准则。此时决策者的指导思想是不放过任何一个可能获得的最好结果的机会，因此这是一个充满冒险精神的决策者。
When the decision maker are optimistic to the objective state estimation, can adopt the optimism criterion. The guiding ideology of makers is to grasp any one opportunity that obtained the best result, so this is an adventurous decision maker.

10 等可能准则又称机会均等法或称拉普拉斯（Laplace）准则，它是 19 世纪数学家 Laplace 提出的。他认为：当决策者面对着 n 种自然状态可能发生时，如果没有充分理由说明某一自然状态会比其他自然状态有更多的发生机会时，只能认为它们发生的概率是相等的，都等于 $1/n$。
May principle, also known equal opportunity laws or Laplace criterion, it is the nineteenth century that mathematician Laplace proposed. He said: when n natural state that the decision maker is facing with may occur, if there is no good reason to illustrate this situation that a natural state than other natural state has more chance of occurrence, can only think of their occurring is equal to the probability, are equal to $1/n$.

例 题 Examples

例 1 某市的自行车厂准备上一种新产品，现有三种类型的自行车可选择：载重车 A_1，轻便车 A_2，山地车 A_3。根据以往的情况与数据，产品在畅销 S_1，一般 S_2 及滞销 S_3 下的益损值如下表

Ex. 1 the bicycle factory of some city ready on a new product, the existing three types of bicycle to choose from: trucks A_1, the light car A_2, mountain bike A_3. Based on previous data, the values of product gains and losses is given as following table in selling well S_1, general S_2 and unmarketable S_3.

自然状态 (Natural state) 决策(decision)	S_1 畅销 Sell well	S_2 一般 General	S_3 滞销 Unmarketable
A_1 生产载重车 Production of trucks	70	60	15
A_2 生产轻便车 Production of light cars	80	80	25
A_3 生产山地车 Production of mountain bikes	55	45	40

问该厂应如何选择方案可使该厂获得的利润最大？
Asked the plant should be how to choose the program will enable the plant to obtain the maximum profit?

解 这本是一个面临三种自然状态和三个行动方案的决策问题,该厂通过对市场进行问卷调查及对市场发展趋势分析,得出的结论是:今后 5 年内,该市场急需自行车,销路极好。因此问题就从三种自然状态变为只有一种自然状态(畅销)的确定型问题,且该厂选择新上轻便产品的方案为最佳方案。在未来 5 年内如果产品畅销,年利润为 80 万元。

Solution: This is a decision problem faced by the three natural states, and three action programs, the plant through a questionnaire on the market and market trends analysis concluded that: the next five years, the market needs for bikes with marketability excellent. So the question is changed from three natural states to determine problem that only a natural state (selling well) is considered, and the program of the plant's selection on new lightweight products is the best option. If the products sell well in the next five years, then, the annual profit is 80 million Yuan.

本章重点
1. 重点掌握不确定型决策分析的基本理论与方法;
2. 熟练应用定量分析的方法解决生产和管理中的各类问题。

Key points of this chapter
1. Key master the theory and method in uncertain decision analysis;
2. Skillful apply quantitative analysis method to solve the production and management of various types of questions.

第四部分 常微分方程
Part 4 Ordinary Differential Equations

引 言

常微分方程已有悠久的历史,而且继续保持着进一步发展的活力,其主要原因是它的根源深扎在各种实际问题之中.

牛顿最早采用数学方法研究二体问题,其中需要求解的运动方程是常微分方程.他以非凡的积分技巧解决了它,澄清了当时关于地球将撞上太阳的一种悲观论点.此后,许多著名数学家,例如伯努利、欧拉、高斯、拉格朗日和拉普拉斯等,都遵循历史传统,把数学研究结合于当时许多重大的实际力学问题,在这些问题中通常离不开常微分方程的求解.海王星的发现是通过对常微分方程的近似计算得到的,这曾是历史上的一段佳话.19 世纪在天体力学上的主要成就应归功于拉格朗日对线性常微分方程的工作.

19 世纪中叶柯西给微积分注入了严格性的要素,同时他也为微分方程的理论奠定了一个基石——解的存在性和唯一性定理.到 19 世纪末期,庞加莱和李雅普诺夫分别创立了常微分方程的定性理论和稳定性理论,这些工作代表了当时非线性力学的最新方法.20 世纪初,伯克霍夫继承并发展了庞加莱在天体力学中的分析方法,创立了拓扑动力系统和各态历经的理论,把常微分方程的研究提高到新的水平.

自 20 世纪 20 年代(特别是第二次世界大战)以来,在众多应用数学家的共同努力下,常微分方程的应用范围不断扩大并深入到机械、电讯、化工、生物、经济和其他社会学科的各个领域,各种成功的实例不胜枚举.自 60 年代以后,常微分方程定性理论发展到现代微分动力系统的理论,对研究一些奇异的非线性现象作出了贡献,构成现代大范围分析学中出色的篇章.另外,现代的(最优)控制理论、微分对策论以及泛函微分方程理论的基本思想,都起源于常微分方程,而且在方法上也与后者有密切关系.

Introduction

With a long history, Ordinary Differential Equation is keeping vigorous for further development which is mainly due to the fact that it is deeply rooted in various kinds of practical problems.

Newton firstly adopted the mathematical methods to study the problem of two bodies, among which the motion equations to be solved are ordinary differential equations. He had solved them with remarkable integral skills and clarified the pessimistic view that the earth would crash into the sun. Thereafter, many well-known mathematicians, such as Bernoulli, Euler, Gauss, Lagrange, Laplace etc., all followed the historical tradition and combined the mathematical study with many important practical mechanical problems. Among these problems the solving process of the ordinary differential equations is unavoidable. The discovery of the Neptune is mainly relied on approximate calculation of ordinary differential equations, and this has become a muchtold tale in history. The chief achievements in celestial mechanics in the 19th century ought to be the Lagrange's contribution in linear ordinary differential equations.

In the middle 19th century Cauchy set strict rules to Calculus. Meanwhile, he had laid a foundation for the theory of differential equations — existence and uniqueness theorem of solutions. By the end of 19th century, Poincaré and Liapunov respectively created the qualitative theory and the stability theory of ordinary differential equation. What they have done represented the newest method in non-linear mechanics at that time. In the early 20th century, Birkhoff succeeded and developed the analytical method in celestial mechanics, and created the topological dynamical system and the ergodic theory. Consequently, it brought the study of differential equations up to a new level.

Since 1920s (especially after World War II), with the joint effort of many applied mathematicians, the applications of ordinary differential equation were expanding to machinery, telecom, chemical industry, biology, economy, fields of the social science, and so on. Various successful instances are too numerous to mention. After 1960s, the qualitative theory of ordinary differential equation developed into the theory of modern differential dynamical system, which contributes to the study of some strange nonlinear phenomena and this makes a remarkable page in modern large-scale analytics. In addition, the modern (optimum) control theory, the game theory of differential and the basic thoughts of functional analysis differential equation theory all originated from ordinary differential equations and their methods are all interrelated with the latter.

第一章　基本概念
Chapter 1　Basic Concepts

　　本章首先介绍微分方程及其解的定义并给出它们的几何解释. 对这些内容的理解需要在以后各章中进行反复和加深.

　　In this chapter, the definitions and solutions of differential equations are first presented, as well as their geometric explanations. More understanding of those contents needs to be repeated and depended in the following chapters.

单词和短语　Words and expressions

- ★ 常微分方程　ordinary differential equation (ODE)
- 阶　order　[ˈɔːdə]
- 线性　linear　[ˈliniə]
- 非线性　nonlinear　[ˈnɔnˈliniə]
- 偏微分方程　partial differential equation (PDE)
- ★ 通解　general solution
- ★ 特解　particular solution
- ★ 初值条件　initial value condition
- 初值问题　initial value problem (IVP)
- 柯西问题　Cauchy problem

- 几何解释　geometry explanation
- 积分曲线　integral curve
- 线素　line element
- 线素场　field of line element
- 方向场　direction field
- ★ 等斜线　isocline　[ˌaisəuˈklain]
- 对称形式　symmetric form
- 奇异点　singularity　[ˌsingjuˈlæriti]
- 通积分　general integral

基本概念和性质　Basic concepts and properties

1 关于一个未知函数和它的一个或者多个导数的方程,称为一个微分方程.

An equation relating an unknown function and one or more of its derivatives is called a differential equation.

2 如果导数 $u', u'', \cdots, u^{(n)}$ 在 I 上存在,且对于所有 $x \in I, F(x, u, u', u'', \cdots, u^{(n)}) \equiv 0$,则称方程 $u = u(x)$ 是微分方程 $F(x, y, y', y'', \cdots, y^{(n)}) = 0$ 在区间 I 上的解.

The function $u = u(x)$ is called a solution of the differential equation $F(x, y, y', y'', \cdots, y^{(n)}) = 0$ on the interval I, provided that the derivatives $u', u'', \cdots, u^{(n)}$ exist on I and $F(x, u, u', u'', \cdots, u^{(n)}) \equiv 0$ for all $x \in I$.

3 如果微分方程中的未知函数依赖于且仅依赖于一个独立变量,就称其为常微分方程;如果未知函数是依赖于两个或两个以上独立变量的方程,则该微分方程中将出现偏导数,就称这种方程为偏微分方程.

If the unknown function of a differential equation (dependent variable) depends only on a single independent variable, the equation is called an ordinary differential equation; If the unknown function of a differential equation is a function of two or more independent variables, then partial derivatives are likely to be involved, and so the equation is called a partial differential equation.

例　题　Examples

例 1　证明:对于所有 $x > 0$,函数 $y(x) = 2x^{1/2} - x^{1/2} \ln x$ 满足微分方程
$$4x^2 y'' + y = 0.$$

解 首先计算导数得到
$$y'(x) = -\frac{1}{2}x^{-1/2}\ln x, \quad y''(x) = \frac{1}{4}x^{-3/2}\ln x - \frac{1}{2}x^{-3/2}.$$

将上式代入方程 $4x^2 y'' + y = 0$ 得到
$$4x^2 y'' + y = 4x^2\left(\frac{1}{4}x^{-3/2}\ln x - \frac{1}{2}x^{-3/2}\right) + 2x^{1/2} - x^{1/2}\ln x = 0,$$

即当 $x > 0$ 时,函数 $y(x) = 2x^{1/2} - x^{1/2}\ln x$ 满足微分方程 $4x^2 y'' + y = 0$.

Ex. 1 Verify that the function $y(x) = 2x^{1/2} - x^{1/2}\ln x$ satisfies the differential equation
$$4x^2 y'' + y = 0 \text{ for all } x > 0.$$

Solution First we compute the derivatives
$$y'(x) = -\frac{1}{2}x^{-1/2}\ln x, \quad y''(x) = \frac{1}{4}x^{-3/2}\ln x - \frac{1}{2}x^{-3/2}.$$

Then on substitution them into the differential equation $4x^2 y'' + y = 0$ yields
$$4x^2 y'' + y = 4x^2\left(\frac{1}{4}x^{-3/2}\ln x - \frac{1}{2}x^{-3/2}\right) + 2x^{1/2} - x^{1/2}\ln x = 0.$$

So the differential equation $4x^2 y'' + y = 0$ is satisfied for all $x > 0$.

例 2 设 $y(x) = 1/(C-x)$ 是微分方程 $dy/dx = y^2$ 的解,求解初值问题
$$\frac{dy}{dx} = y^2, y(1) = 2.$$

解 我们只需找到使得 $y(x) = 1/(C-x)$ 满足初值条件 $y(1) = 2$ 的 C 值. 将 $x = 1, y = 2$ 代入给定解得到 $2 = y(1) = \frac{1}{C-1}$,所以 $2C - 2 = 1$. 因此 $C = \frac{3}{2}$,那么得到所需解为
$$y(x) = \frac{1}{\frac{3}{2} - x} = \frac{2}{3 - 2x}.$$

Ex. 2 Assume that $y(x) = 1/(C-x)$ is a solution of the differential equation $dy/dx = y^2$, solve the initial value problem
$$\frac{dy}{dx} = y^2, y(1) = 2.$$

Solution We need only find a value of C so that the solution $y(x) = 1/(C-x)$ satisfies the initial condition $y(1) = 2$. On substitution of the values $x = 1$ and $y = 2$ into the given solution yields $2 = y(1) = \frac{1}{C-1}$, so $2C - 2 = 1$, and hence $C = \frac{3}{2}$. With this value of C we obtain the desired solution $y(x) = \frac{1}{\frac{3}{2} - x} = \frac{2}{3 - 2x}$.

本章重点

本章主要要求学生理解和掌握微分方程的基本概念及几何意义,为下面各章节的学习作铺垫.

Key points of this chapter

This chapter mainly requires the students a thorough understanding and grasping of the basic concepts and the geometric significance of differential equations, and thus serves as a basis to the following chapters.

第二章 初等积分法
Chapter 2 Elementary integration methods

初等积分法是本章的中心内容,能用初等积分法求解的方程虽属特殊类型,但它们在实际应用中却很常见和重要.

Elementary integration methods are the central contents of this chapter. Even though the equations which can be solved by elementary integration methods belong to a special type, they are widely used in practice and thus important.

单词和短语 Words and expressions

★ 恰当方程　exact equation
可微函数　differentiable function
全微分　total differential
偏导数　partial derivative
隐函数　implicit function
原函数　primitive function
变量可分离的方程　equation of separable variables
★ 一阶线性方程　first-order linear equation

齐次　homogeneous　[ˌhɔməu'dʒi:njəs]
非齐次　non-homogeneous
积分因子法　method of integration factor
★ 常数变易法　method of constants variation
初等变换法　elementary method of transformation
齐次方程　homogeneous equation
伯努利方程　Bernoulli equation
黎卡提方程　Riccati equation

基本概念和性质　Basic concepts and properties

1 对于一阶线性微分方程 $\dfrac{dy}{dx}+P(x)y=Q(x)$，其中函数 $P(x)$ 和 $Q(x)$ 在区间 $I=(a,b)$ 内连续。

(1)当 $Q(x)\not\equiv 0$ 时，则称此方程为非齐次线性方程；(2)当 $Q(x)\equiv 0$ 时，则称方程 $\dfrac{dy}{dx}+P(x)y=0$ 为对应的齐次线性方程.

For the first-order linear differential equation $\dfrac{dy}{dx}+P(x)y=Q(x)$, where the functions $P(x)$ and $Q(x)$ are continuous on the open interval $I=(a,b)$. (1) If $Q(x)\not\equiv 0$, then the equation is called a non-homogeneous linear equation; (2) However, if $Q(x)\equiv 0$, the equation is called a corresponding homogeneous linear equation.

2 定理(一阶线性方程)：如果函数 $P(x)$ 和 $Q(x)$ 在包含点 x_0 的开区间 I 上连续，那么初值问题 $\dfrac{dy}{dx}+P(x)y=Q(x), y(x_0)=y_0$ 在 I 上存在唯一的解 $y(x)$，其形式为

$$y(x)=\dfrac{1}{\rho(x)}\left[\int_{x_0}^{x}\rho(t)Q(t)dt+y_0\right], \rho(x)=\exp(\int_{x_0}^{x}P(t)dt).$$

Theorem (The First-order Linear Equation): If the functions $P(x)$ and $Q(x)$ are continuous on the open interval I containing the point x_0, then the initial value problem

$$\dfrac{dy}{dx}+P(x)y=Q(x), y(x_0)=y_0$$

has a unique solution $y(x)$ on I, given by the following formula

$$y(x)=\dfrac{1}{\rho(x)}\left[\int_{x_0}^{x}\rho(t)Q(t)dt+y_0\right], \rho(x)=\exp(\int_{x_0}^{x}P(t)dt).$$

3 一阶微分方程 $\dfrac{dy}{dx}+P(x)y=Q(x)y^n$ 称为伯努利方程。如果 $n=0$ 或者 $n=1$，则方程是线性的. 当 $n>1$ 时，令 $v=y^{1-n}$，则伯努利方程变为如下的线性方程

$$\dfrac{dv}{dx}+(1-n)P(x)v=(1-n)Q(x).$$

The first-order differential equation of the form $\dfrac{dy}{dx}+P(x)y=Q(x)y^n$ is called a Bernoulli equation. If either $n=0$ or $n=1$, then the corresponding equations are linear. Let $v=y^{1-n}$ as $n>1$, the equation then changes into the following linear equation

$$\dfrac{dv}{dx}+(1-n)P(x)v=(1-n)Q(x).$$

例题 Examples

例 1 求解初值问题 $x^2 \dfrac{dy}{dx} + xy = \sin x, y(1) = y_0$.

解 在方程两边同时除以 x^2，得到 $\dfrac{dy}{dx} + \dfrac{1}{x} y = \dfrac{\sin x}{x^2}$. 令 $P(x) = 1/x, Q(x) = (\sin x)/x^2$. 因为当 $x_0 = 1$ 时，$\rho(x) = \exp\left(\int_1^x \dfrac{1}{t} dt\right) = \exp(\ln x) = x$，所以所求特解为 $y(x) = \dfrac{1}{x}\left[y_0 + \int_1^x \dfrac{\sin t}{t} dt\right]$.

Ex. 1 Solve the initial value problem
$$x^2 \dfrac{dy}{dx} + xy = \sin x, y(1) = y_0.$$

Solution On division by x^2 both sides of the equation yields $\dfrac{dy}{dx} + \dfrac{1}{x} y = \dfrac{\sin x}{x^2}$. Let $P(x) = 1/x$ and $Q(x) = (\sin x)/x^2$. With $x_0 = 1$, the integrating factor is given by $\rho(x) = \exp\left(\int_1^x \dfrac{1}{t} dt\right) = \exp(\ln x) = x$, so the desired particular solution is given by $y(x) = \dfrac{1}{x}\left[y_0 + \int_1^x \dfrac{\sin t}{t} dt\right]$.

例 2 求解齐次线性方程 $\dfrac{dy}{dx} - \dfrac{3}{2x} y = \dfrac{2x}{y}$.

解 这是一个伯努利方程，其中 $P(x) = -3/(2x)$, $Q(x) = 2x, n = -1$ 和 $1 - n = 2$. 令 $v = y^2$, 则得到 $y = v^{1/2}, \dfrac{dy}{dx} = \dfrac{dy}{dv} \dfrac{dv}{dx} = \dfrac{1}{2} v^{-1/2} \dfrac{dv}{dx}$, 将其代入方程, 得到
$$\dfrac{1}{2} v^{-1/2} \dfrac{dv}{dx} - \dfrac{3}{2x} v^{1/2} = 2x v^{-1/2}.$$

两边同乘 $2v^{1/2}$，得到 $\dfrac{dv}{dx} - \dfrac{3}{x} v = 4x$, 其中 $\rho = e^{\int (-3/x) dx} = x^{-3}$. 所以我们得到
$$D_x(x^{-3} v) = \dfrac{4}{x^2}; \quad x^{-3} v = -\dfrac{4}{x} + C; \quad x^{-3} y^2 = -\dfrac{4}{x} + C.$$

由此可得方程的通解为 $y^2 = -4x^2 + Cx^3$.

Ex. 2 Solve the homogeneous linear equation $\dfrac{dy}{dx} - \dfrac{3}{2x} y = \dfrac{2x}{y}$.

Solution Obviously, it is a Bernoulli equation, where $P(x) = -3/(2x)$, $Q(x) = 2x, n = -1$ and $1 - n = 2$.

Let $v = y^2$, we have $y = v^{1/2}$ and $\dfrac{dy}{dx} = \dfrac{dy}{dv} \dfrac{dv}{dx} = \dfrac{1}{2} v^{-1/2} \dfrac{dv}{dx}$. This gives
$$\dfrac{1}{2} v^{-1/2} \dfrac{dv}{dx} - \dfrac{3}{2x} v^{1/2} = 2x v^{-1/2}.$$

Then on multiplication by $2v^{1/2}$ produces the linear equation, it yields $\dfrac{dv}{dx} - \dfrac{3}{x} v = 4x$, with the integrating factor $\rho = e^{\int (-3/x) dx} = x^{-3}$. So we obtain
$$D_x(x^{-3} v) = \dfrac{4}{x^2}; \quad x^{-3} v = -\dfrac{4}{x} + C; \quad x^{-3} y^2 = -\dfrac{4}{x} + C.$$

Consequently, the desired general solution is given by $y^2 = -4x^2 + Cx^3$.

本章重点

理解和掌握用各种初等积分方法求解一阶常微分方程，并能熟练地应用.

Key points of this chapter

Understand and grasp the various methods of elementary integration in solving the first-order differential equations, moreover, apply them proficiently.

第三章 存在和唯一性定理
Chapter 3　Existence and uniqueness theorems

本章主要介绍皮卡定理和皮亚诺存在定理,并介绍解的延伸和解的最大存在区间等有关问题.

This chapter chiefly introduces the Picard theorem, the Peano existence theorem, and the problems of extension and maximum existing interval of solutions, etc.

单词和短语 Words and expressions

★ 皮卡存在和唯一性定理　Picard theorem of existence and uniqueness
★ 李普西兹条件　Lipschitz condition
逐次迭代法　successive iterative method
皮卡序列　Picard sequence
一致收敛　uniform convergence
奥斯古德唯一性条件　Osgood uniqueness condition
★ 皮亚诺存在定理　Peano existence theorem
欧拉折线　Euler's polygonal arc
函数序列　function sequence
一致有界　uniform bound
等度连续　equicontinuous
子序列　sub-sequence
欧拉序列　Euler sequence
矩形区域　rectangular area
解的延伸　extension of solution

有界闭区域　bounded closed region
最大存在区间　maximum interval of existence
边界　boundary　['baundəri]
有限闭区间　finite closed interval
开集　open set
有限半开区间　finite semi-open interval
拉格朗日公式　Lagrange formula
局部　local　['ləukəl]
有限覆盖定理　finite covering theorem
条形域　bar area
单调递减　monotone decreasing
扇形区域　sectorial area
比较定理　comparability theorem
最小解　minimum solution
最大解　maximum solution
斜率　slope

基本概念和性质 Basic concepts and properties

1 李普西兹条件:设函数 $f(x,y)$ 在区域 D 内满足不等式
$$|f(x,y_1)-f(x,y_2)|\leqslant L|y_1-y_2|,$$
其中常数 $L>0$. 则称函数 $f(x,y)$ 在区域 D 内对 y 满足李普西兹条件.

Lipschitz condition: Assume that the function $f(x,y)$ in the range D satisfies the inequality $|f(x,y_1)-f(x,y_2)|\leqslant L|y_1-y_2|$ where the constant $L>0$. Then it is called the function $f(x,y)$ in the range D satisfies the Lipschitz condition with respect to y.

2 存在和唯一性定理:设初值问题 (E): $\dfrac{\mathrm{d}y}{\mathrm{d}x}=f(x,y), y(x_0)=y_0$, 其中 $f(x,y)$ 在矩形区域 R: $|x-x_0|\leqslant a, |y-y_0|\leqslant b$ 内连续,而且对 y 满足李氏条件. 则(E)在区间 $I=[x_0-h,x_0+h]$ 上有并且只有一个解,其中常数 $h=\min\left(a,\dfrac{b}{M}\right)$, 而 $M>\max\limits_{(x,y)\in R}|f(x,y)|$.

Existence and Uniqueness Theorem: For the given IVP (E): $\dfrac{\mathrm{d}y}{\mathrm{d}x}=f(x,y), y(x_0)=y_0$, where $f(x,y)$ is continuous in the rectangle area R: $|x-x_0|\leqslant a, |y-y_0|\leqslant b$, and satisfies the Lipschitz condition with respect to y. Then (E) on the interval $I=[x_0-h,x_0+h]$ has and only has one solution, where the constants $h=\min\left(a,\dfrac{b}{M}\right), M>\max\limits_{(x,y)\in R}|f(x,y)|$.

3 皮亚诺存在定理:设函数 $f(x,y)$ 在矩形区域 R 内连续,则初值问题
$$(E): \frac{\mathrm{d}y}{\mathrm{d}x}=f(x,y), y(x_0)=y_0$$
在区间 $|x-x_0|\leqslant h$ 上至少有一个解 $y=y(x)$, 这里 R,h 的定义同存在和唯一性定理.

Peano Existence Theorem: If the function $f(x,y)$ is continuous in the rectangle area R, then IVP (E): $\frac{dy}{dx} = f(x,y), y(x_0) = y_0$ has at least one solution $y = y(x)$ on the interval $|x-x_0| \leqslant h$, the definitions of R and h are the same as those in the existence and uniqueness theorem.

例 题 Examples

例 设初值问题 $\frac{dy}{dx} = x^2 + (y+1)^2, y(0) = 0$ 的解的右侧最大存在区间为 $[0,\beta)$，试证：
$$\frac{\pi}{4} < \beta < 1.$$

证 由存在和唯一性定理得，初值问题的解存在且唯一，并可延伸到包含坐标原点的任意区域的边界. 下面我们仅给出证明的梗概.

(1) 先证 $\frac{\pi}{4} \leqslant \beta \leqslant 1$.

当 $|x| \leqslant 1$ 时，显然有 $(y+1)^2 \leqslant x^2 + (y+1)^2 \leqslant 1 + (y+1)^2$.

因此，我们可以应用比较定理，把初值问题的解与如下两个可积的初值问题
$$(E_1): \frac{dy}{dx} = (y+1)^2, y(0) = 0$$
和
$$(E_2): \frac{dy}{dx} = 1 + (y+1)^2, y(0) = 0$$
的解分别比较，从而得到 $\frac{\pi}{4} \leqslant \beta \leqslant 1$.

(2) 再证 $\beta < 1$.

在微分方程的积分曲线上取一点 (ξ, η)，其中 $0 < \xi \ll 1$，则初值问题
$$(E_3): \frac{dy}{dx} = (y+1)^2, y(\xi) = \eta$$
是可积的，容易算出它的解的右侧最大存在区间为 $0 \leqslant x < C(\xi)$，其中 $C(\xi) = \xi + \frac{1}{\eta+1}$. 由于
$$\frac{dC}{d\xi} = 1 - \frac{1}{(\eta+1)^2} \frac{d\eta}{d\xi} = 1 - \frac{1}{(\eta+1)^2}[\xi^2 + (\eta+1)^2] < 0,$$
且 $C(0) = 1$，因此，当 $0 < \xi \ll 1$ 时，$C(\xi) < 1$. 再对(1)和 (E_3) 应用比较定理可得 $\beta < 1$.

(3) 最后证 $\beta > \frac{\pi}{4}$.

取正数 λ，使 $0 < 1 - \lambda \ll 1$，则初值问题
$$(E_4): \begin{cases} \frac{dy}{dx} = \lambda^2 + (y+1)^2 \\ y(0) = 0 \end{cases}$$
的解的右侧最大存在区间为 $0 \leqslant x < \widetilde{C}(\lambda)$. 计算表明，$\widetilde{C}(1) = \frac{\pi}{4}$，而且
$$\left. \frac{d\widetilde{C}(\lambda)}{d\lambda} \right|_{\lambda=1} < 0,$$
因此当 $0 < 1 - \lambda \ll 1$ 时，$\widetilde{C}(\lambda) > \frac{\pi}{4}$. 最后，再对(1)和 (E_4) 应用比较定理可得 $\beta > \frac{\pi}{4}$.

Ex Suppose that the right maximum interval of the existence solution of the initial value problem $\frac{dy}{dx} = x^2 + (y+1)^2, y(0) = 0$ is $[0,\beta)$, then $\frac{\pi}{4} < \beta < 1$.

Proof Because of the existence and uniqueness theorem, the existence and uniqueness of the solution of IVP can be extended to the boundary of any domain which goes through $(0,0)$. Here we only present the outline of the proof.

(1) First we proof that $\frac{\pi}{4} \leqslant \beta \leqslant 1$.

When $|x| \leqslant 1$, then $(y+1)^2 \leqslant x^2 + (y+1)^2 \leqslant 1 + (y+1)^2$. So, we can use the comparability theorem to compare the solution of the initial value problem with two integrable initial value problems, i. e.,

$$(E_1): \frac{dy}{dx} = (y+1)^2, y(0) = 0$$

and

$$(E_2): \frac{dy}{dx} = 1 + (y+1)^2, y(0) = 0,$$

then we have $\frac{\pi}{4} \leqslant \beta \leqslant 1$.

(2) Prove that $\beta < 1$.

Choose one point (ξ, η) on the integral curve, for which $0 < \xi \ll 1$, then the initial value problem

$$(E_3): \frac{dy}{dx} = (y+1)^2, y(\xi) = \eta$$

is integrable. It is easy to show that the right maximum interval of existence solution is given by $C(\xi) = \xi + \frac{1}{\eta+1}, 0 \leqslant x < C(\xi)$. Because

$$\frac{dC}{d\xi} = 1 - \frac{1}{(\eta+1)^2} \frac{d\eta}{d\xi} = 1 - \frac{1}{(\eta+1)^2}[\xi^2 + (\eta+1)^2] < 0$$

and $C(0) = 1$, so we have $C(\xi) < 1$ as $0 < \xi \ll 1$. We can use comparability theorem for (1) and (E_3), then $\beta < 1$.

(3) At last we prove $\beta > \frac{\pi}{4}$.

Choose a positive number λ such that $0 < 1 - \lambda \ll 1$, then the initial value problem

$$(E_4): \begin{cases} \frac{dy}{dx} = \lambda^2 + (y+1)^2, \\ y(0) = 0 \end{cases}$$

its right maximum interval of existence solution is $0 \leqslant x < \widetilde{C}(\lambda)$. Then we have $\widetilde{C}(1) = \frac{\pi}{4}$ and $\frac{d\widetilde{C}(\lambda)}{d\lambda}\bigg|_{\lambda=1} < 0$, so we get $\widetilde{C}(\lambda) > \frac{\pi}{4}$ as $0 < 1 - \lambda \ll 1$.

Consequently, we can use comparability theorem for (1) and (E_4), then $\beta > \frac{\pi}{4}$.

本章重点

由于微分方程最重要的理论基础之一是解的存在性和唯一性定理，所以对它们的证明方法及其应用范围都要理解和掌握。

Key points of this chapter

Since one of the fundamental theories concerning differential equations is the theorem of existence and uniqueness of solutions, it is required to understand and grasp the proving methods and their applications.

第四章　奇解
Chapter 4　Singular Solutions

本章先介绍与奇解密切相关的一阶隐式微分方程的解法，然后介绍奇解的概念和判别法，以及奇解和通解的联系。

This chapter first introduces solutions to the first-order implicit differential equations, which are closely related to singular solutions, then studies the concepts, test methods of singular solutions and also the relationship between singular solutions and general solutions.

单词和短语 Words and expressions

一阶隐式微分方程　first-order implicit differential equation
微分法　differentiation
克莱罗方程　Clairaut equation
参数　parameter
★ 奇解　singular solution
P-判别式　P-test formula
P-判别曲线　P-test curve
★ 包络　envelope　['envilǝup]
曲线族　curve family
相切　tangency　['tændʒǝnsi]
C-判别式　C-test formula
非退化　nonsingular　['nɔn'siŋgjulǝ]
切向量　tangent vector

基本概念和性质 Basic concepts and properties

1 设一阶微分方程 $F\left(x, y, \dfrac{dy}{dx}\right) = 0$ 有一特解 $\Gamma: y = \varphi(x)(x \in J)$. 如果对每一点 $Q \in \Gamma$, 且在 Q 点的任何邻域内, 方程有一个不同于 Γ 的解在 Q 点与 Γ 相切, 则称 Γ 是此微分方程的奇解.

Suppose that $\Gamma: y = \varphi(x)(x \in J)$ is a particular solution of the first-order differential equation $F\left(x, y, \dfrac{dy}{dx}\right) = 0$. For every point $Q \in \Gamma$, if the differential equation has a solution which is tangent with Γ, but is not Γ in any neighborhoods of Q, then Γ is called a singular solution of the differential equation.

2 设函数 $F(x,y,p)$ 对 $(x,y,p) \in G$ 是连续的, 而且对 y 和 p 有连续的偏微商 F'_y 和 F'_p, 其中 $p = \dfrac{dy}{dx}$.

若函数 $y = \varphi(x)(x \in J)$ 是微分方程 $F\left(x, y, \dfrac{dy}{dx}\right) = 0$ 的一个奇解, 并且有 $(x, \varphi(x), \varphi'(x)) \in G(x \in J)$, 则奇解 $y = \varphi(x)$ 满足一个称之为 P-判别式的联立方程 $F(x,y,p) = 0, F'_p(x,y,p) = 0$.

Suppose that the function $F(x,y,p)$ is continuous at $(x,y,p) \in G$ and has continuous partial differential quotients F'_y and F'_p with respect to y and p, where $p = \dfrac{dy}{dx}$. If the function $y = \varphi(x)(x \in J)$ is a singular solution of the differential equation and satisfies $(x, \varphi(x), \varphi'(x)) \in G\ (x \in J)$, then the singular solution $y = \varphi(x)$ satisfies an equation set called the P-test formula, i.e., $F(x,y,p) = 0$, $F'_p(x,y,p) = 0$.

例题 Examples

例　求解克莱罗方程 $y = xp + f(p)\left(p = \dfrac{dy}{dx}\right)$, 其中 $f''(p) \neq 0$.

解　利用微分法, 我们得到 $p = p + x\dfrac{dp}{dx} + f'(p)\dfrac{dp}{dx}$, 亦即 $[x + f'(p)]\dfrac{dp}{dx} = 0$.

当 $\dfrac{dp}{dx} = 0$ 时, 我们有 $p = C$. 因此, 得到克莱罗方程的通解为 $y = Cx + f(C)$, 其中 C 是一个任意常数. 当 $x + f'(p) = 0$ 时, 我们得到克莱罗方程的特解为
$$x = -f'(p), y = -f'(p)p + f(p).$$

Ex　Solve the Clairaut equation $y = xp + f(p)\left(p = \dfrac{dy}{dx}\right)$, where $f''(p) \neq 0$.

Solution　Using differentiation, we get $p = p + x\dfrac{dp}{dx} + f'(p)\dfrac{dp}{dx}$, that is $[x + f'(p)]\dfrac{dp}{dx} = 0$. As $\dfrac{dp}{dx} = 0$, we get $p = C$. So the general solution of the Clairaut equation is given by $y = Cx + f(C)$, where

C is an arbitrary constant. Moreover, as $x+f'(p)=0$, the particular solution of the Clairaut equation is given by

$$x=-f'(p), y=-f'(p)p+f(p).$$

本章重点

本章介绍一阶隐式微分方程奇解的概念、求奇解的方法及其判别式。这些内容要求作一般了解。

Key points of this chapter

This chapter mainly introduces the concepts of singular solutions, the methods of finding singular solution and the test formulas of the first-order implicit differential equations. A general understanding of these notions is required.

第五章 高阶微分方程
Chapter 5 Higher-Order Differential Equations

本章将通过一些具体的例子介绍微分方程的降阶技巧，然后讨论一般高阶微分方程的初值问题解的存在性和唯一性，以及解对初值和参数的连续性和可微性。

With some specific examples, this chapter will first present some reduced order skills to the differential equations, and then will discuss the existence and uniqueness of solutions of the higher-order differential equations with initial values, as well as the continuity and differentiability of solution for the given initial values and parameters.

单词和短语 Words and expressions

轨线　　trajectory　[trəˈdʒekətəri]
★ 相平面　　phase plane
★ 相图　　phase diagram
首次积分　　first integral
自变量　　independent-variable
未知函数　　unknown function
拓扑变换　　topological transform
逆变换　　inverse transform
邻域　　neighborhood　[ˈneibəhud]
开区域　　open region
n 维线性空间　　n-dimensional linear space

n 维向量值函数　　n-dimensional vector-valued function
函数组　　function group
标准微分方程　　normal differential equation
欧氏模　　Euler module
解对初值的连续依赖性　　continuous dependence of solution on initial value
解对参数的连续可微性　　continuity and differentiability of solution on parameter
★ 变分方程　　variational equation

基本概念和性质　Basic concepts and properties

如果微分方程组 $\begin{cases} \dfrac{dy_1}{dx} = f_1(x, y_1, y_2, \cdots, y_n) \\ \dfrac{dy_2}{dx} = f_2(x, y_1, y_2, \cdots, y_n) \\ \vdots \\ \dfrac{dy_n}{dx} = f_n(x, y_1, y_2, \cdots, y_n) \end{cases}$ 中的函数 $f_1, f_2, \cdots, f_n$ 都是关于 $y_1, y_2, \cdots, y_n$ 的线性函数，即 $f_k(x, y_1, y_2, \cdots, y_n) = \sum_{i=1}^{n} a_{ik}(x) y_i + e_k(x) (k=1, 2, \cdots, n)$，则称此其为线性微分方程组；否则，叫做非线性微分方程组。

If the functions $f_1, f_2, \cdots, f_n$ in the differential equations set

$$\begin{cases} \dfrac{\mathrm{d}y_1}{\mathrm{d}x} = f_1(x,y_1,y_2,\cdots,y_n) \\ \dfrac{\mathrm{d}y_2}{\mathrm{d}x} = f_2(x,y_1,y_2,\cdots,y_n) \\ \vdots \\ \dfrac{\mathrm{d}y_n}{\mathrm{d}x} = f_n(x,y_1,y_2,\cdots,y_n) \end{cases}$$

are all linear functions with respected to $y_1, y_2, \cdots, y_n$, namely,

$$f_k(x,y_1,y_2,\cdots,y_n) = \sum_{i=1}^{n} a_{ik}(x) y_i + e_k(x) \quad (k=1,2,\cdots,n),$$

then the differential equations is called the linear differential equations, otherwise, is called the nonlinear differential equations.

本章重点

解高阶微分方程的一般方法是降阶法,因此要熟练掌握降阶的一些技巧。对方程的解的性质和求解的方法也要进一步认识和理解。

Key points of this chapter

The most common methods of solving a higher-deferential equation are the methods of reduction of order, and hence a proficient mastering of the skills of reduced order is required. Further understanding and comprehending of properties of a solution to an equation and the methods for finding solutions are also required.

第六章 线性微分方程组
Chapter 6 Linear Differential Equations

本章讨论的线性微分方程组的一般理论和一些解法在应用中有重要的地位,也是进一步研究非线性微分方程组的基础。

This chapter will discuss the general theory of linear differential equations and their solving methods, which are of great importance in applications and also the foundation of nonlinear differential equations.

单词和短语 Words and expressions

★ 存在和唯一性定理　existence and uniqueness theorem
齐次线性微分方程组　homogeneous linear differential equations
基本解组　basic solutions
★ 朗斯基行列式　Wronskian determinant
刘维尔公式　Liouville formula
解矩阵　solution matrix
基解矩阵　basic solution matrix
常数矩阵　constant matrix
非齐次线性微分方程组　non-homogeneous linear differential equations
常数列向量　constant arrange vector
常数变异公式　formula of variation of constant
计算公式　computing formula

结构公式　framework formula
常系数线性微分方程组　constant coefficient linear differential equations
矩阵指数函数　exponential function of matrix
标准基解矩阵　normal basic solution matrix
若尔当标准型　Jordan normal form
矩阵函数　matrix function
实值解　real-valued solution
复值解　complex-valued solution
代数余子式　algebraic cofactor
★ 特征方程　characteristic equation
算子法　operator method
算子多项式　operator polynomial
拉普拉斯逆变换　Laplace inverse transform

基本概念和性质 Basic concepts and properties

1. 对于一阶线性微分方程组 $\dfrac{dY}{dx} = A(x)Y + f(x)$,其中 $Y = (y_1, y_2, \cdots, y_n)^T$, $f(x) = (f_1(x), f_2(x), \cdots, f_n(x))^T$。(1) 当 $f(x) \not\equiv 0$ 时,是非齐次线性微分方程组;(2)当 $f(x) \equiv 0$ 时,即得对应的齐次线性微分方程组。

For the first-order linear differential equations $\dfrac{dY}{dx} = A(x)Y + f(x)$, where $Y = (y_1, y_2, \cdots, y_n)^T$ and $f(x) = (f_1(x), f_2(x), \cdots, f_n(x))^T$. (1) If $f(x) \not\equiv 0$, the equations is called a non-homogeneous linear differential equations; (2) If $f(x) \equiv 0$, then the corresponding equations is called a homogeneous linear differential equations.

2. 存在和唯一性定理:线性微分方程组 $\dfrac{dY}{dx} = A(x)Y + f(x)$ 在区间 $a < x < b$ 上有并且只有一个满足初值条件 $Y(x_0) = Y_0$ 的解 $Y = Y(x)$,其中 $x_0 \in (a,b)$ 和 $Y_0 \in \mathbf{R}^n$ 是任意给定的.

Existence and uniqueness theorem: the linear differential equations $\dfrac{dY}{dx} = A(x)Y + f(x)$ has and only has a unique solution $Y = Y(x)$ on the interval $a < x < b$ which satisfies the initial condition $Y(x_0) = Y_0$, for any $x_0 \in (a,b)$ and for any $Y_0 \in \mathbf{R}^n$.

例 题 Examples

例 1 验证微分方程组 $\dfrac{d}{dx}\begin{pmatrix} y_1 \\ y_2 \end{pmatrix} = \begin{bmatrix} \cos^2 x & \dfrac{1}{2}\sin 2x - 1 \\ \dfrac{1}{2}\sin 2x + 1 & \sin^2 x \end{bmatrix} \begin{pmatrix} y_1 \\ y_2 \end{pmatrix}$ 的通解为

$$\begin{pmatrix} y_1 \\ y_2 \end{pmatrix} = C_1 \begin{pmatrix} e^x \cos x \\ e^x \sin x \end{pmatrix} + C_2 \begin{pmatrix} -\sin x \\ \cos x \end{pmatrix}.$$

证明 不难验证 $\begin{pmatrix} e^x \cos x \\ e^x \sin x \end{pmatrix}, \begin{pmatrix} -\sin x \\ \cos x \end{pmatrix}$ 是齐次线性微分方程组

$$\dfrac{d}{dx}\begin{pmatrix} y_1 \\ y_2 \end{pmatrix} = \begin{bmatrix} \cos^2 x & \dfrac{1}{2}\sin 2x - 1 \\ \dfrac{1}{2}\sin 2x + 1 & \sin^2 x \end{bmatrix} \begin{pmatrix} y_1 \\ y_2 \end{pmatrix}$$

在区间 $-\infty < x < \infty$ 上的两个解;而且它们的朗斯基行列式 $W(x)$ 在 $x = 0$ 处的值为 $W(0) = \begin{vmatrix} 1 & 0 \\ 0 & 1 \end{vmatrix} = 1 \neq 0$. 所以 $\begin{pmatrix} e^x \cos x \\ e^x \sin x \end{pmatrix}, \begin{pmatrix} -\sin x \\ \cos x \end{pmatrix}$ 是一个基本解组,从而 $\begin{pmatrix} y_1 \\ y_2 \end{pmatrix} = C_1 \begin{pmatrix} e^x \cos x \\ e^x \sin x \end{pmatrix} + C_2 \begin{pmatrix} -\sin x \\ \cos x \end{pmatrix}$ 是通解.

Ex. 1 Show that the general solution of the differential equations

$$\dfrac{d}{dx}\begin{pmatrix} y_1 \\ y_2 \end{pmatrix} = \begin{bmatrix} \cos^2 x & \dfrac{1}{2}\sin 2x - 1 \\ \dfrac{1}{2}\sin 2x + 1 & \sin^2 x \end{bmatrix} \begin{pmatrix} y_1 \\ y_2 \end{pmatrix}$$

is given by $\begin{pmatrix} y_1 \\ y_2 \end{pmatrix} = C_1 \begin{pmatrix} e^x \cos x \\ e^x \sin x \end{pmatrix} + C_2 \begin{pmatrix} -\sin x \\ \cos x \end{pmatrix}.$

Proof It is easy to show that $\begin{pmatrix} e^x \cos x \\ e^x \sin x \end{pmatrix}$ and $\begin{pmatrix} -\sin x \\ \cos x \end{pmatrix}$ are two solutions of the homogeneous linear differential equations $\dfrac{\mathrm{d}}{\mathrm{d}x}\begin{pmatrix}y_1\\y_2\end{pmatrix}=\begin{pmatrix}\cos^2 x & \dfrac{1}{2}\sin 2x-1\\ \dfrac{1}{2}\sin 2x+1 & \sin^2 x\end{pmatrix}\begin{pmatrix}y_1\\y_2\end{pmatrix}$ on the interval $-\infty<x<\infty$, moreover the Wronskian determinant of the homogeneous linear differential equations $W(x)$ at $x=0$ is that $W(0)=\begin{vmatrix}1 & 0\\0 & 1\end{vmatrix}=1\neq 0$.

And thus $\begin{pmatrix}\mathrm{e}^x\cos x\\ \mathrm{e}^x\sin x\end{pmatrix}$ and $\begin{pmatrix}-\sin x\\ \cos x\end{pmatrix}$ are the basic solutions of the homogeneous linear differential equations, that is to say, $\begin{pmatrix}y_1\\y_2\end{pmatrix}=C_1\begin{pmatrix}\mathrm{e}^x\cos x\\ \mathrm{e}^x\sin x\end{pmatrix}+C_2\begin{pmatrix}-\sin x\\ \cos x\end{pmatrix}$ is the general solution.

本章重点

理解线性微分方程组解的存在唯一性定理,进一步熟悉和掌握逐步逼近法.了解方程组的所有解的代数结构问题.

Key points of this chapter

Understand the existence and uniqueness theorem of solution to a system of linear differential equations, moreover, familiarize and grasp successive approximation method. Understand the algebraic structure problems of all solutions.

第七章 微分方程的幂级数解法
Chapter 7 Power Series Solutions of Differential Equations

本章的主要内容是,用幂级数解法求解勒让德方程和用广义幂级数解法求解贝塞尔方程,并讨论所得到的两个重要的特殊函数:勒让德多项式和贝塞尔函数.这些内容对于进一步学习数学物理方程是不可缺少的.

The main contents of this chapter are to solve the Legendre equation by using the method of power series and the Bessel equation by using the generalized method of power series, and then discuss the two special functions, i. e., Legendre polynomial and Bessel function. These are indispensable in further study of equations of mathematical physics.

单词和短语 Words and expressions

柯西定理 Cauchy theorem	正交 orthogonal [ɔː'θɔgnəl]
★ 解析解 analytic solution	广义傅里叶级数 generalized Fourier series
优级数 excellent series	广义幂级数 generalized power series
优函数 excellent function	指标 index ['indeks]
强函数 strong function	正则奇点 regular singular point
★ 幂级数 power series	指标方程 index equation
★ 幂级数解法 method of power series	共轭复根 conjugate complex root
奇点 singularity	贝塞尔方程 Bessel equation
勒让德方程 Legendre equation	★ 第一类贝塞尔函数 Bessel function of the first kind
勒让德多项式 Legendre polynomial	
罗德里格斯公式 Rodrigues formula	渐近式 asymptotic expression

本章重点

用幂级数解法求解勒让德方程和用广义幂级数解法求解贝塞尔方程,并讨论所得到的勒让德多项式和贝塞尔函数.

Key points of this chapter

Apply power series solution to solve the Legendre equation and apply generalized power series solution to solve the Bessel equation, and discuss all the obtained Legendre polynomials and Bessel functions.

第八章　定性理论与分支理论
Chapter 8　Qualitative Theory and Bifurcation Theory

微分方程定性理论的应用已深入到许多自然科学和社会科学领域,我们有必要对它的一些基本概念和基本方法作一个初步的介绍.

The qualitative theory of differential equations is applied to many disciplines such as natural science and social science. It is necessary to give a brief introduction on its basic concepts and methods.

单词和短语 Words and expressions

动力系统　dynamical system
相空间　phase space
增广相空间　augmented phase space
速度场　velocity field
向量场　vector field
拓扑结构图　topological structure graph
相图　phase diagram
几何理论　geometrical theory
平衡点　equilibrium point
奇点　singular point
闭轨　closed orbit
轨线的唯一性　uniqueness of orbit
抽象动力系统　abstract dynamical system
拓扑动力系统　topological dynamical system
微分动力系统　differential dynamical system
★ 解的稳定性　stability of solution
李雅普诺夫稳定性　Liapunov stability
渐近稳定　asymptotic stability
渐近稳定域　asymptotic stability field
吸引域　attraction domain
全局渐近稳定　globe asymptotic stability
线性近似　linear approximate
临界情形　critical state
★ 李雅普诺夫第一方法　first method of Liapunov
★ 李雅普诺夫第二方法　second method of Liapunov
直接方法　direct method
李雅普诺夫函数　Liapunov function
正定函数　positive definite function
负定函数　negative definite function
平面上的动力系统　dynamical system on plane
极限环　limit cycle

常点　constant point
初等奇点　elementary singular point
高阶奇点　higher-order singular point
星形结点　star node
临界结点　critical node
鞍点　saddle point
单向结点　one-way node
退化结点　improper node
稳定焦点　stable focus
不稳定焦点　unstable focus
中心点　center point
拓扑等价　topological equivalence
双曲奇点　hyperbolic singularity
稳定极限环　stable limit cycle
不稳定极限环　unstable limit cycle
半稳定极限环　semi-stable limit cycle
轨道稳定性　stability of orbit
庞加莱-班迪克环域定理　Poincaré-Bendixson annulus theorem of theorem
Liénard 方程　Liénard equation
Liénard 作图法　Liénard construction
Liénard 映射　Liénard mapping
后继函数法　subsequence function method
单极限环　single limit cycle
K 重极限环　K-fold limit cycle
鞍-结点　saddle-node
Hopf 分支　Hopf bifurcation
Poincaré 分支　Poincaré bifurcation
开折　unfolding　[ʌnˈfəuldiŋ]
普适开折　universal unfolding

基本概念和性质 Basic concepts and properties

1. 方程 $\dfrac{dx}{dt} = f(x)$ 称为一个一阶自治微分方程，若自变量 t 不明显出现。方程 $f(x) = 0$ 的解，称为自治微分方程 $dx/dt = f(x)$ 的临界点。

   The equation $\dfrac{dx}{dt} = f(x)$ is called a first-order autonomous differential equation, if the independent variable t does not appear explicitly. The solutions of the equation $f(x) = 0$ are called critical points of the autonomous differential equation $dx/dt = f(x)$.

2. 如果初始点 x_0 充分接近于 c，那么对于所有的 $t > 0$，$x(t)$ 仍充分接近于 c，则一阶自治微分方程的临界点 $x = c$ 称为稳定的。更确切地，临界点 $x = c$ 称为稳定的，如果对于每个 $\varepsilon > 0$，存在 $\delta > 0$ 使得：如果 $|x_0 - c| < \delta$，则 $|x(t) - c| < \varepsilon$，对于所有的 $t > 0$。否则，临界点 $x = c$ 称为不稳定的。

   A critical point $x = c$ of an autonomous first-order differential equation is said to be stable provided that if the initial value x_0 is sufficiently close to c, then $x(t)$ remains close to c for all $t > 0$. More precisely, the critical point $x = c$ is stable if, for each $\varepsilon > 0$, there exists $\delta > 0$ such that $|x_0 - c| < \delta$ implies that $|x(t) - c| < \varepsilon$ for all $t > 0$. Otherwise, the critical point $x = c$ is unstable.

3. 李雅普诺夫稳定性定理：假设 $f \in C(D)$，其中 $f(0) = 0$ 且存在一个对于 f 的李雅普诺夫函数 V，那么

   (a) 在 D 中 $\dot{V} \leqslant 0 \Rightarrow Y' = f(Y)$ 的零解稳定.

   (b) 在 $D \backslash \{0\}$ 中 $\dot{V} < 0 \Rightarrow Y' = f(Y)$ 的零解渐近稳定.

   (c) 在 $D(\alpha, \beta, b > 0)$ 中 $\dot{V} \leqslant -\alpha V$ 且 $V(x) \geqslant b|x|^\beta \Rightarrow$ 零解指数渐近稳定.

   Stability Theorem of Liapunov: Assume that $f \in C(D)$ with $f(0) = 0$ and there exists a Liapunov function V for f, Then

   (a) $\dot{V} \leqslant 0$ in $D \Rightarrow$ The zero solution of $Y' = f(Y)$ is stable.

   (b) $\dot{V} < 0$ in $D \backslash \{0\} \Rightarrow$ The zero solution of $Y' = f(Y)$ is asymptotically stable.

   (c) $\dot{V} \leqslant -\alpha V$ and $V(x) \geqslant b|x|^\beta$ in $D(\alpha, \beta, b > 0) \Rightarrow$ The zero solution is exponentially stable.

本章重点

初步介绍了微分方程定性理论的一些基本概念和基本方法.

Key points of this chapter

This chapter preliminarily introduces the fundamental concepts and methods in the qualitative theory of differential equations.

第九章 边值问题
Chapter 9 Boundary Value Problems

本章将讨论某些二阶微分方程的边值问题，而以施图姆-刘维尔边值问题为重点，因为它在数学物理中有重要的应用.

This chapter will discuss the boundary value problems of some second-order differential equations, in particular, the Sturm-Liouville boundary value problem is an emphasis for its important applications in mathematical physics.

单词和短语 Words and expressions

施图姆比较定理 Sturm comparison theorem	孤立点 isolated point
零点 zero ['ziərəu] point	非振动 non-oscillation

振动　vibration　[vaiˈbreiʃən] / oscillation
无限振动　infinite vibration
施图姆-刘维尔边值问题　Sturm-Liouville boundary problem

非零解　nonzero solution
特征函数系　characteristic function series
周期边值问题　periodic boundary value problem

本章重点

边值问题在数学和物理中有重要应用，特别是二阶微分方程的边值问题．

Key points of this chapter

Boundary value problems, especially the boundary value problems of second-order differential equation, are significantly applied in mathematics and physics.

练　习

1. 解初值问题 $\dfrac{dy}{dx} - y = \dfrac{11}{8} e^{-x/3}$，$y(0) = -1$．

   答案　$y(x) = \dfrac{1}{32} e^x - \dfrac{33}{32} e^{-x/3} = \dfrac{1}{32}(e^x - 33 \cdot e^{-x/3})$．

2. 解微分方程 $(6xy - y^3)dx + (4y + 3x^2 - 3xy^2)dy = 0$．

   答案　$3x^2 y - xy^3 + 2y^2 = C$．

3. 解初值问题 $y'' + 2y' + y = 0$，$y(0) = 5, y'(0) = -3$．

   答案　$y(x) = 5e^{-x} + 2xe^{-x}$．

4. 试求齐次系统的通解 $\dot{x}_1 = 2x_1 + x_2, \dot{x}_2 = -3x_1 + 6x_2$．

   答案　$x_1 = c_1 e^{3t} + c_2 e^{5t}$, $x_2 = c_1 e^{3t} + 3c_2 e^{5t}$．

5. 解下面 IVP 系统 $\dot{x}_1 = -3x_1 + 4x_2, \dot{x}_2 = -2x_1 + 3x_2$；$x_1(0) = -1, x_2(0) = 3$．

   答案　$x_1(t) = -\cosh t + 15\sinh t, x_2(t) = 3\cosh t + 11\sinh t$ 或 $x_1(t) = 7e^t - 8e^{-t}$, $x_2(t) = 7e^t - 4e^{-t}$．

6. 解下面 BVP 系统 $y'' + y = x, y(0) = 2, y(\pi) = 1$，其中 $0 \leqslant x \leqslant \pi$．

   答案　此 BVP 系统无解．

7. 证明非线性系统 $\begin{cases} \dot{x} = -3x + 4y + x^2 - y^2 \\ \dot{y} = -2x + 3y - xy \end{cases}$ 在临界点 $(0,0)$ 是不稳定的．

   证明　这里 $a = -3, b = 4, c = -2, d = 3$ 满足 $ad - bc = -1 \neq 0$．
   $F(x,y) = x^2 - y^2, G(x,y) = -xy$，并且 $F(0,0) = G(0,0) = 0$ 对 x 和 y 作极坐标变换：$x = r\cos\theta, y = r\sin\theta$ $(x \to 0, y \to 0 \leftrightarrow r \to 0)$．

   当 $r \to 0$ 时，有
   $$\frac{F(x,y)}{\sqrt{x^2+y^2}} = \frac{r^2(\cos^2\theta - \sin^2\theta)}{r} = r\cos 2\theta \to 0$$

   当 $r \to 0$ 时，有
   $$\frac{G(x,y)}{\sqrt{x^2+y^2}} = -\frac{r^2 \cos\theta\sin\theta}{r} = -r\cos\theta\sin\theta \to 0$$

   因此满足条件．对应的线性系统是
   $$\dot{x} = -3x + 4y, \quad \dot{y} = -2x + 3y \tag{1}$$

   系统(1)的特征方程为 $\lambda^2 - 1 = 0$，特征根为 $\lambda_1 = 1, \lambda_2 = -1$．由于 λ_1 和 λ_2 中有一个是正的，点 $(0,0)$ 是 (1) 的不稳定的临界点．根据定理，点 $(0,0)$ 是非线性系统的不稳定临界点．

Exercises

1. Solve the initial value problem $\dfrac{dy}{dx} - y = \dfrac{11}{8} e^{-x/3}$, $y(0) = -1$.

Answer $y(x) = \frac{1}{32}e^x - \frac{33}{32}e^{-x/3} = \frac{1}{32}(e^x - 33e^{-x/3})$.

2. Solve the differential equation $(6xy - y^3)dx + (4y + 3x^2 - 3xy^2)dy = 0$.

Answer $3x^2y - xy^3 + 2y^2 = C$.

3. Solve the initial value problem $y'' + 2y' + y = 0, y(0) = 5, y'(0) = -3$.

Answer $y(x) = 5e^{-x} + 2xe^{-x}$.

4. Find the general solution of the homogeneous system
$$\dot{x}_1 = 2x_1 + x_2, \quad \dot{x}_2 = -3x_1 + 6x_2.$$

Answer $x_1 = c_1 e^{3t} + c_2 e^{5t}, \quad x_2 = c_1 e^{3t} + 3c_2 e^{5t}$.

5. Solve the IVP $\dot{x}_1 = -3x_1 + 4x_2, \dot{x}_2 = -2x_1 + 3x_2 ; x_1(0) = -1, x_2(0) = 3$.

Answer $x_1(t) = -\cosh t + 15\sinh t, x_2(t) = 3\cosh t + 11\sinh t$ or $x_1(t) = 7e^t - 8e^{-t}, x_2(t) = 7e^t - 4e^{-t}$.

6. Solve the BVP $y'' + y = x, y(0) = 2, y(\pi) = 1$, where $0 \leqslant x \leqslant \pi$.

Answer The BVP has no solution.

7. Show that the critical point $(0,0)$ of the nonlinear system
$$\dot{x} = -3x + 4y + x^2 - y^2$$
$$\dot{y} = -2x + 3y - xy$$
is unstable.

Proof Here $a = -3, b = 4, c = -2$, and $d = 3$ with $ad - bc = -1 \neq 0$.
$F(x,y) = x^2 - y^2$, $G(x,y) = -xy$, with $F(0,0) = G(0,0) = 0$. We express x and y in polar coordinates: $x = r\cos\theta, y = r\sin\theta$. (Then $x \to 0$ and $y \to 0$ is equivalent to $r \to 0$).

$$\frac{F(x,y)}{\sqrt{x^2+y^2}} = \frac{r^2(\cos^2\theta - \sin^2\theta)}{r} = r\cos 2\theta \to 0 \text{ as } r \to 0$$

and

$$\frac{G(x,y)}{\sqrt{x^2+y^2}} = -\frac{r^2\cos\theta\sin\theta}{r} = -r\cos\theta\sin\theta \to 0 \text{ as } r \to 0$$

Hence, the conditions are satisfied. The linearized system is
$$\dot{x} = -3x + 4y, \quad \dot{y} = -2x + 3y \tag{1}$$

The characteristic equation of system (1) is $\lambda^2 - 1 = 0$. Its roots are $\lambda_1 = 1$ and $\lambda_2 = -1$. Since one of them is positive, point $(0,0)$ is an unstable critical point of (1). By Theorem, the point $(0,0)$ is an unstable critical point of the nonlinear system.

第五部分 实变函数与泛函分析
Part 5 Real Variable Function and Functional Analysis

引 言

实变函数论在 19 世纪末 20 世纪初,主要由法国数学家勒贝格(Lebesgue,1875~1941)创立的. 它是普通微积分学的推广,其目的是想克服牛顿和莱布尼茨所建立的微积分学的缺点,使得微分与积分的运算更加对称与完美.

我们以前学习的微积分有一个明显不足:黎曼意义下的可积函数类太少,例如狄利克雷函数看上去很简单却不可积.

那么黎曼积分究竟有什么缺陷? 让我们回顾一下黎曼积分的定义. 对一个由 $y = f(x)$ 围成的曲边梯形来说,要求它的面积总是用内填外包法. 首先将定义区间 $[a,b]$ 分割为小区间 Δ_i,然后以小区间 Δ_i 的长度为底、函数在 Δ_i 上的下确界 m_i 为高的矩形内填,并且以相同的底, Δ_i 的上确界 M_i 为高的矩形外包(其中 $x_{i-1} \leqslant \varepsilon_i \leqslant x_i$). 当把区间分得很细时,内填外包的矩形面积之差就无限小,彼此都趋向一个定值 L,这就得到了定积分:

$$\sum_i m_i \Delta x_i \leqslant \sum_i f(\varepsilon_i) \Delta x_i \leqslant \sum_i M_i \Delta x_i. \qquad (*)$$

而对于狄利克雷函数 $D(x)$,不管把 $[0,1]$ 区间划分成多么小的 n 个区间,每个小区间里都有无理数和有理数,$D(x)$ 的函数值分别取 0 和 1,它们彼此之差都是 1. ($*$)式的左端恒为 0,右端恒为 1,不会趋于相同的值,于是 $D(x)$ 在黎曼意义下就是"不可积"的.

如上所述,用黎曼积分求曲边梯形的面积,是以 Δx_i 为底边的矩形进行"内填外包"的. 现在,我们换个角度去思考问题:用 Δy_i 为底边的矩形去内填外包. 也就是说:求曲边梯形面积时不要去分定义域,而是去分值域,把函数值差不多的点集放在一起考虑,用横放着的小矩形面积之和加以逼近. 仍以 $D(x)$ 为例,它只取两个函数值 0 和 1, 取 0 的是 $[0,1]$ 中的无理数集 I, 取 1 的是 $[0,1]$ 中的有理数集 Q. 假定 I 的"长度"是 $m(I) = 1, Q$ 的长度 $m(Q) = 0$,不管把 y 轴上的 $[0,1]$ 区间分得如何细,因 $D(x)$ 只有两个值,它和 $[0,1]$ 构成的曲边梯形"面积"是:$1 \cdot m(Q) + 0 \cdot m(I)$. 这样,问题归结为如何来确定 $m(Q)$ 和 $m(I)$ 了. 众所周知,在微积分课程里,Q,I 之类的集合是没有"长度"的. 这要求我们重新制定一套理论,按照勒贝格创立的测度论,$m(Q) = 0, m(I) = 1$, 于是 $D(x)$ 的勒贝格积分该是 0, 问题迎刃而解!

通过建立 Lebesgue 积分理论,微积分的缺点得到了弥补. 为了达到这个目的,我们要首先研究集合的度量方法,进一步再研究测度论,之后给出积分的定义及其性质.

函数是数与数之间的对应关系,泛函是函数与数之间的对应关系,算子是函数空间与函数空间之间的对应关系. 由于函数空间是无限维的,因此泛函分析研究的问题要复杂得多. 同时我们也会发现很多有限维空间的理论和方法仍然可以推广到无穷维空间之中. 泛函分析主要研究:距离空间、完备、范数、赋范线性空间、Banach 空间、内积空间、Hilbert 空间等理论.

Introduction

Theory of Real Variable Function is established by the French mathematician Lebesgue (1875~1941) in the later of 19th century and the earlier of 20th century. It is an extension of general calculus. The goal is that it intends to overcome the defect of calculus which was established by Newton-Leibniz, so as to make the operation between differential and integral more symmetric as well as more perfect.

One obvious defect of calculus we have learned before is that there are the Riemann integrable functions are too little. Riemann integrable functions. For example, Dirichlet function is simple but not integrable.

So what defect does the Riemann integrals have? Let us look back upon its definition. For an curvilinear trapezoid rounded by $y = f(x)$, to obtain its area we always use the method of inner expansion and outer contraction. Firstly, we divide the interval $[a,b]$ into many small intervals written as Δ_i. Then we measure the areas of small rectangles by using the method of inner expansion, namely, take the bottom of Δ_i as length and the infimum m_i of the function in Δ_i as height, and by using the method of outer contraction, namely, take the same bottom of Δ_i as length and the supremum M_i of the function in Δ_i as height. As the interval is partitioned so small that the difference of the areas between the rectangles of inner expansion and outer contraction is infinitesimal, tending to a definite number L, and thus we can get the definite integral:

$$\sum_i m_i \Delta x_i \leqslant \sum_i f(\varepsilon_i) \Delta x_i \leqslant \sum_i M_i \Delta x_i. \ (x_{i-1} \leqslant \varepsilon_i \leqslant x_i) \qquad (*)$$

However, for the Dirichlet function $D(x)$, there always have rational numbers and irrational numbers in Δ_i no matter how small Δ_i is. The value of function $D(x)$ is 0 or 1 respectively. Thus, the left of $(*)$ is 0 and the right of $(*)$ is 1, which means that it is not Riemann integrable.

To sum up, in order to get the area of curvilinear trapezoid, we use the method of inner expansion and outer contraction and use Δx_i as the bottom. Correspondingly, we can think about it in another perspective, namely, we use Δy_i as the bottom to compute the area of curvilinear trapezoid. That is to say, we divide the range instead of the domain, and put the set of points having similar value of function together and use these rectangles laid in horizontal as approximation. Still take $D(x)$ as an example, and its function value is 0 or 1. For any $x \in [0,1]$, if $x \in I$, $f(x) = 0$, while if $x \in Q$, $f(x) = 1$. Let $m(I) = 1$ and $m(Q) = 0$. For $D(x)$ only has two values, the areas of curvilinear trapezoids are always equal to $1 \cdot m(Q) + 0 \cdot m(I)$ no matter how small intervals that $[0,1]$ is divided on axis y. Consequently, the problem becomes how to determine $m(Q)$ and $m(I)$. As we all know, Q and I don't have the length in calculus course. We need a new theorem. According to the measure theorem established by Lebesgue, $m(I) = 1$ and $m(Q) = 0$. The Lebesgue integral of $D(x)$ is 0, so the problem is easy to solve.

The defect of Calculus is remedied by the establishment of Lebesgue integral. For this purpose, we need to study the measure method of set firstly. It means that we can give the definition and properties of integral after studying the measure theory.

Corresponding relation between numbers is called functions, corresponding relation between functions and numbers is called functional, corresponding relation between function spaces is called operator. Because on the dimension of function spacs is infinite, the problems studied by Functional Analysis become more complex. Meanwhile, we also find that many theorems and methods of finite dimensional space can be generalized to infinite dimensional space. In this part, we mainly deal with the theories of metric space, such as completeness, norm, normed linear space, Banach space, inner-product space and Hilbert space.

本课程的基本要求

学习本课程需要用到数学分析、解析几何、高等代数课程的基础知识. 在学习本课程过程中,学生应特别注意培养自己的抽象思维能力与逻辑能力和对基本概念实质的理解.

Basic requirements of this course

The course follows Mathematics Analysis, Analytic Geometry and Higher Algebra. To study this course, we should pay attention to advance the abiling for abstract thinking, logical analysis and understanding to the basic concepts.

第一章 预备知识
Chapter 1 Preliminary Knowledge

实变函数论是在点集论的基础上研究分析数学中的一些最基本的概念和性质. 因此,首先要学习集合的运算，包括代数运算与极限运算，以及基数、可数集、不可数集等概念．

Theory of Real Variable Function is to research some basic concepts and properties of analytic mathematics based on set theory. We should learn the operation of sets including algebra operation and limit operation, the concepts of base, countable set, uncountable set, and so on.

单词和短语 Words and expressions

集合的运算 operation of sets	★ 不可数集 uncountable set
对等的集合 equipotent sets	上限集合 upper limit set
基数 base	下限集合 lower limit set
★ 可数集 countable set	实数域 field of real numbers

基本概念和性质 Basic concepts and properties

1 上限集包含下限集．如果上限集与下限集相等，我们称集列收敛．

Upper limit set includes lower limit set. We call the set sequence is convergent if the upper limit set is equal to the lower limit set.

2 对等的集合具有相同的基数．

Equipotent sets have the same base.

3 有理数集合是可数集合．

Rational numbers set is a countable set.

4 实数集合是不可数集合．

Real numbers set is an uncountable set.

例 题 Examples

例 1 证明 $(-1,1)$ 和 $(-\infty,+\infty)$ 之间存在 1-1 映射，并写出这一对应的解析式．

证明 设 $y = \tan\frac{\pi}{2}x, x \in (-1,1)$．很明显，它是从 $(-1,1)$ 到 $(-\infty,+\infty)$ 的 1-1 映射．

Ex. 1 Please give a one to one mapping from $(-1,1)$ and $(-\infty,+\infty)$.

Proof Let $y = \tan\frac{\pi}{2}x, x \in (-1,1)$. Obviously, it is a one to one mapping from $(-1,1)$ to $(-\infty,+\infty)$.

例 2 证明直线上某些互不相交的开区间组成的集合至多为可数集．

证明 我们记那些集合构成的集合为 $\hbar$．对任何的 $A, B \in \hbar$，有 $A \cap B = \emptyset$．因为有理数集是可数的，分别从 A 和 B 中任取两个有理数，可知，$\hbar$ 是可数集．

Ex. 2 Prove that the sets which consist of some open intervals on real line and do not intersect each other is at most countable.

Proof We denote the sets by $\hbar$. For any $A, B \in \hbar$, we know that $A \cap B = \emptyset$. Because the rational set is countable, we take two rational from A and B respectively. So $\hbar$ is a countable set.

本章重点

1 理解上限集合与下限集合的定义．

2 掌握对等集合映射关系的建立方法．

3 理解可数集与不可数集的定义，记住有理数集合是可数集合，实数集合不可数．

Key points of this chapter:

1 Understand the definitions of upper limit and lower limit sets.

2 Master the method to construct the mapping between the equivalent sets.

3 Understand the definitions of countable set and uncountable set. Remember that rational numbers set is a countable set and the real numbers set is an uncountable set.

第二章 点集的拓扑概念
Chapter 2 Topological Concepts for Sets

本章我们将研究点与点之间的距离. 距离是一种结构, 距离空间也叫度量空间. 我们可以研究邻域、开集、闭集、完备集、Cantor 集、覆盖、有限覆盖、可数覆盖, 也就是将欧氏空间的理论推广到更一般的度量空间, 并给出欧氏空间没有的性质.

In this chapter, we study the distance between points. Distance is a class of structures, and a metric space is also called a distance space. We can study neighborhood, open set, closed set, complete set, Cantor set, cover, finite cover and countable cover, that is to say, the theories of a Euclidean space are generalized to a more general metric space. Meanwhile, we will show the properties which Euclidean space does not exist.

单词和短语 Words and expressions

距离空间	distance space	覆盖	cover
开集	open set	有限覆盖	finite cover
闭集	closed set	★ 可数覆盖	countable cover
★ Cantor 集	Cantor set	等价	equivalence

基本概念和性质 Basic concepts and properties

1 掌握距离空间与欧氏空间的区别.
Master the difference between metric space and Euclidean space.

2 距离空间是欧氏空间的推广, 并且它具有许多欧氏空间没有的性质.
Metric space is a generalization of Euclidean space, and it has many properties that Euclidean space does not process.

3 开集、闭集等概念与数学分析中有所不同.
Some concepts, such as open set and closed set, are different from those in mathematics analysis.

例题 Examples

例 1 可数个开集的并集是开集.

证明 设 $p \in \bigcup_{n=1}^{\infty} A_n$ (A_n 是开集), 那么, 必存在一个 A_n, 使得 $p \in A_n$, 因此, 对于 p 的一个邻域 $U(p)$ 有: $U(p) \subset A_n$, 显然, $U(p) \subset \bigcup_{n=1}^{\infty} A_n$, 所以 $\bigcup_{n=1}^{\infty} A_n$ 是一个开集.

Ex. 1 Show that the union of countable open sets is also an open set.

Proof Let $p \in \bigcup_{n=1}^{\infty} A_n$ (A_n is an open set), then there exists A_n such that $p \in A_n$, thus we have $U(p)$ which is a neighborhood of p with $U(p) \subset A_n$, hence that $U(p) \subset \bigcup_{n=1}^{\infty} A_n$, namely, $\bigcup_{n=1}^{\infty} A_n$ is an open set.

本章重点

1 深刻理解和熟悉掌握内点、极限点、开集、闭集等拓扑概念及其性质, 并能熟练运用这些概念进行逻辑推理.

2 掌握 Cantor 集的构造及其性质.

3 了解连续性、覆盖等概念.

Key points of this chapter

1 Understand deeply and master skillfully the topology concepts and their properties of inner point, limit point, open set, closed set, and so on. Be able to make logical inference by using these concepts.

2 Master the construction and its property of Cantor set.

3 Understand the concepts of continuity, cover, etc..

第三章 测度论
Chapter 3　Measure Theory

　　测度是长度、面积、体积等概念的推广,测度为一个集合函数,满足非负性、可列可加性、正则性,常采用内填外包法,通过内外测度取确界方法来求测度.

　　Measure is generalizations of length, area, volume, and so on. It is a function of set satisfying the properties such as non-negativity, countable additivity, regularity. We can find the measure with the method by taking supremum and infimum of inner measure and outer measure respectively, and the usually used method is interion expansion and exterior contraction.

单词和短语　Words and expressions

★ 勒贝格测度　　Lebesgue measure　　　　　　外测度　　exterior measure
★ G_δ 形集合　　G_δ set　　　　　　　　　　　　可数可加　countable additivity
★ F_σ 形集合　　F_σ set　　　　　　　　　　　　正则性　　regularity
★ Borel 集合　　Borel set

基本概念和性质　Basic concepts and properties

1 空集的测度为零.
　　The measure of an empty set is zero.

2 集合 E 为可测的当且仅当对于任意 $A \subset E, B \subset E^c$,总有:
$$m^*(A \cup B) = m^*(A) + m^*(B).$$
A set E is measurable if and only if for any $A \subset E, B \subset E^c$, we have
$$m^*(A \cup B) = m^*(A) + m^*(B).$$

3 开集、闭集皆可测.
　　Both open set and closed set are measurable.

4 凡 Borel 集合都是 L-可测的.
　　All the Borel sets are L-measurable.

例　题　Examples

例 1　证明:集合 E 为可测当且仅当对于任意 $A \subset E, B \subset E^c$,总有
$$m^*(A \cup B) = m^*(A) + m^*(B).$$
　　证明　**必要性**　取 $T = A \cup B$,则 $T \cap E = A, T \cap E^c = B$,所以
$$m^*(A \cup B) = m^*(T) = m^*(T \cap E) + m^*(T \cap E^c) = m^*(A) + m^*(B).$$
　　充分性　对于任意的 T,令 $A = T \cap E, B = T \cap E^c$,则 $A \subseteq E, B \subseteq E^c$,且 $A \cup B = T$,因此
$$m^*(T) = m^*(A \cup B) = m^*(A) + m^*(B) = m^*(T \cap E) + m^*(T \cap E^c).$$

Ex. 1　The set E is measurable if and only if $m^*(A \cup B) = m^*(A) + m^*(B)$ for any $A \subset E, B \subset E^c$.

Proof　**Necessity**　Let $T = A \cup B$, then $T \cap E = A, T \cap E^c = B$, then
$$m^*(A \cup B) = m^*(T) = m^*(T \cap E) + m^*(T \cap E^c) = m^*(A) + m^*(B).$$

Sufficiency　For any T, let $A = T \cap E, B = T \cap E^c$, then $A \subseteq E, B \subseteq E^c$, and $A \cup B = T$, and thus
$$m^*(T) = m^*(A \cup B) = m^*(A) + m^*(B) = m^*(T \cap E) + m^*(T \cap E^c).$$

例 2　证明:开集、闭集都可测.

　　证明　因为任何非空开集可表示为可数多个互不相交的左开右闭区间的并(在 R^1 则可表示为有限

个或可数多个开区间之并,其中包含无界的区间),而区间是可测的.所以开集可测,则闭集作为开集的余集自然也可测.

Ex. 2 Prove that open sets and closed sets are measurable.

Proof Because any open sets may be represented by the union of countable open intervals that are not intersect each other and these intervals are measurable. So the open sets are measurable and the closed sets are measurable too.

本章重点

1. 深入理解和掌握外测度与测度的概念.
2. 熟悉测度的基本性质.
3. 了解从 R^n 上的测度推广到一般集合上的基本思路.

Key points of this chapter

1. Understand and master the concepts of (exterior) Lebesgue measure.
2. Familiarize the basic properties of Lebesgue measure.
3. Understand the basic thinking that a measure of a general set is made from Lebesgue measure on R^n.

第四章 可测函数
Chapter 4 Measurable Functions

在数学分析中所研究的函数基本上是连续的,而实变函数研究的函数范围宽了许多,本章我们研究可测函数,连续函数当然可测. 我们要研究可测函数的构造和可测函数列的收敛问题等等.

The functions we have studied in Mathematics Analysis are on the basis of continuous, while the study range in real variable functions extends a lot. Many problems such as the measurable function, the construction of measurable function and the convergence of a sequence of measurable function can be studied in this chapter. Of course, continuous function is also measurable.

单词和短语 Words and expressions

* 可测函数　measurable function
* 几乎处处收敛　almost everywhere convergence
  一致收敛　uniform convergence
* 依测度收敛　convergence by measure
  Eorou 定理　Eorou theorem

* Lebesgue 控制收敛定理　Lebesgue control convergence theorem
  Riesz 定理　Riesz theorem
  Lusin 定理　Lusin theorem.

基本概念和性质 Basic concepts and properties

1. 可测集上的连续函数都可测.
   Continuous functions defined on a measurable set are measurable.
2. 设 f 是可测集合 E 上的可测函数,则存在闭集 $F \subseteq E$,在其上 f 是连续的.
   Let f be a measurable function that is defined on the measurable set E, there exists a closed set $F \subseteq E$ such that f is continuous on it.
3. 设在可测集合 E 上 $\{f_n\}$ 依测度收敛于 f,则存在子列 $\{f_{n_i}\}$ 在 E 上几乎处处收敛于 f.
   Let $\{f_n\}$ be defined on the measurable set E that converges by measure, there exists subsequence of $\{f_n\}$ such that $\{f_{n_i}\}$ is almost everywhere convergent to f in E.

例题 Examples

例 1 设 $f(x)$ 是定义在可测集 E 上的实函数,则下列任一条件都是 $f(x)$ 在 E 上可测的充要条件:

(1) 对任何的有限实数 a, $E[f \geqslant a]$ 都是可测的.

(2) 对任何的有限实数 a, $E[f < a]$ 都是可测的.

证明 由于对任何的有限实数 a，$E[f \geqslant a] = \bigcap_{n=1}^{\infty} E\left[f > a - \dfrac{1}{n}\right]$，这样，$E[f \geqslant a]$ 可测，并且由于 $E[f < a]$ 与 $E[f \geqslant a]$ 互为余集，因此可测，反之易证。

Ex. 1 Let $f(x)$ be a function defined on the measurable set E, prove that $f(x)$ is measurable on E if and only if one of the following conditions is true.

(1) For every finite real number a, $E[f \geqslant a]$ is measurable.

(2) For every finite real number a, $E[f < a]$ is measurable.

Proof For any a, $E[f \geqslant a] = \bigcap_{n=1}^{\infty} E\left[f > a - \dfrac{1}{n}\right]$, so $E[f \geqslant a]$ is measurable, and $E[f < a]$ is measurable too. The result of converse is easy to prove.

例 2 证明：设 $f_n(x) \Rightarrow f(x)$，$f_n(x) \Rightarrow g(x)$，则 $f(x) = g(x)$ 在 E 上几乎处处成立。

证明 由于 $|f(x) - g(x)| \leqslant |f(x) - f_k(x)| + |f_k(x) - g(x)|$，故对任何自然数 n，

$$E\left[|f - g| \geqslant \dfrac{1}{n}\right] \subset E\left[|f - f_k| \geqslant \dfrac{1}{2n}\right] \cup E\left[|f_k - g| \geqslant \dfrac{1}{2n}\right],$$

从而 $mE\left[|f - g| \geqslant \dfrac{1}{n}\right] \leqslant mE\left[|f - f_k| \geqslant \dfrac{1}{2n}\right] + mE\left[|f_k - g| \geqslant \dfrac{1}{2n}\right]$.

令 $k \to \infty$，即得 $mE\left[|f - g| \geqslant \dfrac{1}{n}\right] = 0$。但是由 $E[f \neq g] = \bigcup_{n=1}^{\infty} E\left[|f - g| \geqslant \dfrac{1}{n}\right]$ 可知，$mE[f \neq g] = 0$，即 $f(x) = g(x)$ 在 E 上几乎处处成立。

Ex. 2 Let $f_n(x) \Rightarrow f(x)$, $f_n(x) \Rightarrow g(x)$, then $f(x) = g(x)$ is valid almost everywhere on E.

Proof Because of $|f(x) - g(x)| \leqslant |f(x) - f_k(x)| + |f_k(x) - g(x)|$, for every natural number n, we have

$$E\left[|f - g| \geqslant \dfrac{1}{n}\right] \subset E\left[|f - f_k| \geqslant \dfrac{1}{2n}\right] \cup E\left[|f_k - g| \geqslant \dfrac{1}{2n}\right],$$

hence, $mE\left[|f - g| \geqslant \dfrac{1}{n}\right] \leqslant mE\left[|f - f_k| \geqslant \dfrac{1}{2n}\right] + mE\left[|f_k - g| \geqslant \dfrac{1}{2n}\right]$.

Let $k \to \infty$, we get $mE\left[|f - g| \geqslant \dfrac{1}{n}\right] = 0$. But from $E[f \neq g] = \bigcup_{n=1}^{\infty} E\left[|f - g| \geqslant \dfrac{1}{n}\right]$, we have $mE[f \neq g] = 0$, that is, $f(x) = g(x)$ is valid almost everywhere on E.

本章重点

1. 掌握可测函数的定义及其基本性质。
2. 掌握可测函数列的几种不同的收敛概念及其相互关系，了解 Eorou 定理、Lebesgue 定理的证明思路。

Key points of this chapter

1. Master the definition and the basic properties of a measurable function.
2. Master several different kinds of convergence concepts and the correlations of measurable functions, and understand Eorou Theorem, Lebesgue Theorem, the thinking of testifying the theorems.

第五章 积分理论
Chapter 5   Integration Theory

经过了前面的准备工作，我们有了可测集、可测函数的概念。本章的 Lebesgue 积分是实变函数的中心理论，主要内容有 Levi 单调收敛定理、Fatou 引理、Lebesgue 控制收敛定理、有界变差函数、绝对连续函数、Fubini 定理等。

We have mastered the concepts of measurable set and measurable function by former study. Lebesgue integral is the kernel theory of real variable function. In this chapter, we mainly deal with Levi monotony convergence theorem, Fatou Lemma, Lebesgue control convergence theorem, variable function, absolutely continuous function, Fubini theorem and so on.

单词和短语 Words and expressions

积分定理　integral theorem
★ Levi 单调收敛定理　Levi monotony convergence theorem
Fatou 引理　Fatou Lemma
★ Fubini 定理　Fubini theorem
★ 有界变差函数　bounded variation function
绝对连续函数　absolutely continuous function

基本概念和性质 Basic concepts and properties

1 测度为有限的可测集 E 上的有界函数 f 在 E 上 L-可积的充要条件为：对于 $\forall \varepsilon > 0$，存在 E 的分划 D 使得 $S(D,f) - s(D,f) < \varepsilon$.

That the measure of a bounded function f defined on the measurable set E is finite is L-integral iff for $\forall \varepsilon > 0$, there exists dividing D of E such that
$$S(D,f) - s(D,f) < \varepsilon.$$

2 测度为有限的可测集 E 上的有界函数 f L-可积的充要条件为 f 可测.

A bounded function f defined on the measurable set E that is finite measure is L-integral iff f is measurable.

3 $[a,b]$ 上 Riemann 可积的函数 f 一定为 Lebesgue 可积,反之未必成立.

f must be Lebesgue integral if it is Riemann integral on $[a,b]$. Contrarily, it is not true.

4 推广的牛顿-莱布尼茨定理为：如果 $F(x)$ 是 $[a,b]$ 上绝对连续函数,则几乎处处有定义的 $F'(x)$ 在 $[a,b]$ 上可积且 $F(x) = F(a) + \int_a^x F'(t)\mathrm{d}t$.

Generalized Newton-Leibniz theorem: If $F(x)$ is an absolutely continuous function on $[a,b]$, then $F'(x)$ is integral almost every where on $[a,b]$ and $F(x) = F(a) + \int_a^x F'(t)\mathrm{d}t$.

例 题 Examples

例 1 设 f 在 $E = [a,b]$ 上可积,则对于任何 $\varepsilon > 0$,必存在 E 上的连续函数 $\varphi(x)$ 使得
$$\int_a^b |f(x) - \varphi(x)| \mathrm{d}x < \varepsilon.$$

证明 设 $e_n = E[|f| > n]$,由于 $f(x)$ 在 E 上 a.e. 有限,那么 $me_n \to 0, n \to \infty$. 根据积分的绝对连续性,对于任何 $\varepsilon > 0$,存在 N,使得
$$N \cdot me_n \leqslant \int_{e_n} |f(x)| \mathrm{d}x < \frac{\varepsilon}{4}.$$

令 $B_n = E \setminus e_n$,在 B_n 上利用 Lusin 定理,存在闭集 $F_n \subset B_n$ 和 $\mathbf{R}^1$ 上的连续函数 $\varphi(x)$ 使得

(1) $m(B_n \setminus F_n) < \frac{\varepsilon}{4N}$;

(2) $x \in F_n, f(x) = \varphi(x)$ 并且 $\sup\limits_{x \in \mathbf{R}^1} |f(x)| = \sup\limits_{x \in \mathbf{R}^1} |\varphi(x)| \leqslant N$.

这样,
$$\int_a^b |f(x) - \varphi(x)| \mathrm{d}x \leqslant \int_{e_n} |f(x) - \varphi(x)| \mathrm{d}x + \int_{B_n} |f(x) - \varphi(x)| \mathrm{d}x$$
$$\leqslant \int_{e_n} |f(x)| \mathrm{d}x + \int_{e_n} |\varphi(x)| \mathrm{d}x + \int_{B_n \setminus F_n} |f(x) - \varphi(x)| \mathrm{d}x + \int_{F_n} |f(x) - \varphi(x)| \mathrm{d}x$$
$$< \varepsilon + \int_{B_n \setminus F_n} |f(x) - \varphi(x)| \mathrm{d}x + \int_{e_n} |\varphi(x)| \mathrm{d}x$$
$$\leqslant \frac{\varepsilon}{4} + N \cdot me_n + 2N \cdot \frac{\varepsilon}{4N} \leqslant \varepsilon.$$

Ex. 1 Let f be integrable on $E=[a,b]$, then for any $\varepsilon > 0$, there exists a continuous function $\varphi(x)$ on $[a,b]$, such that $\int_a^b |f(x)-\varphi(x)|\,dx < \varepsilon$.

Proof Let $e_n = E[|f|>n]$, since $f(x)$ is finite on E a. e. then $me_n \to 0$, $n \to \infty$. By the absolute continuous, for any $\varepsilon > 0$, there exists N, with

$$N \cdot me_n \leqslant \int_{e_n} |f(x)|\,dx < \frac{\varepsilon}{4}.$$

Let $B_n = E \setminus e_n$, using Lusin theorem on B_n, there exists a closed set $F_n \subset B_n$ and a continuous function $\varphi(x)$ on $\mathbf{R}^1$ such that

(1) $m(B_n \setminus F_n) < \frac{\varepsilon}{4N}$;

(2) $x \in F_n, f(x) = \varphi(x)$ and $\sup\limits_{x \in \mathbf{R}^1}|f(x)| = \sup\limits_{x \in \mathbf{R}^1}|\varphi(x)| \leqslant N$.

Thus,
$$\int_a^b |f(x)-\varphi(x)|\,dx \leqslant \int_{e_n} |f(x)-\varphi(x)|\,dx + \int_{B_n} |f(x)-\varphi(x)|\,dx$$
$$\leqslant \int_{e_n} |f(x)|\,dx + \int_{e_n} |\varphi(x)|\,dx + \int_{B_n\setminus F_n} |f(x)-\varphi(x)|\,dx + \int_{F_n} |f(x)-\varphi(x)|\,dx$$
$$< \varepsilon + \int_{B_n\setminus F_n} |f(x)-\varphi(x)|\,dx + \int_{e_n} |\varphi(x)|\,dx$$
$$\leqslant \frac{\varepsilon}{4} + N \cdot me_n + 2N \cdot \frac{\varepsilon}{4N} \leqslant \varepsilon.$$

例 2 设 $f(x)$ 在 E 上可积,$f(x) \geqslant 0$ 且 $\int_E f(x)dx = 0$,则 $f(x) = 0$ a. e. 于 E.

证明 E 可以表示为 $E = \bigcup\limits_{n=1}^{\infty} E[f \geqslant \frac{1}{n}] \cup E[f=0]$. 令 $E_n = E[f \geqslant \frac{1}{n}]$,则它是可测集. 但是

$$0 = \int_E f(x)dx = \int_{E_n} f(x)dx + \int_{E\setminus E_n} f(x)dx \geqslant \int_{E_n} f(x)dx \geqslant \frac{1}{n}mE_n,$$

故 $mE_n = 0$,从而 $mE[f>0] = 0$.

Ex. 2 Let $f(x)$ be integrable defined on E, $f(x) \geqslant 0$ and $\int_E f(x)dx = 0$, then $f(x) = 0$ a. e. on E

Proof E can be denoted by $E = \bigcup\limits_{n=1}^{\infty} E\left[f \geqslant \frac{1}{n}\right] \cup E[f=0]$. Let $E_n = E\left[f \geqslant \frac{1}{n}\right]$, then it is a measurable set. But

$$0 = \int_E f(x)dx = \int_{E_n} f(x)dx + \int_{E\setminus E_n} f(x)dx \geqslant \int_{E_n} f(x)dx \geqslant \frac{1}{n}mE_n,$$

so, $mE_n = 0$, hence $mE[f>0] = 0$.

本章重点

1 深入理解 L-积分的定义,掌握 L-积分的基本性质.
2 牢固掌握 L-积分的定理包括这些定理的条件和结论,弄懂其证明思路.
3 了解 L-积分与 R-积分的关系,R-可积的充要条件.
4 了解推广的牛顿-莱布尼茨定理.

Key points of this chapter

1 Understand the definition of Lebesgue integral, and master the basic properties of Lebesgue integral.
2 Master the theorems of Lebesgue integral, including conditions and conclusions for these theorems,

and make clear of the thinking of testifying the theorems.

3 Understand the relationship between L-integral and R-integral, and the sufficient and necessary conditions for R-integral.

4 Understand the generalized Newton-Leibniz theorem.

第六章 抽象空间论
Chapter 6　Theory of Abstract Spaces

　　本章开始我们将进入"泛函分析"领域. 泛函是函数概念的推广, 函数是数与数之间的对应关系, 泛函是函数与数之间的对应关系, 算子是函数空间与函数空间之间的对应关系. 由于函数空间是无限维的, 因此我们研究的问题要复杂得多. 同时我们也会发现很多有限维空间的理论和方法仍然可以推广到无穷维空间之中. 我们主要研究: 距离空间、完备、范数、赋范线性空间、Banach 空间、内积空间、Hilbert 空间.

　　This chapter will enter into the field of Functional Analysis. Functional is a generalization of function. Corresponding relation between figures is called a function, corresponding relation between function and figure is called a functional, corresponding relation between function spaces is called an operator. Because of the infinite dimensional function space, the problems we should study become more complex. Meanwhile, we also find that many theorems and methods of finite dimensional space can be generalized into infinite dimensional space. In this chapter, we mainly deal with metric space, complete, norm, normed linear space, Banach space, inner space and Hilbert space.

单词和短语　Words and expressions

　三角不等式　　triangle inequality　　　　　★ C-空间　　continuous space
★ 度量空间　　metric space　　　　　　　　紧集　　compact set
★ 范数　　norm　　　　　　　　　　　　　完备集合　　complete set
★ 赋范空间　　normed space　　　　　　　　不动点　　fixed point
　凸集　　convexity　　　　　　　　　　　★ 拓扑空间　　topological space
★ L_p-空间　　p-normed integral space　　　　向量空间　　vector space
★ L_∞-空间　　∞-normed integral space　　　锥　　core

基本概念和性质　Basic concepts and properties

1 度量空间 X 到 Y 之间映射 T 连续当且仅当 Y 中任意开集 M 的原像 $T^{-1}M$ 是 X 中的开集.

Let X and Y are metric space, T is a mapping from X to Y, T is continuous if and only if for every open set $M \subseteq Y$, its inverse image $T^{-1}M$ is also an open set defined in X.

2 $l^\infty, C[a,b]$ 等空间为完备的度量空间, 也为 Banach 空间.

$l^\infty, C[a,b]$ are called a complete metric space, also are a Banach space.

3 完备空间上的压缩映射 T 有且只有一个不动点.

Contraction mapping T defined in a complete space has and only has a fixed point.

例题　Examples

例 1　证明: 设 T 是度量空间 (X,d) 到度量空间 (Y,d) 中的映射, 那么 T 在 $x_0 \in X$ 连续当且仅当 $x_n \to x_0 (n \to \infty)$ 时, 必有 $T_{x_n} \to T_{x_0} (n \to \infty)$.

　　证明　**必要性**　如果 T 在 $x_0 \in X$ 连续, 那么对任意给定的正数 $\varepsilon > 0$, 存在正数 $\delta > 0$, 使当 $d(x, x_0) < \delta$ 时, 有 $d(T_{x_0}, T_x) < \varepsilon$, 因为 $x_n \to x_0 (n \to \infty)$, 所以存在正整数 $N > 0$, 当 $n > N$ 时, $d(x_n, x_0) < \delta$. 因此 $d(T_{x_n}, T_{x_0}) < \varepsilon$. 这就证明了结论.

　　充分性　用反证法. 如果 T 在 $x_0 \in X$ 不连续, 那么存在正数 ε_0, 使对任何正数 $\delta > 0$, 总有 $x \ne x_0$, 满足 $d(x, x_0) < \delta$, 但 $d(T_x, T_{x_0}) \geq \varepsilon_0$, 特取 $\delta = \dfrac{1}{n}$, 则有 x_n, 使 $d(x_n, x_0) < \dfrac{1}{n}$, 但 T_{x_n} 不收敛

于 T_{x_0}，这与假设矛盾，证毕.

Ex. 1 Let T be a mapping from the metric space (X,d) to (Y,d), then T is continuous at $x_0 \in X$ if and only if $T_{x_n} \to T_{x_0} (n \to \infty)$ as $x_n \to x_0 (n \to \infty)$.

Proof   Necessity If T is continuous at $x_0 \in X$, then for any $\varepsilon > 0$, there exists $\delta > 0$, such that $d(T_{x_0}, T_x) < \varepsilon$ as $d(x,x_0) < \delta$. Because of $x_n \to x_0 (n \to \infty)$, so there exists $N > 0$, when $n > N$, there is $d(x_n, x_0) < \delta$. Thus $d(T_{x_n}, T_{x_0}) < \varepsilon$, and so $T_{x_n} \to T_{x_0} (n \to \infty)$.

Sufficiency If T is not continuous at $x_0 \in X$, there exists ε_0, such that for any $\delta > 0$, always have $x \neq x_0$ and it satisfies $d(x,x_0) < \delta$. But $d(T_x, T_{x_0}) \geqslant \varepsilon_0$. Take $\delta = \dfrac{1}{n}$, there is x_n such that $d(x_n, x_0) < \dfrac{1}{n}$. That is to say, although $x_n \to x_0$ as $n \to \infty$, T_{x_n} is not convergence to T_{x_0}, which is a contradiction with the assumption.

例 2 证明：设 X 为赋范线性空间，$X \neq \{0\}$，试证明 X 完备的充要条件是单位球面 $S(X)$ 完备.

证明 必要性 设 $\{x_n\}$ 为基本列，因为 X 完备，则 $x_n \to x$，$\|x\| = 1$，故 $S(X)$ 完备.

充分性 设 $S(X)$ 完备，任取 X 中基本列 $\{x_n\}$，令 $x'_n = \dfrac{x_n}{\|x_n\|}$，则 $\dfrac{x_n}{\|x_n\|}$ 为 $S(X)$ 中的基本列，$S(X)$ 完备，所以 X 完备.

Ex. 2 Let X be a normed space, $X \neq \{0\}$, then X is complete space if and only if $S(X)$ is complete.

Proof   Necessity Let $\{x_n\}$ is a basic sequence. Since X is complete, then we have $x_n \to x$, $\|x\| = 1$, and so $S(X)$ is complete.

Sufficiency Let $S(X)$ be complete, for any basic sequence $\{x_n\}$, let $x'_n = \dfrac{x_n}{\|x_n\|}$, then $\dfrac{x_n}{\|x_n\|}$ is one of basic sequences in $S(X)$, and $S(X)$ is complete, this implies that X is complete..

本章重点

1. 深入理解和熟练掌握距离空间、赋范线性空间、内积空间及其中的完备性、列紧性、可分性等基本概念及其相互联系.
2. 熟悉 $C(\Omega)$、$L_p(\Omega)$ 空间的性质.
3. 深刻理解闭球套定理和 Baire 纲定理.
4. 熟练掌握内积，正交标准交系，Fourier 级数等基本概念.

Key points of this chapter

1. Understand and master the basic concepts and the correlation of metric space, normed linear space, inner product space, and their completeness, sequence compact, separable.
2. Familiarize the properties of spaces of $C(\Omega)$ and $L_p(\Omega)$.
3. Understand closed ball theorem and Baire category theorem.
4. Master the basic concepts of inner product, orthogonal standard relationship and Fourier series.

第七章　抽象空间之间的映射
Chapter 7　Mapping between Abstract Spaces

本章我们开始研究赋范空间之间的映射，即有界线性算子和有界线性泛函、算子空间与对偶空间、有界线性泛函的表示、共鸣定理、开映射定理、闭图像定理、有界线性泛函的延拓、共轭空间与共轭算子等.

In this chapter, we begin to study the mapping between normed spaces. It mainly includes the bounded linear operator and the bounded linear functional, the operator space and the dual space, the representation of bounded linear functional, resonance theorem, open mapping theorem, closed graph theorem, extension of bounded linear functional, conjugate space and conjugate operator, and so on.

单词和短语 Words and expressions

★ Cauchy 序列   Cauchy sequence
★ 完备化空间   complete space
★ 完备子空间   complete sub-space
  收敛序列   convergent sequence
  稠密集合   dense set
  连续变换   continuous transformation
  微分   differential
  凸组合   convex combination.
  凸泛函   convex functional
  阵列   array
  弱紧性   weak compactness
  弱连续性   weak continuous
  弱收敛   weak convergence
  弱 ∗ 收敛   weak ∗ convergence
  内积   inner product
  内积空间   inner product space

  内点   inner point
  同构   isomorphism
  雅可比矩阵   Jacobi matrix
  正交投影   orthogonal projection
  正交补   orthogonal complement
  算子   operator
  最佳逼近   best approximation
  投影   projection
  可分空间   separable space
  支撑泛函   support functional
  共鸣定理   resonance theorem
★ 开映射定理   open mapping theorem
★ 闭图像定理   closed graph theorem
★ 共轭空间与共轭算子   conjugate space and conjugate operator

基本概念和性质 Basic concepts and properties

1 赋范线性空间 X 到 Y 之间的线性算子. T 为有界算子的充要条件是 T 在 X 上是连续的.
Let X, Y be normed spaces and T be a linear operator from X to Y, T is a bounded operator if and only if T is continuous on X.

2 完备的内积空间称为 Hilbert 空间.
A complete inner space is called a Hilbert space.

3 可分的 Hilbert 空间必和某个 $\mathbf{R}^n$ 或 l^2 同构.
Separable Hilbert space must be isomorphic with some $\mathbf{R}^n$ or l^2.

例 题   Examples

例 证明:设 T 是赋范线性空间 X 到赋范线性空间 Y 中的线性算子,则 T 为有界算子当且仅当 T 是 X 上连续算子.

证明 若 T 有界,由式 $\|T_{x_n}\| \leqslant c\|x\|$ 可知,当 $x_n \to x, (n \to \infty)$ 时, $\|T_{x_n} - T_x\| \leqslant c\|x_n - x\|$,所以 $\|T_{x_n} - T_x\| \to 0$,即 $T_{x_n} \to T_x(n \to 0)$,因此 T 连续.

反之,若 T 在 X 上连续,但无界,这时在 X 中必有一列向量 $x_1, x_2, x_3, \cdots$,使得 $\|x_n\| \neq 0$,但 $\|T_{x_n}\| \geqslant n\|x_n\|$. 令 $y_n = \dfrac{x_n}{n\|x_n\|}, n = 1, 2, \cdots$,则 $\|y_n\| = \dfrac{1}{n} \to 0(n \to \infty)$,所以 $y_n \to 0(n \to \infty)$,由 T 的连续性,得到 $T_{y_n} \to T_0 = 0(n \to \infty)$,但由于 T 是线性算子,又可得到对一切正整数 $n > 0$,有 $T_{y_n} = \|T_{x_n}\|/n\|x_n\| \geqslant n\|x_n\|/n\|x_n\| = 1$,与假设矛盾.

所以 T 是有界算子. 证毕.

Ex Let T be a linear operator from normed linear space X to Y, then T is a bounded operator if and only if T is a continuous operator in X.

Proof If T is bounded, according to $\|T_{x_n}\| \leqslant c\|x\|$, when $x_n \to x(n \to \infty)$, because $\|T_{x_n} - T_x\| \leqslant c\|x_n - x\|$, so $\|T_{x_n} - T_x\| \to 0$, that is, $T_{x_n} \to T_x(n \to 0)$. So T is continuous.

Conversely, if T is continuous, but not bounded in X, there must exist a row vectors $x_1, x_2, x_3, \cdots$, such that $\|x_n\| \neq 0$ but $\|T_{x_n}\| \geqslant n\|x_n\|$. Let $y_n = \dfrac{x_n}{n\|x_n\|}, n = 1, 2, \cdots$, then $\|y_n\| = \dfrac{1}{n} \to 0(n$

$\to \infty$), so $y_n \to 0 (n \to \infty)$, because T is continuous and $T_{y_n} \to T_0 = 0, (n \to \infty)$, but T is a linear operator, we can also get for all $n > 0$, there is $T_{y_n} = \|T_{x_n}\|/n\|x_n\| \geqslant n\|x_n\|/n\|x_n\| = 1$. This is a contradiction. So T is a bounded operator.

本章重点

1 深刻理解和熟练掌握有界线性算子与泛函，共轭空间和共轭算子等基本概念及其基本性质．

2 掌握线性算子与泛函的范数的求法．

3 掌握泛函延拓定理，共鸣定理，开映射定理是 Banach 空间中三大基本定理，也是泛函分析的核心理论，应熟悉这些定理的条件概论及其证明思路．

4 掌握赋范线性空间中算子和泛函列的几种基本的收敛性及其关系．

Key points of this chapter

1 Understand and master the basic concepts and the basic properties of bounded linear operator and functional, conjugate space and conjugate operator.

2 Master the methods that define norms of linear operator and functional.

3 Master three basic theorems in Banach spaces, namely, the functional dual operator theorem, the resonant theorem, the open mapping theorem. They are also the kernel theory of functional analysis. Familiarize the conditions and the thinking of testifying the theorems.

4 Master several kinds of the basic convergence and the relationships of operators in normed linear spaces and functional sequences.

练 习

一、判断题（下列各题，你认为正确的，请在题干的括号内打"√"，错的打"×"）

1. $A = (A - B) \cup (A \cap B)$； （ ）
2. 有理数集 **Q** 是闭集； （ ）
3. $\bigcup (A_i - B_i) = \bigcup A_i - \bigcup B_i$； （ ）
4. 设 A,B 为 $\mathbf{R}^n$ 中两集合，它们之间的距离 $d(A,B) > 0$，则 $r^*(A \cup B) = r^*(A) + r^*(B)$； （ ）
5. 设 $\{E_n\}$ 为可测集列，$\lim E_n = E$，则 E 是可测集且 $r(E) = \lim r(E_n)$； （ ）
6. 设 $|f|$ 在 E 上可测，则 f 在 E 上可测； （ ）
7. 设 $f \in C[a,b]$（即 f 在 $[a,b]$ 上连续），$g = f$ 几乎处处成立于 $[a,b]$，则 g 在 $[a,b]$ 上几乎处处连续； （ ）
8. 距离空间 (X,d) 中有界点列必有柯西子点列； （ ）
9. 设 A 为距离空间 (X,d) 中的完备子空间，则 A 为 X 中闭集； （ ）
10. 设 X,Y 为 Banach 空间，$\{T_n\}$ 是从 X 到 Y 的有界线性算子列，若 $\{T_n\}$ 点态有界，则 $\{T_n\}$ 必一致有界． （ ）

二、填空题

1. 设 $f: X \to Y, f$ 在 X 上连续的充要条件是：$\forall$ 闭集 $F \subset Y, f^{-1}(F)$ 为 X 中 _____ 集；
2. 设 f_n 与 f 是 E 上几乎处处有限的可测函数，若 $f_n \to f$ a.e. 于 E，且 _____，则 $f_n \xrightarrow{\gamma} f$ 于 E；
3. _____，则 $f \in L(E)$，且 $\lim\limits_{n\to\infty}\int_E f_n d\mu = \int_E f d\mu$；
4. $L^p(E)$ 空间中的元素是 _____；
5. 设 $|f| \in L(E)$，且 _____，则 $f \in L(E)$；
6. 设 E 是 **R** 上的可测集，f 是 E 上的有限可测函数，则由 Lusin 定理，$\forall \delta > 0$，存在 E 中的集合 F，使得 $f \in C(F)$，及 $\gamma(E - F) < $ _____；

7. A 为距离空间 (X,d) 中列紧集是指_____;
8. 开映射定理的内容是:设 X,Y 为 Banach 空间,若 $T \in B(X,Y)$,且_____,则 T 为开映射;

三、选择题(在下列各小题的备选答案中,请把你认为正确的答案的题号,填入题干的横线上)

1. f 在 E 上可测的充要条件是_____;
(1) $\forall \alpha \in \mathbf{R}^1, f^{-1}$ 为 E 中可测集
(2) $\forall$ 可测集 $A \subset \mathbf{R}^1, f^{-1}(A)$ 为可测集
(3) $\mathbf{R}^1$ 中所有闭集 $F, f^{-1}(F)$ 为可测集
(4) 所有有理数 $r, \{f > r\}$ 为可测集

2. 设 f 在 $[a,b]$ 上 (L) 可积,则_____;
(1) $f \in R[a,b]$
(2) $|f| \in R[a,b]$
(3) $|f| \in L[a,b]$
(4) $r\{|f| = \infty\} = 0$

3. 设 $f(x) = \begin{cases} 1, & x \in p \\ 0 & x \notin p \end{cases}$, p 为 Cantor 集,则_____;
(1) $f \in R[0,1]$
(2) $f \in L[0,1]$
(3) f 在 $[0,1]$ 上可测
(4) f 在 $[0,1]$ 上 a.e. 连续

4. 设 A,B 为内积空间 X 中非空子集,$A \perp B, x \in A, y \in B$,则_____;
(1) $\|x+y\|^2 = \|x\|^2 + \|y\|^2$
(2) $A^\perp$ 是 X 的闭子空间
(3) $A^{\perp\perp} = A$
(4) $X^\perp = \{0\}$

四、完成下列各题

1. 设 $f(x)$ 是闭区间 $[a,b]$ 上的实值连续函数,且对任意常数 a,记 $E = \{x \mid f(x) \geqslant a\}$,证明:$E$ 为闭集。
2. 设 $\mathbf{R}^n = \mathbf{R}^2, E = \{(x,y) \mid x^2 + y^2 < 1\}$,求 E 的导集,闭包,开核,边界$(E', \bar{E}, \dot{E}, \partial E)$。
3. 设函数列 $\{f_n(x)\}$ 在 E 上依测度收敛于 $f(x), f_n(x) = g_n(x), n = 1,2,\cdots$ 几乎处处成立于 E,证明:$g_n(x)$ 依测度收敛于 $f(x)$。

Exercises

I. **Estimation**: True($\checkmark$), False($\times$)

1. $A = (A - B) \cup (A \cap B)$. ( )
2. Rational numbers set $\mathbf{Q}$ is a closed set. ( )
3. $\bigcup (A_i - B_i) = \bigcup A_i - \bigcup B_i$. ( )
4. Let A, B be sets of $\mathbf{R}^n$, the distance of them is $d(A,B) > 0$, then $r^*(A \cup B) = r^*(A) + r^*(B)$. ( )
5. Let $\{E_n\}$ be sequences of measurable set, $\lim E_n = E$, then E is a measurable set, and $r(E) = \lim r(E_n)$. ( )
6. Suppose that $|f|$ is measurable on E, then f is measurable on E. ( )
7. Let $f \in C[a,b]$, $g = f$ a.e on $[a,b]$, then g is continuous a.e. on $[a,b]$. ( )
8. The bounded points set on metric space (X,d) must have a Cauchy sequence. ( )
9. Let A be a complete subspace in metric space (X,d), then A is a closed set in X. ( )
10. Let X, Y be Banach spaces, $\{T_n\}$ is a bounded linear operator sequence from X to Y, if $\{T_n(x)\}$ is bounded for any x in X, then $\{T_n\}$ is necessary uniform bounded. ( )

II. Fill in the blanks

1. Let $f: X \to Y$, f be continuous on X if and only if for any closed set $F \subset Y$, $f^{-1}(F)$ is _____ set.
2. Let f_n and f be finite measurable functions on E a. e., if $f_n \to f$ a. e. on E, and _____, then $f_n \xrightarrow{r} f$ in E.
3. _____, then $f \in L(E)$ and $\lim_{n \to \infty} \int_E f_n \, d\mu = \int_E f \, d\mu$.
4. The definition of $L^p(E)$ is _____.
5. Let $|f| \in L(E)$, and _____ then $f \in L(E)$.
6. Let E be a measurable set in $\mathbf{R}$, f is a finite measurable on E a. e, then by Lusin Theorem, $\forall \delta > 0$, there exists a subset F in E, such that $f \in C(F)$, and $\gamma(E - F) <$ _____.
7. Let A be a sequence compact sets in metric space (X, d), then _____.
8. Open mapping theorem: Let X, Y be Banach spaces, if $T \in B(X, Y)$, and _____, then T is a open mapping.

III. Choose question (please choose the correct answer and take it in the follow blank)

1. f is measurable on E if and only if _____.
   (1) $\forall a \in \mathbf{R}^1$, f^{-1} is a measurable set on E
   (2) For any $A \subset \mathbf{R}^1$, $f^{-1}(A)$ is a measurable set
   (3) All closed set in $\mathbf{R}^1$, $f^{-1}(F)$ is a measurable set
   (4) For any rational number r, $\{f > r\}$ is a measurable set

2. Let f be L-integral in $[a, b]$, then _____.
   (1) $f \in R[a, b]$ (2) $|f| \in R[a, b]$ (3) $|f| \in L[a, b]$ (4) $r\{|f| = \infty\} = 0$

3. Let $f(x) = \begin{cases} 1, & x \in p \\ 0, & x \notin p \end{cases}$, p is a Cantor set, then _____.
   (1) $f \in R[0, 1]$ (2) $f \in L[0, 1]$ (3) f is measurable on $[0, 1]$
   (4) f is continuous on $[0, 1]$ a. e..

4. Let A, B be a non-empty subset in the inner product space X, $A \perp B$, $x \in A$, $y \in B$, then _____.
   (1) $\|x + y\|^2 = \|x\|^2 + \|y\|^2$
   (2) $A^\perp$ is a closed subspace in X
   (3) $A^{\perp\perp} = A$
   (4) $X^\perp = \{0\}$

IV. Complete the following questions:

1. Let $f(x)$ be a continuous function defined in the closed interval $[a, b]$, for every real number a, $E = \{x \mid f(x) \geq a\}$, prove that E is a closed set.

2. Let $\mathbf{R}^n = \mathbf{R}^2$, $E = \{(x, y) \mid x^2 + y^2 < 1\}$, then E', $\bar{E}$, E, $\partial E = ?$.

3. Let $f_n(x)$ be a converge to $f(x)$ by measure and $f_n(x) = g_n(x)$, a. e. on E, $n = 1, 2, 3, \cdots$, then $g_n(x)$ converges to $f(x)$ by measure.

信息与计算科学专业课程

第一部分 数值分析
Part 1 Numerical Analysis

引 言

计算问题可以说是现代社会各个领域普遍存在的共同问题,如工业、农业、交通运输、医疗卫生、文化教育等,这些领域都面临着处理大量数据的问题。人们通过对问题的求解和必要的数据分析,掌握事物发展的规律.

计算数学被称为科学计算的数学,也叫做数值计算方法或数值分析,它是研究计算问题的解决方法及其有关数学理论的一门学科. 计算数学的内容十分丰富,它在科学技术中正发挥着越来越大的作用. 计算数学一般可分为数值逼近、数值代数和微分方程的数值解法三大类,它的主要内容包括代数方程(组)、微分方程的数值解法,函数的数值逼近问题,矩阵特征值的求法,最优化计算问题等等. 此外,它还包括解的存在性、唯一性、收敛性、稳定性和误差分析等理论问题.

本部分包括数值逼近和微分方程的数值解两方面内容.

数值逼近是一门历史悠久、内容丰富而且实践性很强的课程,它的思想和方法渗透于几乎所有的学科,如自然科学和人文科学等;同时它又是诸多数值方法的理论基础和主要依据,如函数值的计算、数值微分、数值积分、微分方程的数值解法、曲线和曲面的拟合等等. 数值逼近不仅是函数逼近理论和数值分析方法的结合,也是应用数学和计算机科学相结合的产物,计算机技术和数学软件的飞速发展也为数值分析的实践提供了物质条件. 随着计算机强大的运算功能、图形功能的开发,以及应用数学软件的不断升级,使师生不仅能在很短的时间内自主地选择软件、比较算法,而且能在屏幕上通过数值的几何观察、联想去发现解决问题的线索,探讨规律性的结果.

作为数值逼近思想的一个重要应用,微分方程的数值解法也是一种近似解法,包括常微分方程的数值解法和偏微分方程的数值解法. 进一步地,常微分方程的数值解法主要包括解初值问题的显式欧拉方法、隐式欧拉方法、修正的欧拉方法、龙格-库塔方法等单步方法、亚当斯型的线性多步方法,以及解边值问题的差分方法、样条函数法和打靶法等等;对于偏微分方程的初值问题或边值问题,目前常用的是有限差分法、Galerkin 方法、有限元素法等.

Introduction

Computational problems may be viewed as the common problems of contemporary society existing in various fields, such as industry, agriculture, traffic, medical treatment, culture education, etc. These fields are all confronted with the problems of handling plentiful data. Through solving problems and analyzing necessary data, one can grasp the developing rules of these problems.

Computational Mathematics is a subject of studying the solving methods of computational problems and the corresponding mathematical theory, which is called mathematics of scientific computing and is also called Numerical Computing Method or Numerical Analysis. Computational Mathematics has so abundant contents that it is exerting more and more action in technology. In general, Computational Mathematics may be divided into three parts, i. e., Numerical Approximation, Numerical Algebra and Numerical Solution of Differential Equations, and its contents mainly involves numerical methods of algebraic equation(s) and differential equations, numerical approximation problems of functions, solving methods of eigenvalues of a matrix, computational problems of optimization, and so on. Moreover, it also involves some theoretical problems, such as existence, uniqueness, convergence, stability and error analysis, etc..

This part involves two classes, namely, Numerical Approximation and Numerical Solutions of Differential Equations.

Numerical Approximation is a course with centuries-old history, abundant contents and powerful practice. Its idea and method almost run through all subjects, such as natural science and humanities. Simultaneously, it is also a main and theoretical basis of many numerical methods, such as computation of functional value, numerical differentiation, numerical integration, numerical solutions of differential equations, approximation of curves and surfaces, and so on. Moreover, Numerical Approximation is not only a combination between approximation of function and methods of numerical analysis, but also a combining result between applied mathematics and computer science. The rapid development of computer technology and mathematical softwares supplies the substance condition for practising numerical analysis. Along with the powerful computational function, exploit of graph function and continuous update of applied mathematical softwares, teachers and students can not only choose softwares freely in short time and compare algorithms, but also discover the clues of solving problems and discuss the disciplinarian results by observing geometrically and associating these problems on the computer screen.

As an important application of idea of numerical approximation, Numerical Solutions of Differential Equations involve Numerical Solutions of Ordinary Differential Equations and Partial Differential Equations, which is also an approximate method. Further, Numerical Solutions of Ordinary Differential Equations mainly contain, for initial problems the single step methods, such as explicit Euler method, implicit Euler method, modified Euler method, Runge-Kutta method, etc. and multistep methods, such as methods of Adams types; for boundary problems, we have some difference methods, spline function methods, shooting methods, and so on. For Numerical Solutions of Partial Differential Equations, the common methods are finite difference methods, Galerkin methods, finite element methods, and so on.

第一章 数值计算中的误差分析
Chapter 1　Error Analysis in Numerical Calculation

本章包括浮点数体系及可能会造成计算机运算失真的舍入误差的一些基本结论。本章也将讨论其他类型的误差和失真问题。例如,当两个几乎相等的数相减的时候,失真就可能会出现. 最后考虑某些稳定和不稳定算法及病态问题.

This chapter contains the system of floating-point numbers and some basic conclusions about round-off errors, which may cause the computer calculations to distort. This chapter also discusses other types of errors and distortions. For example, a distortion may appear as two nearly equal numbers are subtracted. Finally, some problems such as stable, unstable algorithms and ill-conditioned problems will be considered.

单词和短语　Words and expressions

浮点数　floating-point number	二进制数　binary number
初始误差　initial error	尾数　mantissa　[mænˈtisə]
模型误差　model error	指数　exponent　[eksˈpəunənt]
舍入误差　round-off error	舍入　rounds off
截断误差　truncation error	舍入法　rounding method
★绝对误差　absolute error	上溢　overflow　[əuvəˈfləu]
★相对误差　relative error	标准化浮点形式　normalized floating-point form
稳定的计算　stable computation	机器数　machine number
十进制系统　decimal system	双精度　double-precision

差异　discrepancy　[dis'krepənsi]
失真　loss of significance / distortion
函数求值　evaluation of function
周期性　periodicity　['piərə'disiti]
差分方程　difference equation

★ 适定问题　well-posed problem
条件数　condition number
病态的(良态的)问题　ill (well)-conditioned problem
希尔伯特矩阵　Hilbert matrix
指数积分　exponential integral

基本概念和性质　Basic concepts and properties

在任何计算机内,四种基本算术运算满足下面的关系式
$$fl(x \cdot y) = (x \cdot y)(1+\delta), \ |\delta| \leqslant \varepsilon.$$
In any computer, the four arithmetic operations satisfy the following equation
$$fl(x \cdot y) = (x \cdot y)(1+\delta), \ |\delta| \leqslant \varepsilon.$$

本章重点

1. 理解误差的概念及其分类.
2. 明确误差产生的根源和分析误差的一些基本方法.

Key points of this chapter

1. Understand the concept of error and its classification.
2. Make clear of the roots to produce errors and some basic methods to analyze errors.

第二章　多项式插值方法
Chapter 2　Methods of Polynomial Interpolation

本章主要讨论基于等距结点下的多项式插值问题中的各种方法,如拉格朗日插值、内维尔插值、牛顿插值、埃尔米特插值等等;并且引入了一些重要的概念及其性质,如,等距结点下的差分、差商、插值误差等等.

This chapter is to mainly discuss some methods on polynomial interpolation problems based on the equivalent nodes, such as Lagrange interpolation, Neville interpolation, Newton interpolation, Hermite interpolation, and so on. This chapter also introduces some important concepts and their properties based on the equidistant nodes, such as difference, difference quotient, error of interpolation, and so on.

单词和短语　Words and expressions

逼近/近似　approximation
★ 插值方法　interpolation method
插值结点　interpolation node
插值函数　interpolation function
插值基函数　basic function of interpolation
代数插值多项式　algebra interpolation polynomial
三角插值多项式　triangle interpolation polynomial
★ 多项式插值　polynomial interpolation
拉格朗日插值　Lagrange interpolation
内维尔插值　Neville interpolation
牛顿插值　Newton interpolation

等距结点　equidistant node
差分　difference
向前差分　forward difference
向后差分　backward difference
中心差分　centered difference
高阶差分　higher-order difference
差商　difference quotient
高阶差商　higher-order difference quotient
埃尔米特插值　Hermite interpolation
插值误差(余项)　error of interpolation
误差估计　estimation of error
收敛性　convergence
稳定性　stability

基本概念和性质　Basic concepts and properties

1. 多项式插值误差定理

设 f 为 $C^{n+1}_{[a,b]}$ 中的一个函数,设 p 为最高次数是 n 的多项式,使得在 $n+1$ 个不同点 $x_0, x_1, \cdots, x_n$ 处

插入函数 f 的值,则对于 $[a,b]$ 中每个 x,相应于 (a,b) 中的点 ξ_x,使得

$$f(x)-p(x)=\frac{1}{(n+1)!}f^{(n+1)}(\xi_x)\prod_{i=0}^{n}(x-x_i).$$

Theorem on Error of Polynomial Interpolation
Let f be a function in $C_{[a,b]}^{n+1}$, and let p be a polynomial of degree at most n that interpolates the function f at $n+1$ distinct points $x_0, x_1, \cdots, x_n$ in the interval $[a,b]$. To each x in $[a,b]$ there corresponds a point ξ_x in (a,b) such that

$$f(x)-p(x)=\frac{1}{(n+1)!}f^{(n+1)}(\xi_x)\prod_{i=0}^{n}(x-x_i).$$

2 牛顿插值误差定理
设 p 是一个次数不超过 n 的插值多项式,且已知 f 在 $[a,b]$ 上 $n+1$ 个离散点集 $x_0,x_1,\cdots,x_n$ 处的函数值,那么存在 (a,b) 内的一个点 ξ,使得

$$f(t)-p(t)=f[x_0,x_1,\cdots,x_n,\xi]\prod_{j=0}^{n}(t-x_j).$$

Theorem on Error of Newton Interpolation
Let p be a polynomial of degree at most n that interpolates a function f at a set of $n+1$ discrete nodes $x_0, x_1, \cdots, x_n$, then there exists a value ξ in (a,b) such that

$$f(t)-p(t)=f[x_0,x_1,\cdots,x_n,\xi]\prod_{j=0}^{n}(t-x_j).$$

3 高阶差商

$$f[x_0,x_1,\cdots,x_n]=\frac{f[x_1,x_2,\cdots,x_n]-f[x_0,x_1,\cdots,x_{n-1}]}{x_n-x_0}$$

Higher-Order Difference Quotient

$$f[x_0,x_1,\cdots,x_n]=\frac{f[x_1,x_2,\cdots,x_n]-f[x_0,x_1,\cdots,x_{n-1}]}{x_n-x_0}$$

4 导数和差商的关系:
如果 f 在区间 $[a,b]$ 上是 n 阶连续可微的,并且如果 $x_0,x_1,\cdots,x_n$ 是 $[a,b]$ 上不同的点,则在 (a,b) 内存在一个点 ξ 使得

$$f[x_0,x_1,\cdots,x_n]=\frac{1}{n!}f^{(n)}(\xi).$$

Relation between Derivatives and Difference Quotient
If f is n times continuously differentiable on $[a,b]$ and if $x_0,x_1,\cdots,x_n$ are distinct points on $[a,b]$, then there exists a point ξ in (a,b) such that

$$f[x_0,x_1,\cdots,x_n]=\frac{1}{n!}f^{(n)}(\xi).$$

本章重点

1 掌握等距结点下的差分、差商等概念及其性质。
2 掌握各类插值方法公式及其误差估计,即对于给定的函数表,能够计算出相应的拉格朗日插值公式、牛顿插值公式、埃尔米特插值公式等。

Key points of this chapter

1 Master the concepts and properties of difference, difference quotient based on the equidistant nodes.
2 Master the formulae of all methods of interpolation and their estimation of errors, namely, for the given function table, one can compute the corresponding formulae of Lagrange interpolation, Newton interpolation, Hermite interpolation, and so on.

第三章 样条插值
Chapter 3　Spline Interpolation

用多项式插值的方法去逼近函数,虽然有许多优点,但经常会出现诸如龙格现象的情形.为此,本章首先引入分段低次插值方法,然后介绍样条函数及其基本性质、三次样条插值、B-样条插值等内容.

That approximating a function by using the methods of polynomial interpolation has many merits, however, the known Runge phenomenon always appears. Therefore, this chapter is to firstly introduce the method of piecewise lower-order interpolation. Secondly, the spline function and its properties, the cubic spline interpolation and the B-spline interpolation are then introduced.

单词和短语　Words and expressions

龙格现象　Runge phenomenon
伯恩斯坦定理　Bernstein theorem
分段低次插值　piecewise lower-order interpolation
分段线性插值　piecewise linear interpolation
分段抛物插值　piecewise parabola interpolation
分段 3 次埃尔米特插值　piecewise cubic Hermite interpolation
样条插值　spline interpolation
三次样条插值　cubic spline interpolation
样条函数　spline function
自然样条　natural spline
周期样条　periodic spline
B-样条　B-spline

基本概念和性质　Basic concepts and properties

三次样条插值函数的定义

给定区间 $[a,b]$ 的一个分划 $\Delta: a = x_0 < x_1 < \cdots < x_n < x_{N+1} = b$,若分段函数 $s_3(x)$ 满足如下条件:

(1) 在每个子区间 $[x_j, x_{j+1}]$ $(j=0,1,\cdots,N)$ 上,$s_3(x)$ 是一个次数不超过 3 的实系数代数多项式;

(2) $s_3(x)$ 在区间 $[a,b]$ 上具有 2 阶连续导数.则称 $y = s_3(x)$ 为三次样条函数,$x_i(i=0,1,\cdots,N+1)$ 称为样条结点.进一步地,设函数 $f(x)$ 在样条结点上的函数值给定,如果三次样条函数 $s_3(x)$ 还满足条件

$$s_3(x_i) = f(x_i) \ (i = 0,1,\cdots,N+1),$$

则称 $y = s_3(x)$ 为关于分划 Δ 的三次样条插值函数.

Definition of cubic spline interpolation function

For a partition $\Delta: a = x_0 < x_1 < \cdots < x_n < x_{N+1} = b$ of the given interval $[a,b]$, if the subsection function $s_3(x)$ satisfies the following conditions:

(1) On each subinterval $[x_j, x_{j+1}]$ $(j=0,1,\cdots,N)$, $s_3(x)$ is an algebra polynomial with real coefficients and its order does not exceed 3;

(2) $s_3(x)$ has continuous derivatives with degree 2 on the interval $[a,b]$. Then $y = s_3(x)$ is called a cubic spline function, and $x_i(i=0,1,\cdots,N+1)$ are called the spline nodes. Further, assume that the values of $f(x)$ at the spline nodes are given and the cubic spline function $s_3(x)$ also satisfies the following conditions

$$s_3(x_i) = f(x_i) \ (i = 0,1,\cdots,N+1),$$

then $y = s_3(x)$ is called a cubic spline interpolation function about the partition Δ.

本章重点

■1 掌握三次样条插值函数的定义、求解方法、误差估计以及端点条件的提法.
■2 熟悉各类样条插值函数的定义和性质.

Key points of this chapter

1 Master the concept, solving methods, error estimation of cubic spline interpolation function and conditions at the extremal points.
2 Be familiar with the concepts and properties of all spline interpolation functions.

第四章 最佳逼近
Chapter 4　Best Approximation

　　本章通过在空间中引入范数的概念,建立了一种全新的整体截断误差估计理论,即最佳逼近理论.本章的主要内容包括正交多项式及其性质、最佳一致逼近和最佳平方逼近中的一些重要结论.

　　In this chapter, by introducing the concept of norm in a space, a new type theory on estimation of global truncation error is founded, i. e., theory of best approximation. The main contents of this chapter are composed of some important conclusions such as orthogonal polynomials and their properties, the best uniform approximation and the best quadratic approximation.

单词和短语　Words and expressions

★ 最佳逼近　best approximation
范数　norm
最佳一致逼近　best uniform approximation
正线性算子　positive linear operator
魏尔斯特拉斯逼近定理　Weierstrass approximation theorem
最佳逼近多项式　best approximation polynomial
伯恩斯坦多项式　Bernstein polynomial
瓦勒-布松定理　Valle-Poussin theorem
切比雪夫定理　Chebyshev theorem
收敛速度　convergence velocity

连续模　continuous modulus
杰克生定理　Jackson theorem
最佳平方逼近　best quadratic approximation
最小二乘法　method of least squares
正交多项式　orthogonal polynomial
加权函数　weighted function
★ 勒让德多项式　Legendre polynomial
切比雪夫多项式　Chebyshev polynomial
埃尔米特多项式　Hermite polynomial
拉盖尔多项式　Laguerre polynomial

基本概念和性质　Basic concepts and properties

1 魏尔斯特拉斯定理

　　设 $f(x) \in C[a,b]$,则对于任意给定的 $\varepsilon > 0$,都存在这样的多项式 $p(x)$,使得
$$\max_{a \leqslant x \leqslant b} | f(x) - p(x) | < \varepsilon.$$

Weierstrass Theorem

　　Let $f(x) \in C[a,b]$, then for any given $\varepsilon > 0$, there always exists a polynomial $p(x)$ such that
$$\max_{a \leqslant x \leqslant b} | f(x) - p(x) | < \varepsilon.$$

2 连续模及其性质

　　设 $f(x) \in C[a,b]$,数量
$$\omega(\delta) = \omega(f,\delta) = \omega \sup_{|x-y| \leqslant \delta} \{| f(x) - f(y) |\}$$

称为函数 $f(x)$ 的连续模,其中 δ 是任意正数.

连续模的一些简单的性质

2.1　函数 $\omega(\delta)$ 是单调递增的,即当 $\delta_1 < \delta_2$ 时,有 $\omega(\delta_1) \leqslant \omega(\delta_2)$;

2.2　$\omega(\delta)$ 是半可加的,即 $\omega(\delta_1 + \delta_2) \leqslant \omega(\delta_1) + \omega(\delta_2)$;

2.3　函数 $f(x)$ 在区间 $[a,b]$ 上一致连续的充要条件是 $\lim_{\delta \to 0}\omega(\delta) = 0$;

2.4　若 n 是一个自然数,则 $\omega(n\delta) \leqslant n\omega(\delta)$;

2.5　对于任意的 $\lambda > 0$ 都有不等式 $\omega(\lambda\delta) \leqslant (\lambda+1)\omega(\delta)$ 成立;

2.6　设 $f(x) \in \text{Lip}_M \alpha$ 表示函数 $f(x)$ 在区间 $[a,b]$ 上恒适合 Lipschitz 条件 $| f(x) - f(y) | \leqslant$

$M\mid x-y\mid^\alpha$，其中正常数 $\alpha(0<\alpha<1)$ 和 M 分别称为指数和系数. 这样下列两个关系式完全等价, 即
$$f(x)\in \text{Lip}_M\alpha \Leftrightarrow \omega(\delta)\leqslant M\delta^\alpha.$$

Continuous modulus and its properties
Let $f(x)\in C[a,b]$, the quantity
$$\omega(\delta)=\omega(f,\delta)=\omega\sup_{|x-y|\leqslant\delta}\{\mid f(x)-f(y)\mid\}$$
is called the continuous modulus of $f(x)$, where δ is an arbitrary positive number.

Some simple properties of continuous modulus
2.1　$\omega(\delta)$ is a monotonically increasing function, i.e., if $\delta_1<\delta_2$, then $\omega(\delta_1)\leqslant\omega(\delta_2)$;
2.2　$\omega(\delta)$ is a semiadditive function, namely, $\omega(\delta_1+\delta_2)\leqslant\omega(\delta_1)+\omega(\delta_2)$;
2.3　$f(x)$ is a uniform continuous function on $[a,b]$ iff $\lim_{\delta\to 0}\omega(\delta)=0$;
2.4　If n is a natural number, then $\omega(n\delta)\leqslant n\omega(\delta)$;
2.5　For any $\lambda>0$, the inequality $\omega(\lambda\delta)\leqslant(\lambda+1)\omega(\delta)$ is valid identically;
2.6　Let $f(x)\in \text{Lip}_M\alpha$ denote that $f(x)$ satisfies the Lipschitz condition $\mid f(x)-f(y)\mid\leqslant M\mid x-y\mid^\alpha$, on the interval $[a,b]$ where $\alpha(0<\alpha<1)$ and M are respectively called the exponent and the coefficient. In this case, the following expressions are equivalent, namely,
$$f(x)\in \text{Lip}_M\alpha \Leftrightarrow \omega(\delta)\leqslant M\delta^\alpha.$$

3 几类常用的正交多项式
3.1　第一类切比雪夫多项式
$$T_n(x)=\cos(n\arccos x), n=0,1,2,\cdots$$
3.2　第二类切比雪夫多项式
$$U_n(x)=\frac{\sin(n+1)\theta}{\sin\theta},(x=\cos\theta,0\leqslant\theta\leqslant\pi), n=0,1,2,\cdots$$
3.3　勒让德多项式
$$P_n(x)=\frac{1}{2^n n!}\left(\frac{d}{dx}\right)^n(x^2-1)^2;$$
3.4　拉盖尔多项式
$$L_n(x)=e^x\left(\frac{d}{dx}\right)^n(x^n e^{-x});$$
3.5　埃尔米特多项式
$$H_n(x)=e^{x^2}\left(\frac{d}{dx}\right)^n(e^{-x^2}).$$

Some orthogonal polynomials in common use:
3.1　Chebyshev polynomial of the first kind
$$T_n(x)=\cos(n\arccos x), n=0,1,2,\cdots$$
3.2　Chebyshev polynomial of the second kind
$$U_n(x)=\frac{\sin(n+1)\theta}{\sin\theta},(x=\cos\theta,0\leqslant\theta\leqslant\pi), n=0,1,2,\cdots$$
3.3　Legendre polynomial
$$P_n(x)=\frac{1}{2^n n!}\left(\frac{d}{dx}\right)^n(x^2-1)^2;$$
3.4　Laguerre polynomial
$$L_n(x)=e^x\left(\frac{d}{dx}\right)^n(x^n e^{-x});$$
3.5　Hermite polynomial
$$H_n(x)=e^{x^2}\left(\frac{d}{dx}\right)^n(e^{-x^2}).$$

本章重点

1. 掌握一些重要且基础的概念及其性质，如最佳一致逼近、最佳平方逼近、连续模、收敛速度、最佳逼近阶等等.
2. 熟悉一些常用的正交多项式及其各种重要的性质.
3. 了解一些重要的定理的应用，如魏尔斯特拉斯逼近定理、瓦勒-布松定理、切比雪夫定理、杰克生定理等等.

Key points of this chapter

1. Master some important and basic concepts and their properties, such as the best uniform approximation, the best quadratic approximation, continuous modulus, convergence velocity, the best approximation order, etc. .
2. Be familiar with some orthogonal polynomials in common use and all kinds of their important properties.
3. Know the applications of some important theorems, such as the Weierstrass approximation theorem, the Valle-Poussin theorem, the Chebyshev theorem, the Jackson theorem, etc. .

第五章 数值微分与数值积分
Chapter 5 Numerical Differentials and Numerical Integrals

作为插值理论的重要应用，本章首先利用插值方法构造一些具体的数值计算微分和积分的方法，如数值计算微分的李查逊外推法，数值计算积分的牛顿-科特斯求积公式、龙贝格算法、高斯求积公式和切比雪夫算法等等，然后给出衡量这些方法精确程度的标准，最后讨论误差估计问题.

As some important applications of interpolation theory, this chapter firstly constructs some concrete methods on numerically computing differentials and integrals by using interpolation methods, such as Richardson extrapolation algorithm for numerical differentials, Newton-Cotes integral formula, Romberg algorithm, Gaussian quadrature formula and Chebyshev algorithm for numerical integrals, and so on. Secondly, the criterions on distinguishing the exact degree of these methods are presented. Finally, the problem of error estimation is discussed.

单词和短语 Words and expressions

- ★ 数值微分　numerical differential
- ★ 数值积分　numerical integral
- 代数精度　algebraic accuracy
- 减消法　subtractive cancellation
- 外推法　extrapolation
- 理查森外推算法　Richardson extrapolation algorithm
- 插值型求积公式　quadrature formula of interpolation type
- 牛顿-科特斯求积公式　Newton-Cotes quadrature formula
- ★ 梯形法规则　trapezoid rule
- ★ 辛普生规则　Simpson rule
- 中点规则　midpoint rule
- ★ 高斯积分　Gaussian integral
- ★ 龙贝格积分　Romberg integral
- 递归梯形规则　recursive trapezoid rule
- 欧拉-麦克劳林公式　Euler-Maclaurin formula
- 米尔恩公式　Milne rule
- 萨德定理　Sard theory
- 皮亚诺核　Peano kernel
- 伯努利多项式　Bernoulli polynomial

基本概念和性质　Basic concepts and properties

带有误差项的高斯求积公式

对任意 $f \in C^{2n}[a,b]$，我们有

$$\int_a^b f(x)w(x)dx = \sum_{i=0}^{n-1} A_i f(x_i) + \frac{f^{(2n)}(\xi)}{(2n)!}\int_a^b q^2(x)w(x)dx$$

其中：$a < \xi < b$ 且 $q(x) = \prod_{i=0}^{n}(x - x_i)$.

Gaussian quadrature Formula with Error Term
For any $f \in C^{2n}[a, b]$, we have
$$\int_a^b f(x)w(x)dx = \sum_{i=0}^{n-1} A_i f(x_i) + \frac{f^{(2n)}(\xi)}{(2n)!}\int_a^b q^2(x)w(x)dx$$
where $a < \xi < b$ and $q(x) = \prod_{i=0}^{n}(x - x_i)$.

本章重点

1. 了解数值积分与数值微分的思想.
2. 掌握数值微分的插值方法及李查逊的外推算法.
3. 掌握基于插值理论研究的数值积分问题，重点掌握三个很有用的数值积分：龙贝格积分、高斯求积公式和切比雪夫算法.

Key points of this chapter

1. Understand the ideas of numerical integrals and numerical differentials.
2. Master the interpolation methods of numerical differentials and Richardson extrapolation algorithm.
3. Master numerical integral problem based on the interpolation theoretical study, especially the following three useful numerical integrals: Romberg algorithm, quadrature formula of Gauss type and Chebyshev algorithm.

第六章 常微分方程(组)数值解
Chapter 6 Numerical Solutions of Ordinary Differential Equation(s)

本章主要研究常微分方程(组)的数值计算问题．解决问题的途径是用带有有限个未知量的差分方程将常微分方程离散．中心问题是研究解单一变量的一阶方程，它的解曲线上的一个点是已知的．后面的几节研究方程组、高阶方程和两点边值问题．

This chapter mainly concerns with numerical calculation problems involving ordinary differential equation(s). The way of solving problems is to make the ordinary differential equation discretely by a difference equation with finite unknown numbers. The central problem is to solve a single first-order equation when one point on the solution curve is given. Later sections are devoted to systems of equations, higher order equations and two-point boundary-value problems.

单词和短语 Words and expressions

存在性　existence [igˈzistəns]	Heun 法　Heun method
唯一性　uniqueness [juːˈniːknis]	单步方法　single-step method
初值问题　initial value problem	★线性多步方法　linear multi-step method
李普希兹条件　Lipschitz condition	★亚当斯前推公式　Adams explicit formula
整体截断误差　global truncation error	待定系数　undetermined coefficient
延迟微分方程　delay differential equation	★亚当斯隐式公式　Adams implicit formula
延迟变量　retarded argument	变分方程　variational equation
椭圆积分　elliptic integral	边值问题　boundary value problem
欧拉方法　Euler method	格林函数　Green function
显式欧拉方法　explicit Euler method	★打靶法　shooting method
隐式欧拉方法　implicit Euler method	割线法　scant method
★修正的欧拉方法　modified Euler method	有限差分　finite difference
★龙格-库塔方法　Runge-Kutta method	配点法　collocation

施图姆-刘维尔问题	Strum-Liouville problem	若尔当块	Jordan block
指数矩阵	matrix exponential	标准形式	canonical form
可对角化	diagonalizable	刚度方程	stiff equation

本章重点

1. 了解常微分方程数值解的思想.
2. 掌握单步法的构造思想,重点掌握龙格-库塔方法.
3. 掌握多步法的构造思想,重点掌握亚当斯递推公式.
4. 了解边值问题的打靶法、有限差分思想和有限元思想及方法.

Key points of this chapter

1. Understand the idea of the numerical solutions of ordinary differential equations.
2. Master the construction idea of single-step methods, emphatically the Runge-Kutta methods.
3. Master the construction idea of multi-step methods, emphatically the Adams recursion formulas.
4. Understand shooting method, the idea and methods of finite difference, finite element for solving the BVP.

第七章 偏微分方程(组)数值解
Chapter 7 Numerical Solutions of Partial Differential Equation(s)

本章开始偏微分方程(组)数值解的讨论. 在这一领域所产生的数值计算很可能会给容量大、运行速度快的计算机资源带来沉重的负担. 由于巨大的计算负荷通常来自于解偏微分方程(组),所以这方面的数值分析当前有着广泛的研究前景.

This chapter introduces the subject of numerical solutions of partial differential equation(s). The numerical calculations that arise in this area can easily tax the resources of the largest and fastest computers. Because of the immense computing burden usually involved in solving partial differential equation(s), this branch of numerical analysis is one in which there is widespread current research activity.

单词和短语 Words and expressions

★ 热传导方程	heat equation	希尔伯特空间	Hilbert space
★ 扩散方程	diffusion equation	泊松方程	Poisson equation
离散化	discretization	★ 瑞利-里茨方法	Rayleigh-Ritz method
网格点	mesh point	拉普拉斯算子	Lapacian operator
显式方法	explicit method	有限元	finite element
隐式方法	implicit method	特征曲线	characteristic curve
★ 克兰克-尼科尔森方法	Crank-Nicolson method	交替方向法	alternative approach
★ 拉普拉斯方程	Laplace equation	规范形式	canonical form
★ 伽辽金方法	Galerkin method	迎风格式	upwind scheme
狄利克雷问题	Dirichlet problem	拉克斯格式	Lax method
离散问题	discrete problem	多网格方法	multigrid method
自然排序	natural ordering	插值阶段	interpolation phase
字典排序	lexicographic ordering	衰减影响	damping effect
		V 循环	V-cycle [v-'saikl]

本章重点

1. 掌握差分法的思想,了解有限差分法的显示格式、隐式格式的收敛性和稳定性.
2. 掌握解偏微分方程的有限元法思想、构造方法以及刚度矩阵的解法.
3. 重点掌握几种常见的格式:如迎风格式、拉克斯格式等.

Key points of this chapter

1. Master the idea of difference method, understand the convergence and stability of explicit and implicit finite difference methods.
2. Master the solutions of the idea of finite element, construction method, and rigid matrix of partial differential equations.
3. Master emphatically several methods in common use, such as upwind scheme, Lax method, and so on.

练 习

1. 已知函数 $y = f(x)$ 的函数表

x	$f(x)$	x	$f(x)$
0.0	1.00	0.3	2.08
0.1	1.32	0.4	2.52
0.2	1.68	0.5	3.0

计算相应的向前差分,并写出 Newton 向前插值公式.

解 向前差分表如下:

x_i	$f(x_i)$	$\Delta f(x_i)$	$\Delta^2 f(x_i)$	$\Delta^3 f(x_i)$
0.0	1.00	0.32	0.04	0.00
0.1	1.32	0.36	0.04	0.00
0.2	1.68	0.40	0.04	0.00
0.3	2.08	0.44	0.04	
0.4	2.52	0.48		
0.5	3.00			

$f(x)$ 的 Newton 向前插值公式为
$$N_5(x) = N_5(0.1 + 0.1t) = 0.02t^2 + 0.30t + 1.00, \quad t \in [0,1].$$

2. 设 $f(x) \in C^2[a,b]$ 且 $f(a) = f(b) = 0$,证明
$$\max_{a \leqslant x \leqslant b} |f(x)| \leqslant \frac{1}{8}(b-a)^2 \max_{a \leqslant x \leqslant b} |f''(x)|.$$

证明 以 a, b 为插值结点,$f(x)$ 为被插函数的线性插值多项式为
$$L_1(x) = f(a) \frac{x-b}{a-b} + f(b) \frac{x-a}{b-a}.$$

由已知条件 $f(a) = f(b) = 0$ 知, $L_1(x) \equiv 0$,因此插值余项为
$$R(x) = f(x) - L_1(x) = f(x) - 0 = \frac{f''(\xi)}{2!}(x-a)(x-b), \text{ 其中 } \xi \in (a,b),$$

进而有 $\max_{a \leqslant x \leqslant b} |f(x)| = \max_{a \leqslant x \leqslant b} \left| \frac{f''(\xi)}{2!}(x-a)(x-b) \right| \leqslant \frac{1}{2} \left(\frac{b-a}{2} \right)^2 \max_{a \leqslant x \leqslant b} |f''(x)|$

$\leqslant \frac{1}{8}(b-a)^2 \max_{a \leqslant x \leqslant b} |f''(x)|.$

3. 建立下面公式
$$\int_{-\pi}^{\pi} f(x) \cos x \, dx \approx A_0 f\left(-\frac{3}{4}\pi\right) + A_1 f\left(-\frac{1}{4}\pi\right) + A_2 f\left(\frac{1}{4}\pi\right) + A_3 f\left(\frac{3}{4}\pi\right),$$

当 f 是 3 阶多项式时,公式成立.

证明 显然,3 阶多项式是 $1, x, x^2, x^3$ 的线性组合. 于是,令 $f(x) = x^j (0 \leqslant j \leqslant 3)$,我们就能够通过解一个四阶线性方程组来确定系数. 由对称性, $A_0 = A_3, A_1 = A_2$. 于是求解可以简化为:

$$0 = \int_{-\pi}^{\pi} 1 \cos x \, dx = 2A_0 + 2A_1$$

$$-4\pi = \int_{-\pi}^{\pi} x^2 \cos x \, dx = 2A_0 \left(\frac{3}{4}\pi\right)^2 + 2A_1 \left(\frac{1}{4}\pi\right)^2$$

$A_1 = A_2 = \frac{4}{\pi}, A_0 = A_3 = \frac{4}{\pi}$，与 $A_0 = A_3 = -\frac{4}{\pi}$ 不符。

得到解 $A_1 = A_2 = -A_0 = -A_3 = 4/\pi$ 和公式

$$\int_{-\pi}^{\pi} f(x) \cos x \, dx \approx \frac{4}{\pi}\left[-f\left(-\frac{3}{4}\pi\right) + f\left(-\frac{1}{4}\pi\right) + f\left(\frac{1}{4}\pi\right) - f\left(\frac{3}{4}\pi\right)\right].$$

4. 由特征曲线的方法解这个问题

$$\begin{cases} u_x + y u_x = 0 \\ u(0, y) = f(y) \end{cases}$$

解 这是一个在整个 xy-平面上待解的偏微分方程，而且解函数在 y-轴上的值被确定，对应它的特征曲线所满足的常微分方程是

$$\frac{dy}{dx} = -\frac{u_x}{u_y} = y.$$

过 (x_0, y_0) 的解是

$$y = y_0 e^{x-x_0}.$$

这条特征曲线与 y-轴在 $y = y_0 e^{-x_0}$ 相交，有

$$u(x_0, y_0) = f(y_0 e^{-x_0}).$$

作为解. 当然，我们通常把它写成

$$u(x, y) = f(y e^{-x}).$$

可以很容易地验证这是原问题的解.

Exercises

1. The function table of $y = f(x)$ is given as follows

x	$f(x)$	x	$f(x)$
0.0	1.00	0.3	2.08
0.1	1.32	0.4	2.52
0.2	1.68	0.5	3.0

Compute the corresponding forward differences and carry out the forward interpolation formula of Newton type.

Solution Table of forward differences is given by

x_i	$f(x_i)$	$\Delta f(x_i)$	$\Delta^2 f(x_i)$	$\Delta^3 f(x_i)$
0.0	1.00	0.32	0.04	0.00
0.1	1.32	0.36	0.04	0.00
0.2	1.68	0.40	0.04	0.00
0.3	2.08	0.44	0.04	
0.4	2.52	0.48		
0.5	3.00			

The forward interpolation formula of Newton type of is given by

$$N_5(x) = N_5(0.1 + 0.1t) = 0.02t^2 + 0.30t + 1.00, \quad t \in [0, 1].$$

2. Let $f(x) \in C^2[a, b]$ and $f(a) = f(b) = 0$, prove that

$$\max_{a \leq x \leq b} |f(x)| \leq \frac{1}{8}(b-a)^2 \max_{a \leq x \leq b} |f''(x)|.$$

Proof The linear interpolation polynomial of $f(x)$ based on a, b as the interpolation knots is given by
$$L_1(x) = f(a)\frac{x-b}{a-b} + f(b)\frac{x-a}{b-a}.$$
From the known condition $f(a) = f(b) = 0$, we have $L_1(x) \equiv 0$, and thus the interpolation error is given by
$$R(x) = f(x) - L_1(x) = f(x) - 0 = \frac{f''(\xi)}{2!}(x-a)(x-b), \text{ where } \xi \in (a,b),$$
furthermore, $\max\limits_{a \leqslant x \leqslant b} |f(x)| = \max\limits_{a \leqslant x \leqslant b} |\frac{f''(\xi)}{2!}(x-a)(x-b)| \leqslant \frac{1}{2}\left(\frac{b-a}{2}\right)^2 \max\limits_{a \leqslant x \leqslant b} |f''(x)|$
$$\leqslant \frac{1}{8}(b-a)^2 \max\limits_{a \leqslant x \leqslant b} |f''(x)|.$$

3. Find a formula
$$\int_{-\pi}^{\pi} f(x)\cos x \, dx \approx A_0 f\left(-\frac{3}{4}\pi\right) + A_1 f\left(-\frac{1}{4}\pi\right) + A_2 f\left(\frac{1}{4}\pi\right) + A_3 f\left(\frac{3}{4}\pi\right),$$
such that it is exact when f is a polynomial of degree 3.

Proof Obviously, a polynomial of degree 3 is a linear combination of $1, x, x^2, x^3$. Thus, we can determine the coefficients by substituting $f(x) = x^j (0 \leqslant j \leqslant 3)$ and solving the four resulting linear equations. By symmetry we have, $A_0 = A_3$, $A_1 = A_2$. And thus, the work can be simplified:
$$0 = \int_{-\pi}^{\pi} 1\cos x \, dx = 2A_0 + 2A_1,$$
$$-4\pi = \int_{-\pi}^{\pi} x^2 \cos x \, dx = 2A_0 \left(\frac{3\pi}{4}\right)^2 + 2A_1\left(\frac{1\pi}{4}\right)^2. \quad A_1 = A_2 = \frac{4}{\pi}, A_0 = A_3 = -\frac{4}{\pi}$$
The solution is $A_1 = A_2 = -A_0 = -A_3 = 4/\pi$, and the formula is
$$\int_{-\pi}^{\pi} f(x)\cos x \, dx \approx \frac{4}{\pi}\left[-f\left(-\frac{3}{4}\pi\right) + f\left(-\frac{1}{4}\pi\right) + f\left(\frac{1}{4}\pi\right) - f\left(\frac{3}{4}\pi\right)\right].$$

4. Solve this problem by the methods of characteristic curves
$$\begin{cases} u_x + yu_y = 0 \\ u(0, y) = f(y) \end{cases}$$

Solution This is a partial differential equation to be solved at all points in the whole xy-plane, and the values of the solution function are prescribed on the y-axis. The ordinary differential equation that describes the corresponding characteristic curves will be
$$\frac{dy}{dx} = -\frac{u_x}{u_y} = y.$$
The solution passing through (x_0, y_0) is $y = y_0 e^{x-x_0}$.

The intersection of this characteristic curve with the y-axis occurs at $y = y_0 e^{-x_0}$, then we have
$$u(x_0, y_0) = f(y_0 e^{-x_0}).$$
as the solution. Of course, we would usually write this as
$$u(x, y) = f(ye^{-x}).$$
It is easily to show that this is the solution of the original problem.

第二部分 信息论
Part 2 Information Theory

引言

我们生活在信息时代,网络已经成为我们生活的一个不可或缺的部分,使得这个太阳系的第三颗行星变成了一个小村庄. 甚至在电影院中,人们通过有孔电话谈话都是司空见惯的,电影可以以 DVD 碟片的形式被租用. 电子邮件地址、网址和商业信用卡一样普遍. 很多人更喜欢用电子邮件或电子卡片而不是传统的信件给朋友们寄信. 仓储报价可由移动电话来进行检查.

信息已经成为成功的钥匙(虽然它一直都是,但如今它变成了唯一的钥匙). 所有这些信息及其交换均依赖其底层的 1 和 0(这两个无所不在的比特),它们彼此相邻地存在以保存信息. 今天我们生活的信息时代基本归功于它的存在,1948 年出版的美国电气工程师 Claude E. Shannon 撰写的重要文献"通信的数学理论"奠定了信息论重要的理论基础.

在广义上,信息包括任意标准通信媒体所涉及的内容,例如电报、无线电收音机、电视、电子计算机的信号、伺服机构的机械系统和其他数据处理设备. 此理论甚至可以应用于人和动物的神经网络的信号.

信息论主要关注发现控制系统的数学规律以传达或控制信息. 它建立了信息及运输、储存和其他信息过程的定量测量. 所处理的问题是找到使用各种可用的联系系统的最佳方法和从不相干的信息或噪音中分离出想要的信号的最佳方法. 另一个问题是给定信息-传播介质(通常称信道)容量的上界的设定. 结论主要会使信息工程师们感兴趣,有些内容在诸如心理学或语言学方面被采用或者说被发现有用.

信息理论的界线十分模糊,它和通信理论重叠的部分很多,但是在信息的联系和过程上更倾向于基础的局限,较少涉及所用设备的详细操作.

信息论是人们在长期通信工程的实践中,由通信技术与概率论、随机过程和数理统计相结合而逐步发展起来的一门科学. 近半个世纪以来,以通信理论为核心的经典信息论,正以信息技术为手段,向高精尖方向迅猛发展,把人类社会推入了信息时代. 随着信息理论的迅猛发展和信息概念的不断深化,信息论所涉及的内容已经超越了狭义的通信工程的范畴,进入了信息科学这一更广阔、更新兴的领域.

Introduction

Today we are living in the information age. Internet has become an indispensable part of our lives, which makes this, the third planet from the sun, a global village. People talking over the cellular phones is a common sight, sometimes even in theatres. Movies can be rented in the form of a DVD disk. Email addresses and web addresses are common on business cards. Many people prefer to send emails and e-cards to their friends rather than the regular snail mail. Stock quotes can be checked by the mobile phone.

Information has become the key to success (it has always been a key to success, but in today's world it is the only key). And behind all this information and its exchange lie the tiny 1's and 0's (the omnipresent bits) that hold information by merely the way they sit next to one another. Yet the information age that we live in today owes its existence, primarily, to a seminal paper published in 1948 that laid the foundation of the wonderful field of Information Theory initiated by one man, the American electrical engineer Claude E. Shannon, whose ideas appeared in the article "The Mathematical Theory of Communication" (1948).

In its broadest sense, information includes the contents of any standard communication media, such as telegraphy, telephone radio, or television, and the signals of electronic computers, serve-mechanism systems, and other data-processing devices. The theory is even applicable to the signals of the neural networks of humans and other animals.

The chief concern of Information Theory is to discover mathematical laws governing systems designed to communicate or manipulate information. It sets up quantitative measures of information and transmit, store, and otherwise process information. Some of the problems treated are related to finding the best methods of using various available communication systems and the best methods for separating wanted information or signals, from extraneous information or noise. Another problem is the setting of upper bounds on the capacity of a given information-carrying medium (often called an information channel). While the results are of interest to communication engineers, some of the concepts have been adopted and found usefully in such fields such as psychology and linguistics.

The boundaries of Information Theory are not quite clear. The theory overlaps heavily with communication theory but is more oriented towards the fundamental limitations on the processing and communication of information and less towards the detailed operation of the devices employed.

Information Theory is a subject of science which developed step by step combining with communication technique probability, random process and mathematics & physics statistics after long-term communication engineering practice. From nearly half of a century on, the classical information theory whose core is the communication theory developed rapidly towards high, praise, top, with the help of communication technique, pushing the human being society into information time. With impetuous development of the theory of information and the continuously deepening of the idea of information, the content involved by the theory of information have exceeded the narrow sense of the category of communication engineer, and entered the more widely, newly field of information science.

第一章 介绍
Chapter 1 Introduction

本章主要介绍了信息的定义,信息论的起源、发展及研究内容. 通过本章学习,要求学生了解人类利用信息的历史,了解信息论的发展历史,了解信息论研究的内容,掌握信息的本体论层次和认识论层次的定义以及香农信息论中信息的定义,理解信息的分类.

This chapter mainly discusses the definition of information, the origin of Information Theory and its development including its future field of research. Through the study of this chapter, students are required to browse the brief history of information, the stages of Information Theory in its development and the research contents of Information Theory. Moreover, students are required to master the definitions of information in ontological and cognitive levels and also in Shannon theory, comprehend its classification.

单词和短语 Words and expressions

★ 信息论　information theory
★ 编码理论　coding theory
发出　emit
比特　bit
二进制　binary
二进制对称源　binary symmetric source
二进制对称信道　binary symmetric channel
原始字节错误率　raw bit error probability
编码　encode
字节错误率　bit error probability
噪音　noise

冗余　redundant
相互校验　cross check
★ 编码算法　coding algorithm
错误模式　error pattern
综合　synthesis
汉明码　Hamming code
单独错误校正码　single-error-correcting code
速率　rate
二进制熵函数　binary entropy function
能量　capacity
★ 信道编码理论　channel coding theory

本章重点

1. 掌握信息的本体论层次和认识论层次的定义以及香农信息论中信息的定义.
2. 理解信息的分类.

Key points of this chapter

1. Master the definitions of ontological level and cognitive level, the definition of information in Shannon theory.
2. Understand the classification of information.

第二章 信息理论
Chapter 2 Information Theory

本章提到了香农信息论中信息度量的方法、互信息、条件熵、微分熵、严格熵、离散记忆信道等概念. 本章是基础, 但是首先阅读它是错误的, 因为它仅仅是关于熵、互信息等概念的简单罗列. 本章最好看成是一参考章节, 在理解了第三～六章的基础上再来阅读.

This chapter introduces the method of measuring the information in Shannon's theory, the definitions of mutual information, conditional entropy, differential entropy, absolute entropy, discrete memory channel. This chapter is fundamental. However, it is probably a mistake to read it first, since it is really just a collection of technical results about entropy, mutual information, and so forth. It is better to regard as a reference section, and should be consulted as necessity to understand Chapter 3-6.

单词和短语　Words and expressions

比特　bits (binary digits)	操作　manipulation
自然码　natural digits	速记法　shorthand
熵函数　entropy function	通信系统　communication system
可能向量　probability vector	连续信源输出　continuous source outputs
条件熵　conditional entropy	编码员　coder
★离散记忆信道　discrete memory channel	映射　map
过渡可能性　transition probability	目标　destination
产出　output	数据过程定理　data-processing theorem
边际分布　marginal distribution	离散量化　discrete quantization
互信息　mutual information	改进/精炼　refinement
启发式　heuristic	密度　density
联合熵　joint entropy	中值定理　mean value theorem
维恩图　Venn diagram	表面相似　superficial resemblance
马尔可夫链　Markov chain	网格　mesh
限定函数　definite function	★微分熵　differential entropy
串联　tandem	詹森不等式　Jensen inequality
数据过程配置　data-processing configuration	★确定信道　determinate channel
凸组合　convex combination	

本章重点

本章重点掌握信息和熵的概念以及计算公式, 掌握互信息和平均互信息的概念以及计算公式, 掌握单符号离散信源、多符号离散平稳信源以及连续信源熵的计算公式, 掌握无失真信源编码定理.

Key points of this chapter

The emphases of this chapter are to master the definitions and calculation formulas of information and entropy, the definitions and calculation formulas of mutual information and average mutual information, the definitions and calculation formulas of mono-symbol discrete source, multi-symbol discrete stable source and continuous source entropy, the no distortion source coding theorem.

第三章 离散无记忆信道和容量成本方程
Chapter 3 Discrete Memoryless Channels and their Capacity-Cost Equations

本章是基础知识，介绍了离散无记忆信道和容量成本方程、输入符号系统、输出符号系统、香农信息论中信道容量的计算方法、信道编码定理、编码规则、随机编码、弱大数定律.

As the basic knowledge, this chapter introduces the discrete memoryless channels and their capacity-cost equations, the systems of input sign and output sign, the calculating method of capacity of channel in Shannon theory, the channel coding theorem, the coding rule, the random coding and the law of weak large numbers.

单词和短语 Words and expressions

输入符号系统　input sign system
输出符号系统　output sign system
想象　imagine
无记忆假设　memoryless assumption
平均成本　average cost
★ 容量成本方程　capacity-cost equation
验证源头　test-source
n 维容许验证源头　n-dimensional admissible test sources
容许成本　admissible cost
r 对称　r-symmetry
系统比率　rate of system

超过信道容量率　rates above channel capacity
长　length
每个符号的比特　bits per symbol
编码规则　decoding rule
区别代码　distinct code
指示函数　indicator function
随机编码　random coding
期望值　expected value
★ 弱大数定律　weak law of large numbers
编码范围　decoding sphere

第四章 离散无记忆信源和扭曲率方程
Chapter 4 Discrete Memoryless Sources and their Rate-Distortion Equations

本章介绍了离散无记忆信源及其扭曲率方程，源编码定理、扭曲度、汉明扭曲度、错误扭曲率、数据压缩定理、罚函数、无限制和等基本内容.

This chapter introduces some basic contents such as discrete memoryless sources and their distortion rate equations, source coding theorem, distortion measure, Hamming distortion measure, error probability distortion rate, data-compression theory, penalty function, unrestricted sum, and so on.

单词和短语 Words and expressions

源字母表　source alphabet
★ 离散无记忆信源　discrete memoryless sources
统计源　source statistics
目标符号　object sign
扭曲　distortion
扭曲度　distortion measure
平均扭曲　average distortion
★ 测试通道　test channel

扭曲率　distortion rate
★ 源编码定理　source coding theorem
向后测试通道　backwards test channel
Hamming 扭曲度　Hamming distortion measure
错误扭曲率　error probability distortion rate
★ 数据压缩定理　data-compression theorem
目的符号　destination symbols
数据压缩系统　data compression scheme

| 无理数 | irrational |
| 罚函数 | penalty function |

| 无限制和 | unrestricted sum |
| 内部和 | inner sum |

第五章　高斯信道和信源
Chapter 5　Gaussian Channel and Source

　　本章介绍了带有随机错误信息约束平均功率的时间离散无记忆添加的高斯信道、白高斯噪声过程、噪声谱密度、总的容量成本函数噪声密度、离散时间无记忆高斯信源、高斯信源每个符号均值平方扭曲等内容.

　　This chapter is to introduce the Gaussian channel and source discrete-time memoryless additive Gaussian channel with average power constraint error-free information, white Gaussian noise process, noise spectral density, overall capacity-cost function density, discrete-time memoryless Gaussian source, Gaussian source per-symbol mean-squared distortion and so on.

单词和短语　Words and expressions

伏特	Voltage
传送	transmit
信号	signal
瓦	Watts
耗散	dissipate
焦耳	Joule
白高斯噪声过程	white Gaussian noise process
噪声错误密度	noise spectral density
带宽	bandwidth
波段限制	band-limited
功率限制	power-limited
第 n 项容量成本函数	n-th capacity-cost function

平方错误	squared-error
总的容量成本函数噪声密度	overall capacity-cost function oink density
算术-几何均值	arithmetic-geometric average value
★离散时间无记忆高斯信源	discrete-time memoryless Gaussian source
均值平方错误标准	mean-squared error criterion
高斯分布	Gaussian distribution
★高斯信源每个符号均值平方扭曲	Gaussian source per-symbol mean-squared distortion

第六章　信源-信道编码理论
Chapter 6　Source-Channel Coding Theory

　　香农信息论中的无失真信源编码定理只给出了理想信源编码存在的论述. 本章给出了具体信源编码的构造方法. 通过本章的学习,要求学生掌握香农编码、费诺编码、哈夫曼编码的具体编码方法,了解游程编码、冗余位编码,掌握各种编码适用的场合及它们的优缺点.

　　No distortion source coding theorem in Shannon theory only gives an existence assertion of the ideal situation of source coding. This chapter provides some constructing methods for concrete source coding. Through a careful study of this chapter, the students can hopefully grasp some concrete methods of Shannon coding, Fano coding and Huffman coding. Understand run length coding and redundancy encoding. Master the suitable situation of all coding and their merits and defects.

单词和短语　Words and expressions

★信息源	information source
★噪声源	noisy sourse
数据处理	data processing
量子化	quantization
调节	modulation
连续块	successive block

发出信道输出符号	emit channel output symbols
一对一通信	one-to-one correspondence
实验来源	experiment source
信源序列	source sequence
目的序列	destination sequence
信源统计	source statistics

传送率　rate of transmission
冲突　conflict
★信源-信道编码理论　source-channel coding theorem
★数据传输定理　data-processing theorem
中间向量　intermediate vector
最坏扭曲　worst-case distortion

每个符号基础　per-symbol basis
分解　decomposition
传送编码　transmitted coding
负担　afford
交易　tradeoff
可实现区域　realizable region
观点　viewpoint

本章重点

掌握香农编码、费诺编码、哈夫曼编码的定义.

Key points of this chapter

Master the concepts of Shannon coding, Fano coding and Huffman coding.

练 习

1. 一阶 Markov 的标准图形如下图. 信源 X 的符号集为 $\{0, 1, 2\}$.
(1) 计算信源平稳后其概率分布；
(2) 计算信源的熵 H_∞.

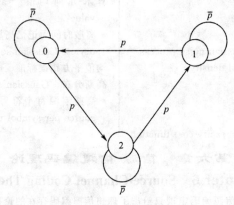

解 (1)
$$\begin{cases} p(e_1) = p(e_1)p(e_1/e_1) + p(e_2)p(e_1/e_2) \\ p(e_2) = p(e_2)p(e_2/e_2) + p(e_3)p(e_2/e_3) \\ p(e_3) = p(e_3)p(e_3/e_3) + p(e_1)p(e_3/e_1) \end{cases}$$

$$\begin{cases} p(e_1) = \bar{p} \cdot p(e_1) + p \cdot p(e_2) \\ p(e_2) = \bar{p} \cdot p(e_2) + p \cdot p(e_3) \\ p(e_3) = \bar{p} \cdot p(e_3) + p \cdot p(e_1) \end{cases}$$

$$\begin{cases} p(e_1) = p(e_2) = p(e_3) \\ p(e_1) + p(e_2) + p(e_3) = 1 \end{cases}$$

$$\begin{cases} p(e_1) = 1/3 \\ p(e_2) = 1/3 \\ p(e_3) = 1/3 \end{cases}$$

$$\begin{cases} p(x_1) = p(e_1)p(x_1/e_1) + p(e_2)p(x_1/e_2) = \bar{p} \cdot p(e_1) + p \cdot p(e_2) = (\bar{p}+p)/3 = 1/3 \\ p(x_2) = p(e_2)p(x_2/e_2) + p(e_3)p(x_2/e_3) = \bar{p} \cdot p(e_2) + p \cdot p(e_3) = (\bar{p}+p)/3 = 1/3 \\ p(x_3) = p(e_3)p(x_3/e_3) + p(e_1)p(x_3/e_1) = \bar{p} \cdot p(e_3) + p \cdot p(e_1) = (\bar{p}+p)/3 = 1/3 \end{cases}$$

$$\begin{pmatrix} X \\ p(X) \end{pmatrix} = \begin{pmatrix} 1 & 2 & 3 \\ 1/3 & 1/3 & 1/3 \end{pmatrix}$$

(2)

$$H_\infty = -\sum_{i=1}^{3}\sum_{j=1}^{3} p(e_i)p(e_j/e_i)\log p(e_j/e_i)$$

$$= -\left[\frac{1}{3}p(e_1/e_1)\log_2 p(e_1/e_1) + \frac{1}{3}p(e_2/e_1)\log_2 p(e_2/e_1) + \right.$$

$$\frac{1}{3}p(e_3/e_1)\log_2 p(e_3/e_1) + \frac{1}{3}p(e_1/e_2)\log_2 p(e_1/e_2) +$$

$$\frac{1}{3}p(e_2/e_2)\log_2 p(e_2/e_2) + \frac{1}{3}p(e_3/e_2)\log_2 p(e_3/e_2) +$$

$$\frac{1}{3}p(e_1/e_3)\log_2 p(e_1/e_3) + \frac{1}{3}p(e_2/e_3)\log_2 p(e_2/e_3) +$$

$$\left.\frac{1}{3}p(e_3/e_3)\log_2 p(e_3/e_3)\right]$$

$$= -\left[\frac{1}{3}\bar{p}\cdot\log_2\bar{p} + \frac{1}{3}p\cdot\log_2 p + \frac{1}{3}p\cdot\log_2 p + \right.$$

$$\left.\frac{1}{3}\bar{p}\cdot\log_2\bar{p} + \frac{1}{3}p\cdot\log_2 p + \frac{1}{3}\bar{p}\cdot\log_2\bar{p}\right]$$

$$= -(\bar{p}\cdot\log_2\bar{p} + p\cdot\log_2 p)$$

2. 试证 $H(X_1X_2\cdots X_N) \leqslant H(X_1) + H(X_2) + \cdots + H(X_N)$.

证

$H(H_N;X_1X_2\cdots X_{N-1}) = H(X_1) + H(X_2/X_1) + H(X_3/X_1X_2) + \cdots + H(H_N/X_1X_2\cdots X_{N-1})$

$I(X_2;X_1) \geqslant 0 \Rightarrow H(X_2) \geqslant H(X_2/X_1)$

$I(X_3;X_1X_2) \geqslant 0 \Rightarrow H(X_3) \geqslant H(X_3/X_1X_2)$

...

$I(X_N;X_1X_2\cdots X_{N-1}) \geqslant 0 \Rightarrow H(X_N) \geqslant H(X_N/X_1X_2\cdots X_{N-1})$

所以 $H(X_1X_2\cdots X_N) \leqslant H(X_1) + H(X_2) + H(X_3) + \cdots + H(X_N)$

3. 假定二元素对称信道的传输矩阵是 $\begin{bmatrix} \frac{2}{3} & \frac{1}{3} \\ \frac{1}{3} & \frac{2}{3} \end{bmatrix}$，若 $p(0) = 3/4, p(1) = 1/4$，计算 $H(X)$，$H(X/Y), H(Y/X)$ 和 $I(X;Y)$.

解

$$H(X) = -\sum_i p(x_i) = -\left(\frac{3}{4}\times\log_2\frac{3}{4} + \frac{1}{4}\times\log_2\frac{1}{4}\right) = 0.811$$

$$H(Y/X) = -\sum_i\sum_j p(x_i)p(y_j/x_i)\log p(y_j/x_i)$$

$$= -\left(\frac{3}{4}\times\frac{2}{3}\lg\frac{2}{3} + \frac{3}{4}\times\frac{1}{3}\lg\frac{1}{3} + \frac{1}{4}\times\frac{1}{3}\lg\frac{1}{3} + \frac{1}{4}\times\frac{2}{3}\lg\frac{2}{3}\right)\times\log_2 10$$

$$= 0.918$$

$$p(y_1) = p(x_1y_1) + p(x_2y_1) = p(x_1)p(y_1/x_1) + p(x_2)p(y_1/x_2)$$

$$= \frac{3}{4}\times\frac{2}{3} + \frac{1}{4}\times\frac{1}{3} = 0.583\ 3$$

$$p(y_2) = p(x_1y_2) + p(x_2y_2) = p(x_1)p(y_2/x_1) + p(x_2)p(y_2/x_2)$$

$$= \frac{3}{4} \times \frac{1}{3} + \frac{1}{4} \times \frac{2}{3} = 0.4167$$

$H(Y) = -\sum_j p(y_j) = -(0.5833 \times \log_2 0.5833 + 0.4167 \times \log_2 0.4167) = 0.980$

$I(X;Y) = H(X) - H(X/Y) = H(Y) - H(Y/X)$

$H(X/Y) = H(X) - H(Y) + H(Y/X) = 0.811 - 0.980 + 0.918 = 0.749$

$I(X;Y) = H(X) - H(X/Y) = 0.811 - 0.749 = 0.062$

Exercises

1. The standard figure of first-order Markov is shown as the following figure. The code set of source X is $\{0, 1, 2\}$.

(1) Compute the probability distribution of the source after it is steady;

(2) Compute the entropy H_∞ of the source.

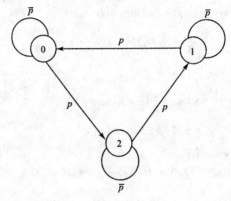

Solution (1)

$$\begin{cases} p(e_1) = p(e_1)p(e_1/e_1) + p(e_2)p(e_1/e_2) \\ p(e_2) = p(e_2)p(e_2/e_2) + p(e_3)p(e_2/e_3) \\ p(e_3) = p(e_3)p(e_3/e_3) + p(e_1)p(e_3/e_1) \end{cases}$$

$$\begin{cases} p(e_1) = \bar{p} \cdot p(e_1) + p \cdot p(e_2) \\ p(e_2) = \bar{p} \cdot p(e_2) + p \cdot p(e_3) \\ p(e_3) = \bar{p} \cdot p(e_3) + p \cdot p(e_1) \end{cases}$$

$$\begin{cases} p(e_1) = p(e_2) = p(e_3) \\ p(e_1) + p(e_2) + p(e_3) = 1 \end{cases}$$

$$\begin{cases} p(e_1) = 1/3 \\ p(e_2) = 1/3 \\ p(e_3) = 1/3 \end{cases}$$

$$\begin{cases} p(x_1) = p(e_1)p(x_1/e_1) + p(e_2)p(x_1/e_2) = \bar{p} \cdot p(e_1) + p \cdot p(e_2) = (\bar{p}+p)/3 = 1/3 \\ p(x_2) = p(e_2)p(x_2/e_2) + p(e_3)p(x_2/e_3) = \bar{p} \cdot p(e_2) + p \cdot p(e_3) = (\bar{p}+p)/3 = 1/3 \\ p(x_3) = p(e_3)p(x_3/e_3) + p(e_1)p(x_3/e_1) = \bar{p} \cdot p(e_3) + p \cdot p(e_1) = (\bar{p}+p)/3 = 1/3 \end{cases}$$

$$\begin{pmatrix} X \\ p(X) \end{pmatrix} = \begin{pmatrix} 1 & 2 & 3 \\ 1/3 & 1/3 & 1/3 \end{pmatrix}$$

(2)

$$H_\infty = -\sum_i^3 \sum_j^3 p(e_i)p(e_j/e_i) \log p(e_j/e_i)$$

$$= -\left[\frac{1}{3}p(e_1/e_1)\log_2 p(e_1/e_1) + \frac{1}{3}p(e_2/e_1)\log_2 p(e_2/e_1) + \right.$$
$$\frac{1}{3}p(e_3/e_1)\log_2 p(e_3/e_1) + \frac{1}{3}p(e_1/e_2)\log_2 p(e_1/e_2) +$$
$$\frac{1}{3}p(e_2/e_2)\log_2 p(e_2/e_2) + \frac{1}{3}p(e_3/e_2)\log_2 p(e_3/e_2) +$$
$$\frac{1}{3}p(e_1/e_3)\log_2 p(e_1/e_3) + \frac{1}{3}p(e_2/e_3)\log_2 p(e_2/e_3) +$$
$$\left.\frac{1}{3}p(e_3/e_3)\log_2 p(e_3/e_3)\right]$$
$$= -\left[\frac{1}{3}\bar{p}\cdot\log_2\bar{p} + \frac{1}{3}p\cdot\log_2 p + \frac{1}{3}p\cdot\log_2 p + \right.$$
$$\left.\frac{1}{3}\bar{p}\cdot\log_2\bar{p} + \frac{1}{3}p\cdot\log_2 p + \frac{1}{3}\bar{p}\cdot\log_2\bar{p}\right]$$
$$= -(\bar{p}\cdot\log_2\bar{p} + p\cdot\log_2 p)\,\text{bit/symbol}$$

2. Prove that $H(X_1 X_2 \cdots X_N) \leqslant H(X_1) + H(X_2) + \cdots + H(X_N)$.

Proof

$$H(H_N; X_1 X_2 \cdots X_{N-1}) = H(X_1) + H(X_2/X_1)$$
$$+ H(X_3/X_1 X_2) + \cdots + H(H_N/X_1 X_2 \cdots X_{N-1})$$

$I(X_2; X_1) \geqslant 0 \Rightarrow H(X_2) \geqslant H(X_2/X_1)$

$I(X_3; X_1 X_2) \geqslant 0 \Rightarrow H(X_3) \geqslant H(X_3/X_1 X_2)$

...

$I(X_N; X_1 X_2 \cdots X_{N-1}) \geqslant 0 \Rightarrow H(X_N) \geqslant H(X_N/X_1 X_2 \cdots X_{N-1})$

so $H(X_1 X_2 \cdots X_N) \leqslant H(X_1) + H(X_2) + H(X_3) + \cdots + H(X_N)$

3. Suppose the deliver matrix of two elements symmetric channel is $\begin{bmatrix} \frac{2}{3} & \frac{1}{3} \\ \frac{1}{3} & \frac{1}{3} \end{bmatrix}$.

If $p(0) = 3/4$, $p(1) = 1/4$, compute $H(X)$, $H(X/Y)$, $H(Y/X)$ and $I(X;Y)$;

Solution

$$H(X) = -\sum_i p(x_i) = -\left(\frac{3}{4}\times\log_2\frac{3}{4} + \frac{1}{4}\times\log_2\frac{1}{4}\right) = 0.811\,\text{bit/symbol}$$

$$H(Y/X) = -\sum_i\sum_j p(x_i)p(y_j/x_i)\log p(y_j/x_i)$$
$$= -\left(\frac{3}{4}\times\frac{2}{3}\lg\frac{2}{3} + \frac{3}{4}\times\frac{1}{3}\lg\frac{1}{3} + \frac{1}{4}\times\frac{1}{3}\lg\frac{1}{3} + \frac{1}{4}\times\frac{2}{3}\lg\frac{2}{3}\right)\times\log_2 10$$
$$= 0.918\,\text{bit/symbol}$$

$$p(y_1) = p(x_1 y_1) + p(x_2 y_1) = p(x_1)p(y_1/x_1) + p(x_2)p(y_1/x_2)$$
$$= \frac{3}{4}\times\frac{2}{3} + \frac{1}{4}\times\frac{1}{3} = 0.5833$$

$$p(y_2) = p(x_1 y_2) + p(x_2 y_2) = p(x_1)p(y_2/x_1) + p(x_2)p(y_2/x_2)$$
$$= \frac{3}{4}\times\frac{1}{3} + \frac{1}{4}\times\frac{2}{3} = 0.4167$$

$$H(Y) = -\sum_j p(y_j) = -(0.5833\times\log_2 0.5833 + 0.4167\times\log_2 0.4167) = 0.980\,\text{bit/symbol}$$

$I(X;Y) = H(X) - H(X/Y) = H(Y) - H(Y/X)$

$H(X/Y) = H(X) - H(Y) + H(Y/X) = 0.811 - 0.980 + 0.918 = 0.749\,\text{bit/symbol}$

$I(X;Y) = H(X) - H(X/Y) = 0.811 - 0.749 = 0.062\,\text{bit/symbol}$

第三部分　数据结构与算法
Part 3　Data Structures and Algorithms

引　言

计算机程序的主要目标不仅是完成运算,而是存储信息和尽快地检索信息。因此,研究数据结构和算法就成了计算机科学的核心问题。

简单地说,数据结构就是一类数据的表示及相关操作,即使是存储在计算机中的一个整数或者一个浮点数,也是一个简单的数据结构。

算法是指解决问题的一种方法或者一个过程。如果将一个问题看做函数,那么算法就是把输入转换为输出。一个问题可能用多种算法解决,一种给定的算法解决一个特定的问题。

估算一种算法或者一个计算机程序的效率的方法,称为算法分析。算法分析还可以度量一个问题的内在复杂程度。

Introduction

The primary purpose of computer programs is not only to perform calculations, but also to store and retrieve information as fast as possible. For this reason, the study of data structures and the algorithms that manipulate them is at the heart of computer science.

In the most general sense, a data structure is any data representation and its associated operations. Even an integer or float in point number stored on the computer is a simple data structure.

An algorithm is a method or a process followed to solve a problem. If the problem is viewed as a function, then an algorithm is an implementation for the function that transforms an input to the corresponding output. A problem can be solved by many different algorithms. A given algorithm solves only one problem.

A method for evaluating the efficiency of an algorithm or computer program, called algorithm analysis. Algorithm analysis also allows you to measure the inherent difficulty of a problem.

第一章　线性表
Chapter 1　Lists

线性表是由称为元素的数据项组成的一种有限且有序的序列。线性表中不包含任何元素时,我们称之为空表。当前存储的元素数目称为线性表的长度。线性表的开始结点称为表头,结尾结点称为表尾。线性表的实现有两种标准方法:顺序表和链表。

We define a list to be a finite, ordered sequence of data items known as elements. A list is said to be empty when it contains no elements. The number of elements currently stored is called the length of the list. The beginning of the list is called the head , the end of the list is called the tail. There are two standard approaches to implementing lists, the array-based list, and the linked list.

单词和短语 Words and expressions

算法	algorithm	元素	element
数据结构	data structure	空表	empty list
线性表	list	表头	head
栈	stack	表尾	tail
队列	queue	有序线性表	sorted list

无序线性表　unsorted list	表头结点　header node
顺序表　array-based list 或 sequential list	动态数组　dynamic array
链表　linked list	双链表　doubly linked list
结点　node	

第二章　栈和队列
Chapter 2　Stack and Queue

栈是限定仅在一端进行插入或删除操作的线性表,也称栈为"LIFO 线性表"(Last In First Out)。一般来讲,称栈的可访问元素为栈顶元,元素插入栈称为入栈,删除元素时称为出栈。实现栈的方法主要有两种:顺序栈和链式栈。

同栈一样,队列也是一种受限的线性表。队列元素只能从队尾插入从队首删除,也称队列为"FIFO 线性表"(First In First Out)。本章将主要介绍队列的两种实现方法:顺序队列和链式队列。

The stack is a list-like structure in which elements may be inserted or removed from only one end. They also called the stack a "LIFO"(Last In First Out) list. It is traditional to call the accessible element of the stack the top element. Elements are said to be inserted; instead they are pushed onto the stack. When removed, an element is said to be popped from the stack. The two approaches presented here are array-based and linked stacks.

Like the stack, the queue is a list-like structure that provide restricted access to its elements. Queue elements may only be inserted at the back (called an enqueue operation) and removed from the front (called a dequeue operation). They call a queue a "FIFO"list, which stands for "First-In First-Out". This section presents two implementations: the array-based queue and the linked queue.

单词和短语 Words and expressions

栈	stack	队列	queue
栈顶	top	入队	enqueue
入栈	push	出队	dequeue
出栈	pop	顺序队列	array-based queue
顺序栈	array-based stack	链队列	linked queue
链栈	linked stack		

第三章　树和二叉树
Chapter 3　Trees and Binary Trees

树 T(Tree)是由一个或一个以上结点组成的有限集,其中有一个特定的结点 R 称为 T 的根结点。集合(T-{R})中的其余结点可被划分为 n(n≥0)个不相交的子集 $T_1, T_2, \cdots, T_n$,其中每个子集都是树,并且其相应的根结点 $R_1, R_2, \cdots R_n$ 是 R 的子结点。子集 Ti(0≤i<n)称为树 T 的子树。结点的出度定义为该结点的子结点数目。森林定义为一棵或更多棵树的集合。

A tree T is a finite set of one or more nodes such that there is one designated node R, called the root of T. The remaining nodes in (T-{R}) are partitioned into n(n≥0) disjoint subsets $T_1, T_2, \cdots, T_n$, each of which is a tree, and whose roots $R_1, R_2, \cdots, R_n$, respectively, are children of R. The subsets Ti(0≤i<n) are said to be subtrees of T. node's out degree is the number of children for that node. A forest is a collection of one or more trees.

二叉树是由结点的有限集合组成,这个集合或为空,或由一个根结点以及两棵不相交的二叉树组成,这两棵二叉树分别称为这个根的左子树和右子树。这两棵子树的根即为根结点的子结点,而根结点则被称为父结点。

A binary tree is made up of a finite set of elements called nodes. This set either is empty or consists of a node called the root together with two binary trees, called the left and right subtrees, which are disjoint from each other and from the root. The roots of these subtrees are children of the root. There is an edge from a node to each of its children, and a node is said to be the parent of its children.

二叉树的遍历是指按一定顺序访问二叉树的结点。主要可分为前序遍历、中序遍历和后序遍历。

Any process for visiting all of the nodes in some order is called a traversal. There are three approaches for traversal, they are preorder traversal, inorder traversal, postorder traversal.

Huffman树的每个叶结点对应于一个字母,叶结点的权重就是它对应字母的出现频率。其目的在于按照最小外部路径建立一棵树。

Each leaf of the Huffman tree corresponds to a letter, and we say that the weight of the leaf node is the weight of its associated letter. The goal is to build a tree with the minimum external path weight.

单词和短语 Words and expressions

树　　tree	边　　edge
二叉树　binary tree	前序遍历　preorder traversal
根结点　root	中序遍历　inorder traversal
父结点　parent	后序遍历　postorder traversal
子结点　children	祖先　ancestor
左子树　left tree	子孙　descendant
右子树　right tree	叶结点　leaf
度　degree	满二叉树　full binary tree
入度　in degree	完全二叉树　complete binary tree
出度　out degree	森林　forest

第四章　图
Chapter 4　Graph

本章介绍图数据结构以及一些通用的图算法。一个图由两个集合定义,一个是由顶点组成的集合。另一个是由顶点之间的边组成的集合。主要内容包括:

1. 定义图的两种基本表示方法:相邻矩阵和邻接表。
2. 给出图的两种最常用遍历算法:深度优先搜索和广度优先搜索。
3. 解决寻找图的最短路径问题的算法。
4. 介绍寻找最小代价树算法。

This chapter introduces the graph data structure and a sampling of some typical graph algorithms. A graph is defined by two sets. The first is a set of vertices. The second is a set of connections linking parts of vertices, called edges. The section consists of four parts.

1. Define two fundamental representations for graphs, the adjacency matrix and adjacency list.
2. Presents the two most commonly used graph traversal algorithms, called depth-first and breadth-first search.
3. Presents algorithms for solving some problems related to finding shortest routes in a graph.
4. Presents algorithms for finding the minimum-cost spanning tree.

单词和短语 Words and expressions

图　　graph	邻接点　neighbors
顶点　vertex	加权图　weighted graph
稀疏图　sparse graph	路径　path
密集图　dense graph	回路　cycle
完全图　complete graph	子图　subgraph
有向图　directed graph	连通　connected
无向图　undirected graph	无环图　acyclic graph

邻接矩阵　adjacency matrix
邻接表　adjacency list
深度优先搜索　depth-first search(DFS)

广度优先搜索　breadth-first search(BFS)
最小代价权　minimum-cost spanning tree(MST)

第五章　内排序
Chapter 5　Internal Sorting

排序就是将给定的一组记录重新排列成按关键字有序的序列。即给定一组记录 $r_1, r_2, \cdots, r_n$，其关键码分别为 $k_1, k_2, \cdots, k_n$，排序问题就是要将这些记录排成顺序为 $rs_1, rs_2, \cdots, rs_n$ 的一个序列 s，满足条件 $ks_1 \leqslant ks_2 \leqslant \cdots \leqslant ks_n$。

本章主要介绍插入排序、起泡排序、选择排序、希尔排序、快速排序、归并排序，堆排序和基数排序等几中常用的内排序方法。

The sorting problem is to arrange a set of records so that the values of their key fields are in order. In another words, given a set of records $r_1, r_2, \cdots, r_n$, with key values $k_1, k_2, \cdots, k_n$, the sorting problem is to arrange the records into any order s such that records $rs_1, rs_2, \cdots, rs_n$ have keys obeying the property $ks_1 \leqslant ks_2 \leqslant \cdots \leqslant ks_n$.

This chapter mainly discuss several approaches about sorting, they are insert sort, bubble sort, selection sort, shell sort, quick sort, merge sort, heap sort and radix sort.

单词和短语 Words and expressions

插入排序　insert sort
逆置　Inversion
起泡排序　Bubble sort
选择排序　Selection sort
希尔排序　shell sort

快速排序　quick sort
归并排序　merge sort
堆排序　heapsort
基数排序　radix sort

第六章　查找
Chapter 6　Searching

查找就是确定一个具有某特定值的元素是否是某集合的成员。检索算法可以分成三类：
1. 顺序表和线性表方法
2. 根据关键字直接访问方法(散列法)
3. 树索引方法

Search can be viewed abstractly as a process to determine if an element with a particular value is a member of a particular set. We can categorize search algorithms into three general approaches:
1. Sequential and list methods.
2. Direct access by key value(hashing).
3. Tree indexing methods.

练习

1. 编写一个 C++ 程序，建立一个能存放 20 个元素而实际只存储了下列结构的线性表。
   <2,23|15,5,9>
2. 编写一个函数，倒置线性表中元素的顺序。
3. 循环链表是指链表中最后一个结点的 next 域指向链表的第一个结点。请用函数实现循环单链表。
4. 编写一个算法，用二叉树的根结点指针做为参数，按照层次顺序将结点的值打印出来。
5. 编写一个递归函数计算二叉树的高度。
6. 一个高度为 h 的堆最大和最小元素数目各为多少？

7. 画出对下列存储于数组中的值执行 buildheap 后的最大值堆：
10 5 12 3 2 1 8 7 9 4

8. 根据下面的字母和权建立 Huffiman 编码树，并给出各字母的代码。

A	B	C	D	E	F	G	H	I	J	K	L
2	3	5	7	11	13	17	19	23	31	37	41

9. 给出一个可导致快速排序的最坏性能的 0 到 7 的排列。

10. 编写算法尽可能快地排序 5 个元素，在最好、最坏、平均情况下的交换和比较次数各为多少？

Exercises

1. Write a C++ program, that creates a list capable of holding twenty elements and which actually stores the list with the following configuration:
<2,23|15,5,9>.

2. Write a function to reverse the order of the elements on the list.

3. A circular linked list is one in which the next field for the last link node of list points to the first link node of the list. Write a function to implement circular singly linked lists.

4. Write an algorithm that takes as input a pointer to the root of a binary tree and prints the node values of the tree in level order.

5. Write a recursive function that returns the height of a binary tree.

6. What are the minimum and maximum number of elements in a heap of height h?

7. Show the max-heap that results from running buildheap on the following values stored in an array:
10 5 12 3 2 1 8 7 9 4

8. Build the Huffman coding tree and determine the codes for the following set of letters and weights:

A	B	C	D	E	F	G	H	I	J	K	L
2	3	5	7	11	13	17	19	23	31	37	41

9. Give a permutation for the values 0 through 7 that will cause Quicksort to have its worst case behavior.

10. Devise an algorithm to sort five numbers. It should make as few comparisons as possible. How many comparisons and swaps are required in the best, worst, and average cases?

第四部分　离散数学
Part 4　Discrete Mathematics

引言

离散数学是数学的几个分支的总称,研究基于离散空间而不是连续空间的数学结构。与光滑变化的实数不同,离散数学的研究对象——例如整数、图和数学逻辑中的命题——不是光滑变化的,而是拥有不等、分立的值。因此离散数学不包含微积分和分析等"连续数学"的内容。离散对象经常可以用整数来枚举。更一般地,离散数学被视为处理可数集合(与自然数集子集基数相同的集合,包括有理数集但不包括实数集)的数学分支。但是,"离散数学"不存在准确且普遍认可的定义。事实上,离散数学经常被定义为不包含连续变化量及相关概念的数学,甚少被定义为包含什么内容的数学。

离散数学中的对象集合可以是有限或者是无限的。有限数学一词通常指离散数学处理有限集合的那些部分,特别是与商业相关的领域。

二十世纪后半叶以来,离散数学由于它在数字计算机上的应用而备受关注。这是因为数字计算机用离散的步骤运算,用离散的点存储数据。人们会使用离散数学里面的概念和表示方法,来研究和描述计算机科学下所有分支的对象和问题,如电脑运算、编程语言、密码学、自动定理证明和软件开发等。相反地,计算机的应用使离散数学的概念得以应用于日常生活当中(如运筹学)。

虽然离散数学的主要研究对象是离散对象,但是连续数学的分析方法往往也可以采用。譬如,数论就是离散和连续数学的交叉学科。

历史上,离散数学涉及了各个领域的一系列挑战性问题。在图论中,大量研究的动机是企图证明四色定理。这些研究虽然从 1852 年开始,但是直至 1976 年四色理论才得到证明,是由肯尼斯·阿佩尔(Kenneth Appel)和沃尔夫冈·哈肯(Wolfgang Haken)大量使用计算机辅助来完成的。

在逻辑领域,大卫·希尔伯特(David Hilbert)于 1900 年提出的公开问题清单的第二个问题是要证明算术的公理是一致的。1931 年,库尔特·哥德尔的第二不完备定理证明这是不可能的——至少算术本身不可能。大卫·希尔伯特的第十个问题是要确定某一整系数多项式丢番图方程是否有一个整数解。1970 年,尤里·马季亚谢维奇证明这不可能做到。

第二次世界大战时盟军有破解纳粹德军密码的需要,带动了密码学和理论计算机科学的发展。英国的布莱切利园因而发明出第一部数位电子计算机——巨像电脑。与此同时,军事上的需求亦带动了运筹学的发展。直至冷战时期,密码学的地位依然重要,其后的几十年间更发展出如公开密钥加密等根本性的长进。随着 1950 年代关键路径方法的创立,运筹学则为商业和项目管理上愈趋重要。电讯工业的出现亦助长了离散数学,特别是图论和信息论上的发展。数理逻辑上叙述的形式验证至今已经成为安全关键系统的软件开发中必不可少的一环,自动定理证明的技术也因此而提高。

离散数学的若干领域,尤其是理论计算机科学、图论和组合数学,在迎接与生命树理解相关的生物信息学的挑战中具有重要作用。

当今,理论计算机科学中最著名的开放问题之一是 P/NP 问题,P/NP 问题中包含了复杂度类 P 与 NP 的关系。克雷数学研究所为此及其他 6 个千禧年大奖难题的第一个正确证明各悬赏 100 万美元。

一门离散数学课有多个目标。学生应该学会特定的一些数学事实并知道怎样应用;更重要的是,这样一门课应教会学生怎样作数学思维。为达到这些目标,本教材强调数学推理及用不同的方法解题。本课程有 5 个重要的主题交织在一起:数学推理、组合分析、离散结构、算法思考以及应用和建模。一门成功的离散数学课应该细心地使这五部分内容交融和取得平衡。

数学推理:学生必须理解数学推理以便阅读、理解和构造数学证明。本课程以数理逻辑开篇,因为数理逻辑是随后讨论的证明方法的基础。数学归纳技术是通过许多例子来重点介绍的。通过这些例子还详细地说明了为什么数学归纳是有效的证明技术。

组合分析:解题的一项重要技巧是计数或枚举对象的能力,本课程中对枚举的讨论就从基本的计数技术着手。重点是用组合分析来解决计数问题而不使用公式。

离散结构：一门离散数学课应该教学生如何使用离散结构。离散结构是抽象的数学结构，用来表示离散对象及离散对象之间的关系。离散结构包括集合、置换、关系、图、树和有限状态机。

算法思考：有几类问题是从给出算法说明入手求解的。描述了算法以后就可构造计算机程序来实现它，这一过程中的数学部分包括算法说明，证实它能正确执行，以及分析执行这一算法所需要的计算机内存和时间。所有这些内容均在课程中介绍。算法是用文字陈述和易于理解的一种伪码这样两种方式描述的。

应用与建模：离散数学已被应用到几乎所有研究领域。本课程既有许多计算机科学和数据网络的应用实例，也有各式各样领域中的应用实例，包括化学、植物学、动物学、语言学、地理、商业以及因特网。这些实例均是离散数学自然而重要的应用，不是编造的。用离散数学建模是十分重要的解题技巧，本课程的练习使学生有机会通过构造自己的模型来发展这一技巧。

Introduction

Discrete mathematics is the ageneraldesignation of some branches of mathematics and the study of mathematical structures that are fundamentally discrete rather than continuous. In contrast to real numbers that have the property of varying "smoothly", the objects studied in discrete mathematics – such as integers, graphs, and statements in logic – do not vary smoothly in this way, but have distinct, separated values. Discrete mathematics therefore excludes topics in "continuous mathematics" such as calculus and analysis. Discrete objects can often be enumerated by integers. More formally, discrete mathematics has been characterized as the branch of mathematics dealing with countable sets (sets that have the same cardinality as subsets of the natural numbers, including rational numbers but not real numbers). However, there is no exact, universally agreed, definition of the term "discrete mathematics". Indeed, discrete mathematics is described less by what is included than by what is excluded: continuously varying quantities and related notions.

The set of objects studied in discrete mathematics can be finite or infinite. The term finite mathematics is sometimes applied to parts of the field of discrete mathematics that deals with finite sets, particularly those areas relevant to business.

Research in discrete mathematics increased in the latter half of the twentieth century partly due to the development of digital computers which operate in discrete steps and store data in discrete bits. Concepts and notations from discrete mathematics are useful in studying and describing objects and problems in branches of computer science, such as computer algorithms, programming languages, cryptography, automated theorem proving, and software development. Conversely, computer implementations are significant in applying ideas from discrete mathematics to real-world problems, such as in operations research.

Although the main objects of study in discrete mathematics are discrete objects, analytic methods from continuous mathematics are often employed as well. For example, number theory is a cross subject of discrete mathematics and continuous mathematics.

The history of discrete mathematics has involved a number of challenging problems which have focused attention within areas of the field. In graph theory, much research was motivated by attempts to prove the four color theorem, first stated in 1852, but not proved until 1976 (by Kenneth Appel and Wolfgang Haken, using substantial computer assistance).

In logic, the second problem on David Hilbert's list of open problems presented in 1900 was to prove that the axioms of arithmetic are consistent. Gödel's second incompleteness theorem, proved in 1931, showed that this was not possible – at least not within arithmetic itself. Hilbert's tenth problem was to determine whether a given polynomial Diophantine equation with integer coefficients has an integer solution. In 1970, Yuri Matiyasevich proved that this could not be done.

The need to break German codes in World War II led to advances in cryptography and theoretical computer science, with the first programmable digital electronic computer being developed at England's Bletchley Park. At the same time, military requirements motivated advances in operations research. The Cold War meant that cryptography remained important, with fundamental advances such as public-key cryptography being developed in the following decades. Operations research remained important as a tool in business and project management, with the critical path method being developed in the 1950s. The telecommunication industry has also motivated advances in discrete mathematics, particularly in graph theory and information theory. Formal verification of statements in logic has been necessary for software development of safety-critical systems, and advances in automated theorem proving have been driven by this need.

Several fields of discrete mathematics, particularly theoretical computer science, graph theory, and combinatorics, are important in addressing the challenging bioinformatics problems associated with understanding the tree of life.

Currently, one of the most famous open problems in theoretical computer science is the P / NP problem, which involves the relationship between the complexity classes P and NP. The Clay Mathematics Institute has offered a $1 million USD prize for the first correct proof, along with prizes for six other millennium mathematical problems.

A discrete mathematics course has more than one purpose. Students should learn a particular set of mathematical facts and how to apply them; more importantly, such a course should teach students how to think mathematically to achieve these goals, this course stresses mathematical reasoning and the different ways problems are solved. Five important themes are interwoven in this course: mathematical reasoning, combinatorial analysis, discrete structures, algorithmic thinking, and applications and modeling. A successful discrete mathematics course should carefully blend and balance all five themes.

Mathematical Reasoning: Students must understand mathematical reasoning in order to read, comprehend, and construct mathematical arguments. This course starts with a discussion of mathematical logic, which serves as the foundation for the subsequent discussions of methods of proof. The technique of mathematical induction is stressed through many different types of examples of such proofs and a careful explanation of why mathematical induction is a valid proof technique.

Combinatorial Analysis: An important problem-solving skill is the ability to count or enumerate objects. The discussion of enumeration in this book begins with the basic techniques of counting. The stress is on performing combinatorial analysis to solve counting problems, not on applying formulae.

Discrete Structures: A course in discrete mathematics should teach students how to work with discrete structures, which are the abstract mathematical structures used to represent discrete objects and relationships between these objects. These discrete structures include sets, permutations, relations, graphs, trees, and finite-state machines.

Algorithmic Thinking: Certain classes of problems are solved by the specification of an algorithm. After an algorithm has been described, a computer program can be constructed implementing it. The mathematical portions of this activity, which include the specification of the algorithm, the verification that it works properly, and the analysis of the computer memory and time required to perform it, are all covered in this course. Algorithms are described using both English and an easily understood form of pseudocode.

Applications and Modeling: Discrete mathematics has applications to almost every conceivable area of study. There are many applications to computer science and data networking in this course, as well as applications to such diverse areas as chemistry botany, zoology, linguistics, geography business, and the Internet. These applications are natural and important uses of discrete mathematics and are not contrived. Modeling with discrete mathematics is an extremely important problem-solving skill, which students have the opportunity to develop by constructing their own models in some of the exercises.

第一章 命题逻辑
Chapter 1 Proposition Logic

逻辑是对有效推理和推理原则，及其连续性、合理性和完整性的研究。本章的内容主要包括命题的概念、命题运算、合式公式、真值表、等值演算、范式和命题逻辑的推理理论。

Logic is the study of the principles of valid and inference, as well as of consistency, soundness, and completeness. The contents of this chapter mainly include the definition of proposition, proposition operations, the well-formed formula, the truth value table, the equivalent calculation, the normal form and the reasoning theory of proposition logic.

单词和短语 Words and expressions

命题　Proposition
命题真值　truth, true value
真命题　true proposition
假命题　fault proposition
联结词　propositional connectives
否定　negation
合取　conjunction
析取　disjunction
蕴含　implication
等价　equivalence
合式公式　well-formed formula
命题常项　propositional constant
命题变项　propositional variable
命题翻译　proposition translation
命题符号化　symbolic notation of a proposition
命题解释　proposition explain
真值表　truth table
重言式 永真式　tautology
永假式　contradiction
还原律　reductive law
幂等律　idempotent law
交换律　commutative law
结合律　associative law
分配律　distributive law
德摩根律　De Morgan law
吸收律　absorbing law
排中律　excluded middle law
矛盾律　contradiction law
等值演算　equivalent calculation
置换规则　equivalent replacement rule
真值函数　truth function
范式　normal form
简单析取式　simple disjunctive form
简单合取式　simple conjunctive form
析取范式　disjunctive normal form
合取范式　conjunctive normal form
主析取范式　principal disjunctive normal form
主合取范式　principal conjunctive normal form
自然推理系统 P　Natural Deduction System P
前提　precondition
假设　hypothesis
有效推理　efficient consequence
有效结论　efficient conclusion
前提引入规则　rule of introducing precondition
结论引入规则　rule of introducing conclusion
基本概念和性质　basic concepts and properties

基本概念和性质　Basic concepts and properties

1 判断结果唯一的陈述句称作命题。

A declarative sentence with only judgment is termed proposition.

2 合式公式（命题公式，公式）递归定义如下：
(1)单个命题常项或变项是合式公式，并称作原子合式公式；
(2)若 A 是合式公式，则 $(\neg A)$ 也是合式公式；
(3)若 A, B 是合式公式，则 $(A \wedge B)$, $(A \vee B)$, $(A \rightarrow B)$, $(A \leftrightarrow B)$ 也是合式公式；
(4)只有有限次地应用(1)～(3)形成的符号串才是合式公式。

The well-formed formula (proposition formula, formula) is defined recursively as follows:
(1) A single propositional constant or a single propositional variable is a well-formed formula, which is called an atomic well-formed formula;
(2) If A is a well-formed formula, then $(\neg A)$ is also a well-formed formula;
(3) If A, B are well-formed formulas, then $A \wedge B$, $A \vee B$, $A \rightarrow B$, $A \leftrightarrow B$ are also well-formed formulas;
(4) Only the symbol string being formed by use (1)～(3) limited times is called a well-formed formula.

③ 命题公式在所有可能的赋值下的取值的列表称作真值表
The truth table is values list of a propositional formula in all possible assignments.
④ 由极小(大)项构成的析取范式称作主析取(合取)范式
The principal disjunctive (conjunctive) normal form is the disjunctive (conjunctive) normal form which is composed of minimum (maximum) terms.
⑤ 任何命题公式都存在着与之等值的主析取范式和主合取范式，并且是惟一的.
There is an only principal disjunctive normal form and an only principal conjunctive normal form to be equivalent to a propositional formula.
⑥ 若对于每组赋值，$A_1 \wedge A_2 \wedge \cdots \wedge A_k$ 为假，或者当 $A_1 \wedge A_2 \wedge \cdots \wedge A_k$ 为真时，C 也为真，则称由前提 $A_1, A_2, \cdots, A_k$ 推 C 的推理有效或推理正确，并称 C 是有效的结论.
If for each assignment, $A_1 \wedge A_2 \wedge \cdots \wedge A_k$ is false, or when $A_1 \wedge A_2 \wedge \cdots \wedge A_k$ is true, C is also true, then the reasoning process from preconditions $A_1, A_2, \cdots, A_k$ to C is called a efficient consequence or a correct reasoning, and C is said a efficient conclusion.
⑦ 由前提 $A_1, A_2, \cdots, A_k$ 推出 C 的推理正确当且仅当 $A_1 \wedge A_2 \wedge \cdots \wedge A_k \rightarrow C$ 为重言式.
The reasoning process from preconditions $A_1, A_2, \cdots, A_k$ to C is correct if and only if $A_1 \wedge A_2 \wedge \cdots \wedge A_k \rightarrow C$ is tautology.

例 题 Example

例 1 求 $\neg(p \rightarrow q) \vee \neg r$ 的主析取范式.

解 $\neg(p \rightarrow q) \vee \neg r \Leftrightarrow \neg(\neg p \vee q) \wedge \neg r \Leftrightarrow (p \wedge \neg q) \vee \neg r$
$\Leftrightarrow ((p \neg \wedge q) \wedge (\neg r \vee r)) \vee ((\neg p \vee p) \wedge (\neg q \vee q) \neg \wedge r)$
$\Leftrightarrow (p \neg \wedge q \neg \wedge r) \vee (p \neg \wedge q \wedge r) \vee (\neg p \neg \wedge q \neg \wedge r) \vee (\neg p \wedge q \neg \wedge r) \vee (p \neg \wedge q \neg \wedge r) \vee (p \wedge q \neg \wedge r)$
$\Leftrightarrow m_0 \vee m_2 \vee m_4 \vee m_5 \vee m_6$
$\Leftrightarrow \sum(0, 2, 4, 5, 6)$

Example 1 Find the principal disjunctive normal form of the formula $\neg(p \rightarrow q) \vee \neg r$.

Solution $\neg(p \rightarrow q) \vee \neg r \Leftrightarrow \neg(\neg p \vee q) \wedge \neg r \Leftrightarrow (p \wedge \neg q) \vee \neg r$
$\Leftrightarrow ((p \neg \wedge q) \wedge (\neg r \vee r)) \vee ((\neg p \vee p) \wedge (\neg q \vee q) \neg \wedge r)$
$\Leftrightarrow (p \neg \wedge q \neg \wedge r) \vee (p \neg \wedge q \wedge r) \vee (\neg p \neg \wedge q \neg \wedge r) \vee (\neg p \wedge q \neg \wedge r) \vee (p \neg \wedge q \neg \wedge r) \vee (p \wedge q \neg \wedge r)$
$\Leftrightarrow m_0 \vee m_2 \vee m_4 \vee m_5 \vee m_6$
$\Leftrightarrow \sum(0, 2, 4, 5, 6)$

例 2 在自然推理系统 P 中构造下面推理的证明：
如果王小红努力学习，她一定取得好成绩。若贪玩或不按时完成作业，她就不能取得好成绩。所以，如果王小红努力学习，她就不贪玩并且按时完成作业。

解 设 p：王小红努力学习；q：王小红取得好成绩；r：王小红贪玩；s：王小红按时完成作业。
前提：$p \rightarrow q$；$(r \vee \neg s) \rightarrow \neg q$.
结论：$p \rightarrow (\neg r \vee s)$.

证明：(1) p P 前提(附加)
(2) $p \rightarrow q$ P 前提
(3) q T(1)(2) 假言推理
(4) $(r \vee \neg s) \rightarrow \neg q$ P 前提
(5) $q \rightarrow \neg(r \vee \neg s)$ T(4) 拒取式
(6) $\neg(r \vee \neg s)$ T(3)(5) 假言推理
(7) $\neg r \vee s$ T(6) 德摩根律，还原律

Example 2 Construct the following reasoning proof in natural deduction system P:
If Wang Xiaohong studies hard, she must get good grades. If more fun or not to complete the work

on time, she can't get good grades. So, if Wang Xiaohong worked hard, she would not have more fun and finish the homework on time.

Solution Let p: Wang Xiaohong studies hard; q: Wang Xiaohong gets good grades; r: Wang Xiaohong has more fun; s: Wang Xiaohong finishes the homework on time.
Precondition: $p \to q$; $(r \vee \neg s) \to \neg q$.
Conclusion: $p \to (\neg r \vee s)$.

Proof:
(1) p Precondition (Subjoined)
(2) $p \to q$ Precondition
(3) q T(1)(2) Modus Ponens
(4) $(r \vee \neg s) \to \neg q$ Precondition
(5) $q \to \neg(r \vee \neg s)$ T(4) Modus Tollendo Ponens
(6) $\neg(r \vee \neg s)$ T(3)(5) Modus Ponens
(7) $\neg r \vee s$ T(6) De Morgan Law, Reductive Law

本章重点

1. 理解命题和命题公式的概念,能判别一个句子是否是命题。能判别命题尤其是复合命题的真假。能熟练进行命题运算和等值演算。
2. 能熟练列出命题公式的真值表。能熟练求出命题公式的主析(合)取范式,并利用其解决某些实际问题。
3. 能熟练在自然系统 P 内构造推理证明。

Key points of this chapter

1. Understand the concept of proposition and propositional formula. Be able to determine whether a sentence is a proposition. Be able to distinguish true and false propositions especially the compound proposition. Be skilled in propositional operation and equivalent calculation.
2. Can skilled list the truth table of a propositional formula. Can skilled find the principal disjunctive (conjunctive) normal form of a propositional formula, and use it to solve some practical problems.
3. Can skilled construct the reasoning proof in natural deduction system P.

第二章 一阶逻辑
Chapter 2 First-order Logic

虽然命题逻辑能够构造一些形式的逻辑证明,但是简单的三段论在命题逻辑中不能证明其正确性。为突破命题逻辑的这种局限性,我们引进谓词逻辑。本章的主要内容包括个体词、谓词、量词、谓词合式公式、等值演算、前束范式和谓词逻辑的推理理论。

Although the propositional logic constructs some logical proof in form, it can't prove the correctness of simple syllogism. In order to break through the limitation of propositional logic, we introduce a predicate logic. The main content of this chapter includes the individual words, the predicate, the quantifier, the predicate well-formed formula, equivalent calculations, the prenex normal form and the reasoning theory of predicate logic.

单词和短语 Words and expressions

个体词 individual word	谓词常项 predicate constant
个体域 individual field	谓词变项 predicate variable
个体常项 individual constant	量词 quantifier
个体变项 individual variable	全称量词 universal quantifier
谓词 predicate	存在量词 existential quantifier

谓词逻辑	predicate logic	换名规则	renaming rule
一阶逻辑	first-order logic	代替规则	instead of rules
指导变元	bound variable	前束范式	prenex normal form
约束出现	bound occurrence		
自由出现	free occurrence		

谓词逻辑　predicate logic
一阶逻辑　first-order logic
指导变元　bound variable
约束出现　bound occurrence
自由出现　free occurrence
辖域　scope
闭式　closed formula
成假解释　explanation with false value
成真解释　explanation with truth value
代换实例　substitution examples
量词否定规则　quantifier negative rule
量词辖域的收缩　contraction of the quantifier scope
量词辖域的扩张　extension of the quantifier scope

换名规则　renaming rule
代替规则　instead of rules
前束范式　prenex normal form
全称量词消去规则，全称指定　universal index (UI)，universal specification (US)
全称量词引入规则　universal generalization (UG)
存在量词消去规则，存在指定　existential index (EI)，existential specification(ES)
存在量词引入规则　existential generalization (EG)
一阶逻辑推理理论　inference theory by the First-order Logic
自然推理系统 F　Natural Deduction System F

基本概念和性质　Basic concepts and properties

1 所研究对象中可以独立存在的具体或抽象的客体称作个体词。
The concrete or abstract objects that can exist independently in research are called individual words.

2 表示个体词性质或相互之间关系的词称作谓词。一元谓词表示个体的性质，多元谓词表示个体间的关系。
The word representing the character or relations of individuals is called predicate. The unary predicate represents the individual character, and the multiple predicate represents relationships between individuals.

3 表示数量的词称作量词。
The word representing the number of individual is said quantifier.

4 在命题翻译中，相对于全称量词，特性谓词作为蕴含式的前件，相对于存在量词，特性谓词作为合取式的一项。
The identity predicate is as the antecedent of the implication respect to universal quantifier and a term of the conjunction respect to existential quantifier in propositional translation.

5 谓词合式公式（谓词公式，公式）递归定义如下：
(1)原子公式是合式公式；
(2)若 A 是合式公式，则($\neg A$)也是合式公式；
(3)若 A, B 是合式公式，则$(A \wedge B)$，$(A \vee B)$，$(A \rightarrow B)$，$(A \leftrightarrow B)$也是合式公式；
(4)若 A 是合式公式，则$\forall xA$，$\exists xA$ 也是合式公式；
(5)只有有限次地应用(1)~(4)形成的符号串才是合式公式。
The predicate well-formed formula (predicate formula, formula) is defined recursively as follows：
(1)An atomic formula is a well-formed formula；
(2)If A is a well-formed formula, then ($\neg A$) is also a well-formed formula；
(3)If A, B are well-formed formulas, then $(A \wedge B)$，$(A \vee B)$，$(A \rightarrow B)$，$(A \leftrightarrow B)$ are also well-formed formulas；
(4)If A is a well-formed formula, then $\forall xA$ and $\exists xA$ are also well-formed formula；
(5)Only the symbol string being formed by use (1)~(4) limited times is called a well-formed formula.

6 设 A 为一个一阶逻辑公式，若具有 $Q_1 x_1 Q_2 x_2 \cdots Q_k x_k B$ 形式，则称 A 为前束范式，其中 $Q_i (1 \leqslant i \leqslant k)$ 为或 $\forall$ 或 $\exists$，B 为不含量词的公式。

Let A be a first-order logic formula. If A possesses a form of $Q_1 x_1 Q_2 x_2 \cdots Q_k x_k B$, then A is said a prenex normal form, where $Q_i (1 \leqslant i \leqslant k)$ is $\forall$ or $\exists$ and B is a predicate formula with no quantifier.

定理 一阶逻辑中的任何公式都存在与之等值的前束范式。

For any first-order logic formula, there are prenex normal forms equivalent to it.

例题 Example

例 1 求 $\forall x F(x) \exists \neg \forall x G(x)$ 的前束范式。

解 $\forall x F(x) \exists \neg \forall x G(x)$
$\Leftrightarrow \forall x F(x) \vee \forall x \neg G(x)$ 量词否定规则
$\Leftrightarrow \forall x F(x) \vee \forall y \neg G(y)$ 换名规则
$\Leftrightarrow \forall x (F(x) \vee \forall y \neg G(y))$ 量词辖域扩张
$\Leftrightarrow \forall x \forall y (F(x) \vee \neg G(y))$ 量词辖域扩张

Example 1 Find a prenex normal form of the formula $\forall x F(x) \exists \neg \forall x G(x)$.

Solusion $\forall x F(x) \exists \neg \forall x G(x)$
$\Leftrightarrow \forall x F(x) \vee \forall x \neg G(x)$ Quantifier Negative Rules
$\Leftrightarrow \forall x F(x) \vee \forall y \neg G(y)$ Renaming Rule
$\Leftrightarrow \forall x (F(x) \vee \forall y \neg G(y))$ Extension of the Quantifier Scope
$\Leftrightarrow \forall x \forall y (F(x) \vee \neg G(y))$ Extension of the Quantifier Scope

例 2 在自然推理系统 F 中构造下面推理的证明：
人都是要死的，苏格拉底是人，所以苏格拉底是要死的．

解 令 $F(x): x$ 是人，$G(x): x$ 是要死的，a：苏格拉底。
前提：$\forall x (F(x) \rightarrow G(x)), F(a)$.
结论：$G(a)$.
证明：① $F(a)$ P 前提
② $\forall x (F(x) \rightarrow G(x))$ P 前提
③ $F(a) \rightarrow G(a)$ T② UI
④ $G(a)$ T①③ 假言推理

Example 2 Construct the following reasoning proof in natural deduction system F：
Every person is doomed to die. Socrates is a person. Thus Socrates is (doomed to die).

Solution Let $F(x)$ be x is a person, $G(x)$ be x is doomed to die and a is Socrates.
Precondition：$\forall x (F(x) \rightarrow G(x)), F(a)$.
Conclusion：$G(a)$.
Proof：① $F(a)$ Precondition
② $\forall x (F(x) \rightarrow G(x))$ Precondition
③ $F(a) \rightarrow G(a)$ T② UI
④ $G(a)$ T①③ Modus Ponens

本章重点

1 理解个体词、谓词、量词、谓词公式等概念，能熟练将自然语言翻译成逻辑符号。能熟练进行谓词逻辑的等值演算。

2 能熟练求出谓词公式的前束范式。

3 能熟练在自然系统 F 内构造推理证明。

Key points of this chapter

1 Understand the concepts of individual word, predicate, quantifier and predicate formula. Be able to translate natural language into logical symbols. Be Skilled in equivalent calculation in predicate logic.

2 Can skilled find a prenex normal form of a predicate formula.

3 Can skilled construct the reasoning proof in natural deduction system F.

第三章 关系
Chapter 3 Relation

集合元素间的某种关联在数学上以关系的形式来表现。数据库中的数据存储就是关系的一个重要应用。本章主要内容包括关系的概念、关系的运算以及两个重要的关系：等价关系和偏序关系；还包括等价关系和偏序关系在实际问题中的应用。

A relevancy between elements in a set is expressed as relation in mathematics. The data storage in the database is an important application of relationship. The main content of this chapter includes the concept of relation, the operation of relations and two important relations, which are the equivalence relation and the partial order relation. This chapter also includes applications in practice of the equivalence relation and the partial order relation.

单词和短语 Words and expressions

集合　set
（集合的）基数，势　cardinality, cardinal numbers, radix
有限集　finite set
非空有限集　nonempty finite set
无限集　infinite set
可数集 可列集　Countable Set
不可数集 不可列集　Uncountable Set
连续统　continuum
有序对　ordered pair
笛卡尔积　Cartesian Product
有序 n 元组，n 元有序组　Ordered n-tuple, Ordered n-ple, fordered N-ple
关系　relation
二元关系　binary relation
N 元关系　N-ple relation
全域关系　universal relation
恒等关系　identity relation
关系矩阵　relational matrix
关系图　relational graph
定义域　domain
值域　range
域　field
逆关系　inverse of a relation
合成关系 复合关系　composition of relations
自反性　reflexivity

反自反性　irreflexivity
自反关系　reflexive relation, reflexivity relation
对称性　symmetry
反对称性　antisymmetry
对称关系　symmetric relation
传递性　transitivity
传递关系　transitive relation
等价关系　equivalence relation
自反（对称，传递）闭包　reflexive (symmetric, transitive) closure
沃舍尔算法　Warshall algorithm
等价类　equivalence class
商集　quotient set
集合的划分　partition of a set
偏序关系　partial-ordered relation
偏序集　partial-ordered set (poset)
全序　total order
链　chain
反链　anti-chain
覆盖　cover
哈斯图　Hass graph
最大（小）元　absolute maximum (minimum) element
极大（小）元　locale maximum (minimum) element

基本概念和性质 Basic concepts and properties

设 A, B 为集合，称 $R \subseteq A \times B$ 为 A 到 B 的一个二元关系，简称为关系，记作 R。若 $\langle x, y \rangle \in R$，可记作 xRy。如果 $R \subseteq A \times A$，称 R 为 A 上的一个二元关系。

Let A, B be sets. R is referred to as a binary relation from A to B, or simply "relation" for short, if

$R \subseteq A \times B$, noted by R. We may note xRy if $\langle x,y \rangle \in R$. R is called a binary relation on A if $R \subseteq A \times A$.

2 设 $|A|=n$, $|B|=m$，那么从 A 到 B 的关系共有 2^{mn} 个，A 上关系共有 2^{n^2} 个。
If $|A|=n$ and $|B|=m$, then the number of relations from A to B is 2^{mn} and then the number of relations on A is 2^{n^2}.

3 若 $A=\{x_1, x_2, \cdots, x_m\}$, $B=\{y_1, y_2, \cdots, y_n\}$，$R$ 是从 A 到 B 的关系，R 的关系矩阵是布尔矩阵 $M_R = [r_{ij}]_{m \times n}$，其中 $r_{ij}=1 \Leftrightarrow \langle x_i, y_j \rangle \in R$。
Let $A=\{x_1, x_2, \cdots, x_m\}$, $B=\{y_1, y_2, \cdots, y_n\}$ and R be a binary relation from A to B. The Boolean matrix $M_R = [r_{ij}]_{m \times n}$ is said the relation matrix of R if $r_{ij}=1 \Leftrightarrow \langle x_i, y_j \rangle \in R$.

4 若 $A=\{x_1, x_2, \cdots, x_m\}$，$R$ 是从 A 上的关系，R 的关系图是 $G_R=\langle A, R \rangle$，其中 A 为结点集，R 为边集。如果 $\langle x_i, x_j \rangle$ 属于关系 R，在图中就有一条从 x_i 到 x_j 的有向边。
Let $A=\{x_1, x_2, \cdots, x_m\}$ and R be a binary relation on A. $G_R=\langle A, R \rangle$, where A is the set of vertexes and R is the set of edges, is said the graph of relation R if there is a directed edge from x_i to x_j when $\langle x_i, x_j \rangle \in R$.

5 $R^{-1}=\{\langle y,x \rangle | \langle x,y \rangle \in R\}$ 称作 R 的逆关系。$R \circ S = \{\langle x,z \rangle | \exists y (\langle x,y \rangle \in R \wedge \langle y,z \rangle \in S)\}$ 称作 R 与 S 的合成或复合。$M_{R^{-1}} = M_R^T$, $M_{R \circ S} = M_R \circ M_S$。
$R^{-1}=\{\langle y,x \rangle | \langle x,y \rangle \in R\}$ is called inverse relation of R. $R \circ S = \{\langle x,z \rangle | \exists y (\langle x,y \rangle \in R \wedge \langle y,z \rangle \in S)\}$ is called the compositive relation of R and S. $M_{R^{-1}} = M_R^T$, $M_{R \circ S} = M_R \circ M_S$.

6 设 R 为非空集合上的关系。如果 R 是自反的、对称的和传递的，则称 R 为 A 上的等价关系。
Let R be a relation on a nonempty set. R is said an equivalence relation if R is reflexive, symmetric and transitive.

7 设 R 为非空集合上的关系。如果 R 是自反的、反对称的和传递的，则称 R 为 A 上的偏序关系。
Let R be a relation on a nonempty set. R is said a partial-ordered relation if R is reflexive, antisymmetric and transitive.

例 题 Example

例 1 设 $A=\{a, b, c, d\}$, $R=\{\langle a,a \rangle, \langle a,b \rangle, \langle a,c \rangle, \langle b,a \rangle, \langle d,b \rangle\}$。求 R 的自反闭包和对称闭包。

解 R 的自反闭包 $r(R) = R \cup I_A = \{\langle a,a \rangle, \langle a,b \rangle, \langle a,c \rangle, \langle b,a \rangle, \langle d,b \rangle\} \cup \{\langle a,a \rangle, \langle b,b \rangle, \langle c,c \rangle, \langle d,d \rangle\} = \{\langle a,a \rangle, \langle a,b \rangle, \langle a,c \rangle, \langle d,d \rangle, \langle a,b \rangle, \langle a,c \rangle, \langle b,a \rangle, \langle d,b \rangle\}$。

R 的对称闭包 $s(R) = R \cup R^{-1} = \{\langle a,a \rangle, \langle a,b \rangle, \langle a,c \rangle, \langle b,a \rangle, \langle d,b \rangle\} \cup \{\langle a,a \rangle, \langle b,a \rangle, \langle c,a \rangle, \langle a,b \rangle, \langle b,d \rangle\} = \{\langle a,a \rangle, \langle a,b \rangle, \langle a,c \rangle, \langle b,a \rangle, \langle d,b \rangle, \langle c,a \rangle, \langle b,d \rangle\}$。

Example 1 Let $A=\{a, b, c, d\}$ and $R=\{\langle a,a \rangle, \langle a,b \rangle, \langle a,c \rangle, \langle b,a \rangle, \langle d,b \rangle\}$. Find the reflexive closure and symmetric closure of R.

Solution The reflexive closure of R is found as follows:
$r(R) = R \cup I_A = \{\langle a,a \rangle, \langle a,b \rangle, \langle a,c \rangle, \langle b,a \rangle, \langle d,b \rangle\} \cup \{\langle a,a \rangle, \langle b,b \rangle, \langle c,c \rangle, \langle d,d \rangle\} = \{\langle a,a \rangle, \langle b,b \rangle, \langle c,c \rangle, \langle d,d \rangle, \langle a,b \rangle, \langle a,c \rangle, \langle b,a \rangle, \langle d,b \rangle\}$.

And symmetric closure of R is found as follows:
$s(R) = R \cup R^{-1} = \{\langle a,a \rangle, \langle a,b \rangle, \langle a,c \rangle, \langle b,a \rangle, \langle d,b \rangle\} \cup \{\langle a,a \rangle, \langle b,a \rangle, \langle c,a \rangle, \langle a,b \rangle, \langle b,d \rangle\} = \{\langle a,a \rangle, \langle a,b \rangle, \langle a,c \rangle, \langle b,a \rangle, \langle d,b \rangle, \langle c,a \rangle, \langle b,d \rangle\}$.

例 2 设 A 是一个集合，2^A 是 A 的幂集，即 A 的所有子集的集合。证明：集合的包含关系 $\subseteq$ 是 2^A

上的偏序关系。

证明：(1)对任意的 $S\in 2^A$ 有 $S\subseteq S$，$\subseteq$是自反的。

(2)对任意的 S，$T\in 2^A$，如果有 $S\subseteq T$ 和 $T\subseteq S$，那么 $S=T$。所以，$\subseteq$是反对称的。

(3)对任意的 R，S，$T\in 2^A$，如果有 $R\subseteq S$ 和 $S\subseteq T$，那么 $R\subseteq T$。所以，$\subseteq$是传递的。

综上，集合的包含关系$\subseteq$是 2^A 上的偏序关系。

Example 2 Let A be a set. 2^A is the power set of A, i.e., the set of all subsets of A. Prove that the inclusion relation of subsets $\subseteq$ is a partial-ordered relation on 2^A.

Proof (1)For any $S\in 2^A$, we have that $S\subseteq S$. So $\subseteq$ is reflexive.

(2)For any S, $T\in 2^A$, if $S\subseteq T$ and $T\subseteq S$, then $S=T$. So $\subseteq$ is antisymmetric.

(3)For any R, S, $T\in 2^A$, if $R\subseteq S$ and $S\subseteq T$, then $R\subseteq T$. So $\subseteq$ is transitive.

本章重点

1. 理解关系、自反关系、反自反关系、对称关系、反对称关系、传递关系等概念，掌握关系的矩阵表示法和图示法。能熟练进行关系的复合运算和逆运算。能求关系的自反闭包、对称闭包和传递闭包。

2. 理解等价关系和偏序关系的概念。理解一个集合上的等价关系和该集合划分间的对应关系。能利用等价关系和偏序关系解决实际问题。

Key points of this chapter

1. Understand the concepts of relation, reflexive relation, irreflexive relation, symmetric relation, anti-symmetric relation and transitive relation. Master the representation of a relation by matrix or graph. Can skilled find the composite relation of relations and the inverse of a relation. Be able to find the reflexive closure, the symmetric closure and the transitive closure of a relation.

2. Understand the concepts of equivalence relation and partial-ordered relation. Understand the correspondence relation between the equivalence relation of a set and the partition of the set. Be able to apply the equivalence relation and the partial-ordered relation in practice.

第四章　函数
Chapter 4　Function

本章我们把函数的概念由定义在数集上推广到定义在一般的集合上，并以关系的形式来表述。这一推广是必要的，因为在自动控制系统、计算机执行的程序、通讯信号的传输、编译程序等中都是把这广义的函数作为工具的。本章主要内容包括函数、单射函数、满射函数、双射函数等概念、函数的复合运算和逆运算。

In this chapter the concept of function defined in number sets is generated to the function defined in general set, and is expressed with a relation. This generalization is necessary because the general function is taken as a tool in the automatic control system, computer program, communication signal transmission, compiler program, etc.. The main content of this chapter includes the concept of function, injective function, surjective function, bijective function and etc., and also includes composite operation and inverse operation of functions.

单词和短语 Words and expressions

函数　function
自变量　argument, independent variable
因变量　dependent variable
函数 f 在 x 的值　the value of a function at x
前陪域　front codomain
后陪域　behind codomain

常值函数　constant function
恒等函数　identity function
特征函数　eigenfunction
单调递减（增）函数　monotony decrease (increase) function
严格单调函数　strictly monotonic function

自然映射　natural map
子集 A 在函数 f 下的像　image of a subset A in the function f
原像　inverse image, preimage
满射　surjection
单射　injection
双射　bijection
复合函数　composite function
反函数　inverse function

基本概念和性质　Basic concepts and properties

1 设 $f \subseteq A \times B$ 为二元关系，若 $\forall\, x \in \text{dom}\,f$ 都存在唯一的 $y \in \text{ran}\,f$ 使 $x\,f\,y$ 成立，则称 f 为函数。对于函数 f，如果有 $x\,f\,y$，则记作 $y = f(x)$，并称 y 为 f 在 x 的值。

Let $f \subseteq A \times B$ be a binary relation. If for $\forall\, x \in \text{dom}\,f$, there is only $y \in \text{ran}\,f$ such that $x\,f\,y$, then f is termed a function. For a function f, $x\,f\,y$ is noted by $y = f(x)$ and y is called the value of f in x.

2 设 A, B 为集合，如果(1) f 为函数；(2) $\text{dom}\,f = A$；(3) $\text{ran}\,f \subseteq B$，则称 f 为从 A 到 B 的函数，记作 $f : A \to B$。

Let A, B be sets. f is referred to as a function from A to B, noted by $f : A \to B$, if that (1) f is a function; (2) $\text{dom}\,f = A$; (3) $\text{ran}\,f \subseteq B$.

3 若 $\text{ran}\,f = B$，则称 $f : A \to B$ 是满射的。若 $\forall\, y \in \text{ran}\,f$ 都存在唯一的 $x \in A$ 使得 $f(x) = y$，则称 $f : A \to B$ 是单射的。若 $f : A \to B$ 既是满射又是单射的，则称 $f : A \to B$ 是双射的。

$f : A \to B$ is a surjection if $\text{ran}\,f = B$. $f : A \to B$ is an injection if for $\forall\, y \in \text{ran}\,f$, there is only $x \in A$ such that $f(x) = y$. $f : A \to B$ is a bijection is $f : A \to B$ is not only a injection but also a surjection.

4 如果 $f : A \to B$，$g : B \to C$ 都是满射的，则 $f \circ g : A \to C$ 也是满射的。如果 $f : A \to B$，$g : B \to C$ 都是单射的，则 $f \circ g : A \to C$ 也是单射的。如果 $f : A \to B$，$g : B \to C$ 都是双射的，则 $f \circ g : A \to C$ 也是双射的。

If $f : A \to B$, $g : B \to C$ are all surjections, then $f \circ g : A \to C$ is a surjection. If $f : A \to B$, $g : B \to C$ are all injections, then $f \circ g : A \to C$ is a injection. If $f : A \to B$, $g : B \to C$ are all bijections, then $f \circ g : A \to C$ is a bijection.

5 如果 $f : A \to B$ 是双射的，则 $f^{-1} : B \to A$ 也是双射的。

If $f : A \to B$ is a bijection, then $f^{-1} : B \to A$ is also a bijection.

例题　Example

例 1　证明：(1) 如果 $f : A \to B$，$g : B \to C$ 都是满射的，则 $f \circ g : A \to C$ 也是满射的。
(2) 如果 $f : A \to B$，$g : B \to C$ 都是单射的，则 $f \circ g : A \to C$ 也是单射的。
(3) 如果 $f : A \to B$，$g : B \to C$ 都是双射的，则 $f \circ g : A \to C$ 也是双射的。

证明：(1) 任取 $c \in C$，由 $g : B \to C$ 的满射性，$\exists\, b \in B$ 使得 $g(b) = c$。
对于这个 b，由 $f : A \to B$ 的满射性，$\exists\, a \in A$ 使得 $f(a) = b$。
由合成定理，有 $f \circ g(a) = g(f(a)) = g(b) = c$。从而证明了 $f \circ g : A \to C$ 是满射的。
(2) 假设存在 $x_1, x_2 \in A$ 使得 $f \circ g(x_1) = f \circ g(x_2)$，由合成定理有 $g(f(x_1)) = g(f(x_2))$。因为 $g : B \to C$ 是单射的，故 $f(x_1) = f(x_2)$。又由于 $f : A \to B$ 也是单射的，所以 $x_1 = x_2$。
从而证明 $f \circ g : A \to C$ 是单射的。
(3) 由(1)、(2)即得(3)的结论。

Example 1　Prove that:
(1) $f : A \to B$ is a surjection, $g : B \to C$ is a surjection, then $f \circ g : A \to C$ is a surjection;
(2) $f : A \to B$ is a injection, $g : B \to C$ is a injection, then $f \circ g : A \to C$ is a injection;
(3) $f : A \to B$ is a bijection, $g : B \to C$ is a bijection, then $f \circ g : A \to C$ is a bijection.

Proof　(1) For any $c \in C$, there is $b \in B$ such that $g(b) = c$ since $g : B \to C$ is surjective. For the previous b, there is $a \in A$ such that $f(a) = b$ since $f : A \to B$ is surjective. By the composite theorem, we have that $fog(a) = g(f(a)) = g(b) = c$. Thus, the surjection of $fog: A \to C$ is proved.

(2) Supposing that there are $x_1, x_2 \in A$ such that $fog(x_1) = fog(x_2)$, we have that $g(f(x_1)) = g(f(x_2))$ by the composite theorem. Since $g: B \to C$ is a injection, $f(x_1) = f(x_2)$. And since $f: A \to B$ is also a injection, $x_1 = x_2$. Thus, the injection of $fog: A \to C$ is proved.

(3) The conclusion of (3) is proved by (1) and (2).

例 2　证明：如果 $f: A \to B$ 是双射的，则 $f^{-1}: B \to A$ 也是双射的.

证明：因为 f 是函数，所以 f^{-1} 是关系，且 $\mathrm{dom}\, f^{-1} = \mathrm{ran}\, f = B$，$\mathrm{ran}\, f^{-1} = \mathrm{dom}\, f = A$。

对于任意的 $x \in B$，假设有 $y_1, y_2 \in A$ 使得 $<x, y_1> \in f^{-1} \wedge <x, y_2> \in f^{-1}$ 成立，则由逆的定义有 $<y_1, x> \in f \wedge <y_2, x> \in f$。根据 f 的单射性可得 $y_1 = y_2$，从而证明了 f^{-1} 是函数，且是满射的。

若存在 $x_1, x_2 \in B$ 使得 $f^{-1}(x_1) = f^{-1}(x_2) = y$，从而有

$$<x_1, y> \in f^{-1} \wedge <x_2, y> \in f^{-1} \Rightarrow <y, x_1> \in f \wedge <y, x_2> \in f \Rightarrow x_1 = x_2$$

从而证明了 f^{-1} 的单射性。

Example 2　Prove that if $f: A \to B$ is bijection, then $f^{-1} B \to A$ is also bijection.

Proof　Since f is a function, f^{-1} is a relation and $\mathrm{dom}\, f^{-1} = \mathrm{ran}\, f = B$, $\mathrm{ran}\, f^{-1} = \mathrm{dom}\, f = A$. For any $x \in B$, if there are $y_1, y_2 \in A$ such that $<x, y_1> \in f^{-1} \wedge <x, y_2> \in f^{-1}$, then $<y_1, x> \in f \wedge <y_2, x> \in f$ by the definition of inverse. So, we have that $y_1 = y_2$ since f is a injection. Therefore, we have proved that f^{-1} is a function and a surjection.

If there are $x_1, x_2 \in B$ such that $f^{-1}(x_1) = f^{-1}(x_2) = y$, then we have that

$$<x_1, y> \in f^{-1} \wedge <x_2, y> \in f^{-1} \Rightarrow <y, x_1> \in f \wedge <y, x_2> \in f \Rightarrow x_1 = x_2$$

Therefore, we have proved that f^{-1} is a injection.

本章重点

1. 理解函数、单射函数、满射函数、双射函数等概念。会判别一个函数是否为单射函数、满射函数和双射函数。能构造两个基数相等的集合间的双射函数。
2. 能求某子集在一函数下的像和完全原像。
3. 熟练掌握函数的复合运算和逆运算。

Key points of this chapter

1. Understand the concepts of function, injection, surjection and bijection. Be able to determine whether a function is a injection, a surjection, or a bijection. Be able to structure a bijection from a set to another if the two sets are with same cardinality.
2. Be able to find the image or complete preimage of a subset in a function.
3. Can skilled find the composite function of functions and the inverse of a function.

第五章　图论
Chapter 5　Graph Theory

研究图形和网络的图论通常被认为是组合学的一部分，但它已成长的足够强大和特色鲜明，拥有自己的研究问题，已经被当作一门独立的学科。图是离散数学研究的基本科目之一。它存在于最广泛的自然和人造的模型中。它能对许多物理的、生物的和社会的系统建立关系模型和过程动态模型。在计算科学中，它可以表述通信网络、数据结构、计算装置、计算流量等等。在数学中，它对几何学、拓扑学某些部分(例如纽结理论)的研究相当有用。代数图论与群论有密切的联系。虽然还有连续图，但图论的大部分研究属于离散数学的范畴。

本章主要包括图及其相关的基本概念，如关联性、邻接性、连通性、可达性、顶点度数、图的关联矩

阵、邻接矩阵、子图等。还包括二部图、欧拉图、哈密尔顿图、平面图及其应用。

Graph theory, the study of graphs and networks, is often considered part of combinatorics, but has grown large enough and distinct enough, with its own kind of problems, to be regarded as a subject in its own right. Graphs are one of the prime objects of study in discrete mathematics. They are among the most ubiquitous models of both natural and human-made structures. They can model many types of relations and process dynamics in physical, biological and social systems. In computer science, they can represent networks of communication, data organization, computational devices, the flow of computation, etc. In mathematics, they are useful in geometry and certain parts of topology, e.g. knot theory. Algebraic graph theory has close links with group theory. There are also continuous graphs, however, for the most part research in graph theory falls within the domain of discrete mathematics.

The main content of this chapter includes the concept of graph and the concepts associated with graphs, such as incidence, adjacency, connectivity, reachability, degrees of a vertex, incidence matrix, adjacent matrix, subgraph, *etc.*. This chapter also includes bipartite graph, Euler graph, Hamilton graph, planar graph and their applications in practice.

单词和短语 Words and expressions

无序对　unordered pair	补图　complement graph
无序积　unordered product	图的同构　isomorphism of graphs
多重集　multiple set	通路　path
重复度　multiplicity	可达矩阵 路径矩阵　path matrix
图　graph	连通图　connected graph
顶点,结点　vertex	连通分支　connected component
边　edge, line	短程线　geodesic
有向图　directed graph, digraph	点割集　cut points set
有向边　directed edge, arc	割点　cut point
环　loop	边割集　cut edges set
孤立点　isolated vertex	割边,桥　cut edge, bridge
关联矩阵　incidence matrix	点(边)连通度　vertex(edge) connectivity
邻接点(边)　adjacent vertices(edges)	可达的　reachable
邻接矩阵　adjacent matrix	弱连通的　weakly connected
顶点 v 的度数　degrees of the vertex v	单向连通　unilaterally connected
完全图　complete graph	强连通　strongly connected
正则图　regular graph	二部图,(两)偶图　bipartite graph
圈图　cycle graph	完全二部图　complete bipartite graph
轮图　wheel graph	欧拉图　Euler graph
方体图　quadratic graph	哈密尔顿图　Hamilton Graph
子图　subgraph	平面图　planar graph
母图　supergraph	平面嵌入　planar embedding
生成子图　spanning subgraph	欧拉公式　Euler formula
导出子图　generated subgraph	对偶图　dual graph

基本概念和性质　Basic concepts and properties

1 $G=<V,E>$ 称作一个无向图,其中 $V\neq\emptyset$ 称为顶点集,其元素称为顶点或结点;E 是 $V\&V$ 的多重子集,称为边集,其元素称为无向边,简称边.

$D=<V,E>$ 称作一个有向图吗,其中 $V\neq\emptyset$ 称为顶点集,其元素称为顶点或结点;E 是 $V\&V$ 的多重子集,称为边集,其元素称为有向边,简称边.

$G=<V,E>$ is termed a undirected graph, where $V\neq \varnothing$ is said the vertex set that its elements is called vertices and E is a multiple subset of $V\&V$ and is said the edge set that its elements is called undirected edges, for short edges or lines.

$D=<V,E>$ is termed a directed graph, where $V\neq \varnothing$ is said the vertex set that its elements is called vertexes and E is a multiple subset of $V\&V$ and is said the edge set that its elements is called directed edges, for short edges or arcs.

2 与顶点 v 关联的边数称作顶点 v 的度数。任何图（无向图和有向图）的所有顶点度数之和都等于边数的 2 倍，此性质称作握手定理，还可以表述为任何图（无向图和有向图）都有偶数个奇度顶点。有向图所有顶点的入度之和等于出度之和。

The number of edges being incidence to vertex v is called the degree of vertex v. The sum of degrees of all vertexes in any graph, either directed graph or undirected graph, is equal to 2 times the number of edges in the graph. This property is said Handshake Theorem and may also be expressed as that the number of the odd degree vertexes is even in any graph, either directed graph or undirected graph. The sum of in-degrees of all vertexes is equal to that of out-degrees in any directed graph.

3 对无向图 $G=<V,E>$ 的两个顶点 $u,v\in V$，如果 u 与 v 之间有通路，则称 u 与 v 连通的。任意两点都连通的无向图称作连通图。

For two vertexes $u,v\in V$ in the undirected graph $G=<V,E>$, u and v are termed connected if there is a path between u and v. An undirected graph is called a connected graph if any two vertexes in the graph connect.

4 对有向图 $D=<V,E>$ 的两个顶点 $u,v\in V$，如果有 u 到 v 有通路，则称 u 可达 v。如果略去各边的方向所得无向图为连通图，则称 D 是弱连通。如果对 $\forall\ u,v\in V$，u 可达 v 或 v 可达 u，则称 D 是单向连通的。如果对 $\forall\ u,v\in V$，u 与 v 相互可达，则称 D 是强连通的。

For two vertexes $u,v\in V$ in the directed graph $D=<V,E>$, that u can reachable to v means there is a path from u to v. D is called weakly connected if the undirected graph obtained from D omitting the edge direction is connected. D is called unilaterally connected if u can reachable to v or v can reachable to u. D is called strongly connected if u and v can reachable each other.

5 设无向图 $G=<V,E>$，若能将 V 分成 V_1 和 V_2 使得 $V_1\cup V_2=V$，$V_1\cap V_2=\varnothing$，且 G 中的每条边的两个端点都一个属于 V_1，另一个属于 V_2，则称为二部图，记为 $<V_1,V_2,E>$，称 V_1 和 V_2 为互补顶点子集。一个无向图是二部图当且仅当图中无奇长度的回路。

Let $G=<V,E>$ be an undirected graph. G is termed a bipartite graph, noted as $<V_1,V_2,E>$, if the V may be divided into V_1 and V_2 such that $V_1\cup V_2=V, V_1\cap V_2=\varnothing$, for the two vertexes of any edge in G, a vertex is belong to V_1 and the other is belong to V_2. V_1 and V_2 are called Complementary vertex subsets. An undirected graph is a bipartite graph if and only if there is no circuit with odd length.

6 每条边恰好经过一次的回路称作欧拉回路。存在欧拉回路的图称作欧拉图。无向图 G 欧拉图当且仅当 G 是连通的且无奇度顶点。有向图 D 是欧拉图当且仅当 D 是连通的且所有顶点的入度等于出度。

A circuit is called an Euler circuit if it goes through every edge in the graph just once. A graph is called an Euler graph if there are Euler circuits in it. An undirected graph G is an Euler graph if and only if G is connected and there is no odd degree vertex in it. A directed graph D is an Euler graph if and only if D is connected and the in-degree is equal to the out-degree for any vertex in D.

7 经过图中所有顶点一次且仅一次的回路称作哈密尔顿回路。存在哈密尔顿回路的图称作哈密尔顿图。若无向图 $G=<V,E>$ 是哈密顿图，则对于 V 的任意非空真子集 V_1 均有 $p(G-V_1)\leqslant |V_1|$。设 G 是 $n(n(3)$ 阶无向简单图，若图 G 的任意两个不相邻的顶点的度数之和大于等于 n，则 G 为哈密顿图。

A circuit is called a Hamilton circuit if it goes through every vertex in the graph just once. A graph is called a Hamilton graph if there are Hamilton circuits in it. If an undirected graph $G=<V,E>$ is a Hamilton graph, then we have that $p(G-V_1) \leqslant |V_1|$ for any non-empty subset V_1 of V. Let G is a simple undirected graph with order $n(n \geqslant 3)$. If the sum of degrees of any two vertexes which is not adjacent in G is greater than or equal to n, then G is a Hamilton graph.

8 如果能将图 G 除顶点外，边不相交地画在平面上，则称 G 是平面图。这个画出的无边相交的图称作 G 的平面嵌入。设 G 为 n 阶 m 条边 r 个面的连通平面图，则有 $n-m+r=2$。

A graph G is referred to as a planar graph if we are able to draw the G in plan such that edges are non-intersecting except for in vertices. The drawn graph with no intersecting edges is said a planar embedding of G. If G is a planar graph, n is its order, m is the number of edges and r is the number of faces, then $n-m+r=2$.

例题 Example

例 1 已知图 G 有 10 条边，4 个 3 度顶点，其余顶点的度数均小于 2，问 G 至少有多少个顶点？

解 设 G 有 n 个顶点. 由握手定理，
$$4(3+2\times(n-4)) \geqslant 2\times 10$$
解得
$$n \geqslant 8.$$

Example 1 Given that a graph G has 10 edges, 4 vertices with 3 degrees and the degrees of other vertices is less than 2. How many vertices G has at least?

Solution Let G have n vertices. By Handshake Theorem, we have that
$$4\times 3 + 2\times(n-4) \geqslant 2\times 10$$
Solving the previous inequality, we obtained that $n \geqslant 8$.

例 2 证明：若无向图 $G=<V,E>$ 是哈密顿图，则对于 V 的任意非空真子集 V_1 均有 $p(G-V_1) \leqslant |V_1|$.

证明：设 C 为 G 中一条哈密顿回路，有 $p(C-V_1) \leqslant |V_1|$.
又因为 $C \subseteq G$，故 $p(G-V_1) \leqslant p(C-V_1) \leqslant |V_1|$.

Example 2 Prove that if an undirected graph $G=<V,E>$ is a Hamilton graph, then $p(G-V_1) \leqslant |V_1|$ for any non-empty subset V_1 of V.

Proof Suppose C be a Hamilton circuit in G. We have that $p(C-V_1) \leqslant |V_1|$.
Because of $C \subseteq G$, so $p(G-V_1) \leqslant p(C-V_1) \leqslant |V_1|$.

例 3 有 7 个人，A 会讲英语，B 会讲英语和汉语，C 会讲英语、意大利语和俄语，D 会讲日语和汉语，E 会讲德语和意大利语，F 会讲法语、日语和俄语，G 会讲法语和德语. 问能否将他们沿圆桌安排就坐成一圈，使得每个人都能与两旁的人交谈？

解 作无向图如下，每人是一个顶点，2 人之间有边（他们有共同的语言.

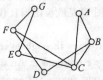

如图，ACEGFDBA 是一条哈密顿回路，按此顺序就坐即可.

Example 3 There are 7 people. A can speak English. B can speak English and Chinese. C speaks English, Italian and Russian. D speaks Japanese and Chinese. E can speak German and Italian. F speaks French, Japanese and Russian. G speaks French and German. Whether will they seated into in a circle along the table, so that everyone can talk to its neighbor persons?

Solution Draw an undirected graph as follows, where every person is a vertex and there is a edge between 2 person if and only if they can speak a common language.

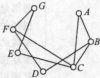

As shown in the graph, ACEGFDBA is a Hamilton circuit. And according to this sequence was then.

本章重点

1. 理解图及其相关的概念。掌握握手定理。会判别一个图是否为连通图。能求出割点、割边（桥）、点割集和边割集。能求出一个图的关联矩阵、邻接矩阵和可达矩阵。
2. 能判断一个图是否为二补图、欧拉图、哈密尔顿图和平面图，掌握这些图的性质，并能利用这些图解决实际问题。

Key points of this chapter

1. Understand the concept of graph and the related concepts. Master Handshake Theorem. Be able to determine whether a graph is connected. Be able to find cut point, cut edge (bridge), cut points set and cut edges set. Be able to find the incidence matrix, the adjacent matrix and the path matrix of a graph.
2. Be able to determine whether a graph is a bipartite graph, a Euler graph, a Hamilton Graph or a planar graph and to apply these graphs in practice.

第六章　树及其应用
Chapter 6　Tree and Its Applications

　　树结构是层级性结构的一种图示方式。之所以称作树结构是因为它的典型表示像一棵树，尽管相对于实际的树这种图示通常是颠倒的，"树根"朝上，"树叶"朝下。

　　树结构被广泛地应用在计算机科学中（参阅树（数据结构）和电信学。树结构用来描述所有类型的分类学知识。例如家谱树、生物进化树、语言谱系进化树、语言的语法结构等。

　　本章主要内容包括无向树的概念和性质、根数的概念和性质、最小生成树的求法、最优二元树的求法及其在编码学中的应用、根数的遍历问题等。

　　A tree structure is a way of representing the hierarchical nature of a structure in a graphical form. It is named a "tree structure" because the classic representation resembles a tree, even though the chart is generally upside down compared to an actual tree, with the "root" at the top and the "leaves" at the bottom.

　　Tree structures are used extensively in computer science (see Tree (data structure) and telecommunications.) Tree structures are used to depict all kinds of taxonomic knowledge, such as family trees, the biological evolutionary tree, the evolutionary tree of a language family, the grammatical structure of a language *etc.*.

　　The main content of this chapter includes the concept and properties of undirected trees, the concept and properties of rooted trees, how to find a minimum spanning tree of a weighted graph, how to find a optimal binary tree and the applications in the coding theory, the problem walking through a binary tree, *etc.*.

单词和短语 Words and expressions

（无向）树　tree	森林　forest
平凡树　trivial tree	树叶　leaf

树枝 branch	树根 root
分支点 branch vertex	内部结点 inner vertex
无向树的性质 properties of trees	根树的层数 layer of a rooted tree
生成树 spanning tree	根树的高度 height of a rooted tree
弦 string	家族树 family tree
余树 cotree	有序树 ordered tree
权重 weight	完全树 complete tree
带权图,加权图 weighted graph	正则树 regular tree
最小生成树 minimum spanning tree	二元树,二叉树 binary tree
避圈法 avoiding-circle method	最优二元树 optimal binary tree
破圈法 breaking-circle method	前缀码 prefix code
克鲁斯卡尔算法 Kruskal Algorithm	最佳前缀码 optimal prefix code
有向树 directed tree	遍历,行遍,周游 walking through, traversal
根树 rooted tree	波兰符号法 Poland symbolic method

基本概念和性质　Basic concepts and properties

1 连通无回路的无向图称作无向树。在无向树中，$m = n-1$，其中，m 是其边数，n 是其顶点数。

An undirected graph is referred to as an undirected tree if it is connected and there is no circuit in it. In any undirected tree, $m = n-1$, where m is the number of edges and n is the number of vertices.

2 设 G 是无向连通图，若 G 的生成子图 T 是一棵树，则称 T 是 G 的生成树。任何无向连通图都有生成树。带权图中权最小的生成树称作最小生成树。

Let G be a connected undirected graph. A spanning subgraph T of G is called a spanning tree of G if T is a tree. Any connected undirected graph has spanning trees. The spanning tree with minimum weights is called a minimum spanning tree in a weighted graph.

3 略去方向后为无向树的有向图称作有向树。有一个顶点入度为 0，其余的入度均为 1 的非平凡的有向树称作根树。每个分支点至多有 2 个儿子的根树称作二元树或二叉树。

A directed graph is termed a directed tree if the graph is a tree when omitting the direction. A nontrivial directed tree is called a rooted tree if there is a vertex with 0 in-degree and the rest vertexes are all with 1 in-degree. A rooted tree is called a binary tree if there are no more than 2 sons of every branch vertex.

例题　Example

例 1 已知无向树 T 中，有 1 个 3 度顶点，2 个 2 度顶点，其余顶点全是树叶。试求树叶数。

解 设有 x 片树叶，则
$$2 \times (2+x) = 1 \times 3 + 2 \times 2 + x$$
解得 $x = 3$，故 T 有 3 片树叶。

Example 1 Given that there is 1 vertex with 3 degree, there are 2 vertexes with 2 degree and the rest vertexes are all leaves in an undirected tree T. Try to find the number of leaves.

Solution Let the number of leaves be x. Then
$$2 \times (2+x) = 1 \times 3 + 2 \times 2 + x$$
Solving the previous equation, we obtained $x = 3$. So, there are 3 leaves in T.

例 2 将如图所示的二叉树表示的算式分别用中序行遍、前序行遍和后续行遍的方式书写出来。

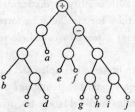

解 中序行遍:$((b+(c+d))*a)\div((e-f)-(g+h)*(i*j))$
前序行遍:$\div*+b+cda-*ef*+gh*ij$
后续行遍:$bcd++a*ef*gh+ij**-\div$

Example 2 Write out the calculation expressed by the following binary tree using the in order, pre order and post order traversal, respectively.

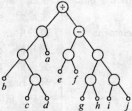

Solution Using the in order traversal:$((b+(c+d))*a)\div((e-f)-(g+h)*(i*j))$
Using the pre order traversal:$\div*+b+cda-*ef*+gh*ij$
Using the post order traversal:$bcd++a*ef*gh+ij**-\div$

本章重点

1. 理解无向树、生成树、根树、二元树等概念。
2. 熟练掌握求最小生成树的方法。
3. 熟练掌握利用最优二元树编制最佳前缀码的方法。
4. 掌握二元树的遍历。

Key points of this chapter

1. Understand the concepts of undirected graph, spanning tree, rooted tree, binary tree, etc..
2. Can skilled find the minimum spanning tree.
3. Can skilled encode the optimal prefix code using an optimal binary tree.
4. Master the traversal of a binary tree.

第七章 代数系统
Chapter 7 Algebraic Systems

抽象代数是研究诸如群、环、域、模、格等代数结构的数学分支(群、环、域、格就是本章研究的对象)。术语抽象代数是在二十世纪初为将这一研究领域与常规代数区分开而杜撰出来的。常规代数,现在通常称作基础代数,是研究运算包括未知量、实数或复数在内的公式和代数表达式的规则的。这种区分很少出现在近来的文献中了。

抽象代数即以离散的又以连续的例子出现。离散的代数包括:用于逻辑门电路库和程序编制的布尔代数、用于数据库的关系代数、在代数编码理论中有重要应用的离散的有限型的群环域、用于形式语言学的离散半群和独异点。

Abstract algebra is the subject area of mathematics that studies algebraic structures such as groups, rings, fields, modules, vector spaces and lattices, which are studied in this chapter. The phrase abstract algebra was coined at the turn of the 20th century to distinguish this area from what was normally referred to as algebra, the study of the rules for manipulating formulae and algebraic expressions involving unknowns and real or complex numbers, often now called *elementary algebra*. The distinction is rarely made in more recent writings.

Algebraic structures occur as both discrete examples and continuous examples. Discrete algebras include: Boolean algebra used in logic gates and programming; relational algebra used in databases; discrete and finite versions of groups, rings and fields are important in algebraic coding theory; discrete semigroups and monoids appear in the theory of formal languages.

单词和短语 Words and expressions

二元运算　binary operation
一元运算　unary operation
S 对运算 f 封闭　closed for the operation f on S
运算表　operation table
单位元，幺元　identity element, unit element
零元　zero element
可逆元　inverse element
代数系统　algebraic systems
同类型　same type
同种的　conspecific
子代数　subalgebra
平凡子代数　trivial subalgebra
真子代数　proper subalgebra
积代数　product algebra
同态映射　homomorphic mapping
同态　homomorphism
同构映射　isomorphic mapping
同构　isomorphism
同余关系　congruence relationship
群　group
子群　subgroup
半群　semi-group
子半群　sub-semigroup
独异点　monoid
子独异点　sub-monoid

阿贝尔群　Abel group
阶　rank
生成子群　generated subgroup
生成元　generator
群的中心　center of a group
循环群　cyclic group
置换群　permutation group
环　ring
整环　integral domain
体　body
域　field
格　lattice
上确界　supremum, least upper bound
下确界　infimum, greatest lower bound
对偶命题　Dual Proposition
对偶原理　Duality Principle
分配格　distributive lattice
钻石格　diamond lattice
五角格　pentagonal lattice
有界格　bounded lattice
补元　complement
有补格　complemented lattice
Algebra 布尔格，布尔代数　Boolean Lattice, Boolean

基本概念和性质　Basic concepts and properties

1. 设 S 为集合，函数 $f: S \times S \rightarrow S$ 称为上的二元运算，简称为二元运算，也称 S 对 f 封闭。函数 $f: S \rightarrow S$ 称为 S 上的一元运算，简称为一元运算，同样也称 S 对 f 封闭。

Let S be a set. The function $f: S \times S \rightarrow S$ is termed binary operation on S, for short binary operation, which is also referred to as f is closed on S. The function $f: S \rightarrow S$ is termed unary operation on S, for short unary operation, as same as binary operation, which is also referred to as f is closed on S.

2. 非空集合 S 和 S 上 k 个一元或二元运算 $f_1, f_2, \cdots, f_k$ 组成的系统称为一个代数系统简称代数，记作 $<S, f_1, f_2, \cdots, f_k>$。

A system composed of nonempty set S and k unary operations or binary operations on S is referred as algebraic system, for short algebra, noted by $<S, f_1, f_2, \cdots, f_k>$.

3 设 $V=<S,\circ>$ 是代数系统，$\circ$ 为二元运算，如果运算 $\circ$ 是可结合的，则称 V 为半群。如果①运算 $\circ$ 是可结合的；② $e\in S$ 是关于运算 $\circ$ 的单位元，则称 V 是含幺半群，也叫做独异点。

Let $V=<S,\circ>$ be a algebraic system and $\circ$ be a binary operation. V is referred to as a semigroup if the operation $\circ$ is associative. V is referred to as a semigroup containing unit element, which also referred to as monoid, if ① the operation $\circ$ is associative; ② $e\in S$ is the identity element for the operation $\circ$.

4 设 $<G,\circ>$ 是代数系统，$\circ$ 为二元运算. 如果

① 运算 $\circ$ 是可结合的，
② $e\in S$ 是关于运算 $\circ$ 的单位元，
③ $\forall\, x\in G$，都有 $x^{-1}\in G$，

则称 G 为群.

Let $<G,\circ>$ be a algebraic system and $\circ$ be a binary operation on G. G is referred to as group if the following satisfied:

①The operation $\circ$ is associative;
② $e\in G$ is the identity element for the operation $\circ$;
③ $\forall\, x\in G$, there exists $x^{-1}\in G$.

5 G 为群，$\forall\, a,b\in G$，方程 $ax=b$ 和 $ya=b$ 在 G 中有解且仅有惟一解.

Let G be a group. For all a, b $\in G$, the equations $ax=b$ and $ya=b$ have one and only one solution in G, respectively.

6 设 G 是群，H 是 G 的非空子集，如果 H 关于 G 中的运算构成群，则称是 G 的子群，记作 $H\leqslant G$.

Let G be a group and H is a nonempty subset of G. H is called a subgroup of G if H constitute a group for the binary operation on G, noted by $H\leqslant G$.

7 判定定理 I：设 G 为群，H 是 G 的非空子集. H 是 G 的子群当且仅当

（ⅰ） $\forall\, a,b\in H$ 有 $ab\in H$，
（ⅱ） $\forall\, a\in H$ 有 $a^{-1}\in H$.

Judgment Theorem I Let G be a group and H is a nonempty subset of G. H is a subgroup of G if and only if the following satisfied:

（ⅰ） $\forall\, a, b\in H$, $ab\in H$ is tenable;
（ⅱ） $\forall\, a\in H$, $a^{-1}\in H$ is tenable.

8 判定定理 II：设 G 为群，H 是 G 的非空子集. H 是 G 的子群当且仅当 $\forall\, a,b\in H$ 有 $ab^{-1}\in H$.

Judgment Theorem II Let G be a group and H is a nonempty subset of G. H is a subgroup of G if and only if for $\forall\, a,b\in H$, $ab^{-1}\in H$ is tenable.

9 设 G_1, G_2 是群，$f:G_1\to G_2$，若 $\forall\, a,b\in G_1$ 都有 $f(ab)=f(a)f(b)$，则称 f 是群 G_1 到 G_2 的同态映射，简称同态.

设 f 是群 G_1 到 G_2 的同态映射，则

(1) $f(e_1)=e_2$，e_1 和 e_2 分别是 G_1 和 G_2 的单位元，
(2) $\forall\, x\in G_1, f(x^{-1})=f(x)^{-1}$，
(3) 设 $H\leqslant G_1$，则 $f(H)\leqslant G_2$.

Let G_1, G_2 be groups and $f:G_1\to G_2$. f is said a homomorphic mapping from G_1 to G_2, for short homomorphism, if $f(ab)=f(a)f(b)$ for $\forall\, a,b\in G_1$.

If f is a homomorphic mapping from group G_1 to group G_2, then

(1) $f(e_1)=e_2$, where e_1, e_2 is identity element of G_1 and G_2, respectively;
(2) $\forall\, x\in G_1, f(x^{-1})=f(x)^{-1}$;

(3) if $H \leqslant G_1$, then $f(H) \leqslant G_2$.

10 设 $<L, \leqslant>$ 是偏序集,如果 $\forall\ x, y \in L, \{x, y\}$ 都有最小上界和最大下界,则称 L 关于偏序 $\leqslant$ 作成一个格。

Let $<L, \leqslant>$ be a partial order set. L is referred $\leqslant$ as a lattice if $\{x, y\}$ has least upper bound and greatest lower bound for $\forall\ x, y \in L$.

11 如果一个格是有补分配格,则称它为布尔格或布尔代数。

A lattice is referred as a Boolean lattice or Boolean algebra if it is a complemented distributive lattice.

例 题 Example

例 1 设 G 为群,$a, b \in G$ 是有限阶元,

证明:(1) $|b^{-1}ab| = a$;   (2) $|ab| = |ba|$.

证明:(1) 设 $|a| = r$, $|b^{-1}ab| = t$, 那么,
$$(b^{-1}ab)^r = b^{-1}a^r b = b^{-1}b = e,$$
所以, $t \mid r$. 另一方面, 由
$$a = (b^{-1})^{-1}(b^{-1}ab) b^{-1},$$
则有 $r \mid t$. 因此, $|b^{-1}ab| = |a|$.

(2) 设 $|ab| = r$, $|ba| = t$, 那么
$$(ab)^{t+1} = a (ba)^t b = ab.$$
由消去律,可得 $(ab)^t = e$.
所以 $r \mid t$. 类似地可证 $t \mid r$.
因此, $|ab| = |ba|$.

Example 1 Let G be a group and $a, b \in G$ be with finite order.

Prove that (1) $|b^{-1}ab| = a$;   (2) $|ab| = |ba|$.

Proof: (1) Let $|a| = r$, $|b^{-1}ab| = t$, then
$$(b^{-1}ab)^r = b^{-1}a^r b = b^{-1}b = e,$$
So, $t \mid r$. On the other hand, from
$$a = (b^{-1})^{-1}(b^{-1}ab) b^{-1},$$
We can obtain that $r \mid t$. Thus, $|b^{-1}ab| = |a|$.

(2) Let $|ab| = r$, $|ba| = t$, then
$$(ab)^{t+1} = a (ba)^t b = ab.$$
By the cancellation law, we get that $(ab)^t = e$.
So $r \mid t$. Similarly, we can prove that $t \mid r$.
Thus, $|ab| = |ba|$.

例 2 设 $<L, \wedge, \vee, 0, 1>$ 是有界分配格. 若 L 中元素 a 存在补元,则存在惟一的补元.

证明:假设 b, c 是 a 的补元,则有
$$a \vee c = 1, \quad a \wedge c = 0, \quad a \vee b = 1, a \wedge b = 0$$
从而得到 $a \vee c = a \vee b, a \wedge c = a \wedge b,$
由于 L 是分配格,因此有 $b = c$.

Example 2 Let $<L, \wedge, \vee, 0, 1>$ be a bounded distributive lattice. Prove that if there is complement element of a in L, then the complement is unique.

Proof Suppose that b, c are complement elements of a. Then
$$a \vee c = 1, \quad a \wedge c = 0, \quad a \vee b = 1, a \wedge b = 0$$
So, $a \vee c = a \vee b, a \wedge c = a \wedge b,$
Since L is distributive, $b = c$.

本章重点

1 理解代数系统的概念,能根据运算公式或运算表判别其满足的运算规律,求出其单位元、零元和逆元

（如果存在）。
2. 理解半群、独异点、和群的概念。掌握群的性质。会判断群的一个非空子集是否位子群。
3. 了解环、域的概念。
4. 理解格和布尔代数的概念，能解相关的一些问题。

Key points of this chapter

1. Understand the concept of algebraic system. Be able to find the operation laws which is satisfied by an operation and the identity, zero and inverse (if there exist) respect to a operation according to the operation formula or operation table.
2. Understand the concepts of semigroup, monoid and group. Master the properties of group. Be able to determine whether a nonempty subset of a group is a subgroup.
3. Know the concepts of ring and field.
4. Understand the concepts of lattice and Boolean algebra. Be able to solve some problem relating to the two algebras.

第五部分　计算机组成原理
Part 5　Computer Organization Principles

引　言

　　世界上第一台计算机基于冯·诺依曼原理,其基本思想是:存储程序与程序控制.存储程序是指人们必须事先把计算机的执行步骤序列(即程序)及运行中所需的数据,通过一定方式输入并存储在计算机的存储器中.程序控制是指计算机运行时能自动地逐一取出程序中指令,加以分析并执行规定的操作.到目前为止,虽然计算机发展了五代,但其基本工作原理没有变.

　　计算机组成原理这门课就是讲述计算机基本组成及工作原理的课程.通过对这门课的学习,使学生能够初步了解整个计算机的基本工作原理及运行机制,掌握计算机各个组成部分的基本工作原理.

Introduction

　　The world's first computer was based on the von Neumann principle whose main idea is the stored procedures and process control. Stored procedures require people must put and store the program and required data into computer memory through some way in advance. Process control is the operation that computer can get, analyze and run the instructions automatically one by one. Although computer has developed five generations, its basic principle has not changed.

　　The course of computer composition principle is about the basic computer components and its working principle. Students can understand the basic working principles and the operating mechanism of computer, and preliminary master the basic working principles of the various parts of the computer by learning this course.

第一章　计算机系统概论
Chapter 1　Introduction for Computer System

　　本章我们主要介绍计算机的分类、计算机发展简史、计算机的硬件组成、软件组成以及计算机系统的层次结构.

　　In this chapter, we mainly introduce classification of computer, computer development history, computer hardware parts, computer software parts and hierarchy architecture of computer system.

单词和短语　Words and expressions

专用计算机	dedicated computer	芯片	chip
通用计算机	general-purpose computer	运算器	arithmetic unit
硬件	hardware	输入设备	input device
存储器	memory	软件	software
输出设备	output device	机器语言	machine language
控制器	controller	汇编语言	assemble language
适配器	adapter	高级程序	advanced program
总线	bus	操作系统	operating system

基本概念和性质　Basic concepts and properties

■ 计算机硬件是指计算机系统中由电子,机械和光电元件等组成的各种物理装置的总称.

Computer hardware is a general term for a variety of physical devices such as electronic, mechanical and optical components and so on.

2. 计算机软件(也称软件),是指计算机系统中的程序及其文档.
Computer software (also known as software) refers to computer systems and procedures documentation.

例 题 Example

例 1 什么是计算机系统、计算机硬件和计算机软件？硬件和软件哪个更重要？
解 计算机系统：由计算机硬件系统和软件系统组成的综合体.
计算机硬件：指计算机中的电子线路和物理装置.
计算机软件：计算机运行所需的程序及相关资料.
硬件和软件在计算机系统中相互依存,缺一不可,因此同样重要.

例 2 冯·诺依曼计算机的特点是什么？
解 (1)计算机由运算器、控制器、存储器、输入设备、输出设备五大部件组成；
(2)指令和数据以同等地位存放于存储器内,并可以按地址访问；
(3)指令和数据均用二进制表示；
(4)指令由操作码、地址码两大部分组成,操作码用来表示操作的性质,地址码用来表示操作数在存储器中的位置；
(5)指令在存储器中顺序存放,通常自动顺序取出执行；
(6)机器以运算器为中心.

Ex. 1 What is computer system, computer hardware and computer software? Hardware and software which is more important?
Solution Computer system: it is complex composed of computer hardware system and software system.
Computer hardware: electronic circuit and physical devices in computer.
Computer software: program and corresponding data needed by computer when running.
Hardware and software are interdependent and indispensable, therefore they are equally important.

Ex. 2 What are the characteristics of the von Neumann computer?
Solution (1)Computer is composed of five components: calculation device, controller, memory, input devices and output devices;
(2)Instructions and data can be stored in memory with equal status and can be accessed by address;
(3)Instructions and data are all expressed by binary;
(4)Instruction is composed two parts, operation code and address code. Operation code used to describe the nature of the operation, the address code used to describe the location of the operand in memory;
(5)Instructions are stored in memory in order and usually run automatically in order;
(6)Arithmetic unit is the core of computer.

本章重点
1. 了解计算机系统的基本概念、计算机系统的工作原理及计算机基本组成部分.

Key points of this chapter
1. It is required that one should understand basic concepts of computer systems, computer system's principle of work and basic computer organization parts.

第二章 运算方法和运算器
Chapter 2 Operational Method and Arithmetic Unit

本章我们主要学习的内容包括:数据与文字的表示、定点加法、减法运算、定点乘法运算、定点除法运算、定点运算器的组成、浮点运算与浮点运算器.

In this chapter, we study some new contents, which include data and text representation, fixed-point addition, subtraction, multiplication of fixed point, fixed point division, fixed-point arithmetic unit composition, floating point arithmetic and floating point arithmetic unit.

单词和短语 Words and expressions

符号数据　symbol data
数值数据　numerical data
编码　coding
定点表示　fixed-point representation
浮点表示　floating point representation
格雷码　gray code
原码　original code
反码　anti-code
补码　complement

移码　frame shift
校验码　check code
ASCII　American standard code for information interchange
逻辑运算单元　logical unit
内部总线　internal bus
三态门　tri-state gate
浮点运算　floating-point operations

基本概念和性质　Basic concepts and properties

1 二进制是计算技术中广泛采用的一种数制. 二进制数据是用 0 和 1 两个数码来表示的数. 它的基数为 2, 进位规则是"逢二进一", 借位规则是"借一当二".

Binary is a number system which is widely used in computing technology. It is represented by 0 and 1 two digital numbers and its base is 2. Binary carry rule is "every two further" and borrow rule is "by a two".

2 算术逻辑单元是中央处理器(CPU)的执行单元, 是所有中央处理器的核心组成部分, 主要功能是进行二进制的算术运算, 如加、减、乘等.

The arithmetic logic unit is the execution unit of CPU. It is the core of CPU and its main function is to carry out binary arithmetic such as addition and subtraction to multiplication and so on.

例题　Example

例 1 设浮点数格式为:阶码 5 位(含 1 位阶符), 尾数 11 位(含 1 位数符). 写出 51/128、$-27/1024$ 所对应的机器数. 要求如下:

(1)阶码和尾数均为原码.
(2)阶码和尾数均为补码.
(3)阶码为移码, 尾数为补码.

解 据题意画出该浮点数的格式:

阶符 1 位	阶码 4 位	数符 1 位	尾数 10 位

将十进制数转换为二进制: $x_1 = 51/128 = 0.0110011B = 2^{-1} * 0.110011B$
$x_2 = -27/1024 = -0.0000011011B = 2^{-5} * (-0.11011B)$

则以上各数的浮点规格化数为:

(1) $[x_1]_浮 = 1,0001;0.110\ 011\ 000\ 0$
　　$[x_2]_浮 = 1,0101;1.110\ 110\ 000\ 0$

(2) $[x_1]_浮 = 1,1111;0.110\ 011\ 000\ 0$
　　$[x_2]_浮 = 1,1011;1.001\ 010\ 000\ 0$

(3) $[x_1]_浮 = 0,1111;0.110\ 011\ 000\ 0$
　　$[x_2]_浮 = 0,1011;1.001\ 010\ 000\ 0$

Ex. 1 Let floating-point format is that order code is 5 (including an order operator), mantissa is 11 (with a median operator), try to write the corresponding machine number of 51/128, $-27/1024$. Requirements are as follows:

(1) order code and mantissa are the original code.
(2) order code and mantissa are the complement code.
(3) order code is shift code, mantissa is complement code.

Solution According to the meaning of the questions drawing the floating-point format:

| order character is 1 | order code is 4 | number character is 1 | mantissa is 10 |

Converts a decimal number to binary: $x_1 = 51/128 = 0.0110011B = 2-1 * 0.110\ 011B$
$x_2 = -27/1024 = -0.0000011011B = 2-5 * (-0.11011B)$

the number of normalized floating-point number as follows:

(1) $[x_1]f = 1,0001; 0.110\ 011\ 000\ 0$
$[x_2]f = 1,0101; 1.110\ 110\ 000\ 0$

(2) $[x_1]f = 1,1111; 0.110\ 011\ 000\ 0$
$[x_2]f = 1,1011; 1.001\ 010\ 000\ 0$

(3) $[x_1]f = 0,1111; 0.110\ 011\ 000\ 0$
$[x_2]f = 0,1011; 1.001\ 010\ 000\ 0$

本章重点

运算方法和运算器是本章的重点,要求学生重点掌握数据的表示方法、加法器的基本工作原理、ALU 的基本工作原理、硬件的具体实现。

Key points of this chapter

Computing operation and ALU are the focus of this chapter. We require students master the data representation methods, the basic working principle and specific hardware implementation of ALU and adder.

第三章 存储系统
Chapter 3　Memory System

本章我们主要了解的内容有:存储器概述、SRAM 存储器、DRAM 存储器、只读存储器和闪速存储器、并行存储器、Cache 存储器。

In this chapter, the contents we should understand include overview of the memory, SRAM memory, DRAM memory, read-only memory and flash memory, parallel memory and Cache memory.

单词和短语 Words and expressions

半导体存储器　semiconductor memory　　动态读写存储器　dynamic read-write memory
只读存储器　read-only memory　　　　　地址译码器　address decoder
随机存储器　random access memory　　　闪速存储器　flash memory
高速缓冲存储器　cache memory　　　　　并行存储器　parallel memory
静态读写存储器　static read-write memory　双端口存储器　dual-port memory

基本概念和性质　Basic concepts and properties

■1 字存储单元是指存放一个机器字的存储单元,相应的单元地址叫字地址。
Word storage unit is a storage unit to store a machine word, the corresponding cell address is called word address.

■2 字节存储单元是指存放一个字节的单元,相应的地址称为字节地址。
Byte storage unit is to store a byte unit, the appropriate address is called byte address.

■3 存储器(Memory)是计算机系统中的记忆设备,用来存放程序和数据.计算机中全部信息,包括输入的原始数据、计算机程序、中间运行结果和最终运行结果都保存在存储器中。

Memory is the memory device for store program and data in the computer system. All information including the input original data, program, the intermediate result and the final result are stored in memory.

例 题 Example

例 1 什么是存储器的带宽？若存储器的数据总线宽度为 32 位，存取周期为 200ns，则存储器的带宽是多少？

解 存储器的带宽指单位时间内从存储器进出信息的最大数量。

存储器带宽 = 1/200ns ×32 位 = 160M 位/秒 = 20MB/秒 = 5M 字/秒，

注：字长 32 位，不是 16 位。（注：1ns=10−9s）。

例 2 一个容量为 16K×32 位的存储器，其地址线和数据线的总和是多少？当选用下列不同规格的存储芯片时，各需要多少片？1K×4 位，2K×8 位，4K×4 位，16K×1 位，4K×8 位，8K×8 位。

解 地址线和数据线的总和 = 14 + 32 = 46 根；

选择不同的芯片时，各需要的片数为：

1K×4：(16K×32) / (1K×4) = 16×8 = 128 片；
2K×8：(16K×32) / (2K×8) = 8×4 = 32 片；
4K×4：(16K×32) / (4K×4) = 4×8 = 32 片；
16K×1：(16K×32)/ (16K×1) = 1×32 = 32 片；
4K×8：(16K×32)/ (4K×8) = 4×4 = 16 片；
8K×8：(16K×32) / (8K×8) = 2×4 = 8 片。

Ex. 1 What is the memory bandwidth? If the width of memory data bus is 32, the access cycle is 200ns, how much is the memory bandwidth?

Solution Memory bandwidth is the maximum in-out amount of information in the unit time.

Memory bandwidth = 1/200ns ×32B = 160MB/S = 20MB/S = 5MW/S,

Note：The word length is 32B, not 16, (Note：1ns = 10−9s).

Ex. 2 There is a memory with capacity of 16K × 32×bit, what is the sum of its address bus and data bus? When we select the following different specifications of the memory chips, how many chips do we need? 1K × 4B 2K × 8B, 4K × 4B, 16K × 1 B, 4K × 8B, 8K × 8B.

Solution the sum of address bus and data bus = 14 + 32 = 46.

When choose different chips, the chip number is：

1K×4：(16K×32) / (1K×4) = 16×8 = 128 chip；
2K×8：(16K×32) / (2K×8) = 8×4 = 32 chip；
4K×4：(16K×32) / (4K×4) = 4×8 = 32 chip；
16K×1：(16K×32)/ (16K×1) = 1×32 = 32 chip；
4K×8：(16K×32)/ (4K×8) = 4×4 = 16 chip；
8K×8：(16K×32) / (8K×8) = 2×4 = 8 chip。

本章重点

1. 存储器的扩展、动态随机存储器的刷新。
2. 高速缓存存储器和虚拟存储器的概念及工作原理。
3. 磁记录原理，磁盘存储器的地址格式。

Key points of this chapter

1. The expansion of memory, dynamic random access memory refresh.
2. The concept and working principle of the cache memory and virtual memory.
3. Magnetic recording principle, the address of the disk storage format.

第四章 指令系统
Chapter 4　Instruction System

本章我们要掌握的内容包括：指令格式、操作数类型、指令和数据的寻址方式、典型指令。

In this chapter, we should grasp the contents including instruction format, operand type, the instruction and data addressing mode, typical instruction.

单词和短语 Words and expressions

复杂指令系统计算机　complex instruction system computer	操作数　operation number
精简指令系统计算机　reduced instruction system computer	直接寻址　direct addressing
	间接寻址　indirect addressing
操作码　operation code	寄存器寻址　register addressing
地址码　address code	相对寻址　relative addressing
指令长度　instruction length	基址寻址　based addressing
	变址寻址　indexed Addressing

基本概念和性质　Basic concepts and properties

1. 指令一般是指微机完成规定操作的命令，一条指令通常由操作码和地址码组成。

Instruction generally refers to the operation command completed by computer, a instruction is generally composed of operation code and address code.

2. 指令系统是计算机所能执行的全部指令的集合，它描述了计算机内全部的控制信息和"逻辑判断"能力。

Instruction set is the full set of instructions performed by computer, which describes all the control information and the logical judgment ability of computer.

3. 计算机在运行过程需要的数据称为操作数。

Operand is the data required by computer when running.

4. 存储器中有许多存放指令或数据的存储单元，每一个存储单元都有一个地址的编号，即地址码。

Memory has many storage unit storing instructions or data, every storage unit has a address number which called address code.

例 题　Example

例 1 某机指令字长 16 位，每个操作数的地址码为 6 位，设操作码长度固定，指令分为零地址、一地址和二地址三种格式。若零地址指令有 M 条，一地址指令有 N 种，则二地址指令最多有几种？若操作码位数可变，则二地址指令最多允许有几种？

解　(1)若采用定长操作码时，二地址指令格式如下：

OP(4 位)	A1(6 位)	A2(6 位)

设二地址指令有 K 种，则：$K = 2^4 - M - N$。

当 M=1(最小值)，N=1(最小值)时，二地址指令最多有：$K_{max} = 16 - 1 - 1 = 14$ 种；

若采用变长操作码时，二地址指令格式仍如(1)所示，但操作码长度可随地址码的个数而变。此时，$K = 2^4 - (N/2^6 + M/2^{12})$；

当 $(N/2^6 + M/2^{12} = 1$ 时 $(N/2^6 + M/2^{12}$ 向上取整)，K 最大，则二地址指令最多有：$K_{max} = 16 - 1 = 15$ 种(只留一种编码作扩展标志用)。

例 2 某 CPU 内有 32 个 32 位的通用寄存器，设计一种能容纳 64 种操作的指令系统。假设指令字长等于机器字长，试回答：如果主存可直接或间接寻址，采用寄存器—存储器型指令，能直接寻址的最大存储空间是多少？画出指令格式并说明各字段的含义。

解 （1）如采用 RS 型指令，则此指令一定是二地址以上的地址格式，指令格式如下：

| OP(6 位) | R(5 位) | I(1 位) | A(20 位) |

操作码字段 OP 占 6 位，因为 $2^6 \geq 64$；
寄存器编号 R 占 5 位，因为 $2^5 \geq 32$；
间址位 I 占 1 位，当 I=0，存储器寻址的操作数为直接寻址，当 I=1 时为间接寻址；
形式地址 A 占 20 位，可以直接寻址 2^{20} 字。

Ex. 1 The instruction word length of some computer is 16 bit, each operand address code is 6 bit. Let the operation code length be fixed, instruction is divided into three formats: zero-address, one-address and two-address. If there are M instructions of zero-address, N instructions of one-address, what are the maximum instructions of two-address.

Solution (1) If we use fixed-length operation code, the instruction format of two-address is as follows

| OP(4 bit) | A1(6 bit) | A2(6 bit) |

Assumed that there are k instructions of two-address, $k = 2^4 - M - N$;
When M = 1 (minimum value), N = 1 (minimum), two-address instruction $K_{max} = 16 - 1 - 1 = 14$;

If we use variable-length operation code, two-address instruction format is still like (1) as shown, but the length of operation code changes with the number of address code. So $K = 2^4 - (N/2^6 + M/2^{12})$;

When $(N/2^6 + M/2^{12}) = 1 (N/2^6 + M/2^{12}$ upward rounding), k has the maximum value, $K_{max} = 16 - 1 = 15$ (leaving only one encoding for expansion flag).

Ex. 2 A CPU has 32 general registers with 32 bit, you are required to design an instruction system containing 64 kinds of operation. Assumed that the length of instruction word is equal to machine word, try to answer the following questions: if main memory can address directly or indirectly, when taking register-memory instruction, what is the maximum storage space for direct addressing? Drawing the instruction format and explaining the meanings of each field.

Solution (1) If taking RS type instruction, the instruction must be address format two-address mode, the instruction format is as follows:

| OP(6 bit) | R(5 bit) | I(1 bit) | A(20 bit) |

Operation code field (OP) is 6 bit, because $2^6 \geq 64$;
Register number R is 5 bit, because $2^5 \geq 32$;
Indirect address (I) is 1 bit, when I=0, the operand of memory addressing is direct addressing, when I=1, it is indirect addressing;
The form address (A) is 20 bit, it can direct address 2^{20} word.

本章重点

1. 指令格式.
2. 指令操作码的扩展技术.
3. 寻址方式.

Key points of this chapter

1. Instruction format.
2. Instruction operation code expansion technology.
3. Addressing method.

第五章　中央处理器
Chapter 5　Central Processing Unit

本章要学习的内容包括：CPU 功能和组成、指令周期、时序产生器、微程序控制器及其设计、硬布线控制器及其设计、传统 CPU、流水 CPU、RISC 的 CPU、多媒体 CPU.

In this chapter, we will study the following contents: CPU functions and composition, instruction cycle, timing generator, micro-program controller and its design, hard-wired controller and its design, traditional CPU, water CPU, RISC CPU, multimedia CPU.

单词和短语 Words and expressions

程序计数器	programming counter	微程序控制器	micro program controller
指令寄存器	instruction register	指令周期	instruction cycle
时序发生器	timing generator	微指令	micro instruction
指令译码器	instruction decoder	硬布线控制器	hard wired controller
地址寄存器	address register	流水 CPU	water CPU
机器周期	machine cycle	多媒体 CPU	multimedia CPU

基本概念和性质　Basic concepts and properties

■1 中央处理器(CPU)是一台计算机的运算核心和控制核心,它是由运算器、控制器、寄存器、总线构成,主要功能是解释计算机指令以及处理计算机中的数据.

The central processing unit (CPU) is the calculation and control core of a computer which is composed of arithmetic unit, controller, register and bus, its main functions are to explain the instructions and address the data in computer.

■2 运算逻辑部件,可以执行定点或浮点算术运算操作、移位操作以及逻辑操作,也可执行地址运算和转换.

Arithmetic logic components can perform fixed-or floating-point arithmetic operations, shift and logical operations and address calculation and conversion.

■3 控制部件主要负责对指令译码,并且发出为完成每条指令所要执行的各个操作的控制信号.

The control unit is mainly responsible for instruction decoding, and it issues the operation control signal for the completion of each instruction to be run.

例　题　Example

例 1　什么是指令周期？指令周期是否有一个固定值？为什么？

解　指令周期是指取出并执行完一条指令所需的时间.

由于计算机中各种指令执行所需的时间差异很大,因此为了提高 CPU 运行效率,即使在同步控制的机器中,不同指令的指令周期长度都是不一致的,也就是说指令周期对于不同的指令来说不是一个固定值.

Ex. 1　What is the instruction cycle? whether instruction cycle is a fixed value? Why?

Solution　Instruction circle is the time that computer get and execute one instruction needed.

Due to the difference of a variety of instruction execution time in computer, in order to improve the efficiency of CPU operation, the length of the instruction cycle of the different instructions is inconsistent. That is to say, the cycles for different instructions is not a fixed value.

本章重点

■1 指令的执行.
■2 时序与控制.

③ 组合逻辑控制器。
④ 微程序控制器。

Key points of this chapter

① Instruction execution.
② Timing and control.
③ Combinational logic controller.
④ Micro-program controller.

第六章 系统总线
Chapter 6   System Bus

本章讲述的内容主要有：总线的概念和结构形态、总线接口、总线的仲裁、总线的定时和数据传送模式、HOST 总线和 PCI 总线、InfiniBand 标准。

This chapter discusses the following contents: the concept and structure of the bus, bus interface, bus arbitration, bus timing and data transfer mode, HOST bus and PCI bus, the InfiniBand standard.

单词和短语 Words and expressions

内部总线　internal bus
系统总线　system bus
I/O 总线　I/O bus
集中式仲裁　centralized arbitration
分布式仲裁　distributed arbitration
HOST 总线　HOST bus
PCI 总线　PCI bus
InfiniBand 标准　the InfiniBand standard

基本概念和性质　Basic concepts and properties

① 总线是构成计算机系统的互联机构，是多个系统功能部件之间进行数据传送的公共通路。借助于总线连接，计算机在各系统功能部件之间实现地址、数据和控制信息的交换，并在争用资源的基础上进行工作。

Bus is the interconnected component in computer system, it is the public access to the data transfer between system function components. With the help of bus, the computer function components can exchange their address, data, control information and can work on the basis of competing resource.

② 一次总线操作所需要的时间称为总线周期。

The time for one bus operation is called a bus cycle.

③ 总线带宽是指单位时间内总线上可传输的数据量。

Bus bandwidth is the amount of data which can be transmitted on the bus in unit time.

例题 Example

例 1　为什么要设置总线判优控制？常见的集中式总线控制有几种？各有何特点？哪种方式响应时间最快？哪种方式对电路故障最敏感？

解　总线判优控制解决多个部件同时申请总线时的使用权分配问题；

常见的集中式总线控制有三种：链式查询、计数器定时查询、独立请求；

特点：链式查询方式连线简单，易于扩充，对电路故障最敏感；计数器定时查询方式优先级设置较灵活，对故障不敏感，连线及控制过程较复杂；独立请求方式速度最快，但硬件器件用量大，连线多，成本较高。

例 2　画一个具有双向传输功能的总线逻辑图。

解　在总线的两端分别配置三态门，就可以使总线具有双向传输功能。

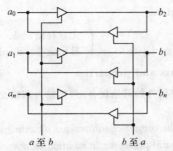

Ex. 1 Why do we set the bus arbiter control? What are the several common centralized bus control modes? What are their characteristics? Which modes have the fastest response time? Which mode is the most sensitive for circuit fault?

Solution Bus arbiter control is aim to deal with the right to use bus when several components applying for bus.

There are three centralized bus control modes: chain query, counter timer query, independent request;

Characteristic: chain query has the following characteristics: simple connection, easy to expand, most sensitive to the circuit failure; counter timer query has the following characteristics: priority setting is flexible, not sensitive to the fault, connection and control process is more complex; independent request has the following characteristics: it has the fastest response time, but it needs more hardware devices with high cost.

Ex. 2 Draw a bus logic diagram with two-way transfer function.

Solution Configure the tri-state gate at both ends of the bus, so you can make the bus with bi-directional transmission function.

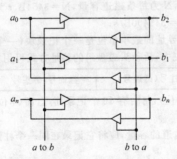

本章重点

系统总线的类型、结构、控制方式和通信方式.

Key points of this chapter

System bus type, structure, control mode and communication mode.

第七章 外围设备
Chapter 7 Peripheral Equipment

本章需要了解的内容主要有:磁盘存储设备、磁带存储设备、显示设备、输入设备和打印设备.

In this chapter, the students should grasp the following contents: disk storage devices, magnetic tape storage devices, display devices, input devices and printing devices.

单词和短语 Words and expressions

磁盘存储设备　disk storage device　　　　光盘存储设备　optical disk storage device
磁带存储设备　magnetic tape storage device　　显示设备　display device

基本概念和性质　Basic concepts and properties

1. 外围设备是指计算机系统中除主机外的其他设备.包括输入和输出设备、外存储器、模数转换器、数模转换器、外围处理机等.

Peripheral equipments refers to the computer equipments including input and output devices, external memory, ADC, DAC and peripheral processor in addition to host.

例题 Example

例 1　磁盘组有 6 片磁盘,每片有两个记录面,最上最下两个面不用.存储区域内径 22cm,外径 33cm,道密度为 40 道/cm,内层位密度 400 位/cm,转速 6000 转/分.问:

(1)共有多少柱面?
(2)盘组总存储容量是多少?
(3)数据传输率多少?
(4)采用定长数据块记录格式,直接寻址的最小单位是什么?寻址命令中如何表示磁盘地址?
(5)如果某文件长度超过一个磁道的容量,应将它记录在同一个存储面上,还是记录在同一个柱面上?

解　(1)有效存储区域 $=16.5-11=5.5(cm)$,因为道密度 $=40$ 道/cm,所以 $40\times 5.5=220$ 道;

(2)内层磁道周长为 $2\pi R=2\times 3.14\times 11=69.08(cm)$;

每道信息量 $=400$ 位/cm$\times 69.08$cm$=27632$ 位 $=3454B$,每面信息量 $=3454B\times 220=759880B$,盘组总容量 $=759880B\times 10=7598800B$;

(3)磁盘数据传输率 $Dr=r\times N$,N 为每条磁道容量,$N=3454B$,r 为磁盘转速,$r=6000$ 转/60 秒 $=100$ 转/秒,$Dr=r\times N=100\times 3454B=345400B/s$;

(4)采用定长数据块格式,直接寻址的最小单位是一个记录块(一个扇区),每个记录块记录固定字节数目的信息,在定长记录的数据块中,活动头磁盘组的编址方式可用如下格式:

台号	柱号(磁道号)	盘面号/磁头号	扇区号

此地址格式表示有 4 台磁盘(2 位),每台有 16 个记录面/盘面(4 位),每面有 256 个磁道(8 位),每道有 16 个扇区(4 位);

(5)如果某文件长度超过一个磁道的容量,应将它记录在同一个柱面上,因为不需要重新找道,数据读/写速度快.

Ex. 1 The disk group has six disks, each disk has two recording surfaces, the two surfaces at the top and at the bottom have no storage area. The inside diameter of bit storage area is 22cm, outside diameter is 33cm, the road density is 40/cm, the inner bit density is 400/cm, the speed is 6000 rev/min. The question is:

(1) What is the total number of cylinders?
(2) What is total storage capacity of disk group?
(3) What is the data transfer rate?
(4) If we use fixed-length data blocks recording format, what is the smallest unit of direct addressing? How to express the disk address with addressing command.
(5) If the length of a file is longer then the capacity a track, it should be recorded in the same storage surface, or recorded in the same cylinder?

Solution (1) The effective storage area = 16.5 − 11 = 5.5(cm), because track density = 40/cm, so the total number of cylinders = 40 × 5.5 = 220;

(2) The inner track circumference = $2\pi R = 2 \times 3.14 \times 11 = 69.08(cm)$;
The amount of information per channel = 400/cm × 69.08cm = 27632/cm = 3454B,
The amount of information per surface = 3454B × 220 = 759880B,
The total capacity of disk group = 759880B × 10 = 7598800B;

(3) Disk data transfer rate $Dr = r \times N$, N is capacity of each track, N = 3454B, r is disk rotational speed, r = 6000 rev seconds = 100 rev/sec, $Dr = r \times N = 100 \times 3454B = 345400$ B/s;

(4) The smallest unit of directly addressing is a record block (a sector) when using fixed-length data block format, each block records a fixed number bytes. In the fixed record data block, the addressing mode of active head disk group is available to the following format:

station number	column number /track number	disk number / head number	sector number

This address format shows that there are 4 disk (2 bit), each disk has 16 record surface (4 bit), each surface has 256 track (8 bit), each track has 16 sector (4 bit);

(5) If the length of a file is longer than the capacity of a track, it should be recorded in the same cylinder. The data read/write is fast because it need not to re-find the track.

本章重点

1. 磁盘记录原理.
2. 显示器显示字符的原理.
3. 打印机打印原理.

Key points of this chapter

1. Disk recording principle.
2. Principle of display characters.
3. Printer principle.

第八章 输入输出系统
Chapter 8 Input/Output System

本章学习的主要内容有：外围设备的定时方式和信息交换方式、程序查询方式、程序中断方式、DMA 方式、通道方式、通用 I/O 标准接口.

In this chapter, students should grasp the following contents: timing mode and information exchange mode of peripherals, program query mode, program interrupt mode DMA mode, channel mode, general I/O standard interface.

单词和短语 Words and expressions

程序查询方式　process query　　　　通道方式　channel mode
程序中断方式　program interrupt　　　中断控制器　interrupt controller
DMA　direct memory access　　　　　I/O 接口标准　I/O interface standard

基本概念和性质 Basic concepts and properties

1. 中断(Interrupt)是指 CPU 暂时中止现行程序,转去处理随机发生的紧急事件,处理完后自动返回原程序的功能和技术.
Interrupt is the function technology that CPU suspend the current program to deal with the emergency events and automatically returned to the original program after finished.

2. 直接存储器访问(Direct Memory Address, DMA)方式是为了在主存储器与 I/O 设备间高速交换批量数据而设置的.基本思想是:通过 DMA 控制器控制实现主存与 I/O 设备间的直接数据传送,在传送过程中无需 CPU 的干预.

DMA (Direct Memory Address) is set for high-speed exchange of data between main memory and I/O devices. The basic idea is that the DMA controller control the direct data transfer between main memory and I/O device without CPU intervention.

例 题   Example

例 1 现有 A、B、C、D 四个中断源,其优先级由高向低按 A、B、C、D 顺序排列.若中断服务程序的执行时间为 $20\mu s$,请根据下图所示时间轴给出的中断源请求中断的时刻,画出 CPU 执行程序的轨迹.

解 A、B、C、D 的响优先级即处理优先级.CPU 执行程序的轨迹图如下:

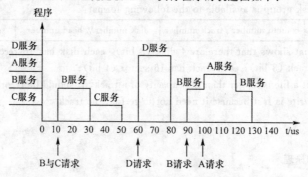

Ex. 1 Existing A, B, C, D four interrupt sources, the priority order from high to low is A, B, C and D. If the interrupt service routine execution time is $20\mu s$, then draw the trace of CPU execution according to the below figure which shows the timeline and the moment of interrupt source requesting interrupt.

Solution The response priority of A, B, C and D is the processing priority. The trace of CPU executing program is as follows:

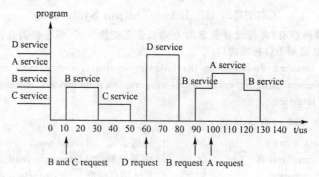

本章重点

1. DMA 的基本接口及程序流程.
2. 程序中断方式的有关概念、中断优先权排队电路、中断方式的接口.
3. 三种 DMA 传送方式,DMA 的周期挪用方式操作过程.

Key points of this chapter

1. The basic interface and the program flow of DMA.

2. Interrupt concept, interrupt priority queuing circuit, interrupt interface.
3. Three DMA transfer, DMA circle shift method procedure.

保险精算专业课程

第一部分 风险理论
Part 1 Risk Theory

第一章 效应理论与保险
Chapter 1 Utility theory and insurance

保险人的存在是可以用期望效用模型进行解释的一个最好的例子。在期望效用模型中,被保险人是一个风险厌恶型的理性决策者,为广争取一个安全的金融地位,由 Jensen 不等式,他会心甘情愿地付给保险人比自己面临的理赔额期望值多的保费。这是一个要在不确定的情况下进行决策的体系,在这一体系下,被保险人做决策凭借的不是直接比较赔付额期望的大小,而是比较与赔付额密切相关的期望效用。

The very existence of insurers can be explained by way of the expected utility model. In this model, an insured is a risk averse and rational decision maker, who by virtue of Jensen's inequality is ready to pay more than the expected value of his claims just to be in a secure financial position. The mechanism through which decisions are taken under uncertainty is not by direct comparison of the expected payoffs of decisions, but rather of the expected utilities associated with these payoffs.

单词和短语 Words and expressions

圣·彼得堡悖论 St. Petersburg paradox 　　风险喜好 risk loving
边际效用递减 diminishing marginal utility 　Jensen 不等式 Jensen's inequality
期望效用 expected utility 　　　　　　　　　绝对风险指数 absolute risk index
风险中性 risk neutral 　　　　　　　　　　　相对风险指数 relative risk index
风险厌恶 risk aversion 　　　　　　　　　　 临界保费 critical premium

基本概念和性质 Basic concepts and properties

① Jensen 不等式:如果 $v(x)$ 是一个凸函数,Y 是一个随机变量,则
$$E[v(Y)] \geqslant v(E[Y]),$$
其中等号成立当且仅当 $v(\cdot)$ 在 Y 的支撑集上是线性的或 $Var(Y) = 0$。

Jensen's inequality: If $v(x)$ is a convex function and Y is a random variable, then
$$E[v(Y)] \geqslant v(E[Y]),$$
with equality if and only if $v(\cdot)$ is linear on the support of Y or $Var(Y) = 0$.

② 停止损失的再保险的最优性:用 $I(X)$ 记当损失为 $X(X \geqslant 0)$ 时,某再保险合同约定的理赔支付。假设 $0 \leqslant I(X) \leqslant x$ 对于任意 $x \geqslant 0$ 成立,则
$$E[I(X)] = E[(X-d)_+] \Rightarrow Var[X - I(X)] \geqslant Var[X - (X-d)_+].$$

Optimality of stop-loss reinsurance: Let $I(X)$ be the payment on some reinsurance contract if the loss is X, with $X \geqslant 0$. Assume that $0 \leqslant I(x) \leqslant x$ holds for all $x \geqslant 0$. Then,
$$E[I(X)] = E[(X-d)_+] \Rightarrow Var[X - I(X)] \geqslant Var[X - (X-d)_+].$$

练 习

1. 证明对厌恶风险的保险人，$P^- \geq E[X]$.

   答案 对 $U(W) = E[U(W + P^- - X)]$ 应用 Jensen 不等式.

2. 一个保险人承保一个风险 X，收保费后，他拥有财富 $w = 100$. 如果他使用的效用函数为 $u(w) = \log(w)$，且 $\Pr[x = 0] = \Pr[x = 36] = 0.5$，那么为了让再保险人承担所有的风险，该保险人乐意支付给保险人的最大保费是多少？

   答案 $P^+ = 20$.

3. 一个保险人使用指数效用函数，以保费 P 承保服从 $N(1000, 100^2)$ 分布的风险. 已知 $P \geq 1250$，求风险厌恶系数 α？如果风险 X 的量纲为"元"，那么 α 的量纲是什么？

   答案 $\alpha \geq 0.5$. α 的量纲为"元$^{-1}$".

Exercises

1. Prove that $P^- \geq E[X]$ for risk averse insurers.

   Answer Apply Jensen's inequality to $U(W) = E[U(W + P^- - X)]$.

2. An insurer undertakes a risk X and after collecting the premium, he owns a capital. What is the maximum premium the insurer is willing to pay to a reinsurer to take over the complete risk, if his utility function is and $\Pr[X = 0] = \Pr[X = 36] = 0.5$ determine the exact value.

   Answer $P^+ = 20$.

3. For the premium P an insurer with exponential utility function asks for a $N(1000, 100^2)$ distributed risk it is known that $P \geq 1250$. What can be said about his risk aversion α if the risk X has *dimension* 'money', then what is the dimension of α?

   Answer $\alpha \geq 0.5$. Dimension of α is (unit of money)$^{-1}$.

第二章 个体风险模型
Chapter 2 The individual risk model

在个体风险模型以及下面要提到的聚合风险模型中，关于保险合同的一个风险组合的总理赔额常常被表示为一个具有一定含义的随机变量. 举例来说，我们需要计算事件"一定的资金足以赔付这些理赔"的概率，或者计算该风险组合的水平 95%. 的在险价值(VaR)，即该风险组合累计分布函数的 95% 的分位点. 在个体模型中，理赔总额被建模为由各保单生成的独立理赔变量之和. 这些理赔往往不可以假设为纯离散或者纯连续的随机变量. 对此我们将提出能够涵盖上述两种情况的一个记号. 尽管个体风险模型是最现实合理的. 但是由于我们可获得的是被取整后的数据. 而且不总是具有密度函数，该模型有时使用起来很不方便. 为了获得这一模型下的许多结果，我们将研究不同于卷积的其它手法，如在一些特殊场合我们采用类似于矩母函数的那样变换会有益于问题的解决. 我们也给出基于分布函数适当阶矩的逼近手法. 由于中心极限定理只涉及到两个矩，对于一些右尾很重的分布函数来说，它不可能提供足够的精度. 于是我们尝试两种涉及到分布函数三个矩的更精确的方法，它们是平移伽玛近似和正态功效近似.

In the individual risk model, as well as in the collective risk model that follows below, the total claims on a portfolio of insurance contracts is the random variable of interest. We want to compute, for instance, the probability that a certain capital will be sufficient to pay these claims, or the value-at-risk at level 95% associated with the portfolio, being the 95% quantile of its cumulative distribution function (cdf). The total claims is modelled as the sum of all claims on the policies, which are assumed independent. Such claims cannot always be modelled as purely discrete random variables, nor as purely continuous ones, and we provide a notation that encompasses both these as special cases. The individual model, though the most realistic possible, is not always very convenient, because the available data is used integrally and not in any way condensed. We study other techniques than convolution to obtain results in this model. Using transforms like the moment generating function helps in some special cases. Also, we present approximations based on fitting moments of the distribution. The Central Limit Theorem, which involves fitting two moments, is not sufficiently accurate in the important righthand tail of the distribution. Hence, we also look at two more refined methods using three moments: the translated gamma approximation and the normal power approximation.

单词和短语 Words and expressions

个体风险模型　individual risk model
卷积　convolution
矩母函数　moment generating function
安全附加保费　safety additional premium
相对安全附加保费　relatively safe additional premium
安全附加保费率　safety additional premium rate
最优再保险　optimal reinsurance

练 习

1. 设理赔概率为 0.1, 求以下两种情形下风险 $X = IB$ 的均值与方差: (1) B 以概率 1 等于 5 (2) $B \sim U(0,10)$.

答案　1) $E[X] = 1/2; Var[X] = 9/4$; 2) $E[X] = 1/2; Var[X] = 37/12$.

2. 设 $I \sim Bernoulli(q)$, b 是固定的实数, 试确定型如 Ib 风险的偏度. 何时是对称的?

答案　$(1-2q)/\sqrt{q(1-q)}$. 如果 X 是对称的, 则 3 阶中心矩等于 0, 所以 $q \in \{0, 0.5, 1\}$. 对 q 的这 3 种取值, I 皆为对称的.

Exercises

1. Determine the expected value and the variance of $X = IB$ if the claim probability equals 0.1. First, assume that B equals 5 with probability 1. Then, let $B \sim U(0,10)$.

Answer　1) $E[X] = 1/2; Var[X] = 9/4$; 2) $E[X] = 1/2; Var[X] = 37/12$.

2. Determine the skewness of a risk of the form Ib where $I \sim Bernoulli(q)$ and is a fixed amount. For which values of q and is the skewness equal to zero, and for which of these values is I actually symmetrical?

Answer　$(1-2q)/\sqrt{q(1-q)}$. If X is symmetrical, then the third central moment equals 0, therefore $q \in \{0, 0.5, 1\}$ must hold. Symmetry holds for all three of these $q-value$.

第三章　聚合风险模型
Chapter 3　Collective risk models

聚合风险模型是把风险组合理解为在随机时间点上产生的理赔全体. 聚合风险模型的主要优点是在计算上它是一个很有效的模型, 该模型也非常接近实际. 聚合模型忽略了一些保单信息. 如果一个保单组合只含有一个可能产生高理赔的保单, 那么该项在个体风险模型至多会出现一次, 而在聚合模型它可以出现若干次. 此外, 在聚合模型中我们要求理赔次数 N 理赔额 X_i 之间相互独立, 这对于汽车保险行业来说就多少有些不妥了, 例如恶劣的天气条件会导致大量的小理赔. 不过, 在实际中这些现象的影响是很小的.

Collective risk models regard the portfolio as a collective that produces a claim at random points in time. The main advantage of a collective risk model is that it is a computationally efficient model, which is also rather close to reality. In collective models, some policy information is ignored. If a portfolio contains only one policy that could generate a high claim, this term will appear at most once in the individual model. In the collective model, however, it could occur several times. Moreover, in collective models we require the claim number N and X_i the claim amounts to be independent. This makes it somewhat less appropriate to model a car insurance portfolio, since for instance bad weather conditions will cause a lot of small claim amounts. In practice, however, the influence of these phenomena appears to be small.

单词和短语 Words and expressions

复合分布　compound distribution
泊松分布　poisson distribution
复合泊松分布　compound possion distribution
Panjer 递推　panjer recursion
聚合风险模型　collective risk model
理赔额分布　claim size distributions
理赔次数分布　distributions for the number of claims
停止损失保险　stop-loss Insurance
停止损失保费　stop-loss premiums

基本概念和性质　Basic concepts and properties

1 理赔次数服从独立泊松分布：设 S 服从复合泊松分布，其中参数为 λ，理赔分布是有个离散型分布，满足

$$\pi_i = p(x_i) = \Pr[X = x_i], i = 1, 2, \ldots, m.$$

如果把 S 写成 $S = x_1 N_1 + x_2 N_2 + \cdots + x_m N_m$ 那样，其中 N_i 表示理赔额 x_i 的发生次数（即 S 的和式里面值 x_i 出现的次数），那么 $N_1, \cdots, N_m$ 那么构成一列独立 $Poisson(\lambda \pi_i)$ 随机变量 $i = 1, \cdots, m$.
Frequencies of claim sizes are independent Poisson: Assume that S is compound Poisson distributed with parameter λ and with discrete claims distribution

$$\pi_i = p(x_i) = \Pr[X = x_i], i = 1, 2, \ldots, m.$$

If S is written as $S = x_1 N_1 + x_2 N_2 + \cdots + x_m N_m$, where N_i denotes the frequency of the claim amount x_i i.e., the number of terms in S with value x_i, then $N_1, \cdots, N_m$ are independent and $Poisson(\lambda \pi_i)$, $i = 1, \cdots, m$, distributed random variables.

2 经验法则（停止损失保费的比）：当自留额 t 大于期望值 $\mu = E[U] = E[W]$ 时，风险变量 U 和 W 的停止损失保费比满足：

$$\frac{E[(U-t)_+]}{E[(W-t)_+]} \approx \frac{Var[U]}{Var[W]}.$$

当 t 的值既不大也不太小的是该法则运行效果很好.
Rule of thumb (Ratio of stop-loss premiums): For retentions t larger than the expectation $\mu = E[U] = E[W]$, we have for the stop-loss premiums of risks U and W:

$$\frac{E[(U-t)_+]}{E[(W-t)_+]} \approx \frac{Var[U]}{Var[W]}.$$

This rule works best for intermediate values of t.

练　习

1. 设一个保单组合包含 100 个有一年期寿险保单，这些保单按照承保额度 1 和 2 以本年死亡概率 0.01 和 0.02 均摊. 试求理赔总额 $\tilde{S}$ 的期望和方差. 选择一个合适的复合泊松随机变量 S 去近似 $\tilde{S}$，并比较它们的期望和方差. 当用适当的平移伽马分布对 S 和 $\tilde{S}$ 作近似时，如何选取参数？

答案　$\tilde{S}: E = 2.25, V = 3.6875, \gamma = 6.41655\sigma^{-3} = 0.906$. $S \sim$ 复合泊松分布，其中 $\lambda = 1.5, p(1) = p(2) = 0.5, E = 2.25, V = 3.75, \gamma = 6.75\sigma^{-3} = 0.930$. $\tilde{S}: \alpha = 4.871, \beta = 1.149, x_0 = -1.988$. S:

$\alpha = 4.630, \beta = 1.111, x_0 = -1.917$.

2. 设一个保单组合由 n 个合同构成，每个合同产生一个大小为1的理赔的概率为 q 试分别在个体模型、聚合模型以及开放聚合模型下求总理赔的分布函数。当 $n \to \infty$ 而 q 固定时，个体模型 S 是否收敛于聚合模型 T，换句话说，概率 $\Pr[(S-E[S])/\sqrt{Var[S]} \leqslant x] - \Pr[(T-E[T])/\sqrt{Var[S]} \leqslant x]$ 是否收敛于0？

答案　　$B(n,q), Possion(nq), Poisson(-n\log(1-q))$，否。

Exercises

1. Consider a portfolio of 100 one-year life insurance policies which are evenly divided between the insured amounts 1 and 2 and probabilities of dying within this year 0.01 and 0.02. Determine the expectation and the variance of the total claims $\tilde{S}$. Choose an appropriate compound Poisson Distribution S to $\tilde{S}$ approximate and compare the expectations and the variances. Determine for both S and $\tilde{S}$ the parameters of a suitable approximating translated gamma distribution.

Answer　$\tilde{S}: E = 2.25, V = 3.6875, \gamma = 6.41655\sigma^{-3} = 0.906$. $S \sim$ compound Poisson with $\lambda = 1.5, p(1) = p(2) = 0.5$, therefore $E = 2.25, V = 3.75, \gamma = 6.75\sigma^{-3} = 0.930$. $\tilde{S}: \alpha = 4.871, \beta = 1.149, x_0 = -1.988$. $S: \alpha = 4.630, \beta = 1.111$ and $x_0 = -1.917$.

2. Consider a portfolio containing n contracts that all produce a claim 1 with probability q. What is the distribution of the total claims according to the individual model, the collective model and the open collective model? If $n \to \infty$ with q fixed, does the individual model S converge to the collective model T, in the sense that the difference of the probabilities
$\Pr[(S-E[S])/\sqrt{Var[S]} \leqslant x] - \Pr[(T-E[T])/\sqrt{Var[S]} \leqslant x]$ converges to 0?

Answer　$B(n,q), Possion(nq), Poisson(-n\log(1-q))$, no.

第四章　破产理论
Chapter 4　Ruin theory

在破产模型中需要研究的问题是保险人运营的稳健性. 设逐年收取的保费固定不变，保险人在时间 $t=0$ 的初始资本为 u，他的资本金会随着时间线性递增，但每当一个理赔发生时，资本金过程就会有一个下跳. 如果在某个时刻该资本金过程取负值，我们便说破产事件发生了. 假设年保费和理赔过程保持不变，当研究保险人的资产与他的负债是否匹配时，破产概率是一个很好的指标. 如果其资产与负债匹配得不好，那么保险人会采取一些措施来加以调整，如增加再保份额. 提高保费或者增加初始资本金.

只有当理赔额分布为指数分布的混合或线性组合时，计算破产概率的解析方法才是有效的. 当赔额服从离散分布而且没有太多的支撑点时，可以编制一些计算程序来实现. 我们还可以给出破产概率精确的上下界估计. 在很多场合下，我们更关心的是破产概率的一个简单的指数型上界，即 Lundberg 界，而不是破产概率本身.

In the ruin model the stability of an insurer is studied. Starting from capital u at time $t = 0$ his capital is assumed to increase linearly in time by fixed annual premiums, but it decreases with a jump whenever a claim occurs. Ruin occurs when the capital is negative at some point in time. The probability that this ever happens, under the assumption that the annual premium as well as the claim generating process remain unchanged, is a good indication of whether the insurer's assets are matched to his liabilities sufficiently well. If not, one may take out more reinsurance, raise the premiums or increase the initial capital.

Analytical methods to compute ruin probabilities exist only for claims distributions that are mixtures and or combinations of exponential distributions. Algorithms exist for discrete distributions with not too many mass points. Also, tight upper and lower bounds can be derived. Instead of looking at the ruin probability, often one just considers an upper bound for it with a simple exponential structure (Lundberg).

单词和短语 Words and expressions

盈余过程　surplus process
破产时刻　ruin time
破产概率　ruin probability
总理陪过程　prime minister with process
理赔次数过程　settling times process
最大损失　maximum loss
调节系数　accommodation coefficient

基本概念和性质　Basic concepts and properties

1 调节系数：设理赔 $X \geqslant 0$ 满足 $E[X] = \mu_1 > 0$，我们称关于 r 的方程
$$1 + (1+\theta)\mu_1 r = m_X(r)$$
的正解 R 为 X 的调节系数。

The adjustment coefficient R for claims $X \geqslant 0$ with $E[X] = \mu_1 > 0$ is the positive solution of the following equation in r: $1 + (1+\theta)\mu_1 r = m_X(r)$.

2 破产概率的 Lundberg 型指数界：设在一个复合泊松风险的过程中，初始资本金为 u，单位时间的保费为 c，理赔分布及矩母函数分别为 $P(\cdot)$ 和 $m_X(t)$ 并且调节系数 R 满足 $1+(1+\theta)\mu_1 r = m_X(r)$，我们有如下关于破产概率的不等式：
$$\varphi(u) \leqslant e^{-Ru}.$$

Lundberg's exponential bound for the ruin probability: For a compound Poisson risk process with an initial capital u, a premium per unit of time c, claims with cdf $P(\cdot)$ and mgf $m_X(t)$ and an adjustment coefficient R that satisfies $1+(1+\theta)\mu_1 r = m_X(r)$, we have the following inequality for the ruin probability:
$$\varphi(u) \leqslant e^{-Ru}$$

3 破产概率：设初始资本金 $u \geqslant 0$ 则破产概率满足
$$\varphi(u) = \frac{e^{-Ru}}{E[e^{-RU(T)} \mid T < \infty]}.$$

Ruin probability: The ruin probability for $u \geqslant 0$ satisfies
$$\varphi(u) = \frac{e^{-Ru}}{E[e^{-RU(T)} \mid T < \infty]}$$

4 破产时刻资金分布：如果初始资本金等于 0，那么对所有的 $y > 0$ 我们有
$$\Pr[U(T) \in (-y-dy, y), T < \infty] = \frac{\lambda}{c}[1-P(y)]dy.$$

Distribution of the capital at time of ruin: If the initial capital u equals 0, then for all $y > 0$ we have:
$$\Pr[U(T) \in (-y-dy, y), T < \infty] = \frac{\lambda}{c}[1-P(y)]dy.$$

练 习

1. 在一个 $\theta = \frac{2}{5}$ 的破产过程中理赔按照如下方式形成：首先，Y 分别以概率 $\frac{1}{2}$ 取两个可能的值 3 和 7. 然后在条件 $Y = y$ 下理赔 X 服从一个 $Exp(y)$ 分布求调节系数 R. 如果对同样的分布有 $R = 2$，那么 θ 会大于还是小于 $\frac{2}{5}$？

答案　$R = 1; \theta = 1.52 > 0.4$.

Exercises

1. Assume that the claims X in a ruin process with $\theta = \dfrac{2}{5}$ arise as follows: first, a value Y is drawn from two possible values 3 and 7, each with probability $\dfrac{1}{2}$. Next, conditionally on $Y = y$, the claim X is drawn from an exponential (y) distribution. Determine the adjustment coefficient R. If $R = 2$ for the same distribution, is θ larger or smaller than $\dfrac{2}{5}$?

   Answer $R = 1; \theta = 1.52 > 0.4$.

第二部分 复利数学
Part 2 Compound Interest Mathematics

引 言

复利数学又称利息理论,是保险精算理论的基础和主要工具。复利数学主要介绍了利率的度量工具,年金的理论和应用,如何计算收益率以及债券、股票的定价原理等内容。这些内容在日常经济生活中有着广泛的应用,在投资理财等方面具有重要的指导作用。

Introduction

Compound interest mathematics is also called interest theory. It's the foundation and major tool of acutrial theory. It introduces the measurement tool of interest rate, the theory and application of annuity, the concept and computing method of yield rate, the pricing principle of bonds and shares and etc.. All the contents are applied widely in daily economic business, and also play important roles in financial investment.

第一章 利息的基本概念
Chapter 1 Basic Concepts of Interest

本章是复利数学的基础,主要介绍了利息度量的相关概念。包括单利、复利、实际利率、名义利率、贴现、实际贴现率、名义贴现率和利息强度等基本的、重要的概念和相关的计算公式。之后,本章也介绍了如何利用价值方程来求解利息基本问题。

This chapter is the foundation of compound interest mathematics. Some conceptions about interest measurement have been introduced that conclude simple interest, compound interest, effective rate of interest, discount, nominal rate of interest, effective rate of discount, nominal rate of discount, force of discount and so on. It's also discussed in this chapter that how to use the equations of value to resolve the interest basic problems.

单词和短语 Words and expressions

单利	simple interest	积累值	accumulation
复利	compound interest	实际利率	effective rate of interest
积累函数	accumulated function	名义利率	nominal rate of interest
贴现	discount	实际贴现率	effective rate of discount
单贴现	simple discount	名义贴现率	nominal rate of discount
复贴现	compound discount	利息强度	force of interest
贴现因子	discount factor	贴现强度	force of discount
利息	interest	价值方程	equation of value
本金	principle	严格单利法	exact simple interest
投资期	term	常规单利法	ordinary simple interest
计息期	period	银行家法则	banker's rule
利率	rate of interest	等时间法	equated time method
现值	present value	72律	rule of 72

基本概念和性质 Basic concepts and properties

■ 若以一个单位本金在初始时刻进行投资,定义该投资在 t 时刻的积累值用积累函数 $a(t)$ 表示,且 $a(0) = 1$.

Denote the accumulation at time t by the accumulated function $a(t)$ with the initial investment 1 and $a(0) = 1$.

2 考虑一个单位本金的投资，
(1)若在时刻 t 投资积累值为 $a(t) = 1 + it$，则称该投资以每期单利 i 计息，以这样方式产生的利息为单利。
(2)若在时刻 t 投资积累值为 $a(t) = (1+i)^t$，则称该投资以每期复利 i 计息，以这样方式产生的利息为复利。

Consider the initial investment 1,
(1) If the accumulated value at time t can expressed as $a(t) = 1 + it$, this investment is paid simple interest at the rate i and this kind of interest is known as simple interest.
(2) If the accumulated value at time t can expressed as $a(t) = (1+i)^t$, this investment is paid compound interest at the rate i and this kind of interest is known as compound interest.

3 第 n 期的实际利率　　$i_n = \dfrac{a(n) - a(n-1)}{a(n-1)} = \dfrac{A(n) - A(n-1)}{A(n-1)}$.

The effective rate of interest of the nth period is
$$i_n = \dfrac{a(n) - a(n-1)}{a(n-1)} = \dfrac{A(n) - A(n-1)}{A(n-1)}.$$

4 第 n 期的实际贴现率　　$d_n = \dfrac{a(n) - a(n-1)}{a(n)} = \dfrac{A(n) - A(n-1)}{A(n)}$.

The effective rate of discount of the nth period is
$$d_n = \dfrac{a(n) - a(n-1)}{a(n)} = \dfrac{A(n) - A(n-1)}{A(n)}.$$

5 一个计息期内计息 m 次的名义利率和实际利率的关系
$$1 + i = \left[1 + \dfrac{i^{(m)}}{m}\right]^m.$$

The relation between the effective rate of interest and nominal rate of interest per unit time payable $m$thly is $1 + i = \left[1 + \dfrac{i^{(m)}}{m}\right]^m$.

6 一个计息期内计息 m 次的名义贴现率和实际贴现的关系
$$1 - d = \left[1 - \dfrac{d^{(m)}}{m}\right]^m.$$

The relation between the effective rate of discount and nominal rate of discount per unit time payable $m$thly is $1 - d = \left[1 - \dfrac{d^{(m)}}{m}\right]^m$.

7 定义时刻 t 的利息强度为 $\delta_t = \dfrac{A'(t)}{A(t)} = \dfrac{a'(t)}{a(t)}$.

Denote the force of interest at time t by $\delta_t = \dfrac{A'(t)}{A(t)} = \dfrac{a'(t)}{a(t)}$.

本章重点
1 掌握单利、复利、单贴现和复贴现等基本概念。
2 掌握等价的实际利率、名义利率、实际贴现率、名义贴现率及利息强度的关系。
3 会用价值方程进行简单的利息基本问题计算。

Key points of this chapter
1 Master some basic concepts such as simple interest, compound interest, simple discount, compound discount and etc..
2 Master the relations about the equal effective rate of interest, nominal rate of interest, effective rate of discount, nominal rate of discount and interest force.

3 Learn to compute some simple interest basic problems by equations of value.

第二章 年金
Chapter 2 Annuity

本章重点介绍了年金的概念及不同类型年金的现值和积累值的计算方法. 年金的分类较多, 按确定的数额进行一系列支付的年金称为确定年金; 在未来相应的时间点上支付是否发生并不确定的年金被称为不确定年金. 按支付时间的不同可以分为期初年金和期末年金. 年金的应用十分广泛, 例如银行存款的零存整取方式, 住房按揭还款, 房租, 养老金的支付方式等都是年金的形式.

This chapter mainly introduces the concept and computing methods of present value and accumulated value of different kinds of annuities. There are many sorts of annuities. It's called annuity-certain if there is exact value of a series of payments and contingency annuity if it can not make sure that there are still payments in the future. In advance, immediate annuity and due annuity can be denoted according to the time of the series of payments. The application of annuity is very wide in the daily lives such as the saving method of fixed deposits by installment, the subsidy, the rent, the old-age pension and etc..

单词和短语 Words and expressions

年金	annuity	递延年金	deferred annuity
确定年金	annuity-certain	永续年金	perpetuity
或有年金	contingency annuity	连续年金	continuous annuity
期末年金	immediate annuity	递增年金	increasing annuity
期初年金	due annuity	递减年金	decreasing annuity

例 题 Example

例 1 在半年名义利率 9% 的条件下, 求每半年末支付 500 美金的 20 年年金的现值.

解 $PV = 500 a_{\overline{40}|0.045} = 500 \times 18.4016 = 9200.80$.

Ex. 1 Find the present value of an annuity which pays MYM500 at the end of every half-year for 20 years if the rate of interest is 9% convertible semiannually.

Solution $PV = 500 a_{\overline{40}|0.045} = 500 \times 18.4016 = 9200.80$.

例 2 季名义利率 8% 时, 按期末年金形式投资 1000 美金, 若投资期是 10 年, 求每季度末应支付多少钱?

解 每季度实际利率 = 2%,

$$1000 = R a_{\overline{40}|0.02}$$

$$R = \frac{1000}{a_{\overline{40}|0.02}} = \frac{1000}{27.3555} = 36.56.$$

Ex. 2 If a person invests MYM1000 at 8% per annum convertible quarterly, how much can be withdrawn at the end of every quarter to use up the fund exactly at the end of 10 years?

Solution The effective rate of interest every quarter-year = 2%, then

$$1000 = R a_{\overline{40}|0.02}$$

$$R = \frac{1000}{a_{\overline{40}|0.02}} = \frac{1000}{27.3555} = 36.56.$$

本章重点

1 掌握确定年金的概念和计算方法.
2 了解各种年金之间的关系.
3 学习年金在实际生活中的应用.

Key points of this chapter

1 Master the concept and computing method of annuity-certain.

2 Make sure about the relations among all kinds of annuities.
3 Learn the applications of annuity in practice.

第三章 收益率
Chapter 3 Yield Rate

在各种复杂的经济活动中,例如银行存款,购买年金保险,投资股票等等,人们都希望获得高收益的回报.因此,如何评价投资方案成为了一个非常重要的问题.本章将主要介绍评估收益率的现金流分析法.

In people's some complicated economic activities, such as saving in the bank, buying annuity insurance, investing stocks and etc., people always hope to obtain high yield awards. So how to estimate an investment becomes a very important problem. This chapter will mainly introduce the discounted cash flow analysis method to estimate the yield rate.

单词和短语 Words and expressions

收益率　yield rate
贴现现金流分析法　discounted cash flow analysis
内部收益率　internal rate of return
再投资收益率　yield rate of reinvestment
投资组合　investment portfolio
资金预算　capital budgeting

净现值　net present value
收益曲线　yield cure
基金　fund
未结投资价值　outstanding balance
资本加权法　money-weighted rate method
时间加权法　time-weighted rate method

例 题 Example

例 1 求利率为何值时,第 2 年末支付 2000 元、第 4 年末支付 3000 元的现值和为 4000 元.

解 这是一个简单的求收益率的问题.现金流是 $R_0 = 4000, R_2 = -2000, R_4 = -3000$. 则

$$4000 - 2000v^2 - 3000v^4 = 0$$

$$v^2 = \frac{-2 \pm \sqrt{52}}{6}$$

因为 $v^2 > 0$,所以 $v^2 = \frac{-2 + \sqrt{52}}{6} = 0.868517$

$$i = 0.0730.$$

Ex. 1 Find the yield rate of the initial investment 4000 yuan which obtains payments 2000 yuan and 3000 yuan at the end of the 2nd year and 4th year, respectively.

Solution This is in fact an simple problem of yield rate. The cash flows is $R_0 = 4000, R_2 = -2000, R_4 = -3000$. Thus,

$$4000 - 2000v^2 - 3000v^4 = 0$$

$$v^2 = \frac{-2 \pm \sqrt{52}}{6}.$$

Because $v^2 > 0$, we have $v^2 = \frac{-2 + \sqrt{52}}{6} = 0.868517$

$$i = 0.0730.$$

例 2 某人在年初贷款 10000 元,本金年利率 9%,为期 10 年,若所还款的再投资利率为 7%,求以下三种方式收回贷款本例的收益率:

(1)以年金方式,每年年末归还一次,每次还款额相等.
(2)每年年末将本年所生成的利息支付给贷款人,本金 10 年末支付.
(3)第 10 年年末一次性归还本利和.

解

(1) 10年末所支付的终值是：
$$\frac{1000}{a_{\overline{10}|0.09}} S_{\overline{10}|0.07} = 1558.2 \times 13.8164 = 21528.8$$

收益率 i 可由价值方程得：
$$10000 \times (1+i)^{10} = 21528.8$$
$$i = 7.97\%$$

(2) 10年末所有支付的终值是：
$$10000 + 900 S_{\overline{10}|0.07} = 10000 + 900 \times 13.8164 = 22434.76$$

收益率 i 可由价值方程得：
$$10000(1+i)^{10} = 22434.76$$
$$i = 8.42\%$$

(3) 10年末全部支付的终值是：
$$10000 \times 1.09^{10} = 23673.6$$

收益率 i 可由价值方程：
$$10000 \times (1+i)^{10} = 23673.6$$
$$i = 9\%$$

Ex. 2 Find the yield rates of the following three payment methods, if someone borrows 10000 yuan at the begging of a year at the effective rate 9% per annum for 10 years and the payments are invested again with the yield rate of reinvestment 7% per annum:

(1) The investor will receive the equal at the end of the year in arrear for 10 years.

(2) The investor will receive the interest at the end of the year in arrear for 10 years and the principle at the end of the 10th year.

(3) The investor will receive the principle and the interest at the end of the 10th year.

Solution

(1) The accumulation for 10 years is:
$$\frac{1000}{a_{\overline{10}|0.09}} S_{\overline{10}|0.07} = 1558.2 \times 13.8164 = 21528.8.$$

From the value equation, we have:
$$10000 \times (1+i)^{10} = 21528.8$$
$$i = 7.97\%$$

(2) The accumulation for 10 years is:
$$10000 + 900 S_{\overline{10}|0.07} = 10000 + 900 \times 13.8164 = 22434.76.$$

From the value equation, we have:
$$10000(1+i)^{10} = 22434.76$$
$$i = 8.42\%$$

(3) The accumulation for 10 years is:
$$10000 \times 1.09^{10} = 23673.6.$$

From the value equation, we have:
$$10000 \times (1+i)^{10} = 23673.6$$
$$i = 9\%$$

本章重点

1. 会利用贴现现金流分析法评估投资的好坏。
2. 理解再投资收益率的概念和应用。
3. 掌握投资额加权和时间加权收益率的计算方法。
4. 了解投资组合法和投资年法。

Key points of this chapter

1. Learn to use the discounted cash flow analysis method to estimate an investment.
2. Understand the concept of yield rate of reinvestment and its application.

3. Master the computing method of money-weighted rate and time-weighted rate of interest.
4. Understand portfolio method and investment year method.

第四章 债务偿还
Chapter 4 Debt Repayment

本章在第二章的基础上，讨论债务偿还问题。债务偿还的方式有三种：满期偿还方法，分期偿还方法，偿债基金法。本章主要介绍后两种方法。另外，本章还讨论偿还频率和计息频率不同时分期偿还表的建立及变动偿还计划等相关计算问题。

Based on Chapter2, this chapter discusses debt repayment. There are three way of repayment: expiration method, amortization method and sinking funds method. The final two methods are introduced in this chapter. Moreover, it's also discussed in this chapter that how to build amortization schedules when the repayment frequency is different from interest frequency and how to compute problems about adjusting repayment plan.

单词和短语 Words and expressions

债务偿还 debt repayment	追溯法 retrospective method
分期偿还法 amortization method	预期法 prospective method
偿债基金法 sinking funds method	偿债基金表 sinking funds schedules

例 题 Example

例 1 某借款人年初贷款，利率为 5%，贷款期限为 10 年，首年末还款额为 2000 元，第二年末为 1900 元，以此类推，第 10 年末还款额为 1100 元，计算：

(1) 贷款本金。
(2) 第 5 次还款中的本金部分和利息部分。

解 (1) 贷款本金为：
$$L = 1000 a_{\overline{10}|} + 100(Da)_{\overline{10}|}$$
$$= 1000 \times (7.72173) + 100 \times \frac{10 - 7.72173}{0.05} = 12278.27$$

(2) $R_5 = 1600$
$$B_5^p = 1000 a_{\overline{6}|} + 100(Da)_{\overline{6}|}$$
$$I_5 = i B_4^p = 1000 \times (1 - v^6) + 100 \times (6 - a_{\overline{6}|})$$
$$= 1000 \times (1 - 0.74622) + 100 \times (6 - 5.0757) = 346.21$$
$$P_5 = R_5 - I_5 = 1600 - 346.21 = 1253.79$$

Ex. 1 Someone has a 10-year loan at the begging of a year and should pay for it at the interest rate of 5% per annum. He pays 2000 yuan at the end of this year, 1900 yuan at the end of the 2nd year, ···, 1100 yuan at the end of the 10th year. Please find:

(1) The principle of the loan.
(2) The principle and the interest in the 5th payment.

Solution (1) The principle of the loan is:
$$L = 1000 a_{\overline{10}|} + 100(Da)_{\overline{10}|}$$
$$= 1000 \times (7.72173) + 100 \times \frac{10 - 7.72173}{0.05} = 12278.27$$

(2) $R_5 = 1600$
$$B_5^p = 1000 a_{\overline{6}|} + 100(Da)_{\overline{6}|}$$
$$I_5 = i B_4^p = 1000 \times (1 - v^6) + 100 \times (6 - a_{\overline{6}|})$$
$$= 1000 \times (1 - 0.74622) + 100 \times (6 - 5.0757) = 346.21$$
$$P_5 = R_5 - I_5 = 1600 - 346.21 = 1253.79$$

例 2 甲借款 2000 元,为期 2 年,贷款年利率为 10%,借款人建立偿债基金并每半年末在偿债基金中存一次,偿债基金利率为每年计息 4 次的年名义利率 8%,建立这一贷款的偿债基金表.

解 借款人每年末付款利息为 $0.1 \times 2000 = 200$,设每半年末的偿债基金存款为 D,则

$$D \frac{S_{\overline{8}|0.02}}{S_{\overline{2}|0.02}} = 2000$$

因此,$D = \dfrac{2000 S_{\overline{2}|0.02}}{S_{\overline{8}|0.02}} = 2000 \times \dfrac{2.02}{8.5830} = 470.70$

偿债基金表如下:

期间	支付利息	偿债基金存款	偿债基金每计息期所得利息	偿债基金额	净贷款余额
0					2000
$\frac{1}{4}$	0	0	0	0	2000
$\frac{1}{2}$	0	470.70	0	470.70	1529.30
$\frac{3}{4}$	0	0	9.41	480.11	1519.89
1	200	470.70	9.60	960.41	1039.59
$1\frac{1}{4}$	0	0	19.21	979.62	1020.38
$1\frac{1}{2}$	470.70	470.70	19.59	1469.91	530.09
$1\frac{3}{4}$	0	0	29.40	1499.31	500.69
2	470.70	470.70	29.99	2000	0

Ex. 2 A person borrows 2000 yuan for 2 years with the effective rate of 10% per annum. He builds the sinking funds that interest four times at the nominal rate of 8% per annum and deposits money at the end of every half year. Please show the sinking funds schedules.

Solution Every year the payment's interest is $0.1 \times 2000 = 200$. Suppose D is the deposit in the sinking funds every half year, thus

$$D \frac{S_{\overline{8}|0.02}}{S_{\overline{8}|0.02}} = 2000.$$

So, we have $D = \dfrac{2000 S_{\overline{8}|0.02}}{S_{\overline{8}|0.02}} = 2000 \times \dfrac{2.02}{8.5830} = 470.70$.

The sinking funds schedules are as following:

term	interest	sinking fund accumulation	Every period interest of the sinking fund	Sinking fund amount	Net loans
0					2000
$\frac{1}{4}$	0	0	0	0	2000
$\frac{1}{2}$	0	470.70	0	470.70	1529.30
$\frac{3}{4}$	0	0	9.41	480.11	1519.89
1	200	470.70	9.60	960.41	1039.59
$1\frac{1}{4}$	0	0	19.21	979.62	1020.38
$1\frac{1}{2}$	470.70	470.70	19.59	1469.91	530.09
$1\frac{3}{4}$	0	0	29.40	1499.31	500.69
2	470.70	470.70	29.99	2000	0

本章重点

① 了解分期偿还和偿债基金两种债务偿还方式.
② 重点掌握偿还频率与计息频率不同时的分期偿还和变动偿还的相关计算.

Key points of this chapter

① Understand the amortization method and sinking funds method.
② Master the amortization method when the repayment frequency is different from interest frequency and the computing method about adjusting repayment.

第五章　债券及其定价理论
Chapter 5　Bond and Its Pricing Theory

本章主要介绍债券的定价和收益率的计算等等. 债券可归属于固定收益类的证券, 其定价方法比较简单. 本章介绍了著名的 Makeham 公式及与其相关的计算方法. 此外, 本章还对股票做了简要的讨论, 最后介绍了债券市场中的新兴产品的定价知识, 这些产品的定价相对复杂.

This chapter mainly discusses the computing methods about pricing and yield of bond. Bond has flat yield so its pricing method is simple. The famous Makeham formula and some computing methods about it have been introduced in this chapter. Moreover, knowledge about stock is also mentioned. Finally, some pricing methods of new products in the bond market are provided that whose pricing methods are more complicated than bond.

单词和短语　Words and expressions

名义收益率　nominal yield rate
当前收益率　current yield rate
到期收益率　yield rate to maturity
溢价　premium
折价　discount
账面价值　book value

应计票息　accrued coupon
系列债券　serial bonds
期权　options
货币市场基金　money of funds
存单　certificate of deposit
期货　futures

例题　Example

例 1　面值 1000 元的 10 年期债券, 票息率为每年计息两次的年名义利率 8.4%, 赎回值为 1050 元, 票息所得税为 20%. 若按每年计息两次的年名义收益率 10% 购买, 求该债券的价格. 分别用四种公式计算.

解　由题意得:

$N = 1000$, $C = 1050$, $r = \dfrac{8.4\%}{2} = 0.042$, $g = \dfrac{Nr}{C} = 1000 \times \dfrac{0.042}{1050} = 0.04$,

$i = \dfrac{10\%}{2} = 0.05$, $n = 10 \times 2 = 20$, $K = Cv^n = 1050 \times (1+0.05)^{-20} = 395.734$,

$G = \dfrac{Nr}{i} = 1000 \times \dfrac{0.042}{0.05} = 840$, $t_1 = 20\%$, $1 - t_1 = 1 - 20\% = 0.8$.

按照四个计算公式, 有

(1) 基本公式:
$P = Nr(1-t_1)a_{\overline{n}|} + K = 42 \times 0.8 a_{\overline{20}|0.05} + 395.734$
$= 33.6 \times 12.4622 + 395.734$
$\approx 814.46.$

(2) 溢价/折价公式:
$P = C + [N_r(1-t_1) - Ci]a_{\overline{n}|}$

$$= 1050 + (42 \times 0.8 - 1050 \times 0.05) \times 12.4622$$
$$\approx 814.46.$$

(3)基础金额公式：
$$P = G(1-t_1) + [C - G(1-t_1)]v^n$$
$$= 840 \times 0.8 + (1050 - 840 \times 0.8) \times (1+0.05)^{-20}$$
$$= 672 + (1050 - 672) \times 0.37689 \approx 814.46.$$

(4)Makeham 公式
$$P = K + \frac{g(1-t_1)}{i}(C - K)$$
$$= 395.734 + \frac{0.04 \times 0.8}{0.05}(1050 - 395.734)$$
$$\approx 814.46.$$

Ex. 1 There's a 10-year-term bond with the face value 1000 yuan, the coupon rate of 8.4 per annum convertible semiannually, the redemption 1050 yuan and coupon tax of 20 per annum. If it is sold at the nominal rate of 10 per annum convertible semiannually, please use classic four formulas to compute the bond price.

Solution According to the problem, we have that.
$$N = 1000, C = 1050, r = \frac{8.4\%}{2} = 0.042, g = \frac{Nr}{C} = 1000 \times \frac{0.042}{1050} = 0.04,$$
$$i = \frac{10\%}{2} = 0.05, n = 10 \times 2 = 20, K = Cv^n = 1050 \times (1+0.05)^{-20} = 395.734,$$
$$G = \frac{Nr}{i} = 1000 \times \frac{0.042}{0.05} = 840, t_1 = 20\%, 1 - t_1 = 1 - 20\% = 0.8.$$

(1) Basic formula:
$$P = Nr(1-t_1)a_{\overline{n}|} + K = 42 \times 0.8 a_{\overline{20}|0.05} + 395.734$$
$$= 33.6 \times 12.4622 + 395.734$$
$$\approx 814.46.$$

(2) Premium formula:
$$P = C + [N_r(1-t_1) - Ci]a_{\overline{n}|}$$
$$= 1050 + (42 \times 0.8 - 1050 \times 0.05) \times 12.4622$$
$$\approx 814.46.$$

(3) Basic amount formula:
$$P = G(1-t_1) + [C - G(1-t_1)]v^n$$
$$= 840 \times 0.8 + (1050 - 840 \times 0.8) \times (1+0.05)^{-20}$$
$$= 672 + (1050 - 672) \times 0.37689 \approx 814.46.$$

(4) Makeham formula:
$$P = K + \frac{g(1-t_1)}{i}(C - K)$$
$$= 395.734 + \frac{0.04 \times 0.8}{0.05}(1050 - 395.734)$$
$$\approx 814.46.$$

本章重点

1. 掌握债券的定价方法.
2. 会使用 Makeham 公式解决问题.
3. 关注债券市场新兴产品及其定价方法.

Key points of this chapter

1. Master the bond pricing methods.
2. Learn to use Makeham formula to solve problems.
3. Pay attention to some new products in bond market and their pricing methods.

第六章 复利数学的应用
Chapter 6 Application of compound interest mathematics

复利数学在实务中有着广泛的应用。本章主要介绍复利在银行的信贷业务实际利率的计算，投资成本计算以及固定资产折旧等实际问题中的应用。

The compound interest mathematics is widely applied in practice. This chapter mainly introduces its application in areas of the effective rate computing in the bank's credit operations, capitalize cost and depreciation of on fixed asset, etc..

单词和短语 Words and expressions

诚实借贷	truth in lending	固定利率抵押贷款	fixed rate mortgage
财务费用	finance charge	最大收益率法	maximum yield rate method
年利率	APR	最小收益率法	minimum yield rate method
交割日	settlement date	折旧	depreciation
可调利率抵押贷款	adjustable rate mortgage	卖空	short sales

例题 Example

例 1 某人以 10% 的利率借款 1000 元，期限为 12 个月。如果借款人在 3 个月末还款 200 元，在 8 个月末还款 300 元，求在 12 个月末必须还款额。分别按(1)实际利率计算。(2)商业规则计算。(3)联邦规则计算。

解

(1) 最后支付额为：

$$1000 \times (1+10\%) - 200 \times (1+10\%)^{\frac{3}{4}} - 300 \times (1+10\%)^{\frac{1}{3}} = 575.50.$$

(2) 最后支付额为：

$$1000 \times (1+10\%) - 200 \times \left(1+10\% \times \frac{3}{4}\right) - 300 \times \left(1+10\% \times \frac{1}{3}\right) = 575.00.$$

(3) 在 3 个月末 1000 元应计息为：

$$1000 \times 10\% \times \frac{1}{4} = 25.00.$$

200 元中 25 元用于支付利息，余下 175 元用于偿还本金，未尝贷款余额变为 825 元。在 8 月末 825 元未尝贷款余额应计息为：

$$825 \times 10\% \times \frac{8-3}{12} = 34.38.$$

300 元先支付利息 34.38 元，余下 300 − 34.38 = 265.62 用于偿还本金，未尝贷款余额变为 825 − 265.62 = 559.38。在 12 个月末，未尝贷款余额为：

$$559.38 \times \left(1+10\% \times \frac{12-8}{12}\right) = 578.03.$$

Ex Someone borrows 1000 yuan at the effective rate of 10 per annum for 12 months. If he pays 200 yuan at the end of the 3rd month and 300 yuan at the end of the 8th month, please find the payment at the end of this year under：(1) Effective rate. (2) Merchant rate. (3) United State rule.

Solution

(1) The final payment is:
$$1000\times(1+10\%)-200\times(1+10\%)^{\frac{3}{4}}-300\times(1+10\%)^{\frac{1}{3}}=575.50.$$

(2) The final payment is:
$$1000\times(1+10\%)-200\times\left(1+10\%\times\frac{3}{4}\right)-300\times\left(1+10\%\times\frac{1}{3}\right)=575.00.$$

(3) The interest of 1000 yuan for 3 months are:
$$1000\times10\%\times\frac{1}{4}=25.00.$$

There is 25 yuan as the payments of interest and 175 yuan as the payments of principle within the 200 yuan. Thus, the outstanding loans balance is 825 yuan whose interest is
$$825\times10\%\times\frac{8-3}{12}=34.38.$$

In a similar way, there is 34.38 yuan as the payments of interest and 265.62 yuan as the payments of principle within the 300 yuan. Thus, the outstanding loans balance is $825-265.62=559.38$ yuan whose interest is
$$559.38\times\left(1+10\%\times\frac{12-8}{12}\right)=578.03.$$

本章重点

1. 了解复利数学有哪些实际应用。
2. 会利用复利知识解决诚实借贷，不动产抵押贷款，APR 近似计算，折旧，成本核定等问题。

Key points of this chapter

1. Master the applications of compound interest mathematics.
2. Solve problems about truth in lending, fixed asset mortgage, APR computing, depreciation and cost measurement with compound interest knowledge.

第七章 金融分析
Chapter 7　Analytical Finance

本章讨论了在特定情况下利率水平是由哪些因素确定的；在预测未来的投资时，利率水平如何估计；资产和负债的未来现金流如何管理。

This chapter discusses that, under certain situations, what the interest rate decided by, how to estimate the future investment interest and some management business about the future cash flow of assets and debts.

单词和短语 Words and expressions

通货膨胀　inflation　　　　　　　　　期望现时值　expected present value
风险和不确定性　risk and uncertainty

基本概念和性质　Basic concepts and properties

1. 通货膨胀：指在纸币流通的条件下，因货币供给大于货币实际需求，也即现实购买力大于供给，导致货币贬值，从而引起的一段时间内物价持续而普遍上涨现象。扣除通货膨胀后的利率通常称为真实利率，记为 i'，市场利率称为表面利率，记为 i，通货膨胀率记为常数 $r>0$，则有
$$1+i=(1+i')(1+r).$$

Inflation: Under the condition of paper circulation, for the money supply is bigger than the money demand, the money devaluates. And the inflation is a phenomenon of the market prices keep rising during some period because of the above reasons. The real interest rate is the interest rate eliminating the inflation, which is noted by i'. The market interest is noted by i. And the constant $r > 0$ stands for the inflation rate. Then, we have

$$1 + i = (1 + i')(1 + r).$$

本章重点
1. 了解利息的经济原理和决定利率水平的因素.
2. 掌握通货膨胀的概念并会计算通货膨胀率.
3. 了解风险和收益的关系,并培养在投资时规避风险的意识.

Key points of this chapter
1. Understand the interest economic principle and the factors deciding the interest rate.
2. Master the concept about inflation and be able to compute the inflation rate.
3. Understand the relations between risk and profit and train the attitude of avoiding risk in the investment business.

练 习

1. 对于任意的 t, 假设利息强度 $\delta(t) = 0.06(0.9)^t$. 则求 100 元在 $t = 3.5$ 年的现值 $v(t)$ 的表达式.

解
$$v(t) = \exp\left[-\int_0^t 0.06(0.9)^s ds\right]$$
$$= \exp[-0.06(0.9^t - 1)/\log(0.9)]$$

因此, 100 元在 $t = 3.5$ 年的现值 $v(t)$ 的表达式为

$$100\exp[-0.06(0.9^{3.5} - 1)/\log(0.9)] = 83.89.$$

2. 在利率为 i 的情况下, 证明

$$\frac{1}{a_{\overline{n}|}} = \frac{1}{S_{\overline{n}|}} + i.$$

解 $S_{\overline{n}|} = [(1+i)^n - 1]/i$, 则有

$$\frac{1}{S_{\overline{n}|}} + i = \frac{i}{(1+i)^n - 1} + i$$
$$= \frac{i(1+i)^n}{(1+i)^n - 1} = \frac{i}{[1 - (1+i)^{-n}]} = \frac{1}{a_{\overline{n}|}}$$

成立.

3. 一项 100000 元的投资将会以每年支付 10500 元并持续 25 年的年金形式得到回报. 当贷款利率是 9% 的时候多少年后回本. 如果回本后收益率变为 7%, 求 25 年后投资的收益是多少.

解 由条件得, 当利率为 9% 时, 设投资回本时的最短整数年为 t, 满足

$$10500 a_{\overline{t}|} \geq 100000.$$

由复利表可得, 时间 $t = 23$. 则 25 年后年金的收益为

$$P = [-100000(1.09)^{23} + 10500 S_{\overline{20}|0.09}](1.07)^2 + 10500 S_{\overline{2}|0.07}$$
$$= 26656.$$

4. 前 3 年每季度初在银行存入 1000 元, 后 2 年每季度初存入 2000 元, 每月实际利率为 1%, 计算第 5 年年末存款积累值.

解 5 年中共有 60 个计息期, 存款积累值为:

$$1000 \times \frac{S_{\overline{60}|0.01} + S_{\overline{24}|0.01}}{a_{\overline{3}|0.01}} = 1000 \times \frac{81.66967 + 26.973465}{2.940985} = 36941.07$$

即为第五年年末存款积累值.

Exercises

1. Measure time in years from the present, and suppose that $\delta(t) = 0.06(0.9)^t$ for all t. Find a simple expression for $v(t)$, and hence find the discounted present value of 100 yuan due in 3.5 years' time.

Solution

$$v(t) = \exp[-\int_0^t 0.06(0.9)^s ds]$$
$$= \exp[-0.06(0.9^t - 1)/\log(0.9)]$$

Hence the present value of 100 yuan due in 3.5 years' time is
$$100\exp[-0.06(0.9^{3.5} - 1)/\log(0.9)] = 83.89.$$

2. Prove that, at rate of interest i,
$$\frac{1}{a_{\overline{n}|}} = \frac{1}{s_{\overline{n}|}} + i.$$

Solution $s_{\overline{n}|} = [(1+i)^n - 1]/i$, we have
$$\frac{1}{s_{\overline{n}|}} + i = \frac{i}{(1+i)^n - 1} + i$$
$$= \frac{i(1+i)^n}{(1+i)^n - 1} = \frac{i}{[1 - (1+i)^{-n}]} = \frac{1}{a_{\overline{n}|}}$$

as required.

3. An investment of 100000 yuan will produce an annuity of 10500 annually in arrear for 25 years. Find the discounted payback period when the interest rate on borrowed money is 9% per annum. Find also the accumulated profit after 25 years if money may be invested at 7% per annum.

Solution

By conditions, the discounted payback period is the smallest integer t such that
$$10500 a_{\overline{t}|} \geqslant 100000 \text{ at } 9\%$$

From compound interest tables we see that the discounted payback period is 23 years. The accumulated profit after 25 years is
$$P = [-100000(1.09)^{23} + 10500 s_{\overline{23}|0.09}](1.07)^2 + 10500 s_{\overline{2}|0.07}$$
$$= 26656.$$

4. Someone saves 1000 yuan in the bank at every begging of the quarter in the first 3 years and 2000 yuan at every begging of the quarter in the last 2 years. Suppose the year effective rate is 1%, please find the accumulated value at the end of these 5 years.

Solution

There are 60 periods in these 5 years and the accumulated value is
$$1000 \times \frac{s_{\overline{60}|0.01} + s_{\overline{24}|0.01}}{a_{\overline{3}|0.01}} = 1000 \times \frac{81.66967 + 26.973465}{2.940985} = 36941.07.$$

第三部分 寿险精算实务
Part 3 Life Insurance Products and Finance

引 言

"寿险精算实务"是既覆盖理论又涵盖实务的一门课程。主要学习寿险精算理论与技术在实务中的应用,其内容包括:寿险产品设计及其特点、寿险产品定价、负债评估、资本需求、利润分析等。

Introduction

"Life Insurance Products and Finance" is a course that covers theory and practice. The main learning is the applications of life insurance actuarial theory and technology in practice. Its contents include: life insurance product design and its characteristics, life insurance products pricing, liability assessment, capital demand, profit analysis.

第一章 人寿保险的主要类型
Chapter 1 Main types of life insurance

本章讨论人寿保险的主要类型,分为普通型人寿保险(定期寿险、终身寿险、两全保险、年金保险)和新型人寿保险(分红保险、投资连结保险、万能保险)。

This chapter discuss the main types of life insurance which includes common types of life insurance (term insurance、whole life insurance、endowment insurance、annuity insurance) and new types of life insurance (participating insurance、investment-linked insurance、universal insurance).

单词和短语 Words and expressions

定期寿险	term insurance	个人年金	individual annuity
终身寿险	whole life insurance	联合年金	joint annuity
限期交费终身保险	limited payment life insurance	最后生存者年金	last survivor annuity
		联合及生存者年金	joint and survivor annuity
趸交终身保险	single premium whole life insurance	定额年金	level annuity
两全保险	endowment insurance	变额年金	variable annuity
生存保险	survival insurance	分红保险	participating insurance
年金保险	annuity insurance	保单红利	policy dividend
趸交年金	single premium annuity	现金红利	cash dividend
即期年金	immediate annuity	增额红利	incremental dividend
延期年金	deferred annuity	投资连结保险	investment-linked insurance
终身年金	whole life annuity	万能保险	universal insurance
退还年金	refund annuity	结算利率	settlement rates
定期生存年金	temporary life annuity	保证利率	guaranteed interest rate

本章重点

1. 掌握人寿保险的主要类型。
2. 掌握分红保险、投资连结保险和万能保险的特点。

Key points of this chapter

1. Master the main types of life insurance.
2. Master the characteristics of participating insurance and investment-linked insurance and universal insurance.

第二章 保单现金价值与红利
Chapter 2　Cash value and dividend of policy

　　本章介绍了保单现金价值的含义和保单现金价值的不同计算方法，给出了交清保险、展期保险、自动垫交保费等保单选择权的计算公式，介绍了资产份额的基本公式，最后对保单红利的基本计算方法、保单红利分配与红利选择权进行概述。

　　Chapter 2 is devoted to introducing the meaning of the policy cash value and its different calculation methods. The formulas of policy options such as paid-up insurance、extended insurance、automatic premium loan are given, as well as the basic formula of asset share. At last the basic calculation method of policy dividend、policy dividend allocation and dividend accumulation are introduced.

单词和短语 Words and expressions

现金价值　cash value	调整保费法　adjusted premium method
红利　bonus; dividend	资产份额法　asset share method
退保金　surrender benefits	交清保险　paid-up insurance
不丧失价值　non-forfeit value	展期保险　extended insurance
不丧失现金价值　non-forfeit cash value	自动垫交保费　automatic premium loan (APL)
节约费用　surrender charge	保单红利　policy dividend
死亡力　force of mortality	红利分配　dividend allocation
退保力　force of surrender	红利选择权　dividend accumulation
均衡净保费法　net level premium method	现金红利　cash dividend
修正净保费法　modified net premium method	终了红利　terminal dividend

本章重点

① 理解保单现金价值的含义。
② 掌握保单现金价值的不同计算方法。
③ 会计算保单红利。

Key points of this chapter

① Understand the meaning of the policy cash value.
② Master different calculation methods of the policy cash value.
③ Know how to calculate policy dividend.

第三章 特殊年金与保险
Chapter 3　Special annuity and insurance

　　本章将讨论特殊形式的年金与保险的精算现值、净保费、毛保费及责任准备金等，主要包括最低保证年金、分期退还年金、现金退还年金、家庭收入保险、退休收入保单等。

　　This chapter will discuss actuarial present value、net premium、gross premium and liability reserve for special types of annuity and insurance, mainly including minimum guaranteed annuity、installment refunded annuity、cash refund annuity、family income insurance、retirement income policy and so on.

单词和短语 Words and expressions

最低保证年金　minimum guaranteed annuity	退休收入保单　retirement income policy
分期退还年金　installment refunded annuity	变额保险　variable insurance
现金退还年金　cash refund annuity	基金份额　fund share
家庭收入保险　family income insurance	变额年金　variable annuity

完全变额人寿保险　fully variable life insurance
固定保费的变额人寿保险　fixed premium variable life insurance

风险保额　risk premium
残疾收入给付　disability income benefit

本章重点

1. 知道如何计算特殊形式年金的精算现值。
2. 掌握家庭收入保险的计算方法。

Key points of this chapter

1. Know how to calculate the actuarial present value for special types of annuity.
2. Master the calculation method of family income insurance.

第四章　寿险定价概述
Chapter 4　An overview of life insurance pricing

本章首先介绍了寿险定价的基本原则和过程，然后介绍了定价的主要方法，包括净保费加成法、资产份额法和宏观定价法，最后介绍了定价假设及影响定价假设的因素。

This chapter introduces the basic principles and processes of life insurance pricing at first. Then the main pricing methods are presented, including additive method of net premium、asset share method and macro-pricing method. At last the pricing assumptions and factors that affect the pricing assumptions are introduced.

单词和短语 Words and expressions

净保费加成法　additive method of net premium
毛保费等价公式　equivalence formula of gross premium
换算表　translation table
资产份额法　asset share method
利润现值　present value of profits
投资回报率　rate of return on investments
宏观定价法　macro-pricing method

死亡率　mortality rate
利率　interest rate
费用率　expense ratio
失效率　lapse rate
合同初始费　contract initial premium
保单维持费　policy maintenance expenses
平均保额　average amount of insurance

本章重点

1. 掌握寿险定价的原则和主要方法。

Key points of this chapter

1. Master the principles and main methods of life insurance pricing.

第五章　资产份额定价法
Chapter 5　Asset share pricing method

本章重点介绍了用资产份额法进行定价的具体计算过程、利润指标的实现及保费调整等问题。

This chapter is focused on the calculation process of pricing by asset share method、profit indexes to achieve and premium adjustment and other issues.

单词和短语 Words and expressions

资产份额　asset share
有效保单　policies in force

积累盈余　accumulated surplus
利润　profit

账面利润	book profits	红利给付	dividend
贴现因子	discount factor	营运损益	contractors profit and loss
保单费用	policy fee	获利比率	profit ratio
平均保额	average amount of insurance	投资回报率(ROI)	rate of return on investments
期初费用	initial charge	利润现值	present value of profits
死亡给付费用	death benefit charge	保费现值	present value of premiums
退保费用	urrender charge	风险成本现值	present value of risk cost
红利费用	dividend charge	死亡给付现值	present value of death benefits
死亡给付	death benefit		

本章重点

1. 理解资产份额定价的过程。
2. 掌握资产份额法的基本公式。

Key points of this chapter

1. Understand the process of pricing by asset share method.
2. Master the basic formulas of asset share method.

第六章　资产份额法的进一步分析
Chapter 6　Further analysis of asset share method

本章在上一章的基础上对资产份额法的进一步改良、利润的变动以及资产份额法的其他应用进行深入讨论。

Based on Chapter 5, the further improvement of asset share method、variation in profit and asset share method to other applications are discussed in depth.

单词和短语 Words and expressions

现金价值	cash value	免交保费	waiver of premium
通货膨胀	inflation	趸交保费即期年金	single premium immediate annuity
保单贷款	policyloan		
红利	dividend	健康保险	health insurance
投资年方法	investment year method	团体保险	group insurance
选择权	option	准备金	reserve
再保险	reinsurance	法定准备金	statutory reserve
利润变动	profit variation	GAAP 准备金	GAAP reserve
贴现因子	discount factor	自然准备金	nature reserve
利润	profit	毛保费准备金	gross premium reserve
意外死亡给付	accidental death benefit		

本章重点

1. 掌握资产份额法的改良、利润变动以及资产份额法的其他应用。

Key points of this chapter

1. Master the improvement of asset share method、profit variation and asset share method to other applications.

第七章 准备金评估 Ⅰ
Chapter 7　Reserve valuation Ⅰ

　　本章主要讨论传统寿险的准备金评估。首先介绍了准备金的来源与定义，并从不同的角度介绍了几种常见的准备金类型，包括法定责任准备金、盈余准备金和税收准备金；接下来总结了传统寿险产品法定准备金的评估方法；随后则讨论了准备金评估基础的选择；最后介绍了准备金方法在实务中的应用。

　　This chapter mainly discusses reserve assessment of traditional life insurance. Firstly introduces the sources and definitions of reserve, as well as several common types of reserve form different points of view, including legal reserve、surplus reserve、tax reserves. Then the assessment methods of statutory reserve for traditional life insurance productions are summarized, and the choices of reserve valuation basis are discussed. Finally, the applications of the reserve method in practice are introduced.

单词和短语 Words and expressions

准备金　reserve
法定责任准备金　legal reserve
盈余准备金　surplus reserve
税收准备金　tax reserve
未来法　prospective method
过去法　retrospective method
均衡净保费法　net level premium method
修正净保费评估法　modified net premiums valuation method
费用补贴　expense subsidy
Zillmer 方法　Zillmer method

保险监督官准备金评估方法（CRVM）　commissioner's reserve valuation method
保单保费评估法（PPM）　policy premium valuation method
累积法　accumulation method
评估利率　valuation rate of interest
平均准备金　average reserve
期中准备金　mean reserve
延期保费　deferred premiums
准备金调整　reserve adjustment
退还准备金　refund reserve

本章重点
1 掌握传统寿险产品法定准备金的评估方法。
2 掌握准备金方法在实务中的应用。

Key points of this chapter
1 Master the valuation methods of statutory reserve for traditional life insurance productions.
2 Master the applications of the reserve method in practice.

第八章 准备金评估 Ⅱ
Chapter 8　Reserve valuation Ⅱ

　　本章首先介绍了可变动保费万能寿险和固定保费万能寿险等利率敏感型寿险的评估；然后介绍了年金的评估，包括趸交净保费延期年金、年交保费年金、可变动保费年金、即期年金；在变额保险的评估中主要介绍了年交保费变额寿险、趸交保费变额寿险、变额年金、保证最小死亡给付准备金；最后讨论了评估的进一步应用。

　　This chapter firstly introduces the valuation of interest-sensitive life insurance, such as flexible premium universal life insurance and fixed premium universal life insurance. Then introduces annuity valuation, including single net premium deferred annuity、annual premium annuity and flexible premium annuity and immediate annuity. In the valuation of variable insurance, annual premium variable life insurance、single premium premium variable life insurance and variable annuity are mainly introduced. At last we discuss further applications of the valuation.

单词和短语 Words and expressions

可变动保费万能寿险　flexible premium universal life insurance
最小准备金　minimum reserve
固定保费万能寿险　fixed premium universal life insurance
年金评估　annuity valuation
趸交净保费延期年金　single net premium deferred annuity
退保现金价值　cash surrender value
年交保费年金　annual premium annuity
可变动保费年金　flexible premium annuity
即期年金　immediate annuity
变额保险　variable insurance
年交保费变额寿险　annual premium variable life insurance
固定保费变额寿险　fixed premium variable life insurance

变动保费变额寿险　flexible premium variable life insurance
混合保费变额寿险　hybrid premium variable life insurance
趸交保费变额寿险　single premium premium variable life insurance
可变动趸交保费　variable single premium
固定趸交保费　fixed single premium
变额年金　variable annuity
保费不足准备金　premium deficiency reserve
意外死亡给付　accidental death benefit
残疾免交保费给付　waiver of premium for disability benefit
失效支持保单　lapse-supported policy

本章重点

1 掌握利率敏感型寿险的评估。
2 掌握变额保险的评估及评估的进一步应用。

Key points of this chapter

1 Master the valuation of interest-sensitive life insurance.
2 Master the valuation of variable insurance and further applications of the valuation.

第九章　寿险公司内含价值
Chapter 9　embedded value of life insurance company

本章首先介绍了内含价值的基本定义、内含价值计算方法的比较和如何由内含价值得到市场价值。随后介绍了内含价值具体的计算过程，进而介绍了内含价值具体应用及评价。

This chapter firstly introduces the basic definition of embedded value、comparison of calculation methods of embedded value and how to get market value by embedded value. Then the calculation process of embedded value is described，as well as the specific application and evaluation of embedded value.

单词和短语 Words and expressions

内含价值　embedded value
经济价值　economic value
市场价值　market value
锁定资本　capital lock-in
偿付能力资本　capital solvency
要求资本　capital requirement
自由资本　free solvency
自由盈余　free surplus

有效业务价值　value of in-force business
调整净资产　adjusted net assets
风险贴现率　risk discount rate
评估价值　valuation value
死亡率　mortality rate
投资收益率　portfolio interest rate
退保费　surrender charge
费用率　expense ratio

佣金　commission
权益回报率（ROE）　rate of return on equity
内部收益率（IRR）　internal rate of return

本章重点
1. 理解内含价值的基本定义。
2. 掌握内含价值的计算方法。

Key points of this chapter
1. Understand the basic definition of embedded value.
2. Master the calculation methods of embedded value.

第十章　偿付能力监管
Chapter 10　solvency regulation

　　本章介绍了偿付能力、偿付能力额度、偿付能力监管的基本概念，同时对欧盟及北美偿付能力监管实践及其进展，以及偿付能力监管中的资产评估进行了介绍。

　　This chapter introduces the basic concepts of solvency、solvency margin and solvency regulation. Also the practice and progress of solvency regulation in European Union and North American, and the asset valuation in solvency regulation are described.

单词和短语　Words and expressions

偿付能力　solvency
偿付能力监管　solvency regulation
偿付能力额度　solvency margin
最低偿付能力额度　minimum solvency margin
实际偿付能力额度　real solvency margin
认可资产　admitted assets
认可负债　admitted liability
责任准备金　liability reserve
风险资本　risk-based capital
资产风险（C1）　asset risk

定价风险（C2）　pricing risk
利率风险（C3）　interest rate risk
经营风险（C4）　business risk
资产评估　asset valuation
非认可资产　non-admitted assets
现金流测试　cash flow testing
动态资本充足性测试　dynamic capital adequacy testing
经济资本　economic capital

本章重点
1. 理解偿付能力、偿付能力额度、偿付能力监管的基本概念。
2. 了解欧盟及北美偿付能力监管实践及其进展。
3. 掌握偿付能力监管中的资产评估。

Key points of this chapter
1. Understand the basic concepts of solvency、solvency margin and solvency regulation.
2. Understand the practice and progress of solvency regulation in European Union and North American.
3. Master the asset valuation in solvency regulation.

第十一章　养老金概述
Chapter 11　An overview of pension

　　本章首先介绍了养老金计划的基本概念，包括养老金计划的设计、计划成本、精算评估等，然后对人口、经济等精算成本因素进行了分析，主要介绍了给付分配精算成本法和成本分配精算成本法。

　　This chapter presents the basic concepts of pension plan, including the design of pension plan, plan cost and actuarial valuation and so on. Then the factors of the actuarial cost such as population and economy are analyzed, and benefit allocation actuarial cost method and cost allocation actuarial cost method are mainly introduced.

单词和短语 Words and expressions

养老金　pension
养老金计划　pension plan
社会保险　social Insurance
个人储蓄　personal savings
单个雇主计划　single-employer plan
多雇主计划　multi-employer plans
确定给付计划　defined benefit plan
固定给付计划　fixed benefit plan
个人账户　personal Account
混合计划　combination plan
计划成本　plan cost
替代率　replacement rate
养老金给付　pensionable pay
盈余　surplus
精算成本法　actuarial cost method
正常成本　normal cost
精算负债(AL)　actuarial liability
现收现付制　pay-as-you-go plan
精算评估　actuarial valuation
个体成本法　individual cost method
聚合成本法　aggregate cost method
总体精算负债　total actuarial liability
未来法　prospective method
过去法　retrospective method
养老基金　pension fund
负债评估　liability valuation
静态评估法　static assessment method
动态评估法　dynamic assessment method
投资收益率　portfolio interest rate

死亡　death
退休　retirement
终止　termination
终止率(退保率)　termination rate (surrender rate)
给付分配的精算成本法　benefit allocation actuarial cost method
贴现利率　discount rate
正常成本负债　normal cost liability
附加成本负债　additional cost liability
传统应计给付成本法　traditional accrued benefit cost method
计划终止负债　plan termination liability
附加成本　additional cost
利息惟一附加成本　interest only additional cost
均衡货币附加成本　level money additional cost
均衡百分比附加成本　level precentage additional cost
计划终止成本法　plan termination cost method
成本分配精算成本法　cost allocation actuarial cost method
个体成本分配法　individual cost allocation method
均衡货币成本法　level money cost method
个体均衡保费成本法　individual level premium cost method
均衡百分比成本法　level precentage cost method
聚合成本分配法　aggregate cost allocation method
聚合均衡货币成本法　aggregate level level money cost method
个体聚合成本法　individual aggregate cost method

本章重点

1. 掌握养老金计划的基本概念。
2. 掌握精算成本法的分类及其特点。

Key points of this chapter

1. Master the basic concept of pension plan.
2. Master the classifications and characteristics of actuarial cost method.

第十二章　养老金数理及实例
Chapter 12　Pension mathematics and examples

本章主要介绍了递增成本的个体成本法、传统单位信用成本法(TUC)、规划单位信用成本法

(PUC)、均衡成本的个体成本法、进入年龄正常成本法(均衡货币)(EAN)、进入年龄正常成本法(均衡百分比)、个体均衡保费法(ILP)、聚合成本法、个体聚合成本法、冻结初始负债法(进入年龄)、冻结初始负债法(到达年龄)、聚合进入年龄成本法的精算负债、正常成本、获利、损失、附加成本等的计算方法。

This chapter mainly introduces the related calculation methods (such as actuarial liability, normal cost, profit, loss, additional cost and so on) of increasing cost individual cost method, traditional unit credit cost method, level cost individual cost method, entry age normal cost method (level money), entry age normal cost method (level precentage), individual level premium method, aggregate cost method, individual aggregate cost method, frozen initial liability (entry age), frozen initial liability (attained age), aggregate entry age cost method.

单词和短语 Words and expressions

递增成本的个体成本法　increasing cost individual cost method
传统单位信用成本法（TUC）　traditional unit credit cost method
应计年养老金给付　accrued pension benefits
精算负债　Actuarial liability
工资增长　salary inflation
规划单位信用成本法（PUC）　plan unit credit cost method
均衡成本的个体成本法　level cost individual cost method
进入年龄正常成本法(均衡货币)(EAN)　entry age normal cost method (level money)
进入年龄正常成本法（均衡百分比）　entry age normal cost method (level precentage)
个体均衡保费法(ILP)　individual level premium method
聚合成本法　aggregate cost method
个体聚合成本法　individual aggregate cost method
冻结初始负债法(进入年龄)　frozen initial liability (entry age)
冻结初始负债法(到达年龄)　frozen initial liability (attained age)
聚合进入年龄成本法　aggregate entry age cost method

本章重点

掌握递增成本的个体成本法、均衡成本的个体成本法和聚合成本法相关的计算方法。

Key points of this chapter

Master the related calculation methods of increasing cost individual cost method、level cost individual cost method and aggregate cost method.

练　习

1. 已知 3 年期两全保险,保险金额为 1000 元,$G=350$ 元,$i=15\%$,且见表 1 所示。

表 1

K	$p_{x+k}^{(\tau)}$	$q_{x+k}^{(\tau)}$	$q_{x+k}^{(1)}$	$q_{x+k}^{(2)}$	c_k	e_k	$_{k+1}CV$
0	0.5	0.5	0.1	0.4	0.2	8	230
1	0.6	0.4	0.15	0.25	0.06	2	560
2	0.5	0.5	0.5	0	0.06	2	1000

求各年的资产份额。

解　根据资产份额公式
$$_{k+1}AS p_{x+k}^{(\tau)} = (_kAS + G(1-c_k) - e_k)(1+i) - q_{x+k}^{(1)} - {_{k+1}}CV \cdot q_{x+k}^{(2)}, k=0,1,2,3\ldots,有$$

$$_{k+1}AS = \frac{1}{p_{x+k}^{(\tau)}}[_kAS + G(1-c_k) - e_k)(1+i) - 1000 q_{x+k}^{(1)} - {_{k+1}}CV \cdot q_{x+k}^{(2)}]$$

$$_0AS = 0$$

$$_1AS = \frac{1}{0.5}[(0+350(1-0.2)-8)(1+0.15) - 1000 \times 0.1 - 230 \times 0.4]$$

$$= 241.6$$

$$_2AS = \frac{1}{0.6}[(241.6 + 350(1-0.06) - 2)(1+0.15) - 1000 \times 0.15 - 560 \times 0.25]$$
$$= 606.6$$
$$_3AS = \frac{1}{0.5}[(606.6 + 350(1-0.06) - 2)(1+0.15) - 1000 \times 0.5]$$
$$= 1147$$

2.已知计划生效日：1985 年 1 月 1 日；

正常退休给付：每服务 1 年月给付 20 元；

精算假设：$i = 7\%, r = 65$；

退休前无死亡之外终止；

参加者资料：生日：1953 年 1 月 1 日

受雇日：1984 年 1 月 1 日

$\ddot{a}_{65}^{(12)} = 8.5$

换算值：

x	D_x	N_x
31	1540	25240
32	1500	24000
41	900	13050
42	860	12150
65	200	1792

试用个体均衡保费法（ILP）法计算 1994 年 1 月 1 日的未来成本现值。

解 $e = 31, a = 32, x = 41$，则有

$$NC \frac{N_{32} - N_{65}}{D_{32}} B_{65} \cdot \frac{D_{65}}{D_{32}} \cdot \ddot{a}_{65}^{(12)}$$

$$NC\left(\frac{24000 - 1792}{1500}\right) = 20 \times 12 \times (65-31) \times \left(\frac{200}{1500}\right) \times 8.5$$

$$NC = 624.64$$

则现值为

$$pv_{41}NC = NC \cdot \frac{N_{41} - N_{65}}{D_{41}}$$
$$= 624.64 \times \frac{13050 - 1792}{900} = 7815.20$$

Exercises

1. Known three-year endowment insurance, the amount of insurance is 1000 yuan, $G = 350$ yuan, $i = 15\%$, as shown in Table 1.

Table 1

K	$p_{x+k}^{(r)}$	$q_{x+k}^{(r)}$	$q_{x+k}^{(1)}$	$q_{x+k}^{(2)}$	c_k	e_k	$_{k+1}CV$
0	0.5	0.5	0.1	0.4	0.2	8	230
1	0.6	0.4	0.15	0.25	0.06	2	560
2	0.5	0.5	0.5	0	0.06	2	1000

Find the asset share of each year.

Solution According to the asset share formula

$_{k+1}AS p_{x+k}^{(r)} = (_kAS + G(1-c_k) - e_k)(1+i) - q_{x+k}^{(1)} - {_{k+1}CV} \cdot q_{x+k}^{(2)}, k = 0,1,2,3\ldots$, we have

$$_{k+1}AS = \frac{1}{p_{x+k}^{(\tau)}}[_k AS + G(1-c_k) - e_k)(1+i) - 1000 q_{x+k}^{(1)} - {}_{k+1}CV \cdot q_{x+k}^{(2)}]$$

$$_0 AS = 0$$

$$_1 AS = \frac{1}{0.5}[(0 + 350(1-0.2) - 8)(1+0.15) - 1000 \times 0.1 - 230 \times 0.4]$$
$$= 241.6$$

$$_2 AS = \frac{1}{0.6}[(241.6 + 350(1-0.06) - 2)(1+0.15) - 1000 \times 0.15 - 560 \times 0.25]$$
$$= 606.6$$

$$_3 AS = \frac{1}{0.5}[(606.6 + 350(1-0.06) - 2)(1+0.15) - 1000 \times 0.5]$$
$$= 1147$$

2. Known effective date of the plan is January 1, 1985;
Normal retirement benefit: 20 yuan per month for each year service;
Actuarial assumptions: $i = 7\%, r = 65$;
No death terminates before retirement;
Participant information: Birthday: January 1, 1953
  Employment date: January 1, 1984
  $\ddot{a}_{65}^{(12)} = 8.5$

The value of translation:

x	D_x	N_x
31	1540	25240
32	1500	24000
41	900	13050
42	860	12150
65	200	1792

Using individual level premium method(ILP) method to calculate the present value of the future cost of January 1, 1994.

Solution $e = 31, a = 32, x = 41$, we have

$$NC \frac{N_{32} - N_{65}}{D_{32}} B_{65} \cdot \frac{D_{65}}{D_{32}} \cdot \ddot{a}_{65}^{(12)}$$

$$NC \left(\frac{24000 - 1792}{1500}\right) = 20 \times 12 \times (65 - 31) \times \left(\frac{200}{1500}\right) \times 8.5$$

$$NC = 624.64$$

Then the present value is $pv_{41} NC = NC \cdot \frac{N_{41} - N_{65}}{D_{41}}$

$$= 624.64 \times \frac{13050 - 1792}{900} = 7815.20$$

第四部分 多元统计分析
Part 4 Multivariate Statistical Analysis

第一章 矩阵代数基本知识
Chapter 1 A Short Excursion into Matrix Algebra

本章帮助大家回忆多元统计分析中特别有用的矩阵代数的一些基本概念,例如特征值与特征向量在多元统计技术中占据重要地位。同时,分析多元正态分布离不开分块矩阵。多元正态分布的几何表示和多元统计技术的几何理解都将频繁的使用向量夹角、点在向量上的投影以及两点间距离等概念。

This chapter is a reminder of basic concepts of matrix algebra, which are particularly useful in multivariate statistical analysis. It also introduces the notations used in this book for vectors and matrices. Eigenvalues and eigenvectors play an important role in multivariate techniques. In analyzing the multivariate normal distribution, partitioned matrices appear naturally. The geometry of the multinormal and the geometric interpretation of the multivariate techniques intensively uses the notion of angles between two vectors, the projection of a point on a vector and the distances between two points.

单词和短语 Words and expressions

秩	rank	特征向量	eigenvectors
迹	trace	二次型	quadratic forms
行列式	determinant	导数	derivatives
转置	transpose	分块矩阵	partitioned matrices
逆阵	inverse	距离	distance
广义逆矩阵	generalized inverse	范数	norm
特征值	eigenvalues	夹角	angle

基本概念和性质 Basic concepts and properties

1 矩阵 A 的迹 $tr(A)$ 等于 A 的特征值之和。

The trace $tr(A)$ is the sum of the eigenvalues of A.

2 向量 $\vec{X}$ 的长度范数定义为 $\|\vec{X}\| = d(0,x) = \sqrt{x^T x}$。

Consider a vector $\vec{X}$, the norm or length of $\vec{X}$ is defined as $\|\vec{X}\| = d(0,x) = \sqrt{x^T x}$.

3 矩阵 A 的秩等于 A 中线性独立的行(或列)的最大数目。

The rank(A) is the maximal number of linearly independent rows (columns) of A.

4 二次型可以被一个对称矩阵 A 描述。

A quadratic form can be described by a symmetric matrix A.

5 二次型可以对角化。

Quadratic forms can always be diagonalized.

6 二次型的正定性与对应矩阵 A 的特征值的正定性等价。

Positive definiteness of a quadratic form is equivalent to positiveness of the eigenvalues of the matrix A.

7 限制条件下的二次型的最大值与最小值可以用特征值表示。

The maximum and minimum of a quadratic form given some constraints can be expressed in terms of eigenvalues.

8 两个 p 维点 x 和 y 间的距离为随 $(x-y)$ 变化的二次型 $(x-y)^T A(x-y)$。距离定义了向量的范数。

A distance between two p-dimensional points x and y is a quadratic form $(x-y)^T A(x-y)$ in the vectors of differences $(x-y)$. A distance defines the norm of a vector.

本章重点
1. 对线性代数中的向量与矩阵的相关概念要理解掌握。

Key points of this Chapter
1. Understand and grasp the concepts of vector and matrix in linear algebra.

第二章 统计分析
Chapter 2 Statistical analysis

本章作为一个起点，将介绍一些描述数据依赖性的基本工具。其中的基本概念来自概率论和初等统计学。

In this chapter, as a starting point, simple and basic tools are used to describe dependency. They are constructed from elementary facts of probability theory and introductory statistics.

单词和短语 Words and expressions

协方差 covariance	厚尾分布 heavy-tailed distribution
相关系数 correlation	混合模型 mixture model
概括统计量 summary statistics	多元混合模型 multivariate mixture model
线性变换 linear transformation	联结函数 copulae
两变量线性模型 linear model for two variables	威沙特分布 Wishart distribution
多元线性模型 multiple linear model	霍特林 T^2 分布 Hotelling's T^2 distribution
累积分布函数 cumulative distribution function	似然函数 likelihood function
概率密度函数 probability density function	克拉美—拉奥下界 Cramer-Rao lower bound
条件期望 conditional expectations	费希尔信息矩阵 Fisher information matrix
特征方程 characteristic functions	似然比检验 likelihood ratio test
变换 transformation	置信域 confidence region
多元正态分布 the multinormal distribution	线性假设 linear hypothesis

基本概念和性质 Basic concepts and properties

1. 存在非线性依赖的两个随机变量的协方差为 0。
   There are nonlinear dependencies that have zero covariance.

2. 对于小的样本，我们应该用 $\frac{1}{(n-1)}$ 代替因子 $\frac{1}{n}$ 进行计算。
   For small n, we should replace the factor $\frac{1}{(n-1)}$ in the computation of the covariance by $\frac{1}{n}$.

3. 在给定 X 中的一个可能的观察值 x 时，线性回归可以预测 Y。
   A linear regression predicts values of Y given a possible observation x of X.

4. 关于假设 $\beta = 0$ 的 t 检验为 $t = \frac{\hat{\beta}}{SE(\hat{\beta})}$，其中 $SE(\hat{\beta}) = \frac{\hat{\sigma}}{(n \cdot s_{XX})^{\frac{1}{2}}}$。
   The t-test for the hypothesis $\beta = 0$ is $t = \frac{\hat{\beta}}{SE(\hat{\beta})}$, where $SE(\hat{\beta}) = \frac{\hat{\sigma}}{(n \cdot s_{XX})^{\frac{1}{2}}}$.

5. 简单线性回归模型 $y_i = \alpha + \beta x_i + \varepsilon_i$ 的 F 统计量等于斜率系数 t 检验统计量的平方。
   The F-test statistic for the slope of the linear regression model $y_i = \alpha + \beta x_i + \varepsilon_i$ is the square of the t-test statistic.

6. 模型 $y = X\beta + e$ 构建了一维变量 Y 与 p 维变量 X 的线性关系。最小二乘估计量为 $\hat{\beta} = (X^T X)^{-1} X^T y$。
   The relation $y = X\beta + e$ models a linear relation between a onedimensional variable Y and a p-dimensional variable X. The least squares parameter estimator is $\hat{\beta} = (X^T X)^{-1} X^T y$.

7 随机向量 X 的特征方程为 $\varphi_X(t) = E(e^{it^T X})$。
The characteristic function of a random vector X is $\varphi_X(t) = E(e^{it^T X})$.

8 条件期望 $E(X_2 | X_1)$ 是 X_2 的 MSE 最优近似,它是 X_1 的函数。
The conditional expectation $E(X_2 | X_1)$ is the MSE best approximation of X_2 by a function of X_1.

9 在线性关系 $Y = AX + b$ 的情况下,X 与 Y 的密度函数通过 $f_Y(y) = abs(|A|^{-1}) f_X\{A^{-1}(y-b)\}$ 联系起来。
In the case of a linear relation $Y = AX + b$ the pdf's of X and Y are related via $f_Y(y) = abs(|A|^{-1}) f_X\{A^{-1}(y-b)\}$.

10 若 $X_1, X_2, \cdots, X_n$ 为独立同分布的随机向量,$X_i \sim N(\mu, \sigma)$,则由 CLT 可以构建一个渐进的置信区间:$\bar{x} \pm \dfrac{\hat{\sigma}}{\sqrt{n}} u_{1-\frac{\alpha}{2}}$。
If $X_1, X_2, \cdots, X_n$ are i.i.d. random variables with $X_i \sim N(\mu, \sigma)$, then an asymptotic confidence interval can be constructed by the CLT: $\bar{x} \pm \dfrac{\hat{\sigma}}{\sqrt{n}} u_{1-\frac{\alpha}{2}}$.

11 T^2 分布和 F 分布的关系为 $T^2(p, n) = \dfrac{np}{n-p+1} F_{p, n-p+1}$。
The relation between Hotelling's T^2 and Fisher's F-distribution is given by $T^2(p, n) = \dfrac{np}{n-p+1} F_{p, n-p+1}$.

12 得分函数是对数似然函数相对于 θ 的导数 $s(X; \theta) = \dfrac{\partial}{\partial \theta} l(X; \theta)$。$s(X; \theta)$ 的协方差是费希尔信息矩阵。$E\{s(X; \theta)\} = 0$。
The score function is the derivative $s(X; \theta) = \dfrac{\partial}{\partial \theta} l(X; \theta)$ of the loglikelihood with respect to θ. The covariance matrix of $s(X; \theta)$ is the Fisher information matrix. $E\{s(X; \theta)\} = 0$.

例题 Example

经典蓝套衫公司正在分析三个市场策略的影响:
1. 在当地报纸上登广告;
2. 销售助理的存在;
3. 商店橱窗上用豪华装饰进行宣传。

以上三个策略在 10 个不同的商店中进行试验,最终的销售额观察值如表所示。

shop k	marketing strategy factor l		
1	9	10	18
2	11	15	14
3	10	11	17
4	12	15	9
5	7	15	14
6	11	13	17
7	12	7	16
8	10	15	14
9	11	13	17
10	13	10	15

现讨论一下三个市场策略是否具有相同的均值影响或者差异。

解 假设检验为 $H_0: \mu_l = \mu, l = 1, \cdots, 3$,备则假设为市场策略有不同的影响,可以写作 $H_1: \mu_l \neq \mu_{l'}, l = 1, \cdots, 3$,对任意的 l 和 l',这意味着一个市场策略比其余的要好。

利用 $F = \dfrac{\dfrac{\{SS(reduced) - SS(full)\}}{df(r) - df(f)}}{\dfrac{SS(full)}{df(f)}}$，其中 $SS(reduced) = \sum_{l=1}^{p}\sum_{k=1}^{m}(y_{kl} - \overline{y})^2$，$SS(full) = \sum_{l=1}^{p}\sum_{k=1}^{m}(y_{kl} - \overline{y_l})^2$，$df(f)$ 和 $df(r)$ 分别表示完全模型和退化模型中的自由度。

所以 $df(f) = n - p = 30 - 3 = 27$，$df(r) = n - p = 30 - 1 = 29$。

可以解得 $SS(reduced) = \sum_{l=1}^{p}\sum_{k=1}^{m}(y_{kl} - \overline{y})^2 = 260.3$，$SS(full) = \sum_{l=1}^{p}\sum_{k=1}^{m}(y_{kl} - \overline{y_l})^2 = 157.7$。

所以 F 统计量是 $F = 8.78$。

此值需要和 $F_{(2,27)}$ 分布的临界值进行比较。查看 F 分布的临界值可以看出在此检验下统计量是高度显著的。我们因此得出结论：不同的市场策略有不同的影响。

The classic blue pullover company analyzes the effect of three marketing strategies：
1. advertisement in local newspaper；
2. presence of sales assistant；
3. luxury presentation in shop windows.

All of these strategies are tried in 10 different shops. The resulting sale observations are given in Table. There are p = 3 factors and n = mp = 30 observations in the data. The classic blue pullover company wants to know whether all three marketing strategies have the same mean effect or whether there are differences.

Solution The hypothesis to be tested is therefore $H_0 : \mu_l = \mu, l = 1, \cdots, 3$. The alternative hypothesis, that the marketing strategies have different effects, can be formulated as $H_1 : \mu_l \neq \mu_{l'}, l = 1, \cdots, 3$ for any l and l'. This means that one marketing strategy is better than the others.

The method used to test this problem is to compute $F = \dfrac{\dfrac{\{SS(reduced) - SS(full)\}}{df(r) - df(f)}}{\dfrac{SS(full)}{df(f)}}$, in it $SS(reduced) = \sum_{l=1}^{p}\sum_{k=1}^{m}(y_{kl} - \overline{y})^2$, $SS(full) = \sum_{l=1}^{p}\sum_{k=1}^{m}(y_{kl} - \overline{y_l})^2$, $df(f)$ and $df(r)$ denote the degrees of freedom under the full model and the reduced model respectively. The degrees of freedom are essential in specifying the shape of the F-distribution.

They have a simple interpretation：df(_) is equal to the number of observations minus the number of parameters in the model. $df(f) = n - p = 30 - 3 = 27$. Under the reduced model, there is one parameter to estimate, namely the overall mean, $df(r) = n - p = 30 - 1 = 29$. We can compute $SS(reduced) = \sum_{l=1}^{p}\sum_{k=1}^{m}(y_{kl} - \overline{y})^2 = 260.3$ and $SS(full) = \sum_{l=1}^{p}\sum_{k=1}^{m}(y_{kl} - \overline{y_l})^2 = 157.7$.

The F-statistic is therefore $F = 8.78$.

This value needs to be compared to the quantiles of the $F_{(2,27)}$ distribution. Looking up the critical values in a F-distribution shows that the test statistic above is highly signficant. We conclude that the marketing strategies have different effects.

本章重点

1 解均值和协方差的基本性质（在边际及条件分布意义上）。
2 掌握多元正态分布的概率属性，了解多元正态分布的两个同伴分布：威沙特分布和霍特林分布。
3 会使用基本理论工具获得估计量，并基于最大似然理论研究其渐近的最优性质。
4 学会假设检验和置信区间的构建。

Key points of this Chapter

1 Understand the basic properties on means and covariances (on marginal and conditional ones).

2 Grasp the probabilistic properties of the multinormal, understand two "companion" distributions of the multinormal: the Wishart and the Hotelling distributions.
3 Use the basic theoretical tools are developed which are needed to derive estimators and to determine their asymptotic optimal properties properties on maximum likelihood theory.
4 We should concentrate on hypothesis testing and confidence interval issues.

第三章 主成分分析
Chapter 3 Principal Components Analysis

主成分分析就是设法将原来指标重新组合成一组新的互相无关的几个综合指标来代替原来指标。同时根据实际需要从中可取几个较少的综合指标尽可能多地反映原来的指标的信息。

Principal components analysis is to the original index reassembled into a new set of independent several comprehensive index to replace the original index. At the same time, according to the actual need of desirable from the few comprehensive index as much as possible to reflect the original index information.

单词和短语 Words and expressions

主成分分析　principal components analysis　　　旋转变换　rotation transformation
因子载荷量　factor load capacity

基本概念和性质　Basic concepts and properties

1 第 k 个主成分 y_k 的系数向量是第 k 个特征根 λ_k 所对应的标准化特征向量 U_k。
The coefficient vector y_k of the kth principal component is the standard feature vector U_k. of the kth characteristic root λ_k.

2 第 k 个主成分的方差为第 k 个特征根 λ_k，且任意两个主成分都是不相关的，也就是主成分 $y_1, y_2, \cdots, y_p$ 的样本协方差矩阵是对角矩阵。
The variance of the kth principal component is the kth characteristic root λ_k, and any two principal components are not correlated, which means the sample covariance matrix of the principal components $y_1, y_2, \cdots, y_p$ is a diagonal matrix.

3 样本主成分的总方差等于原变量样本的总方差。
The total variance of the principal component is equel to the total variance of the primary variable samples.

4 第 k 个样本主成分与第 j 个变量样本之间的相关系数为：$r(y_k, x_j) = r(y_k, zx_j) = \sqrt{\lambda_k} u_{kj}$。
The correlation of the kth sample principal component and the jth variable samples is: $r(y_k, x_j) = r(y_k, zx_j) = \sqrt{\lambda_k} u_{kj}$.

5 该相关系数又称为因子载荷量。
The correlation coefficient is also called factor load capacity.

6 旋转变换的目的是为了使得 n 个样本点在 y_1 轴方向上的离散程度最大，即 y_1 的方差最大，变量 y_1 代表了原始数据的绝大部分信息，在研究问题时，即使不考虑变量 y_2 也损失不多的信息。y_1 与 y_2 除起了浓缩作用外，还具有不相关性。y_1 称为第一主成分，y_2 称为第二主成分。
Rotation transformation aims to make the biggest discrete degree in the y_1-axis direction of the n sample points, which means the variance in the y_1-axis direction is the biggest, variable y_1 behalf the most information of the original data, you won't lost much information if you don't consider variable y_1. With the exception of the concentration effect between y_1 and y_2, it also has irrelevance. y_1 is called the first principal component, and y_2 is called the second principal component.

例题　Example

例 我国 2000 年各地区大中型工业企业主要经济效益指标见表1，对各地区经济效益作出分析。

解 将数据标准化，并求相关矩阵 R 为：

$$\begin{pmatrix} 1 & & & & & & \\ 0.704 & 1 & & & & & \\ -0.566 & -0.375 & 1 & & & & \\ -0.146 & 0.418 & -0.058 & 1 & & & \\ 0.773 & 0.771 & -0.615 & 0.127 & 1 & & \\ 0.385 & 0.523 & -0.621 & 0.430 & 0.521 & 1 & \\ -0.007 & 0.163 & 0.181 & 0.245 & -0.117 & -0.079 & 1 \end{pmatrix}$$

求 R 的特征根及相应的单位正交特征向量和贡献率。由 R 的特征方程 $|R-\lambda I|=0$ 求得 R 的单位特征根 λ 为:

$$\lambda_1=3.422\ \lambda_2=1.445\ \lambda_3=1.017\ \lambda_4=0.590\ \lambda_5=0.279\ \lambda_6=0.162\ \lambda_7=0.085$$

再由齐次线性方程组求得特征向量 U, 将具体结果整理为下表:

对应的 特征向量							
	U_{1j}	.440	-.250	.414	.016	.157	-.509	.538
	U_{2j}	.460	.228	.241	.398	.082	-.205	-.690
	U_{3j}	-.408	.227	.247	.591	.530	.175	.244
	U_{4j}	.158	.690	-.373	.233	-.354	-.223	.366
	U_{5j}	.487	-.126	.130	.243	-.252	.748	.219
	U_{6j}	.408	.153	-.450	-.285	.708	.158	.037
	U_{7j}	-.022	.566	.592	-.544	.007	.182	.012
特征根		3.422	1.445	1.017	0.590	0.279	0.162	0.085
		48.88%	20.65%	14.52%	8.43%	3.99%	2.32%	1.21%

接下来确定主成分的个数 q:

按 $\lambda \geqslant 1$ 的原则, 取三个主成分就能够对工业企业经济效益进行分析, 且这三个主成分的累计方差贡献率达到 84.06% 的主成分的表达式为:

$y_1=0.440xx1+0.46xx2-0.48xx3+0.158xx4+0.487xx5+0.408xx6-0.022xx7$
$y_2=-0.251xx1+0.228xx2+0.227xx3+0.690xx4-0.126xx5+0.153xx6+0.566xx7$
$y_3=0.414xx1+0.241xx2+0.247xx3-0.373xx4+0.130xx5-0.45xx6+0.592xx7$

主成分的经济意义:

y_1 的含义是在综合其它变量所反映信息的基础上, 突出地反映了企业经营风险的大小。
y_2 在综合其它变量信息的基础上, 突出地反映了企业投入资金的周转速度。
y_3 在综合其它变量信息的基础上, 突出地反映了工业产品满足社会需求的情况。

EX. 1 The main economic indicators of the large and medium-sized industrial enterprises of our country in 2000. Analysis the economic indicators.

Solution Makeing the data standardization, then we get the matrix R:

$$\begin{pmatrix} 1 & & & & & & \\ 0.704 & 1 & & & & & \\ -0.566 & -0.375 & 1 & & & & \\ -0.146 & 0.418 & -0.058 & 1 & & & \\ 0.773 & 0.771 & -0.615 & 0.127 & 1 & & \\ 0.385 & 0.523 & -0.621 & 0.430 & 0.521 & 1 & \\ -0.007 & 0.163 & 0.181 & 0.245 & -0.117 & -0.079 & 1 \end{pmatrix}$$

Get the characteristic roots, eigenvector with normal orthogonal unit and contribution rate. Get the characteristic roots by using the characteristic equation $|R-\lambda I|=0$:

$$\lambda_1=3.422\ \lambda_2=1.445\ \lambda_3=1.017\ \lambda_4=0.590\ \lambda_5=0.279\ \lambda_6=0.162\ \lambda_7=0.085$$

Again, get the eigenvector U by using the homogeneous linear equations, sort the results in list:

对应的特征向量	U_{1j}	.440	−.250	.414	.016	.157	−.509	.538
	U_{2j}	.460	.228	.241	.398	.082	−.205	−.690
	U_{3j}	−.408	.227	.247	.591	.530	.175	.244
	U_{4j}	.158	.690	−.373	.233	−.354	−.223	.366
	U_{5j}	.487	−.126	.130	.243	−.252	.748	.219
	U_{6j}	.408	.153	−.450	−.285	.708	.158	.037
	U_{7j}	−.022	.566	.592	−.544	.007	.182	.012
特征根		3.422	1.445	1.017	0.590	0.279	0.162	0.085
		48.88%	20.65%	14.52%	8.43%	3.99%	2.32%	1.21%

Next, make sure the number of the main component-q:

Because $\lambda \geq 1$, take out three main components will analysis the economic indicators, and the expression of the three main components cumulative variance contribution over 84.06%, expression like this:

$y_1 = 0.440zx1 + 0.46zx2 - 0.48zx3 + 0.158zx4 + 0.487zx5 + 0.408zx6 - 0.022zx7$
$y_2 = -0.251zx1 + 0.228zx2 + 0.227zx3 + 0.690zx4 - 0.126zx5 + 0.153zx6 + 0.566zx7$
$y_3 = 0.414zx1 + 0.241zx2 + 0.247zx3 - 0.373zx4 + 0.130zx5 - 0.45zx6 + 0.592zx7$

Economic significance of the main component:

y_1: base on comprehensive the information which the other variables reflectted, reflect the business risk.

y_2: base on comprehensive the information of other variables, reflect the turnover speed of capital input.

y_3: base on comprehensive the information of other variables, reflect t e situation of meeting social needs.

本章重点

1. 在应用主成分分析研究问题时,通常先将数据标准化,以消除量纲对结果的影响。标准化的常用公式为:$zx_i = \dfrac{x_i - E(x_i)}{\sqrt{D(x_i)}}$,为了求出主成分,只需求样本协方差矩阵 S 或相关系数矩阵 R 的特征根和特征向量就可以。(可以证明,变量 $x_1, x_2, \cdots, x_p$ 标准化以后,其协方差矩阵 S 与相关系数矩阵 R 相等。)

2. 在应用主成分分析研究问题的基本步骤:
   (1)对原变量的样本数据矩阵进行标准化变换;
   (2)求标准化数据矩阵的相关系数矩阵 R;
   (3)求 R 的特征根及相应的特征向量和贡献率等;
   (4)确定主成分的个数;
   (5)解释主成分的实际意义和作用。

Key points of this Chapter

1. When research questions with principal component analysis, we make the data standardization first, in order to eliminate the effect of the dimension, formula of standardization is: $zx_i = \dfrac{x_i - E(x_i)}{\sqrt{D(x_i)}}$, If you want to get the principal component, you just need the covariance matrix S or the characteristic root and the feature vector of coefficient matrix R. (We can prove, if we standardization the variables, then S equel to R)

2. The steps of research questions with principal component analysis:
   (1) make the data standardization first;
   (2) get the coefficient matrix R;

(3) get the characteristic root, the feature vector and the contribution rate of coefficient matrix R;
(4) make sure the number of the principal components;
(5) explane the practical significance and action of the principal components.

表 1

地区	工业增加值率(%)x1	总资产贡献率(%)x2	资产负债率(%)x3	流动资产周转次数 x4	成本费用利润率(%)x5	劳动生产率(元/人年)x6	产品销售率(%)x7
北 京	27.90	5.22	57.23	1.31	2.63	53987.95	98.10
天 津	27.28	8.12	58.84	1.85	6.80	78191.27	99.35
河 北	36.46	8.12	60.28	1.49	5.49	42629.81	98.90
山 西	36.80	5.41	62.59	.88	2.49	24413.53	97.97
内蒙古	38.09	6.09	57.37	1.24	2.31	35129.88	99.02
辽 宁	28.15	7.36	59.49	1.48	4.67	47955.68	98.37
吉 林	28.18	8.28	64.72	1.28	6.20	40141.03	98.89
黑龙江	52.80	24.07	55.81	1.67	37.64	73120.46	99.07
上 海	29.29	9.01	47.48	1.51	7.21	118816.16	99.46
江 苏	26.48	8.58	59.73	1.68	4.48	56044.81	98.26
浙 江	26.48	10.84	55.04	1.85	6.67	66785.21	98.22
安 徽	32.67	7.35	60.96	1.36	2.67	35674.30	99.11
福 建	34.22	10.89	58.52	1.85	6.79	91263.70	97.84
江 西	28.51	6.40	67.53	1.19	1.55	29123.49	98.11
山 东	32.60	11.70	61.54	1.92	8.42	52621.24	98.32
河 南	30.77	7.39	65.02	1.23	4.07	29296.86	98.31
湖 北	33.75	7.46	62.58	1.23	4.58	49374.62	101.23
湖 南	34.48	9.67	66.40	1.29	2.20	38179.94	99.61
广 东	31.23	10.56	56.89	1.76	7.59	120863.34	97.99
广 西	33.70	8.63	69.68	1.39	5.54	44140.89	98.16
海 南	25.93	6.44	66.08	1.18	4.46	58521.50	94.96
重 庆	30.25	6.11	63.82	1.07	2.19	36638.20	99.54
四 川	31.62	6.84	63.93	1.07	4.22	35013.75	98.98
贵 州	34.06	7.99	68.29	.83	3.71	32317.42	99.71
云 南	55.32	20.93	50.04	1.37	11.59	96702.70	99.09
西 藏	57.56	10.67	25.43	.59	29.89	103001.24	95.78
陕 西	37.04	8.28	67.63	1.07	7.67	38102.97	98.00
甘 肃	29.98	5.14	65.87	1.08	1.18	34352.92	97.47
青 海	30.79	5.21	72.47	.58	1.67	51599.98	97.38
宁 夏	30.62	6.32	60.98	1.16	3.07	33620.79	96.65
新 疆	45.02	12.81	60.85	1.78	15.81	118599.82	98.73

第四章 因子分析
Chapter 4 Factor Analysis

因子分析是通过对变量相关系数矩阵内部结构的研究，找出能够控制所有变量的少数几个潜在随机变量去描述多个显著随机变量之间的相关关系。换句话说，因子分析是把每个可观测的原始变量分解为两部分因素，一部分是由所有变量共同具有少数几个公共因子构成的，另一部分是每个原始变量独自具有的，即特殊因子部分，对于所研究的问题就可试图用最少个数的不可观测的公共因子的线性函数与特殊因子之和来描述原来观测的每一分量。

Factor analysis are based on the research of the internal structure of the coefficient matrix, finding a few potential random variable who can control all the variable, to describe the relationship of the remarkable random variable. In other words, factor analysis divide the original variable into two parts, one part is made of a few public factors of all the variable, another part is unique of each variable, that is specific factors, then we use the linear function of the few potential public factors and the sum of the specific factors to discribe every variable.

单词和短语 Words and expressions

因子分析　factor analysis　　　　　　public factor
变量共同度　common degree of variable　　载荷矩阵　loading matrix
公共因子的方差贡献　variance contribution of

基本概念和性质 Basic common factor properties

■ 符号与假定：设有 n 个样本，每个样本观测 p 个变量，记：原始变量矩阵为 $X: X = \begin{bmatrix} x_1 \\ x_2 \\ \vdots \\ x_p \end{bmatrix}$，公共因子变量矩阵为 $F: F = \begin{bmatrix} F_1 \\ F_2 \\ \vdots \\ F_q \end{bmatrix}$，特殊因子矩阵为 $E: E = \begin{bmatrix} e_1 \\ e_2 \\ \vdots \\ e_p \end{bmatrix}$。

假定因子模型具有以下性质：
1. $E(X) = 0, Cov(X) = \Sigma$；
2. $E(F) = 0, Cov(F) = I$；
3. $E(E) = 0, Cov(E) = diag(\sigma_1^2, \cdots, \sigma_p^2)$；
4. $Cov(F, E) = 0$。

数学模型为：
$$\begin{cases} x_1 = a_{11}F_1 + a_{12}F_2 + \cdots\cdots + a_{1p}F_p + e_1 \\ x_2 = a_{21}F_1 + a_{22}F_2 + \cdots\cdots + a_{2p}F_p + e_2 \\ \cdots\cdots \\ x_p = a_{p1}F_1 + a_{p2}F_2 + \cdots\cdots + a_{pp}F_p + e_p \end{cases}$$

若用矩阵形式表示，则为：$X = AF + E$。式中的 A，称为因子载荷矩阵，并且称 a_{ij} 为第 i 个变量在第 j 个公共因子上的载荷，反映了第 i 个变量在第 j 个公共因子上的相对重要性。

1. Symbol and assume: If there are n samples, observe p variables each sample, note:

original variable matrix is $X: X = \begin{bmatrix} x_1 \\ x_2 \\ \vdots \\ x_p \end{bmatrix}$, public factors variable matrix is $F: F = \begin{bmatrix} F_1 \\ F_2 \\ \vdots \\ F_q \end{bmatrix}$, specific factors matrix is E: $E = \begin{bmatrix} e_1 \\ e_2 \\ \vdots \\ e_p \end{bmatrix}$

Assum the model of factor have the basic properties:

1. $E(X) = 0, Cov(X) = \Sigma$;
2. $E(F) = 0, Cov(F) = I$;
3. $E(E) = 0, Cov(E) = diag(\sigma_1^2, \cdots, \sigma_p^2)$;
4. $Cov(F, E) = 0$。

Mathematical model is:

$$\begin{cases} x_1 = a_{11}F_1 + a_{12}F_2 + \cdots\cdots + a_{1p}F_p + e_1 \\ x_2 = a_{21}F_1 + a_{22}F_2 + \cdots\cdots + a_{2p}F_p + e_2 \\ \cdots\cdots \\ x_p = a_{p1}F_1 + a_{p2}F_2 + \cdots\cdots + a_{pp}F_p + e_p \end{cases}$$

If expressed in matrixform, it is $X = AF + E$. In the formula, A is called factor loadig metrix, and aig is the loading of geh jth common factor on the bith varian which reflects the relative dmportance of ith variance on the jth common factor.

2. 因子载荷的统计含义：因子载荷 a_{ij} 为第 i 个变量 x_i 与第 j 个公共因子 F_j 的相关系数，即反映了变量与公共因子的关系密切程度，a_{ij} 越大，表明公共因子 F_j 与变量 x_i 的线性关系越密切。

Statistics meaning of the factor loading matrix: factor loading matrix means the correlation coefficient of the No. i variable x_i and the No. j public factor F_j, indicate the intimate level between the variable and the public factor, the more the a_{ij} is, means the closer linear relationship.

3. 变量共同度：因子载荷矩阵中各行元素的平方和：

$$\begin{cases} h_1^2 = a_{11}^2 + a_{12}^2 + \cdots\cdots + a_{1q}^2 \\ h_2^2 = a_{21}^2 + a_{22}^2 + \cdots\cdots + a_{2q}^2 \\ \cdots\cdots \\ h_p^2 = a_{p1}^2 + a_{p2}^2 + \cdots\cdots + a_{pq}^2 \end{cases}$$

称为变量 $x_1, x_2, \cdots, x_p$ 的共同度。它表示 q 个公共因子对变量 x_i 的方差贡献，变量共同度的最大值为 1，值越接近于 1，说明该变量所包含的原始信息被公共因子所解释的部分越大，用 q 个公共因子描述变量 x_i 就越有效；而当值接近于 0 时，说明公共因子对变量的影响很小，主要由特殊因子来描述。

3. Common degree of variable: sum of squares of each line elements in the factor loading matrix:

$$\begin{cases} h_1^2 = a_{11}^2 + a_{12}^2 + \cdots\cdots + a_{1q}^2 \\ h_2^2 = a_{21}^2 + a_{22}^2 + \cdots\cdots + a_{2q}^2 \\ \cdots\cdots \\ h_p^2 = a_{p1}^2 + a_{p2}^2 + \cdots\cdots + a_{pq}^2 \end{cases}$$

called the common degree of variable $x_1, x_2, \cdots, x_p$. express the variance contribution of the q public factors to the variable x_i, the maximum value of the common degree is one, more close to one, means more original information the public factors can explain and more effective to describe the variable x_i with the q public factors; when it's close to zero, means less effective and described by the specific factors chiefly.

4. 公共因子的方差贡献：因子载荷矩阵中各列元素的平方和：

$$\begin{cases} g_1 = a_{11}^2 + a_{21}^2 + \cdots\cdots + a_{p1}^2 \\ g_2 = a_{12}^2 + a_{22}^2 + \cdots\cdots + a_{p2}^2 \\ \cdots\cdots \\ g_q = a_{1q}^2 + a_{2q}^2 + \cdots\cdots + a_{pq}^2 \end{cases}$$

称为公共因子 $F_1, F_2, \cdots, F_p$ 的方差贡献。它与 p 个变量的总方差之比为：

$$F_j \text{ 的贡献率} = \frac{g_j}{p} = \frac{\sum_{i=1}^{p} a_{ij}^2}{p}$$

它是衡量各个公共因子相对重要程度的一个指标。方差贡献率越大，该因子就越重要。

4. Variance contribution of public factors: Sum of squares of each line elements in the factor loading matrix:

$$\begin{cases} g_1 = a_{11}^2 + a_{21}^2 + \cdots\cdots + a_{p1}^2 \\ g_2 = a_{12}^2 + a_{22}^2 + \cdots\cdots + a_{p2}^2 \\ \cdots\cdots \\ g_q = a_{1q}^2 + a_{2q}^2 + \cdots\cdots + a_{pq}^2 \end{cases}$$

Called the variance contribution of the public factors $F_1, F_2, \cdots, F_p$. The ratio of it to the p variable's total variance is:

$$\text{Contribution rate of } F_j = \frac{g_j}{p} = \frac{\sum_{i=1}^{p} a_{ij}^2}{p}$$

It's the index to judge the relative important extent of every public factor. The biggest the variance contribution is, the more important the factor is.

例题 Example

例 仍以我国 2000 年各地区大中型工业企业主要经济效益指标作为研究对象，试求：(1)正交因子模型；(2)各个变量的共同度以及特殊因子方差；(3)每个因子的方差贡献率以及三个因子的累计方差贡献率；

解 (1)将原始数据标准化后求得其相关系数矩阵 R 为

$$\begin{bmatrix} 1 & & & & & & \\ 0.704 & 1 & & & & & \\ -0.566 & -0.375 & 1 & & & & \\ -0.146 & 0.418 & -0.058 & 1 & & & \\ 0.773 & 0.771 & -0.615 & 0.127 & 1 & & \\ 0.385 & 0.523 & -0.621 & 0.431 & 0.521 & 1 & \\ -0.007 & 0.163 & 0.181 & 0.245 & -0.117 & -0.079 & 1 \end{bmatrix}$$

(2)求出特征根与特征向量：$\lambda_1 = 3.422$ $\lambda_2 = 1.445$ $\lambda_3 = 1.017$.

$$U = \begin{bmatrix} .440 & -.250 & .414 \\ .460 & .228 & .241 \\ -.408 & .227 & .247 \\ .158 & .689 & .373 \\ .487 & -.126 & .130 \\ .408 & .153 & -.450 \\ -.022 & .566 & .592 \end{bmatrix}。$$

(3)因子载荷矩阵为：

$$A = \sqrt{\lambda}U = \begin{bmatrix} .814 & -.301 & .417 \\ .851 & .274 & .243 \\ -.754 & .273 & .249 \\ .293 & .829 & -.376 \\ .901 & -.151 & .131 \\ .754 & .184 & -.454 \\ -.040 & .680 & .597 \end{bmatrix}.$$

(4)因子模型为
$$\begin{cases} x_1 = 0.814F_1 - 0.301F_2 + 0.417F_3 + e_1 \\ x_2 = 0.851F_1 + 0.274F_2 + 0.243F_3 + e_2 \\ x_3 = -0.754F_1 + 0.273F_2 + 0.249F_3 + e_3 \\ x_4 = 0.293F_1 + 0.829F_2 - 0.376F_3 + e_4 \\ x_5 = 0.901F_1 - 0.151F_2 + 0.131F_3 + e_5 \\ x_6 = 0.754F_1 + 0.184F_2 - 0.454F_3 + e_6 \\ x_7 = -0.040F_1 + 0.680F_2 + 0.597F_3 + e_7 \end{cases}.$$

变量	因子载荷			共同度	特殊因子方差
	F1	F2	F3		
X1	814	−.301	417	926	0.074
X2	851	274	243	858	0.142
X3	−754	273	249	705	0.295
X4	293	829	−376	914	0.086
X5	901	−151	131	853	0.147
X6	754	184	−454	808	0.192
X7	4.02E−02	680	597	820	0.180
方差贡献率	48.88%	20.66%	14.52%	—	—
累计方差贡献率	48.88%	69.53%	84.05%	—	—

Exercise: Still choose the main economic indicators of the large and medium-sized industrial enterprises of our country in 2000 as the research object, try to get:(1)the orthogonal factor model;(2)common degree of every variable and the variance of special factors;(3) variance contribution of every factor and cumulative variance contribution of three factors.

Solution(1) Making the data standardization, get the matrix R:

$$\begin{bmatrix} 1 & & & & & & \\ 0.704 & 1 & & & & & \\ -0.566 & -0.375 & 1 & & & & \\ -0.146 & 0.418 & -0.058 & 1 & & & \\ 0.773 & 0.771 & -0.615 & 0.127 & 1 & & \\ 0.385 & 0.523 & -0.621 & 0.431 & 0.521 & 1 & \\ -0.007 & 0.163 & 0.181 & 0.245 & -0.117 & -0.079 & 1 \end{bmatrix}$$

(2)Get the characteristic roots and the eigenvector:

$$\lambda_1 = 3.422 \quad \lambda_2 = 1.445 \quad \lambda_3 = 1.017, U = \begin{bmatrix} .440 & -.250 & .414 \\ .460 & .228 & .241 \\ -.408 & .227 & .247 \\ .158 & .689 & .373 \\ .487 & -.126 & .130 \\ .408 & .153 & -.450 \\ -.022 & .566 & .592 \end{bmatrix}.$$

(3) Factor loading matrix:

$$A = \sqrt{\lambda} U = \begin{bmatrix} .814 & -.301 & .417 \\ .851 & .274 & .243 \\ -.754 & .273 & .249 \\ .293 & .829 & -.376 \\ .901 & -.151 & .131 \\ .754 & .184 & -.454 \\ -.040 & .680 & .597 \end{bmatrix}.$$

(4) Orthogonal factor model $\begin{cases} x_1 = 0.814F_1 - 0.301F_2 + 0.417F_3 + e_1 \\ x_2 = 0.851F_1 + 0.274F_2 + 0.243F_3 + e_2 \\ x_3 = -0.754F_1 + 0.273F_2 + 0.249F_3 + e_3 \\ x_4 = 0.293F_1 + 0.829F_2 - 0.376F_3 + e_4 \\ x_5 = 0.901F_1 - 0.151F_2 + 0.131F_3 + e_5 \\ x_6 = 0.754F_1 + 0.184F_2 - 0.454F_3 + e_6 \\ x_7 = -0.040F_1 + 0.680F_2 + 0.597F_3 + e_7 \end{cases}$

本章重点

1 因子的求解：设相关系数矩阵的特征根为 $\lambda_1 \geqslant \lambda_2 \geqslant \cdots \geqslant \lambda_p$，相应的特征向量为 $U_1, U_2, \cdots, U_p$，设由列向量构成的矩阵有 A 表示，即 $A = (\sqrt{\lambda_1} U_1, \sqrt{\lambda_2} U_2, \ldots, \sqrt{\lambda_p} U_p)$。

一般来说，公共因子的个数 q 要小于等于变量的个数 p。

2 因子分析的基本步骤：

(1) 用公式 $zx = \dfrac{x - E(x)}{\sqrt{D(x)}}$ 对原始数据标准化；

(2) 建立相关系数矩阵 R；

(3) 根据 $|R - \lambda I| = 0$ 及 $(R - \lambda I)U = 0$，求 R 的单位特征根 λ 与特征向量 U；

(4) 根据 $A = \sqrt{\lambda} U$ 求因子载荷矩阵 A；

(5) 写出因子模型 $X = AF + E$。

Key points of this Chapter

1 Calculation of factor: Let the characteristic roots $\lambda_1 \geqslant \lambda_2 \geqslant \cdots \geqslant \lambda_p$, the eigenvector $U_1, U_2, \cdots, U_p$, the matrix of column is $A = (\sqrt{\lambda_1} U_1, \sqrt{\lambda_2} U_2, \ldots, \sqrt{\lambda_p} U_p)$.

Generally speaking, the number q of public factor is less than the number p of variable.

2 Steps of factor analysis:

(1) Making the data standardization with the formula $zx = \dfrac{x - E(x)}{\sqrt{D(x)}}$;

(2) Buliding the coefficient matrix R;

(3) According to $|R-\lambda I|=0$ and $(R-\lambda I)U = 0$, get the characteristic root λ and the eigenvector U;

(4) According to $A=\sqrt{\lambda}U$, get the factor loading matrix A;

(5) Get the factor model $X = AF+E$.

第五章　聚类分析
Chapter 5　Cluster Analysis

聚类分析的基本思想是认为研究的样本或变量之间存在着程度不同的相似性，根据一批样本的多个观测指标，具体找出一些能够度量样本或指标之间相似程度的统计量，以这些统计量为划分类型的依据，把一些相似程度较大的样本（或变量）聚合为一类，把另外一些彼此之间相似程度较大的样本（变量）也聚合为一类，关系密切的聚合到一个小的分类单位，关系疏远的聚合到一个大的分类单位，直到把所有的样本（或变量）都聚合完毕，把不同的类型一一划分出来，形成一个由小到大的分类系统；最后再把整个分类系统画成一张图，将亲疏关系表示出来。

Cluster analysis takes it that there exists similarity between samples and variables in varying degrees. According to various observing index in a number of samples, statistic scores, which can measure the degree of similarity between samples and index, are found specifically as the reason of type classification. Samples or variables with greater degree of similarity are aggregated as one group, while other samples or variables with greater degree of similarity are aggregated as one group; those with close relations are aggregated as one small classification unit, and those with distant relations are aggregated as one large classification unit, until all samples or variables are aggregated. Then, different types are classified one by one to form a classification system from the smallest to the largest. At last, the whole classification system is drawn as a picture, which presents the closeness relationship.

单词和短语　Words and expressions

聚类分析　cluster analysis　　　　　　　　相似系数统计量　statistics of similarity coefficient
距离统计量　distance statistics

基本概念和性质　Basic concepts and properties

1 聚类分析可以分为 Q 型聚类和 R 型聚类两种，Q 型聚类是指对样本进行分类，R 型聚类是指对变量进行分类。通常 Q 型聚类采用距离统计量，R 型聚类采用相似系数统计量。

Cluster analysis can be divided into two kinds: Q cluster and R cluster. Q cluster is to make classification on samples, while R cluster is to make classification on variables. Q cluster usually takes distance statistics, while R cluster takes statistics of similarity coefficient.

2 距离：设有 n 个样本，每个样本观测 p 个变量，数据结构为 $\begin{bmatrix} x_{11} & x_{12} & \cdots\cdots & x_{1p} \\ x_{21} & x_{22} & \cdots\cdots & x_{2p} \\ \cdots & \cdots & \cdots\cdots & \cdots \\ x_{n1} & x_{n2} & \cdots\cdots & x_{np} \end{bmatrix}$。

绝对距离：$d_{ij} = \sum |x_{ik}-x_{jk}|$；

欧氏距离：$d_{ij} = \sqrt{\sum_{k=1}^{p}(x_{ik}-x_{jk})^2}$；

切比雪夫距离：$d_{ij} = \max_{1 \leqslant k \leqslant p}|x_{ik}-x_{jk}|$；

马氏距离：$d_{ij} = [(X_i-X_j)'S^{-1}(X_i-X_j)]^{\frac{1}{2}}$。

Distance: the hypothesis is tested that the number of samples is n, and each sample can observe p variables, data structure
$$\begin{bmatrix} x_{11} & x_{12} & \cdots\cdots & x_{1p} \\ x_{21} & x_{22} & \cdots\cdots & x_{2p} \\ \cdots\cdots & \cdots\cdots & \cdots\cdots & \cdots\cdots \\ x_{n1} & x_{n2} & \cdots\cdots & x_{np} \end{bmatrix}.$$

Absolute distance: $d_{ij} = \sum |x_{ik} - x_{jk}|$;

Euclidean Distance: $d_{ij} = \sqrt{\sum_{k=1}^{p}(x_{ik} - x_{jk})^2}$;

Chebyshev distance: $d_{ij} = \max_{1 \le k \le p} |x_{ik} - x_{jk}|$;

Mahalanobis Distance: $d_{ij} = [(X_i - X_j)' S^{-1} (X_i - X_j)]^{\frac{1}{2}}$.

③ 夹角余弦: $\cos\vartheta_{ij} = \dfrac{\sum_{k=1}^{p} x_{ki} x_{kj}}{\sqrt{\sum_{k=1}^{p} x_{ki}^2 \sum_{k=1}^{p} x_{kj}^2}}$ $i,j = 1,2,\ldots\ldots,p$;

相关系数: $r_{ij} = \dfrac{\sum (x_{ki} - \bar{x}_i)(x_{kj} - \bar{x}_j)}{\sqrt{\sum_{k=1}^{p}(x_{ki} - \bar{x}_i)^2 \sum_{k=1}^{p}(x_{kj} - \bar{x}_j)^2}}$ 。

Cosine: $\cos\vartheta_{ij} = \dfrac{\sum_{k=1}^{p} x_{ki} x_{kj}}{\sqrt{\sum_{k=1}^{p} x_{ki}^2 \sum_{k=1}^{p} x_{kj}^2}}$ $i,j = 1,2,\ldots\ldots,p$;

Similarity coefficient: $r_{ij} = \dfrac{\sum (x_{ki} - \bar{x}_i)(x_{kj} - \bar{x}_j)}{\sqrt{\sum_{k=1}^{p}(x_{ki} - \bar{x}_i)^2 \sum_{k=1}^{p}(x_{kj} - \bar{x}_j)^2}}$ 。

④ 分类的形成:先将所有的样本各自算作一类,将最近的两个样本点首先聚类,再将这个类和其他类中最靠近的结合,这样继续合并,直到所有的样本合并为一类为止。若在聚类过程中,距离的最小值不唯一,则将相关的类同时进行合并。

The forming of classification: at first, each sample is considered as one kind, and two samples with closest distance are first aggregated. This clusters goes on to be aggregated with other sample with closest distance, the whole process will continue until all the samples are aggregated as one cluster. If there is no single smallest distance score on the process, related cluster can be aggregated simultaneously.

⑤ 类与类之间的距离:设两个类 G_1 与 G_2,分别为 n_1 和 n_2 个样本,

最短距离法: $d_{lm} = \min\{d_{ij}, X_i \in G_l, X_j \in G_m\}$;

最长距离法: $d_{lm} = \max\{d_{ij}, X_i \in G_l, X_j \in G_m\}$;

重心法:两类的重心分别为 $\bar{x}_l, \bar{x}_m$,则 $d_{lm} = d_{\bar{x}_l \bar{x}_m}$;

类平均法: $d_{lm} = \dfrac{1}{n_1 n_2} \sum_{X_i \in G_i} \sum_{X_j \in G_j} d_{ij}$。

离差平方和法:首先将所有的样本自成为一类,然后每次缩小一类,每缩小一类离差平方和就要增大,选择使整个类内离差平方和增加最小的两类合并,直到所有的样本归为一类为止。

The distance between clusters: the hypothesis is tested to be G_1 and G_2, each cluster has n_1 and n_2 samples.

The shortest distance method: $d_{lm} = \min\{d_{ij}, X_i \in G_l, X_j \in G_m\}$;

The longest distance method: $d_{lm} = \max\{d_{ij}, X_i \in G_l, X_j \in G_m\}$;

Core method, the cores for each clusters are $\bar{x}_l, \bar{x}_m$, thus, $d_{lm} = d_{\bar{x}_1 \bar{x}_2}$;

Cluster average method $d_{lm} = \dfrac{1}{n_1 n_2} \sum\limits_{X_i \in G_i} \sum\limits_{X_j \in G_j} d_{ij}$.

Sum of deviation squares method: All the samples have their own cluster and reduce one cluster each, accordingly sum of squares becoming larger. Two clusters, increasing the minimum of sum of deviation squares in the whole cluster, are selected. The process will continue until all the samples are aggregated as one cluster.

例题 Example

例 根据距离矩阵式

$$D = (d_{ij})_{9\times 9} \begin{bmatrix} 0 & & & & & & & & \\ 1.52 & 0 & & & & & & & \\ 3.10 & 2.70 & 0 & & & & & & \\ 2.19 & 1.47 & 1.23 & 0 & & & & & \\ 5.86 & 6.02 & 3.64 & 4.77 & 0 & & & & \\ 4.72 & 4.46 & 1.86 & 2.99 & 1.78 & 0 & & & \\ 4.79 & 5.53 & 2.93 & 4.06 & 0.83 & 1.07 & 0 & & \\ 1.32 & 0.88 & 2.24 & 1.29 & 5.14 & 3.96 & 5.03 & 0 & \\ 2.62 & 1.66 & 1.20 & 0.51 & 4.84 & 3.06 & 3.32 & 1.40 & 0 \end{bmatrix}$$

用直接聚类法对某地区的9个农业区进行聚类分析。

解 (1) 在距离矩阵 D 中，除去对角线元素以外，$d_{49} = d_{94} = 0.51$ 为最小者，故将第4区与第9区并为一类，划去第9行和第9列；

(2) 在余下的元素中，除对角线元素以外，$d_{75} = d_{57} = 0.83$ 为最小者，故将第5区与第7区并为一类，划掉第7行和第7列；

(3) 在第2步之后余下的元素之中，除对角线元素以外，$d_{82} = d_{28} = 0.88$ 为最小者，故将第2区与第8区并为一类，划去第8行和第8列；

(4) 在第3步之后余下的元素中，除对角线元素以外，$d_{43} = d_{34} = 1.23$ 为最小者，故将第3区与第4区并为一类，划去第4行和第4列，此时，第3、4、9区已归并为一类；

(5) 在第4步之后余下的元素中，除对角线元素以外，$d_{21} = d_{12} = 1.52$ 为最小者，故将第1区与第2区并为一类，划去第2行和第2列，此时，第1、2、8区已归并为一类；

(6) 在第5步之后余下的元素中，除对角线元素以外，$d_{65} = d_{56} = 1.78$ 为最小者，故将第5区与第6区并为一类，划去第6行和第6列，此时，第5、6、7区已归并为一类；

(7) 在第6步之后余下的元素中，除对角线元素以外，$d_{31} = d_{13} = 3.10$ 为最小者，故将第1区与第3区并为一类，划去第3行和第3列，此时，第1、2、3、4、8、9区已归并为一类；

(8) 在第7步之后余下的元素中，除去对角线元素以外，只有 $d_{51} = d_{15} = 5.86$，故将第1区与第5区并为一类，划去第5行和第5列，此时，第1、2、3、4、5、6、7、8、9区均归并为一类。

根据上述步骤，可以作出聚类过程的谱系图。

Ex According to distance matrix

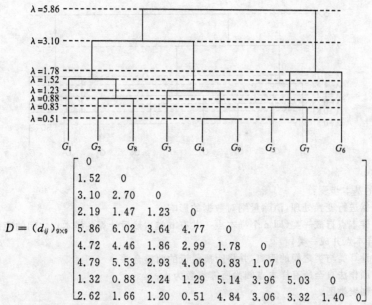

$$D = (d_{ij})_{9\times 9} \begin{bmatrix} 0 & & & & & & & & \\ 1.52 & 0 & & & & & & & \\ 3.10 & 2.70 & 0 & & & & & & \\ 2.19 & 1.47 & 1.23 & 0 & & & & & \\ 5.86 & 6.02 & 3.64 & 4.77 & 0 & & & & \\ 4.72 & 4.46 & 1.86 & 2.99 & 1.78 & 0 & & & \\ 4.79 & 5.53 & 2.93 & 4.06 & 0.83 & 1.07 & 0 & & \\ 1.32 & 0.88 & 2.24 & 1.29 & 5.14 & 3.96 & 5.03 & 0 & \\ 2.62 & 1.66 & 1.20 & 0.51 & 4.84 & 3.06 & 3.32 & 1.40 & 0 \end{bmatrix}$$

Making cluster analysis on 9 agricultural regions in an area with the method of direct cluster.

Solution:

(1) In distance matrix D, apart from Diagonal elements, the smallest one is $d_{49}=d_{94}=0.51$, therefore, the region 4 and 9 are aggregated as one cluster, and delete line 9 and row 9.

(2) In the rest elements, apart from Diagonal elements, the smallest one is $d_{75}=d_{57}=0.83$, therefore, the region 5 and 7 are aggregated as one cluster, and delete line 7 and row 7.

(3) In the rest elements after step 2, apart from Diagonal elements, the smallest one is $d_{82}=d_{28}=0.88$, therefore, the region 2 and 8 are aggregated as one cluster, and delete line 8 and row 8.

(4) In the rest elements after step 3, apart from Diagonal elements, the smallest one is $d_{43}=d_{34}=1.23$, therefore, the region 3 and 4 are aggregated as one cluster, and delete line 4 and row 4, and region 3, 4 and 9 have been aggregated as one cluster.

(5) In the rest elements after step 4, apart from Diagonal elements, the smallest one is $d_{21}=d_{12}=1.52$, therefore, the region 1 and 2 are aggregated as one cluster, and delete line 2 and row 2, and region 1, 2 and 8 have been aggregated as one cluster.

(6) In the rest elements after step 5, apart from Diagonal elements, the smallest one is $d_{65}=d_{56}=1.78$, therefore, the region 5 and 6 are aggregated as one cluster, and delete line 6 and row 6, and region 5, 6 and 7 have been aggregated as one cluster.

(7) In the rest elements after step 6, apart from Diagonal elements, the smallest one is $d_{31}=d_{13}=3.10$, therefore, the region 1 and 3 are aggregated as one cluster, and delete line 3 and row 3, and region 1, 2, 3, 4, 8 and 9 have been aggregated as one cluster.

(8) In the rest elements after step 7, apart from Diagonal elements, the smallest one is $d_{51}=d_{15}=5.86$, therefore, the region 1 and 5 are aggregated as one cluster, and delete line 5 and row 5, and region 1, 2, 3, 4, 5, 6, 7, 8 and 9 have been aggregated as one cluster.

Based on the steps above, the Pedigree chart on cluster process is concluded as follows:

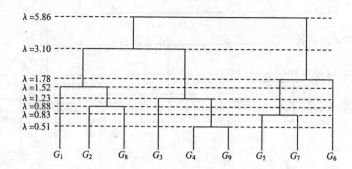

本章重点

1. 聚类分析的基本步骤：
(1) 先对数据进行变换处理，消除量纲对数据的影响；
(2) 认为各样本点自成一类（即 n 个样本点一共有 n 类），然后计算各样本点之间的距离，并将距离最近的两个样本点并成一类；
(3) 选择并计算类与类之间的距离，并将距离最近的两类合并；
(4) 重复上面作法直至所有样本点归为所需类数为止；
(5) 最后绘制聚类图。

Key points of this Chapter

1. Basic steps of cluster analysis:
(1) First make transform on data so as to remove the impact of dimension;
(2) Each sample is considered as one cluster(there are N samples with N clusters). calculate the distance between samples and aggregate two samples with closest distance;
(3) Select and calculate the distance between clusters and aggregate the two ones with closest distance;
(4) The process is repeated until all the samples are completely aggregated as the required clusters;
(5) Draw cluster chart.

第六章　判别分析
Chapter 6　Discriminant Analysis

　　判别分析是在已知研究对象用某种方法已分成若干类的情况下，确定新的观察数据属于已知类别中的哪一类的分析方法。判别分析方法在处理问题时，通常要给出一个衡量新样本与已知组别接近程度的描述指标，即判别函数，同时也指定一种判别规则，用以判定新样本的归属。

　　Discriminant analysis is used to clarify the catalogue of the new observed data from the research objects, which has been known to be classified into several catalogue. When dealing with matters, this kind of analyzing method usually presents a index, dicriminant index, which describes the proximity degree between new data and known ones; meanwhile, it also sets a kind of discriminant rules to clarify the catalogue of new data.

单词和短语 Words and expressions

判别分析　discriminant analysis
距离判别　distance discriminant
费歇尔判别　fishier discriminant
贝叶斯判别　Bayesian discrimination

先验概率　prior probability
超平面　hyperplane
错判率　false discrimination rate

基本概念和性质 Basic concepts and properties

1. 距离判别：先根据已知分类的数据，分别计算各类的重心，然后计算待判样本与各类的距离，与哪一类距离最近，就判待判样本 x 属于哪一类。

判别函数：$W(x) = D(x,G_2) - D(x,G_1)$；

判别准则：$\begin{cases} x \in G_1, \text{当 } W(x) > 0 \\ x \in G_2, \text{当 } W(x) < 0 \\ \text{待判, 当 } W(x) = 0 \end{cases}$

Distance discriminant: To calculate the core of each classification, according to the known data; then to count the distance between the sample and various classifications to find out which group has the shortest distance. The sample belongs to the catalogue which has the shortest distance from the former.

Discriminant function: $W(x) = D(x,G_2) - D(x,G_1)$

Discriminant rules: $\begin{cases} x \in G_1, \text{当 } W(x) > 0 \\ x \in G_2, \text{当 } W(x) < 0 \\ \text{待判, 当 } W(x) = 0 \end{cases}$

2. 费歇尔判别：通过将多维数据投影至某个方向上，投影的原则是将总体与总体之间尽可能分开，然后再选择合适的判别规则，将待判的样本进行分类判别。所谓的投影实际上是利用方差分析的思想构造一个或几个超平面，使得两组间的差别最大，每组内的差别最小。

判别函数：$y = (\overline{X}_1 - \overline{X}_2)'\hat{\Sigma}^{-1} X$；

判别准则：$\begin{cases} x \in G_1 & y_1 > y_2, y > y_0 \\ x \in G_2 & y_1 > y_2, y < y_0 \\ x \in G_2 & y_1 < y_2, y > y_0 \\ x \in G_1 & y_1 < y_2, y < y_0 \end{cases}$

Fishier discriminant: According to the projection principles, various types should be apart as far as possible, to project multidimensional data towards a certain direction, and select a an appropriate rules of discrimination to classify the samples. In fact, so-called projection is to construct one or several hyperplane to extend the difference between two groups to the largest degree, and reduce the difference of each group to the smallest degree by using the analysis of variance.

Discriminant function: $\begin{cases} x \in G_1 & y_1 > y_2, y > y_0 \\ x \in G_2 & y_1 > y_2, y < y_0 \\ x \in G_2 & y_1 < y_2, y > y_0 \\ x \in G_1 & y_1 < y_2, y < y_0 \end{cases}$

3. 贝叶斯判别：设有两个总体，它们的先验概率分别为 q_1、q_2，各总体的密度函数为 $f_1(x)$、$f_2(x)$，在观测到一个样本 x 的情况下，可用贝叶斯公式计算它来自第 k 个总体的后验概率为：$P(G_k/x) = \dfrac{q_k f_k(x)}{\sum_{k=1}^{2} q_k f_k(x)}, k = 1,2$。

判别准则：对于待判样本 x，如果在所有的 $P(G_k/x)$ 中 $P(G_h/x)$ 是最大的，则判定 x 属于第 h 总体。通常会以样本的频率作为各总体的先验概率。

Bayesian discrimination:

For the hypothesis of two types, their prior probability are q_1, q_1, and density functions for each overall are $f_1(x)$, $f_2(x)$. The Bias formula can be used to calculate its posterior probability from the kth overall, based on the situation of knowing a sample x: $P(G_k/x) = \dfrac{q_k f_k(x)}{\sum_{k=1}^{2} q_k f_k(x)}$, $k = 1,2$.

Discriminant principles: as for sample x, if $P(G_h/x)$ is the biggest in $P(G_k/x)$, thus x belongs to h. The sample frequency is usually taken as prior probability of each type.

例 题 Example

例 1 欲用显微分光光度计对病人细胞进行检查以判断病人是否患有癌症。现抽取110例癌症病人和190例正常人。指标：X_1, X_2 和 X_3。X_1：三倍体的得分，X_2：八倍体的得分，X_3：不整倍体的得分。(0—10分)

收集数据，得到训练样本：对于若干已明确诊断为癌症的110个病人和无癌症的190个正常人均用显微分光光度计对细胞进行检测，得到 X_1, X_2 和 X_3 的值。这就是训练样本。

例号	X_1	X_2	X_3	Y(类别)
1	1	2	2	0
2	2	5	6	1
……				
300	3	3	3	0

解 根据实测资料(训练样本)用判别分析方法可建立判别函数，本例用 Fisher 判别分析方法得到：$Y = X_1 + 10X_2 + 10X_3$。

并确定判别准则为：如有某病人的 X_1, X_2 和 X_3 实测值，代入上述判别函数可得 Y 值，Y>100 则判断为癌症，Y<100 则判断为非癌症。

Ex 1 To check whether patients have cancer or not by testing their cells with microscopic spectrophotometer. We extract cells from 100 cancer patients and 190 normal persons. Index: X_1, X_2 and X_3. X_1: the scores of triploid, X_2: the scores of octoploid, X_3: the scores of non-aneuploidy. (0-10points)

Collect data and get training samples: to test cells from 110 cancer patients and normal persons with microscopic spectrophotometer to get X_1, X_2 and X_3. Those are training samples.

number	X1	X2	X3	Y(catalogue)
1	1	2	2	0
2	2	5	6	1
……				
300	3	3	3	0

Solutions Constructing discriminant function by means of disciminant analysis, this examples is to conclude $Y = X_1 + 10X_2 + 10X_3$ and set discriminant principles as such that if X_1, X_2 and X_3 has measured value, puting them into previous function and get Y, and Y>100, that person is sentenced to have cancer, otherwise, he is normal.

第五部分 概率论与数理统计
Part 5 Probability Theory and Mathematical Statistics

引言

概率论是一个非常重要且令人着迷的学科,它起源于17世纪,是 Fermat 和 Pascal 两位数学家在研究赌博几率的基础上发展起来的.直到20世纪,基于公理、定义和定理的严格的数学理论才发展起来.随着时间的推移,不但工程、科学和数学,甚至精算、农业和商业到医学和心理学,概率理论都得到了广泛的应用.在很多情况下,应用本身又促进了理论的发展.

统计学科比概率理论起源得早,主要用来处理怎样搜集、组织、表达表格和图表里的数据.随着概率论的发展,人们逐渐认识到,在很多方面如:抽样理论和预报或预测,在分析数据的基础上,应用统计理论可以得出正确的结果和合理的决策.

Introduction

Probability Theory is a very important and fascinating subject, which began in the 17th century through the efforts of mathematicians such as Fermat and Pascal who study the questions concerning gambling chance. It was not until the 20th century that a rigorous mathematical theory based on axioms, definitions and theorems was developed. As time progressed, Probability Theory found its way into many applications, not only in engineering, science and mathematics but also in fields ranging from actuarial science, agriculture and business to medicine and psychology. In many instances the applications themselves contributes to the further development of theory.

The subject of Statistics originated much earlier than Probability Theory and mainly deals with the collection, organization and presentation of data in tables and charts. With the development of Probability Theory it is realized that Statistics could be used in drawing valid conclusions and making reasonable decisions on the basis of analysis of data, such as in sampling theory and prediction or forecasting.

第一章 概率论的基本概念
Chapter 1 Introduction of Probability Theory

本章我们主要介绍事件之间的关系、概率的定义、古典概率的求法、条件概率的定义、事件的相互独立定义、概率的计算公式,其中包括概率的加法定理、乘法公式、全概率公式、贝叶斯公式.

In this chapter we mainly introduce the relationship among events, the definition of probability, the solution of classical probability, the definitions of conditional probability and pairwise independent of events and some formulas concerned with probability, such as addition theorem, multiplication of probabilities, formula of total probability, Bayes formula.

单词和短语 Words and expressions

不确定性　indeterminacy
必然现象　certain phenomenon
随机现象　random phenomenon
试验　experiment
结果　result
频率数　frequency number
★样本空间　sample space
出现次数　frequency of occurrence
n 维样本空间　n-dimensional sample space

样本点　point in sample space
★随机事件　random event
★基本事件　elementary event
必然事件　certain event
不可能事件　impossible event
等可能事件　equally likely event
事件运算律　operational rules of events
事件的包含　implication of events
并事件　union events

交事件　intersection events
互不相容事件、互斥事件　mutually exclusive events/ incompatible events
互逆的　mutually inverse
逆事件　complementary event
加法定理　addition theorem
★ 古典概率　classical probability
古典概率模型　classical probabilistic model
几何概率　geometric probability
乘法定理　product theorem
概率乘法　multiplication of probabilities
条件概率　conditional probability
★ 全概率公式、全概率定理　formula of total probability
★ 贝叶斯公式、逆概率公式　Bayes formula
后验概率　posterior probability
先验概率　prior probability
独立事件　independent event
独立随机事件　independent random event
独立试验　independent experiment
两两独立　pairwise independence
两两独立事件　pairwise independent events

基本概念和性质　Basic concepts and properties

1 事件 $A \cup B$ 是指 A 和 B 至少有一个发生。
The event $A \cup B$ means that at least one of the events And B occurs.

2 事件 $A \cap B$ 是指 A 和 B 同时发生。
The event $A \cap B$ means that both the event A and the event B occur.

3 事件 A,B 相互独立，即指 $P(AB) = P(A)P(B)$。
That the events A and B are mutually independent is to say $P(AB) = P(A)P(B)$。

例题　Example

例 1 掷硬币

假设掷一枚质地均匀的硬币 10 次，试求：
(a) 恰好出现 3 次正面向上的概率 p；(b) 出现小于等于 3 次正面向上的概率 p'。

解 （a）掷 10 次硬币出现正面和反面的不同结果的总数为 2^{10}，并且每一种结果都是等可能出现的，而恰好出现 3 次正面向上的结果数目等于 3 个正面和 7 个反面的不同的排列数目，由于此数为 C_{10}^3，则恰好出现三次正面向上的概率为 $p = \dfrac{C_{10}^3}{2^{10}}$。

（b）一般的，由于样本空间中正面恰好出现 k 次（$k=0,1,2,3$）的结果数目为 C_{10}^k，则出现等于或少于 3 次正面向上的概率为 $p' = \dfrac{C_{10}^0 + C_{10}^1 + C_{10}^2 + C_{10}^3}{2^{10}} = \dfrac{176}{2^{10}}$。

Ex. 1 Tossing a coin.

Suppose that a fair coin is to be tossed ten times, and it is desired to determine: (a) the probability p of obtaining exactly three heads; (b) the probability p' of obtaining three or fewer heads.

Solution (a) The total number of different arrangements of ten heads and tails is 2^{10}, and it may be assumed that each of these sequences is equally probable. The number of these sequences that contain exactly three heads will be equal to the number of different arrangements that can be formed with three heads and seven tails. Since this number is C_{10}^3, the probability of obtaining exactly three heads is $p = \dfrac{C_{10}^3}{2^{10}}$.

(b) Since, in general, the number of sequences in the sample space that contain exactly k heads ($k = 0,1,2,3$) is C_{10}^k, the desired probability of obtaining three or fewer heads is
$$p' = \dfrac{C_{10}^0 + C_{10}^1 + C_{10}^2 + C_{10}^3}{2^{10}} = \dfrac{176}{2^{10}}.$$

例 2 机器工作

假设在一个工厂机器 1 和 2 相互独立地工作，事件 A 表示在给定的 8 小时内机器 1 发生故障，事件

B 表示在相同的时间内机器 2 发生故障, 并设 $p(A) = \frac{1}{3}$ 和 $p(B) = \frac{1}{4}$, 试求在给定的时间内至少有一台机器发生故障的概率.

解 在给定的时间内两台机器同时发生故障的概率

$$p(AB) = p(A)p(B) = \frac{1}{3} \times \frac{1}{4} = \frac{1}{12}.$$

进而, 在该时间内至少有一台机器发生故障的概率

$$p(A \cup B) = p(A) + p(B) - p(AB) = \frac{1}{3} + \frac{1}{4} - \frac{1}{12} = \frac{1}{2}.$$

Ex. 2 Machine operation.

Suppose that the operations of two machines 1 and 2 in a factory are mutually independent. Let A be the event that the machine 1 will become inoperative during a given 8-hour period, let B be the event that the machine 2 will become inoperative during the same period, and suppose that $p(A) = \frac{1}{3}$ and $p(B) = \frac{1}{4}$. Try to determine the probability that at least one of the machines will become inoperative during the given period.

Solution The probability $p(AB)$ that both machines will become inoperative during the period is

$$p(AB) = p(A)p(B) = \frac{1}{3} \times \frac{1}{4} = \frac{1}{12}.$$

Therefore, the probability $p(A \cup B)$ that at least one of the machines will become inoperative during the period is

$$p(A \cup B) = p(A) + p(B) - p(AB) = \frac{1}{3} + \frac{1}{4} - \frac{1}{12} = \frac{1}{2}.$$

例 3 假设三个机器 M_1, M_2, M_3 用来大批量生产同一种产品, 20% 的产品由机器 M_1 生产, 30% 的产品由机器 M_2 生产, 50% 的产品由机器 M_3 生产. 进一步, 机器 M_1 生产的产品有 1% 是次品, 机器 M_2 生产的产品有 2% 是次品, 机器 M_3 生产的产品有 3% 是次品. 最后, 从制成的产品中随机选取一个, 经检测为次品, 试求此产品是由机器 M_2 生产的概率.

解 设事件 A_i 表示选取的产品由机器 $M_i (i=1,2,3)$ 生产, 事件 B 表示选取的产品是次品, 首先求条件概率 $p(A_i \mid B)$. 任选一个产品, 设它是由机器 M_i 生产的概率为 $p(A_i)(i=1,2,3), p(A_1) = 0.2, p(A_2) = 0.3, p(A_3) = 0.5$. 进而, 由机器 M_i 生产的产品是次品的条件概率 $p(B \mid A_i)$ 分别是 $p(B \mid A_1) = 0.01, p(B \mid A_2) = 0.02, p(B \mid A_3) = 0.03$. 则根据贝叶斯定理得

$$p(A_2 \mid B) = \frac{p(A_2)p(B \mid A_2)}{\sum_{i=1}^{3} p(A_i)p(B \mid A_i)} = \frac{0.3 \times 0.02}{0.2 \times 0.01 + 0.3 \times 0.02 + 0.5 \times 0.03} = 0.26.$$

Ex. 3 Three different machines M_1, M_2, and M_3 were used for producing a large batch of similar manufactured items. Suppose that 20 percent of the items were produced by the machine M_1, 30 percent by the machine M_2, and 50 percent by the machine M_3. Suppose further that 1 percent of the items produced by M_1 are defective, that 2 percent of the items produced by M_2 are defective, and that 3 percent of the items produced by M_3 are defective. Finally, suppose that one item is selected at random from the entire items and it is found to be defective. Try to determine the probability that this defective item was produced by M_2.

Solution Let A_i be the event that the selected item was produced by the machine $M_i (i=1,2,3)$ and let B be the event that selected item is defective. We must evaluate the conditional probability $p(A_i \mid B)$. The probability $p(A_i)$ that an item selected at random from the entire batch was produced by M_i is as follows, for $i=1,2,3$.

$$p(A_1) = 0.2, \ p(A_2) = 0.3, \ p(A_3) = 0.5$$

Furthermore, the probability $p(B \mid A_i)$ that an item produced by M_i will be defective is $p(B \mid A_1) = 0.01$, $p(B \mid A_2) = 0.02$, $p(B \mid A_3) = 0.03$. It now follows from the Bayes theorem that

$$p(A_2 \mid B) = \frac{p(A_2)p(B \mid A_2)}{\sum_{i=1}^{3} p(A_i)p(B \mid A_i)} = \frac{0.3 \times 0.02}{0.2 \times 0.01 + 0.3 \times 0.02 + 0.5 \times 0.03} = 0.26.$$

本章重点

1 理解随机试验、基本事件、样本空间、随机事件的概念.
2 掌握事件的包含、相等、对立、互斥四种关系.
3 掌握事件的并、事件的和的关系及对偶原则.
4 理解概率、条件概率的定义和性质.
5 理解古典概率的定义并会熟练地计算各种古典概率.
6 理解两个或三个事件相互独立性的定义并用此定义解决实际问题.
7 熟练地利用加法公式、乘法公式、全概率公式、贝叶斯公式解决实际问题.

Key points of this chapter

1 Understand the concepts of random experiments, elementary events, sample spaces and random events.
2 Master the four relationships among events, namely, implication, equal, opposition, incompatible.
3 Master the union of events, addition of events and duality principle.
4 Understand the concepts and properties of probability and conditional probability.
5 Understand the concept of classical probability, and operate various classical probabilities proficiently.
6 Understand the concept of mutual independence of two or three events and solve practical problems by using this concept.
7 Solve various practical problems by using proficiently addition formula, product formula, formula of total probability and Bayes formula.

第二章 随机变量及其分布
Chapter 2 Random Variables and Distributions

本章我们主要学习随机变量的定义、分布函数的定义、离散型随机变量的分布律、连续型随机变量概率密度的定义、几种重要分布.

In this chapter, we mainly study the definitions of random variable and distribution functions, the law of probability distribution of discrete random variables, the probability densities of continuum random variable and some important distributions.

单词和短语 Words and expressions

★ 随机变量　random variable
离散随机变量　discrete random variable
概率分布律　law of probability distribution
一维概率分布　one-dimension probability distribution
★ 概率分布　probability distribution
两点分布　two-point distribution
伯努利试验　Bernoulli trials
★ 二项分布、伯努利分布　binomial distribution
超几何分布　hypergeometric distribution

三项分布　trinomial distribution
多项分布　polynomial distribution
★ 泊松分布　Poisson distribution
泊松参数　Poisson parameter
泊松定理　Poisson theorem
分布函数　distribution function
概率分布函数　probability density function
连续随机变量　continuous random variable
概率密度　probability density
★ 概率密度函数　probability density function

概率曲线　probability curve
均匀分布　uniform distribution
指数分布　exponential distribution
指数分布密度函数　exponential distribution density function
正态分布、高斯分布　normal distribution
标准正态分布　standard normal distribution
正态概率密度函数　normal probability density function
正态概率曲线　normal probability curve
标准正态曲线　standard normal curve
柯西分布　Cauchy distribution

基本概念和性质　**Basic concepts and properties**

1 随机变量：定义在样本空间上的实值函数.
A real-valued function that is defined on the sample space is called a random variable.

2 随机变量分为离散型和非离散型(常用到的是连续型).
There are two kinds of random variables: discrete random variable and non-discrete random variable (continuous random variables are often used).

3 正态分布是随机变量的最重要的一种分布.
Normal distribution is one of the most important distribution of random variables.

例 题　Example

例 1 若 $X \sim N(\mu,\sigma^2)$ 则 $Z = \dfrac{X-\mu}{\sigma} \sim N(0,1)$.

证明　$Z = \dfrac{X-\mu}{\sigma}$ 的分布函数为

$$P\{Z \leqslant x\} = P\left\{\dfrac{X-\mu}{\sigma} \leqslant x\right\} = P\{X \leqslant \mu+\sigma x\} = \dfrac{1}{\sqrt{2\pi}\sigma}\int_{-\infty}^{\mu+\sigma x} \exp\left\{-\dfrac{(t-\mu)^2}{2\sigma^2}\right\}dt.$$

令 $\dfrac{t-\mu}{\sigma} = u$，得 $P\{Z \leqslant x\} = \dfrac{1}{\sqrt{2\pi}}\int_{-\infty}^{x} e^{-\frac{u^2}{2}} du = \Phi(x)$. 由此可知

$$Z = \dfrac{X-\mu}{\sigma} \sim N(0,1).$$

Ex. 1 Suppose that $X \sim N(\mu,\sigma^2)$, prove $Z = \dfrac{X-\mu}{\sigma} \sim N(0,1)$.

Proof　The distribution function of $Z = \dfrac{X-\mu}{\sigma}$ is given by

$$P\{Z \leqslant x\} = P\left\{\dfrac{X-\mu}{\sigma} \leqslant x\right\} = P\{X \leqslant \mu+\sigma x\} = \dfrac{1}{\sqrt{2\pi}\sigma}\int_{-\infty}^{\mu+\sigma x} \exp\left\{-\dfrac{(t-\mu)^2}{2\sigma^2}\right\}dt.$$

Let $\dfrac{t-\mu}{\sigma} = u$, then $P\{Z \leqslant x\} = \dfrac{1}{\sqrt{2\pi}}\int_{-\infty}^{x} e^{-\frac{u^2}{2}} du = \Phi(x)$. So we know that

$$Z = \dfrac{X-\mu}{\sigma} \sim N(0,1).$$

本章重点

1 理解随机变量的定义及分类.
2 理解离散型随机变量概率分布的定义、性质.
3 掌握离散型随机变量概率分布的求法及利用性质解决问题.
4 理解随机变量分布函数的定义、性质.
5 利用随机变量分布函数的定义、性质求分布函数中的未知常数和事件的概率.
6 理解连续型随机变量概率密度的定义、性质.
7 知道如何利用随机变量概率密度的定义、性质求概率密度中的未知常数和事件的概率.
8 理解两点分布、二项分布、泊松分布、超几何分布、均匀分布、指数分布和正态分布七个重要分布并能

准确判断.
9. 掌握离散型随机变量函数的概率分布的求法.
10. 知道如何用公式法或分布函数法求连续型随机变量函数的概率密度.
11. 熟练掌握正态分布标准化的过程.

Key points of this chapter

1. Understand the concept and classification of random variables.
2. Understand the concept and properties of law of probability distribution of discrete random variables.
3. Grasp the methods for finding the probability distribution of discrete random variables and apply the properties to solve practical problems.
4. Understand the definition and properties of distribution function of random variables.
5. Apply the definition and properties of distribution function of random variables to find the probability of unknown constant and events in this distribution function.
6. Understand the concept and properties of probability density of continuous random variables.
7. Know how to apply the definition and properties of probability density of random variables to find the probability of unknown constant and events in the same probability density.
8. Understand and identify precisely the seven major distributions, namely, two-point distribution, binomial distribution, Poisson distribution, hypergeometric distribution, uniform distribution, exponential distribution and normal distribution.
9. Master the methods for finding the probability distribution of discrete random variable function.
10. Know how to apply distribution functions or formulae to find the probability density of continuous random variable functions.
11. Thorough understanding of the standardization procedure of normal distribution.

第三章 多维随机变量及其分布
Chapter 3 Multivariate Random Variables and Distributions

本章我们主要研究二维随机变量的定义、联合分布函数的定义、二维连续型随机变量概率密度的定义、二维离散型随机变量的概率分布、边缘概率分布的求法.

In this chapter we study the definitions of two-dimensional random variables, joint distribution function, two-dimensional continuous random variables probability density, and the solving methods of two-dimensional discrete random variable distribution, marginal probability distribution.

单词和短语 Words and expressions

★ 二维随机变量　two-dimensional random variable
联合分布函数　joint distribution function
★ 二维离散型随机变量　two-dimensional discrete random variable
联合分布　joint distribution
联合概率分布　joint probability distribution
★ 二维连续型随机变量　two-dimensional continuous random variable
联合概率密度　joint probability density
n 维随机变量　n-dimensional random variable
n 维分布函数　n-dimensional distribution function
n 维概率分布　n-dimensional probability distribution
★ 边缘分布　marginal distribution
★ 边缘分布函数　marginal distribution function
★ 边缘分布律　law of marginal distribution
边缘概率密度　marginal probability density
二维正态分布　two-dimensional normal distribution
二维正态概率密度　two-dimensional normal

probability density
二维正态概率曲线　two-dimensional normal probability curve
条件分布　conditional distribution
条件分布律　law of conditional distribution
条件概率分布　conditional probability distribution
条件概率密度　conditional probability density
边缘密度　marginal density
★独立随机变量　independent random variables

基本概念和性质　Basic concepts and properties

1 两个随机变量的联合概率分布称为两变量分布.
The joint probability distribution of two random variables is called a bivariate distribution.

2 两个随机变量的相互独立当且仅当联合分布函数等于两个随机变量的边缘分布函数的乘积.
The mutual independence between two random variables is valid if and only if the joint distribution function is equal to the multiplication of two marginal distribution functions of two random variables.

例题　Example

例 设二维随机变量(X,Y)的联合概率密度为
$$f(x,y) = \begin{cases} 15xy^2, & 0 \leqslant x \leqslant 1, 0 \leqslant y \leqslant x \\ 0, & \text{其他} \end{cases}$$
试求X,Y的边缘概率密度.

解　$f_X(x) = \int_{-\infty}^{\infty} f(x,y)\mathrm{d}y = \begin{cases} \int_0^x 15xy^2 \mathrm{d}y = 5x^4, & 0 \leqslant x \leqslant 1 \\ 0, & \text{其他} \end{cases}$

$f_Y(y) = \int_{-\infty}^{\infty} f(x,y)\mathrm{d}x = \begin{cases} \int_y^1 15xy^2 \mathrm{d}x = \frac{15}{2}(y^2 - y^4), & 0 \leqslant y \leqslant 1 \\ 0, & \text{其他} \end{cases}$

Ex Suppose that the joint probability density function of (X,Y) is
$$f(x,y) = \begin{cases} 15xy^2, & 0 \leqslant x \leqslant 1, 0 \leqslant y \leqslant x \\ 0, & \text{others} \end{cases}$$
Find the two marginal probability density of X,Y.

Solution

$f_X(x) = \int_{-\infty}^{\infty} f(x,y)\mathrm{d}y = \begin{cases} \int_0^x 15xy^2 \mathrm{d}y = 5x^4, & 0 \leqslant x \leqslant 1 \\ 0, & \text{others} \end{cases}$

$f_Y(y) = \int_{-\infty}^{\infty} f(x,y)\mathrm{d}x = \begin{cases} \int_y^1 15xy^2 \mathrm{d}x = \frac{15}{2}(y^2 - y^4), & 0 \leqslant y \leqslant 1 \\ 0, & \text{others} \end{cases}$

本章重点

1 理解二维随机变量的概念、联合分布函数的定义及性质.
2 会利用联合分布函数的定义、性质求联合分布函数中的未知常数和事件的概率.
3 会求二维离散型随机变量的联合概率分布、边缘概率分布和条件概率分布.
4 理解二维连续型随机变量的联合概率密度的定义、性质.
5 利用联合概率密度的定义、性质求相同的联合概率密度中未知常数和事件的概率.
6 理解随机变量的独立性的概念.
7 掌握联合分布函数是否等于两个边缘分布函数的乘积、或联合概率分布是否等于两个边缘概率分布的乘积、或联合概率密度是否等于两个边缘概率密度的乘积,判断两个随机变量的相互独立性.
8 了解二维均匀分布、二维正态分布.

> 9 会求两个独立随机变量的简单函数的分布。

Key points of this chapter

1. Understand the concept of two-dimensional random variables and the definition and properties of joint distribution functions.
2. Know how to apply the definition and properties of joint distribution functions to find the probability of unknown constant and events of the same joint distribution function.
3. Grasp the methods for finding joint probability distribution, marginal probability distribution and conditional probability distribution of a two-dimensional discrete random variable.
4. Understand the definition and properties of joint probability density of a two-dimensional continuous random variable.
5. Know how to apply the definition and properties of joint probability density to find the probability of unknown constant and events of the same joint probability density.
6. Understand the concept of independence of random variables.
7. Identify the mutual independence between two random variables by the criteria whether the joint distribution function is equal to the multiplication of two marginal distribution functions, or whether the joint probability distribution is equal to the multiplication of two marginal probability distribution, or whether the joint probability density is equal to the multiplication of the two marginal probability density.
8. Understand two-dimensional uniform distribution and two-dimensional normal distribution.
9. Know how to apply the methods for finding the distribution of simple function of two mutually independent random variables.

第四章 随机变量的数字特征
Chapter 4 Numerical Characteristics of Random Variables

本章我们主要掌握随机变量的数学期望、方差、协方差、相关系数的定义和求法。

In this chapter we primarily acquire the definitions of mathematical expectation, variance, covariance, correlation coefficient of random variables and how to solve them.

单词和短语 Words and expressions

★ 数学期望、均值 mathematical expectation	相关关系 correlation relation
期望值 expectation value	相关系数 correlation coefficient
★ 方差 variance	协方差 covariance
标准差 standard deviation	协方差矩阵 covariance matrix
随机变量的方差 variance of random variables	切比雪夫不等式 Chebyshev inequality
★ 均方差 mean square deviation	

基本概念和性质 Basic concepts and properties

1. 随机变量的方差可被定义为 $Var(X) = E[(X - EX)^2]$.

   The variance of X, denoted by $Var(X)$, is defined as $Var(X) = E[(X - EX)^2]$.

2. 随机变量的方差的正的平方根称为它的标准差。

   The standard deviation of a random variable is defined to be the nonnegative square root of the variance.

例题 Example

例 已知随机变量 X 只可取以下四个值: $-2, 0, 1$ 和 4, 而且

$P(X=-2) = 0.1, \quad P(X=0) = 0.4, \quad P(X=1) = 0.3, \quad P(X=4) = 0.2,$

试求变量 X 的数学期望.

解 $E(X)=-2\times 0.1+0\times 0.4+1\times 0.3+4\times 0.2=0.9$.

Ex Suppose that the random variable X can be only the four different values $-2, 0, 1$ and 4, and that

$$P(X=-2)=0.1, \quad P(X=0)=0.4, \quad P(X=1)=0.3, \quad P(X=4)=0.2,$$

solve the mathematical expectation of X.

Solution $E(X)=-2\times 0.1+0\times 0.4+1\times 0.3+4\times 0.2=0.9$.

本章重点

1. 理解并掌握随机变量的数字特征(数学期望、方差、标准差、协方差、相关系数)的概念,并能运用数字特征的定义和基本性质计算具体分布的数字特征.
2. 牢记常用分布的数字特征.
3. 理解随机变量不相关的概念并会判断随机变量的不相关性.
4. 会求一维、二维随机变量的函数的数学期望.
5. 了解切比雪夫不等式.

Key points of this chapter

1. Understand and grasp the concept of numerical characteristics (mathematical expectation, variance, standard deviation, covariance, coefficient of correlation) of a random variable, the methods for the calculation of the numerical characteristics of a specific distribution through the application of the definition and fundamental properties of numerical characteristics.
2. Remember the numerical characteristics of distribution patterns that are commonly used.
3. Understand the concept of independence of random variables and know how to identify the independence of random variables.
4. Know how to find the mathematical expectations of one-dimensional, two-dimensional random variable functions.
5. Understand the Chebyshev inequality.

第五章 大数定律及中心极限定理

Chapter 5 Law of Large Numbers and Central Limit Theorem

本章我们主要讨论大数定律的真正含义和中心极限定理及其应用.

In this chapter we discuss the basic connotation of law of large numbers, central limit theorem and their applications.

单词和短语 Words and expressions

★ 大数定律 law of great numbers / law of large numbers
切比雪夫定理的特殊形式 special form of Chebyshev theorem
★ 依概率收敛 convergence in probability
伯努利大数定律 Bernoulli law of large numbers
同分布 same distribution

列维-林得伯格定理、独立同分布中心极限定理 Levy-Lindberg theorem
★ 辛钦大数定律 Khinchine law of large numbers
李雅普诺夫定理 Liapunov theorem
棣莫弗-拉普拉斯定理 De Moivre-Laplace theorem

基本概念和性质 Basic concepts and properties

棣莫弗-拉普拉斯定理

设随机变量 $\eta_n \sim b(n,p)$,则对任意的 x,有

$$\lim_{n\to\infty} P\left\{ \frac{\eta_n - np}{\sqrt{np(1-p)}} \leqslant x \right\} = \int_{-\infty}^{x} \frac{1}{\sqrt{2\pi}} e^{-\frac{t^2}{2}} dt = \Phi(x).$$

De Moivre-Laplace theorem

Suppose that the random variables $\eta_n \sim b(n,p)$, then for any x, we get

$$\lim_{n\to\infty} P\left\{\frac{\eta_n - np}{\sqrt{np(1-p)}} \leqslant x\right\} = \int_{-\infty}^{x} \frac{1}{\sqrt{2\pi}} e^{-\frac{t^2}{2}} dt = \Phi(x).$$

例题　Example

例 设随机变量 $V_k(k=1,2,\cdots,20)$ 是相互独立且都在区间 $(0,10)$ 上服从均匀分布，记 $V = \sum_{k=1}^{20} V_k$，试求概率 $P(V > 105)$.

解 易知 $E(V_k) = 5$，$D(V_k) = \frac{100}{12}(k=1,2,\cdots,20)$，则

$$P(V > 105) = P\left(\frac{V - 20\times 5}{(10/\sqrt{12})\sqrt{20}} > \frac{105 - 20\times 5}{(10/\sqrt{12})\sqrt{20}}\right)$$

$$= P\left(\frac{V - 100}{(10/\sqrt{12})\sqrt{20}} > 0.387\right)$$

$$= 1 - P\left(\frac{V - 100}{(10/\sqrt{12})\sqrt{20}} \leqslant 0.387\right)$$

$$\approx 1 - \Phi(0.387) = 0.348.$$

Ex Suppose that the random variables $V_k(k=1,2,\cdots,20)$ are independent and have the same distribution $U(0,10)$. Let $V = \sum_{k=1}^{20} V_k$, please solve $P(V > 105)$.

Solution We can easily know that $E(V_k) = 5$, $D(V_k) = \frac{100}{12}(k=1,2,\cdots,20)$, then

$$P(V > 105) = P\left(\frac{V - 20\times 5}{(10/\sqrt{12})\sqrt{20}} > \frac{105 - 20\times 5}{(10/\sqrt{12})\sqrt{20}}\right)$$

$$= P\left(\frac{V - 100}{(10/\sqrt{12})\sqrt{20}} > 0.387\right)$$

$$= 1 - P\left(\frac{V - 100}{(10/\sqrt{12})\sqrt{20}} \leqslant 0.387\right)$$

$$\approx 1 - \Phi(0.387) = 0.348.$$

本章重点

1. 理解切比雪夫定理的特殊形式、伯努利大数定律、辛钦大数定律、李雅普诺夫定理的内容.
2. 理解独立同分布中心极限定理、棣莫弗-拉普拉斯定理的前提和结论，并会利用两定理解决实际问题.

Key points of this chapter

1. Understand the special form of Chebyshev theorem, Bernoulli law of large numbers, Khinchine law of large numbers, Liapunov theorem.
2. Understand the premises and conclusions of Levy-Lindberg, De Moivre-Laplace theorem and know how to solve practical problems applying these two theorems.

第六章　样本及抽样分布
Chapter 6　Samples and Sampling Distributions

本章我们主要了解数理统计的基本概念，如：总体、样本、容量、统计量和几种经常使用的分布.

In this chapter we mainly acquire some basic concepts of statistics, such as, population, sample, capacity, statistic, and some common distributions.

单词和短语 Words and expressions

- ★ 统计量　statistics
- ★ 总体　population
- ★ 个体　individual
- ★ 样本　sample
- 容量　capacity
- 统计分析　statistical analysis
- 统计分布　statistical distribution
- 统计总体　statistical ensemble
- 随机抽样　stochastic sampling / random sampling
- 随机样本　random sample
- 简单随机抽样　simple random sampling
- 简单随机样本　simple random sample
- 经验分布函数　empirical distribution function
- 样本均值　sample average / sample mean
- 样本方差　sample variance
- 样本标准差　sample standard deviation
- 标准误差　standard error
- 样本 k 阶矩　sample moment of order k
- 样本中心矩　sample central moment
- 样本值　sample value
- 样本大小、样本容量　sample size
- 样本统计量　sampling statistics
- 随机抽样分布　random sampling distribution
- 抽样分布、样本分布　sampling distribution
- 自由度　degree of freedom
- Z 分布　Z-distribution
- U 分布　U-distribution

基本概念和性质　Basic concepts and properties

1 统计量是不含有未知参数的样本函数.
A statistic is defined to be any sample functions without any unknown parameters.

2 统计量的分布称为抽样分布.
The distribution of a statistic is called a sampling distribution.

本章重点

1 理解总体、简单随机样本、统计量、样本均值、样本方差及样本矩的概念.
2 理解 x^2 分布、t 分布和 F 分布的概念及性质,并理解分位数的概念,能查表计算.
3 理解正态总体的常用分布.

Key points of this chapter

1 Understand the concepts of population, simple random sample, statistic, sample average, sample variance and sample moment.
2 Understand the concepts and properties of x^2 distribution, t distribution and F distribution, understand the concept of quantile and know how to calculate according to the chart.
3 Understand some common distributions of normal population.

第七章　参数估计
Chapter 7　Parameter Estimations

　　本章主要学习参数估计中的点估计的两种方法:矩法估计和极大似然估计,主要掌握参数估计中的区间估计.

　　In this chapter we are primarily concerned with two methods of point estimations, namely, estimations of moments and maximum likelihood estimation, as well as interval estimations in parameter estimations.

单词和短语 Words and expressions

- 统计推断　statistical inference
- ★ 参数估计　parameter estimation

分布参数　parameter of distribution
参数统计推断　parametric statistical inference
点估计　point estimate / point estimation
总体中心矩　population central moment
总体相关系数　population correlation coefficient
总体分布　population distribution
总体协方差　population covariance
点估计量　point estimator
★ 估计量　estimator
无偏估计　unbiased estimate / unbiased estimation
估计量的有效性　efficiency of estimator
矩法估计　moment estimation
总体均值　population mean
总体矩　population moment
总体 k 阶矩　population moment of order k
总体参数　population parameter
★ 极大似然估计　maximum likelihood estimation
★ 极大似然估计量　maximum likelihood estimator
极大似然法　maximum likelihood method / maximum-likelihood method
似然方程　likelihood equation
似然函数　likelihood function
★ 区间估计　interval estimation
置信区间　confidence interval
置信水平　confidence level
置信系数　confidence coefficient
单侧置信区间　one-sided confidence interval
置信上限　confidence upper limit
置信下限　confidence lower limit
U 估计　U-estimator
正态总体　normal population
总体方差的估计　estimation of population variance
置信度　degree of confidence
方差比　variance ratio

基本概念和性质　Basic concepts and properties

如果随机变量 $X_1, X_2, \cdots, X_k$ 是相互独立的，且 $X_i \sim x^2(n_i)(i=1,2,\cdots,k)$，则
$$X_1 + X_2 + \cdots + X_k \sim x^2(n_1 + n_2 + \cdots + n_k)$$

If the random variables $X_1, X_2, \cdots, X_k$ are independent and if X_i has a x^2 distribution with degrees of freedom $n_i(i=1,2,\cdots,k)$, then the sum $X_1 + X_2 + \cdots + X_k$ has a x^2 distribution with degrees of freedom $n_1 + n_2 + \cdots + n_k$.

例题　Example

例　表 1 中的一组甜菜根的含糖量百分比数据如下：

表 1

样本 1	17.4	18.6	18.8	18.6	18.4	17.6	17.7	17.8	18.1	18.4	18.1

求给定数据的样本均值和标准差（表 1）；求样本均值的标准差．

解　样本均值为 $\overline{Y} = \dfrac{17.4 + \cdots + 18.1}{11} \approx 18.136$，

样本方差为 $S^2 = \dfrac{1}{10}[(17.4-18.136)^2 + \cdots + (18.1-18.136)^2] \approx \dfrac{2.145}{10} \approx 0.215$，

则标准差为 $S = \sqrt{0.215} \approx 0.463$．

Ex　For the given data in Table 1 on the percentage of sugar content of a set of beets.

Table 1

Sample 1	17.4	18.6	18.8	18.6	18.4	17.6	17.7	17.8	18.1	18.4	18.1

Find the given values for the sample mean and standard deviation (Table 1).
Determine the standard error of the sample mean.

Solution　The sample mean is given by $\overline{Y} = \dfrac{17.4 + \cdots + 18.1}{11} \approx 18.136$．

The sample variance is given by $S^2 = \frac{1}{10}\left[(17.4-18.136)^2 + \cdots + (18.1-18.136)^2\right] \approx \frac{2.145}{10} \approx 0.215$.

Thus, the sample standard deviation is given by $S = \sqrt{0.215} \approx 0.463$.

本章重点
1. 理解参数的点估计、估计量与估计值的概念.
2. 了解矩估计法（一、二阶矩）和极大似然估计法（求偏导数和定义法）.
3. 理解估计量的无偏性、有效性（最小方差性）和一致性（相合性）的概念，并会验证估计量的无偏性.

Key points of this chapter
1. Understand the concepts of point estimation, estimator and estimate value.
2. Know how to estimate by the method of moments (moments of the first and the second order) and maximum-likelihood method. (Finding the partial derivatires and definition method)
3. Understand the concepts of unbiasedness, efficiency (the minimum of variance) and consistency, and know how to test the unbiasedness of an estimator.

第八章 假设检验
Chapter 8 Hypothesis Testings

单词和短语 Words and expressions

★ 参数假设　parametric hypothesis
★ 假设检验　hypothesis testing
两类错误　two types of errors
统计假设　statistical hypothesis
统计假设检验　statistical hypothesis testing
检验统计量　test statistics
显著性检验　test of significance
统计显著性　statistical significance
单边检验、单侧检验　one-sided test
单侧假设、单边假设　one-sided hypothesis

双侧假设　two-sided hypothesis
双侧检验　two-sided testing
显著水平　significant level
★ 拒绝域/否定区域　rejection region
★ 接受区域　acceptance region
U 检验　U-test
F 检验　F-test
方差齐性的检验　homogeneity test for variances
拟合优度检验　test of goodness of fit

基本概念和性质　Basic concepts and properties

在两个统计假设中选择的过程称为假设检验.
A deciding process of two statistical hypothesis is called a problem of hypothesis testing.

例题 Example

例 假设总体 $X \sim N(\mu, 40^2)$，其中 μ 未知，并且得一组容量为 9 的样本的样本均值 $\bar{x} = 780$，检验如下假设：$H_0: \mu = \mu_0 = 800$
$H_1: \mu \neq \mu_0 = 800$. 这里检验水平 $\alpha = 0.05$.

解 首先建立假设 $H: \mu = \mu_0 = 800$，选取统计量 $U = \dfrac{\bar{X} - 800}{40/\sqrt{9}}$.

在 H_0 成立的条件下，$U \sim N(0, 1)$.

因为检验水平 $\alpha = 0.05$，所以临界值 $u_\alpha = 1.96$，而 $|u| = \left|\dfrac{780-800}{40/3}\right| = 1.5 < 1.96$，所以我们接受原假设 $H: \mu = \mu_0 = 800$.

Ex Suppose that the population $X \sim N(\mu, 40^2)$, where μ is unknown, and that the mean of the sample $\bar{x} = 780$ with content 9, then the following hypotheses are to be tested: $H_0: \mu = \mu_0 = 800$
$H_1: \mu \neq \mu_0 = 800$. Here

the size of test will be $\alpha = 0.05$.

Solution First we make hypothesis $H: \mu = \mu_0 = 800$. Then we choose statistic $U = \dfrac{\overline{X} - 800}{40/\sqrt{9}}$. On the condition of H_0 yields $U \sim N(0,1)$.

For $\alpha = 0.05$, so $u_\alpha = 1.96$ and $|u| = \left| \dfrac{780 - 800}{40/3} \right| = 1.5 < 1.96$.

So we will accept the hypothesis $H: \mu = \mu_0 = 800$.

本章重点

1. 理解统计假设和显著性检验的基本思想.
2. 掌握假设检验的基本步骤,会构造简单假设的显著性检验.
3. 理解假设检验可能产生的两类错误;掌握单个和两个正态总体参数的假设检验.
4. 理解 x^2 拟合优度检验方法与秩和检验法.

Key points of this chapter

1. Understand the basic ideas concerning statistical hypothesis and test of significance.
2. Grasp the basic procedures of statistical hypothesis testing and know how to construct simple hypothesis tests of significance.
3. Understand the errors of two types resulting from a testing hypothesis and grasp hypothesis testing of single or two normal population parameters.
4. Understand the methods of x^2 chi-square test of goodness of fit and rank sum test.

练 习

1. 已知事件 A, B 相互独立,且 $P(B) > 0$,试求条件概率 $P(A \mid B)$.

  答案 $P(A)$.

2. 假设一个盒子里有 7 个红球和 3 个蓝球,如果随机不放回抽取 5 个球,试求取出的红球数的概率函数.

  答案 $f(x) = \begin{cases} \dfrac{C_7^x C_3^{5-x}}{C_{10}^5}, & 1 < x \leqslant 5 \\ 0, & \text{其他} \end{cases}$

Exercises

1. If A, B are mutually independent events and $P(B) > 0$, Find the conditional probability $P(A \mid B)$.
Answer $P(A)$

2. Suppose that a box contains 7 red balls and 3 blue balls. If 5 balls are selected at random without replacement, determine the probability function of the number of red balls that will be obtained.

Answer $f(x) = \begin{cases} \dfrac{C_7^x C_3^{5-x}}{C_{10}^5} & \text{for } 1 < x \leqslant 5 \\ 0 & \text{others} \end{cases}$

第六部分 统计学原理
Part 6 Introduction to Statistical Theory

引 言

统计学是论述收集、分析并解释数字信息的科学。统计科学的理论基础是数理统计,数理统计是由抽象的公理、定理和严谨的证明紧密结合而形成的完整的结构。为了使这种理论结构也适用于非数学家,产生了一个称为一般统计学的解释性的学科,其中的描述是最简化的,且常常是非数学的。很多领域,例如,经济学、生物学、农学、社会学、人类学、工程学、心理学等都从这个简化形式中得到了适合于各自数据的分析方法。

Introduction

Statistics is the science which tells us how to collect, analyze and explain the digital information. The theoretical basis of statistics is a discipline called mathematical statistics, mathematical statistics has the complete structure which contains the abstract axioms, theorems and rigorous proofs. In order to make the theory structure is also suitable for non-mathematicians, resulting in explanatory disciplines called the general statistics, which description is the most simplified, and often non-mathematical. Many fields such as economics, biology, agronomy, sociology, anthropology, engineering, psychology, and so on obtain suitable analysis methods for themselves.

第一章 绪论
Chapter 1 Introduction

本章主要介绍了统计的涵义、统计研究对象的特点、统计的作用、统计学中的基本概念、统计研究的基本方法和统计工作的一般过程以及统计工作的任务和组织。

This chapter mainly discusses the definition of statistics, the characteristics of the statistical study, the role of statistics, the basic concepts in statistics, the basic methods of statistical methods and the general process of statistical work, the task and the organization of the statistical work.

单词和短语 Words and expressions

统计调查　statistical survey	统计单位　statistical unit
统计工作　statistical work	品质标志　attributive indication
统计指标　statistical indicator	变量　variables
统计资料　statistical data	变异　variation
大量观察法　mass observation	统计设计　statistical design
分组法　grouping method	标志与指标　signs and indicators
综合指标法　comprehensive index method	同质性　homogeneity
统计分析　statistical analysis	统计监督　statistical supervision
数量标志　quantitative indication	统计设计　statistical design
统计总体　statistical population	

本章重点

1. 统计的涵义;
2. 统计学中的基本范畴。

Key points of this chapter

1. The definition of statistics.
2. Basic aspects of statistics.

第二章 统计设计和统计调查
Chapter 2 Statistical Design and Statistical Survey

本章主要介绍了统计设计的概念和意义，统计设计的种类以及统计设计的内容，统计表及其设计，统计调查的概念和种类，统计调查方案及其组织方式。

This chapter introduces the concept, significance, the types and contents of statistical design, statistical tables and its design, the concepts, types, programs and organizations of statistical survey.

单词和短语 Words and expressions

中文	English	中文	English
整体设计	overall design	简单分组	simple grouping
专项设计	special design	复合分组	composite grouping
全过程设计	whole process design	全面调查	full-scope survey
单阶段	design of single phase	图表	charts
中期设计	medium-term design	类	classes
长期设计	long-term design	调查表	enumeration form
短期设计	short-term design	采访法	gathering method
统计表	statistical table	普查	general census
统计设计	statistical design	重点调查	key-point survey
统计指标	statistical indicator	大量观察法	mass obseruation
总量指标	aggregate indicators	非全面调查	non-full-scope survey
相对指标	relative indicators	经常性调查	regular survey
平均指标	average index	抽样调查	sample survey
数量指标	quantitative indicators	一次性调查	single-round survey
质量指标	quality indicators	统计报表	statistical reports
实体指标	entity index	典型调查	typical survey
行为指标	behavioral indicators	专门调查	special survey
客观指标	objective indicators	统计分析	statistical analysis
主观指标	subjective indicators	统计资料	statistical data
考核指标	assessment indicators	统计调查	statistical survey
非考核指标	non-assessment indicators	调查方法	survey method
时期指标	period indicators	调查对象	survey object
时点指标	point indicators	调查时间	survey time
分析表	analysis form	调查单位	survey unit
汇总表	summary form	直接观察法	direct observation method

本章重点

1. 统计报表和四种专门调查；
2. 数据搜集方案的内容；
3. 数据的质量要求。

Key points of this chapter

1. Statistical reports and four special investigations.

2. The collection program of data.
3. The quality requirements of data.

第三章　统计整理
Chapter 3　Statistical Processing

本章主要介绍了统计整理的意义，统计整理的步骤和方法，统计整理在整个统计工作中的重要作用，统计分组的作用和分组方法，统计汇总的组织与技术，手工汇总与计算机汇总技术。

This chapter focuses on the significance, important role, steps and methods of statistical processing, statistical grouping methods and its roles, the organization and technology of statistical summary; the technologies of manual and computer tabulation summary.

单词和短语 Words and expressions

数量标志	quantitative indication	连续变量	continuous variable
品质标志	attributive indication	离散变量	discrete variable
简单分组	simple grouping	钟形	bell-shaped nature
复合分组	composite grouping	数据	data
频数分布	frequency distribution	累积	accumulative
组距	class interval	条形图	bar charts
组限	class limit	频数百分比	percentage frequency
组全距	class range	累积频数	cumulative frequency
开口组	open-ended class	未分组数据	ungrouped data
曲线图	curvilinear data plots	频数折线图	frequency polygons
直方图	histogram		

本章重点

1. 统计数据分组的原则和方法；
2. 频数分布的概念、种类和编制方法。

Key points of this chapter

1. The principles and methods of statistical data grouping.
2. The concept, types and compilation methods of frequency distribution.

第四章　总量指标和相对指标
Chapter 4　Aggregate Indicators and Relative Indicators

统计指标可以分为总量指标、相对指标和平均指标。本章分别介绍了总量指标和相对指标的概念、作用和种类以及计算和运用总量指标和相对指标的基本原则。

Statistical indicators can be divided into the total index, relative index and average index. This chapter describes the concept, role and types of the total index and relative index, and the basic principles of calculation and application of them.

单词和短语 Words and expressions

基期	base period	水平法	horizontal method
基期水平	base level	指标	indicator
超额完成任务	beat the target	计划期间	in planned period
复名数	composite unit	劳动单位	labour unit
动态相对指标	dynamic relative indicator	价值指标	merit indicator

货币单位　monetary unit
实物单位　physical unit
总体标志总量　population mark total amount
总体单位总量　population size
比例相对指标　ratio relative indicator
相对的　relative
相对指标　relative indicator
计划完成相对指标　relative quantities of fulfillment of plan
报告期　reporting period
增长速度　speed of growth

结构相对指标　structure relative indicator
强度相对指标　intensity relative indicator
比较相对指标　comparison relative indicator
报告期水平　current level
时期指标　time-period indicator
时点指标　time-point indicator
总量指标　total amount indicator
实体指标　entity index
劳动指标　labor indicators
累计法　cumulative Method

本章重点
1. 总量指标的概念和分类；
2. 相对指标的概念及各种相对指标的计算方法；
3. 相对指标的运用原则。

Key points of this chapter
1. The concept and classification of aggregate indicators.
2. The concept of relative indicators and the calculation methods of the relative indicators.
3. The application principles of the relative indicators.

第五章　平均指标和变异指标
Chapter 5　Average Indicators and Variation Indicators

本章阐述了平均指标的概念和作用，各种平均数的计算原则、方法与应用条件，变异指标的作用、计算方法与应用条件，主要的平均指标和变异指标。

In this chapter, the concept and role of the average indicators are elaborated. The calculation principles of average indicators, methods and application conditions of average indicators are introduced. The main average index, calculation principles, methods and application conditions of variation indicators are interpreted. The main average indicators and variation indicators are showed.

单词和短语 Words and expressions

集中趋势　central tendency
变异系数　coefficient of variation
加权算术平均数　weighted arithmetic mean
加权平均　weighted mean
几何平均数　geometric mean
调和平均数　harmonic average
平均绝对误差　mean absolute deviation

平均误差　mean deviation
众数　mode
中位数组　median class
标准差　standard deviation
简单算术平均值　simple arithmetic mean
变量　variables

本章重点
1. 算术平均数；
2. 调和平均数；
3. 众数；
4. 中位数；
5. 标准差与标准差系数。

Key points of this chapter

1. Arithmetic mean
2. Harmonic average
3. Mode
4. Median
5. Standard deviation and coefficient of standard deviation

第六章 动态数列
Chapter 6 Dynamic Series

动态分析是统计分析方法之一，其依据是动态数列。本章阐述了动态数列的概念、作用、种类和编制原则；总量指标、相对指标和平均指标三种动态数列；动态分析的水平和速度指标；测定实物变动长期趋势的主要方法；直线配合法的常用方法；测定季节变动的主要指标。

Dynamic analysis is one of the statistical analysis methods, and it is based on the dynamic series. In this chapter, the concept, role, type and preparation of a dynamic series are introduced. The total index, relative index and average index are elaborated. The level and speed indicators of dynamic analysis are interpreted. Determination of the long-term trends of physical changes, the commonly used method of linear match method and the main determination indicators of the seasonal variation are showed.

单词和短语 Words and expressions

发展水平   development level
动态数列   dynamic series
消除季节性销售量   deseasonalized sales
消除季节性数据   deseasonalizing data
消除季节性数据   with deseasonalizing data
时间序列   in time series
增长水平   growth level

增长量   increment
移动平均法   moving-average method
季节变动   seasonal variation
时间序列和预测   time series and forecasting
时期数列   the series of time period
序时平均数   chronological average

本章重点

1. 时间序列的概念、构成要素及种类；
2. 各种动态分析指标的计算和应用；
3. 长期趋势、季节变动的测定和分析。

Key points of this chapter

1. The concept, elements and types of the time series.
2. The calculation and application of dynamic analysis indicators.
3. The determination and analysis of long-term trends and seasonal variations.

第七章 统计指数
Chapter 7 Statistical Index

统计指数法是统计分析中广为采用的重要方法。本章阐述了统计指数的概念、作用和种类；个体指数和总指数；简单指数和加权指数；定基指数和环比指数；综合指数的编制原则与方法；平均指数的编制方法；指数体系和因素分析；总量指标的两因素分析和多因素分析；平均指标的因素分析。

Statistical index method is an important method widely used in statistical analysis. This chapter describes the concept, role and types of statistical index; individual index and total index, simple index and weighted index, fixed base index and chain index number, the principle and method of aggregate index number, the preparation method of average index, index system and factor analysis, two-factor analysis and multivariate analysis of the total index, factor analysis of the average index.

单词和短语 Words and expressions

综合指数　aggregate index number
算术平均数指数　arithmetic average index
合计指数　aggregate index
价格指数平均　average of the price indexes
环比指数　chain index number
消费者价格指数　consumer price index (CPI)
指数数据　data converted to indexes
因素　factors
指数　index numbers
质量指数　index number of quanlity
数量指数　index number of quantity
固定组成指数　index of fixed formation
可变结构指数　index of variable construction
可变组成指数　index of variable formation
个体指数　individual index number
同度量因素　isometric factor
拉氏价格指数　Laspeyres price index
拉氏指数　Laspeyres index
纽约股票交易指数　New York stock Exchange index
帕氏价格指数　Paasche's price index
帕氏指数　Passche index
季节指数　seasonal index
简单综合指数　simple aggregate index
价格指数的简单平均　simple average of price indexes
简单指数　simple index numbers
加权指数　weighted indexes
未加权指数　unweighted indexes
道—琼斯工业平均指数　Dow jones industrial Average (DJIA)

本章重点

1. 统计指数的概念；
2. 综合指数公式的意义；
3. 掌握算术平均指数的计算；
4. 了解调和平均数的计算以及平均指数的应用；
5. 掌握指数体系的概念，熟练运用指数体系进行统计绝对数变动的因素分析，了解平均数变动的因素分析；
6. 对常用的几个经济指数及其具体应用做一个大体的了解。

Key points of this chapter

1. The concept of statistical index.
2. The significance of aggregate index formula.
3. Master the calculation of arithmetic average index.
4. Understand the calculation of harmonic average and the application of average index.
5. Master the concept of index system, perform the factors analysis of statistical absolute numbers changes based on the index system, and understand the factors analysis of average changes.
6. Understand the concepts and specific application of several commonly used economic index.

例题 Example

例　设某公司生产三种产品的有关资料如下表，试以1990年不变价格为权数，计算各年的产品产量指数。

商品名称	产量			1990年不变价格(p_{90})(元)
	1994年(q_{94})	1995年(q_{95})	1996年(q_{96})	
A	1000	900	1100	50
B	120	125	140	3500
C	200	220	240	300

解 各年的产量指数为

$$q_{95/4} = \frac{\sum p_{90}q_{95}}{\sum p_{90}q_{94}} = \frac{50 \times 900 + 3500 \times 125 + 300 \times 220}{50 \times 1000 + 3500 \times 120 + 300 \times 200} = 103.49\%$$

$$q_{96/5} = \frac{\sum p_{90}q_{96}}{\sum p_{90}q_{95}} = \frac{50 \times 1100 + 3500 \times 140 + 300 \times 240}{50 \times 900 + 3500 \times 125 + 300 \times 220} = 112.49\%$$

$$q_{96/4} = \frac{\sum p_{90}q_{96}}{\sum p_{90}q_{94}} = \frac{50 \times 1100 + 3500 \times 140 + 300 \times 240}{50 \times 1000 + 3500 \times 120 + 300 \times 200} = 116.42\%$$

Ex The relevant information of three products is listed in the following table, compute the output index of products each year based on the weight of constant prices of 1990.

Commodity	Quantities			1990 Constant price(p_{90})(yuan)
	1994 (q_{94})	1995 (q_{95})	1996 (q_{96})	
A	1000	900	1100	50
B	120	125	140	3500
C	200	220	240	300

Solution

Aggregate quantity index numbers of each year are as follows:

$$q_{95/4} = \frac{\sum p_{90}q_{95}}{\sum p_{90}q_{94}} = \frac{50 \times 900 + 3500 \times 125 + 300 \times 220}{50 \times 1000 + 3500 \times 120 + 300 \times 200} = 103.49\%$$

$$q_{96/5} = \frac{\sum p_{90}q_{96}}{\sum p_{90}q_{95}} = \frac{50 \times 1100 + 3500 \times 140 + 300 \times 240}{50 \times 900 + 3500 \times 125 + 300 \times 220} = 112.49\%$$

$$q_{96/4} = \frac{\sum p_{90}q_{96}}{\sum p_{90}q_{94}} = \frac{50 \times 1100 + 3500 \times 140 + 300 \times 240}{50 \times 1000 + 3500 \times 120 + 300 \times 200} = 116.42\%$$

第八章 抽样调查
Chapter 8 Sample Surveys

本章阐述了抽样调查的概念、特点、作用;影响抽样误差的主要因素;抽样调查几种主要组织方式的抽样平均误差的计算;抽样估计;点估计和区间估计;必要抽样数目的确定。

In this chapter, the concept, characteristics and roles of sample surveys are introduced. The main affected factors of the sampling errors and the average sampling error calculation of several sample survey organizations are introduced. The sampling estimation, point and interval estimation and the determination method of necessary sample size are outlined.

单词和短语 Words and expressions

整群抽样　cluster sampling
置信区间　confidence interval
置信限　confidence limits
区间估计　interval estimate
大样本　large samples
概率抽样　probability sampling
比例抽样　proportional sampling

预测区间　prediction intervals
抽样　sample
样本均值　sample mean
样本比例　sample proportion
样本量　sample size
样本量和估计　sample size and estimation
抽样统计量　sample statistic

抽样方差　sample variance
样本均值的抽样方差　sampling distribution of the sample mean
抽样误差　sampling error
抽样方法　sampling method
小样本　small sample

简单随机抽样　simple random sampling
分层随机抽样　stratified random sampling
抽样调查　sample survey
总体标准差　standard deviation in population
估计标准误差　standard error of estimate
系统抽样　systematic random sampling

本章重点

1. 参数的区间估计方法
2. 抽样误差的计算

Key points of this chapter

1. The parameter interval estimation method.
2. The calculation of sample errors.

例 题　Example

例　我们从某个大城市中随机抽取 100 个家庭，他们的平均年收入为 $\bar{x} = 5,565$ 元，根据经验可知，总体服从正态分布 $N(\mu, \sigma^2)$，且标准差 $\sigma = 1,800$ 元。试求 μ 的置信水平为 95% 的置信区间。

解　由题可知，$Z_a = 1.96$，$\mu = \bar{x} \pm Z_a \dfrac{\sigma}{\sqrt{n}} = 5565 \pm 1.96 \times \dfrac{1800}{\sqrt{100}} = 5565 \pm 352.8$

则 μ 的置信水平为 95% 的置信区间为 [5212.2, 5917.8]。

Ex　We assume that a sample of $n = 100$ families is selected from several million families in a big city. The annual income of each family is $\bar{x} = 5,565$. By calculating the sample, we know that the probability distribution of the population is normal and its standard deviation of the population is $\sigma = 1,800$ from the experience. Now, estimate the confidence interval of μ, with 95% confidence level.

Solution　From the diagram of normal-curve areas, $Z_a = 1.96$, the 95% confidence interval estimate is the follow：

$$\mu = \bar{x} \pm Z_a \dfrac{\sigma}{\sqrt{n}} = 5565 \pm 1.96 \times \dfrac{1800}{\sqrt{100}} = 5565 \pm 352.8$$

i.e. $5212.2 \leqslant \mu \leqslant 5917.8$, the 95% confidence interval estimation is [5212.2, 5917.8].

第九章　相关与回归分析
Chapter 9　Correlation and Regression Analysis

本章主要介绍了相关系数的概念与特点；相关关系与函数关系的区别与联系；相关关系的种类与测定方法；回归分析的概念与特点；直线回归方程的求解及其精确度的评价；估计标准误差的计算。

This chapter mainly introduces the concept and characteristics of the correlation coefficient, the differences between function relation and correlation relation, the types and determination method of correlation relation, the concept and characteristics of regression analysis, the solution and accuracy evaluation of linear regression equation, and the calculation of estimate standard error.

单词和短语　Words and expressions

自相关　autocorrelation
相关　correlation
相关分析　correlation analysis
相关系数　coefficient of correlation

线性回归　linear regression
画回归线　drawing the line of regression
线性回归的一般形式　general form of linear regression

线性相关　linear correlation
线性回归　linear regression
线性回归方程　linear regression equation
多元估计标准误差　multiple standard error of estimate
负相关　negative correlation
非线性相关　nonlinear correlation
回归　regression

回归分析　regression analysis
回归系数　regression coefficient
回归方程　regression equation
回归变差　regression variation
回归直线　regression line
回归定义　regression defined
k 个自变量　with k independent variables

本章重点

① 两个变量的直线相关分析；
② 两个变量的线性回归分析。

Key points of this chapter

① Linear correlation analysis of two variables.
② Linear regression analysis of two variables.

例　题　Example

例　某旅游景点历年观光游客资料如下表，试用最小二乘法拟合直线趋势方程 $y_c = a + bt$。

年份	时间 t	游客 y(百人)	t^2	ty
1994	1	100	1	100
1995	2	112	4	224
1996	3	125	9	375
1997	4	140	16	560
1998	5	155	25	775
1999	6	168	36	1008
2000	7	180	49	1260
合计		980	140	4302

解　由上表可知，$\sum t = 28, \sum y = 980, \sum t^2 = 140, \sum ty = 4320$

得到 $b = \dfrac{n\sum ty - \sum t \sum y}{n\sum t^2 - (\sum t)^2} = \dfrac{7 \times 4302 - 28 \times 980}{7 \times 140 - 28 \times 28} = \dfrac{2674}{196} = 13.64$

$a = \bar{y} - b\bar{t} = \dfrac{980}{7} - 13.64 \times 4 = 85.44$

从而求得直线趋势方程为 $y_c = 85.44 + 13.64t$

Ex　The tourists information of a tourist attraction over the years is listed in the following table, compute the linear regression equation $y_c = a + bt$ based on the least squares method.

Year	Time t	Tourist y(Hundred people)	t^2	ty
1994	1	100	1	100
1995	2	112	4	224
1996	3	125	9	375
1997	4	140	16	560
1998	5	155	25	775
1999	6	168	36	1008
2000	7	180	49	1260
合计		980	140	4302

Solution From the above table, we obtain
$$\sum t = 28, \sum y = 980, \sum t^2 = 140, \sum ty = 4320$$

Then $b = \dfrac{n\sum ty - \sum t \sum y}{n\sum t^2 - (\sum t)^2} = \dfrac{7 \times 4302 - 28 \times 980}{7 \times 140 - 28 \times 28} = \dfrac{2674}{196} = 13.64$

$a = \bar{y} - b\bar{t} = \dfrac{980}{7} - 13.64 \times 4 = 85.44$

The linear regression equation is $y_c = 85.44 + 13.64t$.

第七部分 生命表基础
Part 7 Life Table Foundation

引言

人寿保险是以人的寿命为保险标的,被保人的生存与死亡决定保险金的给付与否及给付时间。因此,被保人的生命规律对寿险经营在财务上有直接影响。事实上,在人寿保险中,从保险费率的厘定、责任准备金的提取、保单现金价值的计算到保单红利的分配都必须考虑一个重要的因素——死亡率。而各年龄的死亡率构成一个生命表,也就是说,所有这些计算都建立在生命表的基础之上。

被保险人在未来某个时期的生死是一个不确定性事件,这个事件发生的不确定性对寿险公司和养老金计划造成财务上的影响是精算学研究的一个重要领域。本书即在这一领域阐述精算的基本理论与方法,而生命表是精算研究的基本工具。

Introduction

Life insurance takes people's life as the subject matter of the insurance. Whether to pay the insurance fee or not and the time to pay depend on the insurant's survival or mortality. Therefore, insurant's life has a direct effect on financial affairs. In fact, in the life insurance, many important aspects, raging from the determination of insurance rates, the reserves of extraction, the calculation of the cash value of the policy to the policy of dividend distribution, must consider an important factor - mortality rate. It is well know that the mortality rations for every age produce a life table. It is clear that all of these calculations are based on life table.

As the mortality of the insurant in forthcoming period is an uncertainty event, the research on the uncertainty on the financial implications for the life insurance companies and pension plans is an important area of actuarial science. This book covers the basic theories and methods of actuarial science. Life table is a basic tool for actuarial studies.

第一篇 生存模型及其应用
Part 1 Survival Model and Application

第一章 生存模型及其性质
Chapter 1 Survival Model and Character

本章以概率为基础介绍生存模型的基本概念及形式,着重对生存模型 $S(t) = Pr(T>t)$ 进行研究,给出几种描述生存模型的参数分布形式,进而对死亡率和条件死亡率进行详细的阐述,在此基础上,给出中心死亡率的概念和计算公式。

This chapter introduces the basic concept and form of the survival model based on probability knowledge. We focus on our research on survival model $S(t) = Pr(T>t)$. We first give the parameter distribution functions of several survival models, then show the mortality rate and conditional mortality rate in details. Base on the prior knowledge, the concept and the calculation formula of center of mortality rate are given.

单词和短语 Words and expressions

★ 生存模型　survival models　　　　★ T分布　T distribution

★ 均匀分布　average distribution
★ 指数分布　exponential distribution
Gompertz 分布　Gompertz distribution
Makeham 分布　Makeham distribution
Weibull 分布　Weibull distribution
条件度量　conditional measure
连续运转　continuous running
报废　scrapped
失效　invalidation
未来寿命　future lifetime
年限　fixed number of year
精算生存模型　actuarial survival model
确切生存模型　specific survival model
确切年龄　specific time of life
自然年龄　natural time of life
精算师　actuary
选择模型　selection model
未来寿命随机变量　future life random variable
参数生存模型　parameter survival model
横向研究　transversal study
纵向研究　longitudinal study
可控数据研究　research on controllable data
删截　censored
平均未来寿命　average future lifetime
失效密度　invalidation density

未来预期寿命　future expectant life
寿命期望　life expectation
经验样本数据　empirical sample data
无生命物体　inanimate object
长时间区间段　long-time interval
短时间区间段　short-time interval
死亡时间随机变量　random variable of time of death
单参数分布　single parameter distribution
样本空间子集　subset of sample space
条件危险率函数　conditional dangerous rate function
截尾函数　censored function
双截尾生存分布函数　double cutoff survival distribution function
下界的截尾　low censored bound
上界的截尾　upper censored bound
★ 截尾分布　truncated distribution
★ 死亡率　mortality
★ 条件死亡率　conditional mortality
★ 中心死亡率　central rate of death
累积分布函数　accumulative distribution function
★ 危险率函数　hazard rate function
生存分布函数　survival distribution function

例 题　Example

例　设 X 服从标准指数分布，$F_X(x) = 1 - e^{-x}$，令 $y = g(x) = \dfrac{1}{x}$，求 Y 的分布函数 $S_Y(y)$、概率密度函数 $f_Y(y)$ 和危险率函数 $\lambda_Y(y)$。

解　因为 $x = h(y) = \dfrac{1}{y}$，所以

$$S_Y(y) = F_X[h(y)] = 1 - e^{-\frac{1}{y}}$$

$$f_Y(y) = -\frac{d}{dy} S_Y(y) = y^{-2} e^{-\frac{1}{y}}$$

$$\lambda_Y(y) = \frac{f_Y(y)}{S_Y(y)} = \frac{y^{-2} e^{-\frac{1}{y}}}{1 - e^{-\frac{1}{y}}}$$

Ex　Suppose that X is the standard index distribution, Let $F_X(x) = 1 - e^{-x}$, $y = g(x) = \dfrac{1}{x}$. Please determine distribution function $S_Y(y)$, probability density function $f_Y(y)$, hazard rate function $\lambda_Y(y)$.

Solution　Since $x = h(y) = \dfrac{1}{y}$, then $S_Y(y) = F_X[h(y)] = 1 - e^{-\frac{1}{y}}$

$$f_Y(y) = -\frac{d}{dy} S_Y(y) = y^{-2} e^{-\frac{1}{y}}$$

$$\lambda_Y(y) = \frac{f_Y(y)}{S_Y(y)} = \frac{y^{-2} e^{-\frac{1}{y}}}{1 - e^{-\frac{1}{y}}}.$$

本章重点
1. 掌握生存模型的概念和性质。
2. 会计算生存分布函数，概率密度函数及危险率函数。
3. 掌握条件和无条件分布之间的区别，以及死亡率的度量。

Key points of this chapter
1. Know the concept and character of the survival model.
2. Calculate survival distribution function, probability density function and risk ratio function.
3. Master the distinction between conditional and unconditional distribution, and mortality ratio measurement.

第二章 生命表
Chapter 2 Life Table

本章给出生命表的传统形式和标准精算符号，推导出死亡年龄随机变量的概率密度函数，最后给出暴露数的概念与计算公式，并对死亡率与中心死亡率的关系进行阐述。

This chapter gives the traditional forms and standard actuarial symbols of life table. The probability density function of random variables of age at death is derived. Finally, the concept and calculation formula of exposure number is given. The relationship between the mortality ratio and the central mortality ratio is discussed.

单词和短语 Words and expressions

生命表　life table
★死力　the force of mortality
★暴露数　exposure number
完全预期寿命　total life expectancy
★表格模型　table model
★传统生命表　traditional life table
绝对瞬时变化率　absolute instantaneous changing rate
瞬时死亡率　instantaneous death rate
死力函数　function of the force of mortality
概率分布　probability distribution
初始规模　initial scale
新生群体　new group
瞬时死亡人数　instantaneous mortality amount
死亡风险　mortality risk
黎曼和　Riemann sum
生存人数　survival number

静止人口学　stationary population demography
平均余命　life expectancy
平均年比率　the average annual rate
连续参数模型　Continuous parameter model
l_{x+t} 线性形式　l_{x+t} linear formation
l_{x+t} 指数形式　l_{x+t} exponential formation
l_{x+t} 的双曲线形式　l_{x+t} hyperbolic formation
人寿保险业　life insurance office
双曲线假设　hyperbolic assumption
死亡均匀分布假设(UDD)　the assumption of uniform distribution of death
常死力假设　constant force of mortality
精算学　Actuarial Science
Balducci假设　Balducci assumption
选择—终极表　select—ultimate table
选择期长度　select period length
★取整余命　integral remainder

例题 Example

例 在某一生命表中，生存人数 l_x，已知 $l_x = 1000, l_{x+1} = 900$，分别在UDD假设、指数假设和Balducci假设下估计中心死亡率 m_x。

解 由已知得：x 岁到 $(x+1)$ 岁之间的死亡人数 $d_x = 100$，

x 岁活到 $(x+1)$ 岁的概率 p_x，$\ln p_x = -0.10536$。

在 UDD 假设下，$L_x = l_x - \frac{1}{2}d_x = 950$，则 $m_x = \frac{d_x}{l_x} = 0.10526$；

在指数假设下，直接有 $m_x = -\ln p_x = 0.10536$；

在 Balducci 假设下，$L_x = \frac{-l_{x+1} \cdot \ln p_x}{q_x} = 948.24$，从而有 $m_x = 0.10546$。

Ex In a life table, the number of survival is l_x, let $l_x = 1000, l_{x+1} = 900$. Under the UDD hypothesis, index hypothesis and the Balducci assumption, please estimate central rate of death m_x.

Solution We can easily know that the number of death of aged-x to aged-$(x+1)$ is $d_x = 100$, the probability of aged-x to aged-$(x+1)$ is $p_x, \ln p_x = -0.10536$.

Under the UDD hypothesis, $L_x = l_x - \frac{1}{2}d_x = 950, m_x = \frac{d_x}{l_x} = 0.10526$;

Under the index hypothesis, $m_x = -\ln p_x = 0.10536$;

Under the Balducci assumption, $L_x = \frac{-l_{x+1} \cdot \ln p_x}{q_x} = 948.24, m_x = 0.10546$.

本章重点

1 掌握生命表传统形式和标准精算符号。

2 理解由生存分布函数及其相关函数所表达的生命表和生存模型之间的对应关系。

Key points of this chapter

1 Know the traditional forms of the life table and standard actuarial symbols.

2 Understand the correspondence between the life table and survival model expressed by the survival distribution function and its related functions.

第三章　完整样本数据情况下表格生存模型的估计
Chapter 3　Table Survival Model Estimation in Complete Sample Data

本章以前两章的知识为基础，在已知死亡确切时间的条件下，利用经验分布对完整样本数据情况下表格生存模型进行估计；在死亡确切时间未知时，利用死亡时间分组法，给出完整样本数据情况下表格生存模型的估计。

In this chapter, we use the knowledge introduced in previous chapters to estimate the form survival table with empirical distributions. We limit our discussion to Complete sample data, knowing the extract time of mortality. For the condition without knowing extract time of mortality, we mainly discuss the estimation of the form survival model using the mortality grouping method with the Complete sample date.

单词和短语 Words and expressions

* ★完整样本数据　Complete sample data
* ★不完整的样本数据　incomplete sample data
* 初始群体　initial colony
* 封闭　close down
* 诊断日期　diagnostic date
* 死亡日期　death date
* 实际年龄　actual age
* 小样本　small sample
* 估计量曲线　estimation curve
* 降幅　drop in
* 初始估计量　initial estimation amount
* ★死亡时间　mortality time
* ★死亡确切时间　accrual mortality time
* 生存分布　survival distribution
* ★生存概率　survival probability
* 二项比随机变量　random variable of binomial proportion
* 标准精算符号　standard actuarial symbol
* ★经验生存分布　experienced survival distribution
* N_t 的分布　N_t distribution
* D_t 的分布　D_t distribution
* 死亡分组　mortality grouping

★ 表格生存模型　table survival model
统计微分法　statistical differential method
直接估计　direct estimation
样本结果　sample results

无偏极大似然估计　unbiased maximum likelihood estimation
★ 标准估计法　standard estimation method
多项随机变量　multivariate random variables

本章重点

1. 掌握表格生存模型的有关内容。
2. 能够从样本数据中得出未知生存分布的估计及表格生存模型的估计。

Key points of this chapter

1. Grasp relevant contents of the table survival model.
2. Be able to make the estimation for unknown survival distributes and table survival model from the sample data.

第四章　非完整样本数据情况下表格生存模型的估计
Chapter 4　Table Survival Model Estimation in Incomplete Sample Data

　　本章首先介绍了观察期的进入观察年龄、预计退出年龄等概念，以此为基础，定义了特例A、特例B和特例C三种情形；然后介绍了双风险情况下的精算理论；接着阐述了对非完整样本数据的表格生存模型进行估计的方法。最后介绍了乘积极限估计及Nelson—Aalen估计量的概念及计算公式。

　　This chapter introduces some basic concepts, including the entrance observing age for the period under observation, estimation of the debuting age. Based on these concepts, we define three special cases: case A, case B and case C. The actuary theory on double risk condition is introduced. We also elaborate the estimation method of form table survival model for uncompleted sample data. In the end of this chapter, the concepts and formulas of product-limit estimation and Nelson-Aalen estimation are introduced.

单词和短语 Words and expressions

★ 观察期的年龄　age of the observation period
★ 计划结束年龄　plans to end age
★ 进入年龄　enter age
★ 退出年龄　exit age
单风险环境　single risk environment
双风险环境　double risk environment
退出力　exiting force
死亡密度函数　mortality Death density function
退出密度函数　exit density function
矩方程　moment equation
观察日期　observation date
特例A　special case A
确切暴露数　exact exposure number
预计暴露数　estimated exposure number
期望死亡人数　expected number of death
依靠引出　rely on lead
期望观察值　expected observation
实际观察值　actual observation
期望生存人数　number of expected survival
期望死亡人数　number of expected death
实际结束人数　number of actual end

生存经历　survival experience
完全数据　full data
部分数据　part data
★ 矩估计法　moment estimation method
★ Hoem法　Hoem method
★ 精算法　actuarial method
精算暴露数　actuarial exposure number
统计判断　statistical judgment
精算估计量　actuarial estimator
Greenwood 估计法　Greenwood estimation method
似然函数　likelihood function
★ 极大似然估计法　maximum likelihood estimation method
★ 乘积极限估计量　product-limit estimator
Kaplan - Meier 估计量　Kaplan-Meier estimator
★ Nelson - Aalen 估计量　Nelson - Aalen estimator
风险集合　risk collection
累积风险率函数　cumulative hazard rate function

例题 Example

例1 在某研究样本中三个观察对象,其有序数组均为 $(r_i, s_i) = (0, 0.75), (i = 1, 2, 3)$,对象1在 $x + 0.75$ 时仍在样本中,对象2在 $x + 0.5$ 处退出;对象3在 $x + 0.5$ 处死亡。求(1)精算暴露数;(2) Home 矩估计法中的暴露数。

解 (1)对象1,2,3的暴露数分别为 $0.75, 0.5, 1$,所以总和为 $0.75 + 0.5 + 1 = 2.25$;

(2)对象1,2,的暴露数分别为 $0.75, 0.5$,对象3为 0.75,总和为 $0.75 + 0.5 + 0.75 = 2$。

Ex. 1 There are three observation objects in a study. First is still in the sample at $x + 0.75$. Second exits at $x + 0.5$. Third deaths at $x + 0.5$. Determine

(1) Actuarial exposure number;

(2) Exposure number of Home moment estimation method.

Solution (1) The exposure number of first is 0.75, the exposure number of second is 0.5, and the exposure number of third is 1. so Actuarial exposure number is 2.25.

(2) Under Home moment estimation, the exposure number of first is 0.75, the exposure number of second is 0.5, and the exposure number of third is 0.75, so $0.75 + 0.5 + 0.75 = 2$.

例2 在某一完整数据研究中,初始样本容量 $n = 10$,$S(12)$ 的乘积极限估计为 $\hat{S}(12) = 0.7$,求 $S(12)$ 的 Nelson-Aalen 估计量。

解 因为是完整的数据,所以在 $t = 12$ 时,有7个生存者(因为此时乘积极限估计是一简单的二项比例),这样,在 $t = 12$ 以前发生了3次死亡,于是 Nelson-Aalen 估计量为 $\hat{\Lambda}(12) = \frac{1}{10} + \frac{1}{9} + \frac{1}{8} = \frac{242}{720}$,从而有 $\hat{S}(12) = e^{-\frac{242}{720}} = 0.71454$(注意 Nelson-Aalen 估计量比乘积极限估计量要大)。

Ex. 2 In a complete data research, the initial sample size is $n = 10$, product limit estimation of $S(12)$ is $\hat{S}(12) = 0.7$. Please determine Nelson - Aalen estimator of $S(12)$.

Solution Because the data is complete, so $t = 12$, there are seven survivors. Death happened three times before $t = 12$. Nelson - Aalen estimator is $\hat{\Lambda}(12) = \frac{1}{10} + \frac{1}{9} + \frac{1}{8} = \frac{242}{720}$, then $\hat{S}(12) = e^{-\frac{242}{720}} = 0.71454$ (Note that the Nelson-Aalen estimator is bigger than the product-limit estimator).

本章重点

1. 掌握非完整样本数据的表格生存模型的估计方法,包括矩估计法和极大似然估计法。
2. 熟悉矩估计法中的传统估计法、Hoem 法及精算法。
3. 能够计算出不同特例情形下及一般情形下的极大似然估计。
4. 计算乘积极限估计量和 Nelson-Aalen 估计量。

Key points of this chapter

1. Know the estimation methods of form table survival model for uncompleted sample data, including moment-based estimation method and maximum likelihood estimation method.
2. Be familiar with the traditional estimation method Hoem method and actuarial estimation method in moment estimation method.
3. Be able to calculate the maximum likelihood estimation for some specific cases and some usual cases.
4. Calculate the product-limit estimation and Nelson-Aalen estimation.

第五章 参数生存模型的估计
Chapter 5 Estimation of Parametric Survival Model

本章针对完整样本数据和非完整样本数据,对参数生存模型中的参数进行估计和做出假设检验。估计方法包括矩估计法、极大似然法、分位点法及最小二乘法等。最后介绍了含伴随变量的参数生存模型,并对其中的和模型和积模型进行了阐述。

In this chapter, we consider how to make estimations and hypotheses for the parameters in parameter-survival method with full sample data or non-complete sample data. The estimation methods we use include moment-bases method, maximum likelihood estimation method, fractile method and the least squares method. In the end of this chapter, the parameter-survival model with some random variables is introduced. As two important models, the additive model and product model are addressed.

单词和短语 Words and expressions

★ 参数估计　parameter estimation
临床研究　clinical research
矩方法　moment method
中位数估计法　median estimate method
估计参数　estimate parameter
死亡事件　death event
分组死亡时间　time of group death
单参数均匀分布　uniform distribution of single parameter
理论模型　theoretical model
模型参数　model parameter
死亡对象　death object
样本数据估计　sample data estimation
随机退出　random exit
生存分布符号　survival distribution symbol
随机死亡　random death
乘积常数　product constant
似然方程　Likelihood equation
拟合模型参数　fitting model parameter
参数的最小 χ^2 估计法　minimum χ^2 estimation method of the parameter
修正最小 χ^2 估计法　modified minimum χ^2 estimation method
加权平方和　weighted sum of square
中心率　central rate

参数估计量　parameter estimator
柯尔莫哥洛夫——斯米尔诺夫(K—S)统计量　Kolmogorov - Smirnov statistic
偏差度量　deviation measurement
最大偏差　maximum deviation
A^2 统计量　A^2 statistic
★ 假设检验　hypothesis testing
★ 分位点法　fractile method
★ 最小二乘法　Least squares method
双变量模型　two variable model
伴随变量　adjoint variable
二元模型　binary model
总危险率　total risk rate
基本危险率　fundamental risk rate
附加危险率　additional risk rate
线性指数模型　linear exponent model
数值分析法　numerical analysis method
适宜模型　suitable models
多变量模型　multivariate model
★ 和模型　additive model
★ 积模型　product model.
Cox 模型　Cox model
对数线性指数模型　logarithmic linear exponent model

例题　Example

例 1 由 10 只实验老鼠组成的样本，其死亡时间(以天为单位)为：3,4,5,7,7,8,10,10,10,12。假定适合的生存模型为指数分布，试分别运用矩方法和中位数估计法来估计参数 λ。

解 该样本平均值 $\bar{t} = 7.6$，样本中位数 $\tilde{t} = 7.5$。

则在矩估计法下，$\hat{\lambda} = \dfrac{1}{\bar{t}} = 0.13158$；

在中位数法下，$\hat{\lambda} = \dfrac{-\ln \dfrac{1}{2}}{\tilde{t}} = 0.09242$。

Ex. 1 The sample book is composed of 10 experiment mice, and the time of death (in days) is 3,4,5,7,7,8,10,10,10,12. If the survival model is an exponential distribution, please respectively use the moment method and the median estimate method to estimate the parameter λ.

Solution Mean of sample is $\bar{t} = 7.6$, and median of sample is $\tilde{t} = 7.5$.

So, under moment estimation method, $\hat{\lambda} = \dfrac{1}{\bar{t}} = 0.13158$;

under median method, $\hat{\lambda} = \dfrac{-\ln\dfrac{1}{2}}{\tilde{t}} = 0.09242$.

例2 对于由例1的矩估计法所得的 $\hat{S}(t)$ 的指数分布形式，估计 $Y = (D_n - \dfrac{0.2}{n})(\sqrt{n} + 0.26 + \dfrac{0.5}{\sqrt{n}})$ 的值，并将其与样本数据比较。

解 假定模型为 $\hat{S}(t) = e^{-0.13158t}$，将 $\hat{S}(t)$ 和 $S^o(t)$ 的值进行比较，见表 5-1，此处 $S^o(t^-)$ 和 $S^o(t^+)$ 分别表示在 t 以前的 $S^o(t)$ 的瞬时值和在 t 以后的 $S^o(t)$ 的瞬时值。

表 5-1

t	$\hat{S}(t)$	$S^o(t^-)$	$S^o(t^+)$	D_n
3	0.67387	1.0	0.90	0.32613
4	0.59079	0.90	0.80	0.30921
5	0.51796	0.80	0.70	0.28204
7	0.39812	0.70	0.50	0.30188
8	0.34904	0.50	0.40	0.15096
10	0.26828	0.40	0.10	0.16828
12	0.20621	0.10	0.00	0.20621
>12	<0.20621	0.00	0.00	<0.20621

最大偏差在 $t = 3^-$ 处发生（3^- 表示从 3 的左边趋近 3 的数值），则有 $D_{10} = 0.32613$，于是 $y = \left(D_{10} - \dfrac{0.2}{10}\right)\left(\sqrt{10} + 0.26 + \dfrac{0.5}{\sqrt{10}}\right) = 1.09607$。Stephens 也发现对应于显著性水平 0.05 的临界值 $y = 1.094$，由于 $y = 1.09607$ 非常接近 1.094，因此以 95% 的置信度拒绝 $\hat{S}(t)$ 这一假设。

Ex. 2 The exponential distribution $\hat{S}(t)$ is estimated by the moment estimation method of Ex. 1. Please estimate $Y = \left(D_n - \dfrac{0.2}{n}\right)\left(\sqrt{n} + 0.26 + \dfrac{0.5}{\sqrt{n}}\right)$, and compare it with sample data.

Solution Assume that the model is $\hat{S}(t) = e^{-0.13158t}$, and will compare $\hat{S}(t)$ with $S^o(t)$ (See table 5-1), in which, $S^o(t^-)$ represents the instantaneous value before t, $S^o(t^+)$ represents the instantaneous value after t.

Table 5-1

t	$\hat{S}(t)$	$S^o(t^-)$	$S^o(t^+)$	D_n
3	0.67387	1.0	0.90	0.32613
4	0.59079	0.90	0.80	0.30921
5	0.51796	0.80	0.70	0.28204
7	0.39812	0.70	0.50	0.30188
8	0.34904	0.50	0.40	0.15096
10	0.26828	0.40	0.10	0.16828
12	0.20621	0.10	0.00	0.20621
>12	<0.20621	0.00	0.00	<0.20621

The maximum deviation is in $t = 3^-$. So, $D_{10} = 0.32613$, $y = \left(D_{10} - \dfrac{0.2}{10}\right)\left(\sqrt{10} + 0.26 + \dfrac{0.5}{\sqrt{10}}\right) =$ 1.09607. Stephens also found that the critical value corresponding to a significance level 0.05 is $y =$ 1.094. Because $y = 1.09607$ is very close to 1.094, the confidence factor with 95% declines the assumption t.

本章重点

1. 在完整数据情形下，分别对确切死亡时间和分组死亡时间条件下的单变量生存模型的参数进行估计和做出假设检验。
2. 在非完整样本数据下，给出单变量生存模型参数估计和假设检验。

Key points of this chapter

1. For complete sample data, make estimation and hypothesis testing for the parameter in single-variable survival model under the extract mortality time and time of group death.
2. For non-completer sample data, make estimation and hypothesis testing for the parameter in single-variable survival model.

第六章 大样本数据下年龄的处理及暴露数的计算
Chapter 6　Process of Age and Calculation of Exposure Number in Large Sample Data

本章在大样本数据假设条件下，介绍了实际年龄、保险年龄和会计年龄的基本概念和计算公式，给出暴露数的计算方法；进一步介绍了计算暴露数的表格估值法。

In this chapter, we introduce the basic concepts and formulas of actual age, insurance age and accounting age under the assumption for large sample data. We give the calculation methods for exposure number. Furthermore, we pay more attention on the table estimation method.

单词和短语 Words and expressions

个人保单　personal policy	★会计年龄(FA)　fiscal age
团体保单　group policy	日历持续期分组　calendar duration grouping
养老金计划　pension plans	观察期限估计　observation period estimates
★小数年　decimal years	计划周年日　plan anniversary
★计划终止年龄　plan termination age	计划评估日期　scheduled valuation date
观察期　observation period	T 日期　T date
计划暴露数　program exposure number	出生评估年(VYB)　valuation year of birth
精算暴露数　actuarial exposure number	假设生日　assume birthday
确切暴露数　exact exposure number	非整数退出年龄　non-integer exit age
指数假设　index hypothesis	确切实际年龄　exact actual age
终止期限日　termination date	整数会计年龄　integer accounting age
期限向量　term vectors	★表格估值　table valuation
★保险年龄　insurance age	个体记录法　individual records method
周年日方法　anniversary date method	净迁移　net migration
日历保险年龄　calendar insurance age	日历保险年龄分组法　calendar insurance age grouping
事件进入　event access	
保单持续期　policy duration	

例题 Example

例 1 考察年龄向量 $v'_i = (39.85, 40.75, 0.25, 0)$，求：
(1)（确切暴露数）$_{i,40}$；(2)（计划暴露数）$_{i,40}$；(3)（精算暴露数）$_{i,40}$。

解 首先，将 v'_i 转换为 $u'_{i,40} = (0, 0.75, 0.25, 0)$，则可得：
(1)（确切暴露数）$_{i,40} = 0.25(\tau_i < s_i)$；
(2)（计划暴露数）$_{i,40} = 0.75$；
(3)（精算暴露数）$_{i,40} = 1$。

Ex. 1 Study the age vector $v'_i = (39.85, 40.75, 0.25, 0)$, please solve:
(1) (exact exposure number)$_{i,40}$;
(2) (number of program exposure)$_{i,40}$;
(3) (actuarial exposure number)$_{i,40}$.

Solution We will replace v'_i with $u'_{i,40} = (0, 0.75, 0.25, 0)$, then
(1) (exact exposure number)$_{i,40} = 0.25(\tau_i < s_i)$;
(2) (number of program exposure)$_{i,40} = 0.75$;
(3) (actuarial exposure number)$_{i,40} = 1$.

例 2 运用以下值估计 q_{50} 和 q_{51}。
$I_{50}^{1985} = 1000, I_{50}^{1986} = 1010, d_{51}^{1985} = 990, I_{51}^{1986} = 1030, d_{51}^{1986} = 4$.

解 由于 $_ad_x^z$ 的值没有给出，用 $\frac{1}{2}d_x^z$ 来近似它。运用（精算暴露数）$_x \approx \frac{1}{2}(I_x^{z-1} + I_x^z) + _aI_x^z$，有

$$\hat{q}_{50} = \frac{2}{\frac{1}{2}(1000 + 1010) + 1} = \frac{2}{1006}$$

$$\hat{q}_{50} = \frac{4}{\frac{1}{2}(990 + 1030) + 1} = \frac{4}{1012}.$$

Ex. 2 Using the following value estimate q_{50} and q_{51}.
$I_{50}^{1985} = 1000, I_{50}^{1986} = 1010, d_{51}^{1985} = 990, I_{51}^{1986} = 1030, d_{51}^{1986} = 4$.

Solution Because the value $_ad_x^z$ is unknown, it is approximated with $\frac{1}{2}d_x^z$. (actuarial exposure number)$_x \approx \frac{1}{2}(I_x^{z-1} + I_x^z) + _aI_x^z$, then

$$\hat{q}_{50} = \frac{2}{\frac{1}{2}(1000 + 1010) + 1} = \frac{2}{1006}$$

$$\hat{q}_{50} = \frac{4}{\frac{1}{2}(990 + 1030) + 1} = \frac{4}{1012}.$$

本章重点

1. 在大样本数据假设条件下，掌握实际年龄、保险年龄和会计年龄的基本概念。
2. 能够利用保险公司或养老金基金所得数据计算暴露数。

Key points of this chapter

1. Under the assumption of large sample data, grasp the basic concepts and formulas of the actual age, insurance age and accounting age.
2. Be able to calculate the number of exposed with data from insurance companies or pension funds.

第二篇 人口统计
Part 2 Population Statistics

第七章 死亡和生育测度
Chapter 7 Death Measure and Birth Measure

本章首先介绍人口统计学的基本概念、起源,对人口统计采样的方法进行了阐述;并对人口统计误差来源及相关指标的概念和计算方法进行了介绍。其次,对死亡的测度指标和生育的测度指标的概念与计算方法进行了阐述。

This chapter introduces the basic concepts and origins of the population statistics as well as the statistical sampling method. We also discuss the origins of the population statistics error. The relevant indicators are introduced in both basic concepts and calculation methods. The concepts as well as calculation methods for the measuring index of mortality and birth are presented.

单词和短语 Words and expressions

★ 抽样误差　sampling error
数据来源　data source
误差分析　error analysis
人口统计学　demography
人口理论　population theory
人口政策　population policy
精算报告　actuarial report
人口普查　census
人口动态统计　dynamic statistical of population
抽样调查　sampling investigation
常住人口　resident population
人口迁移　population migration
动态统计率　dynamic statistical rate
户口管理　management of registered permanent residence
动态事件　dynamic events
国民生命表　national life tables
★ 总误差比率　total error rate
误差来源　error sources
误差校正　error correction
终止误差　closure error
年龄误报　age misreporting

混合技术　mixing technology
★ 净误差比率　net error ratio
★ 粗死亡率　crude mortality rate
★ 粗出生率　crude birth rate
性别死亡率　sex mortality
★ 人口自然增长率　natural growth rate of population
★ 婴儿死亡率　infant mortality rate
婴儿死亡测度　infant mortality measure
分年龄死亡率　age mortality
★ 直接调整死亡率　directly adjusted mortality
直接调查法　direct adjustment method
间接调查法　indirect adjustment method
标准死亡比率　standardized mortality ratio, SMR
分年龄生育率　age-specific fertility rate
最小生育年龄　minimal fertility age
最晚生育年龄　late fertility age
★ 总和生育率　total fertility rate
性别生育率　sex fertility rate
★ 毛再生育率　gross reproduction rate, GRR
★ 净再生育率　net reproduction rate, NRR

例题 Example

例1　某 $\frac{1}{6}$ 采样估计有如下结果:250 人在一个有 5000 人居住的地区受过大学教育,计算抽样标准误差。

解 $SE_{250} = \sqrt{(6-1)X\left(1-\dfrac{X}{N}\right)} = \sqrt{5 \times 250 \times \left(1-\dfrac{250}{5000}\right)} = 34.46$.

Ex. 1 The sampled-data estimation of $\dfrac{1}{6}$ has the following result: there is 250 college-educated people in a 5000 people. Please calculate sample standard error.

Solution
$$SE_{250} = \sqrt{(6-1)X\left(1-\dfrac{X}{N}\right)} = \sqrt{5 \times 250 \times \left(1-\dfrac{250}{5000}\right)} = 34.46$$

例 2 假设 2000 人的地区,有 20% 是移民人口,假设问卷调查导致了 −10% 的净误差,15% 的总误差,求隐瞒了自己是移民人口的人数。

解 设 $n = 2000$,b 代表隐瞒自己是移民的人数,c 代表误报自己是移民的人数,那么
$$\left.\begin{array}{l} b+c = 2000 \times 15\% \\ b-c = -2000 \times 10\% \end{array}\right\} \Rightarrow b = 50$$

Ex. 2 Assumed that there is 20% migrant of 2000 people. Suppose that questionnaire leads to −10% net error and 15% total error. Determine the number of population that conceal himself immigrant.

Solution let $n = 2000$, b indicates the number of population that conceal himself migrant, c indicates the number of population that misreport himself migrant. So
$$\left.\begin{array}{l} b+c = 2000 \times 15\% \\ b-c = -2000 \times 10\% \end{array}\right\} \Rightarrow b = 50$$

本章重点

1. 掌握人口统计学的基本概念。
2. 掌握人口统计误差来源及相关指标的概念和计算方法。

Key points of this chapter

1. Grasp the basic concepts of demography
2. Know the error sources of demographic and grasp the concept and calculation method for some related indicators.

第八章 人口模型
Chapter 8 Population Model

本章阐述三个人口模型,首先对静止人口模型的假设、特征及计算方法进行了详细阐述,其次介绍稳定人口模型,对该模型的主要性质、相关指标的计算方法给出详细的分析,最后介绍了拟稳定人口模型。

This chapter discusses three population models, firstly, we introduce the assumptions, characteristics and computational methods for stationary population model in details. Secondly, we introduce stable population model with some main characters and computational method for some related indicators. Finally, the quasi-stable population model is given.

单词和短语 Words and expressions

★ 静止人口模型　stationary population model
生存组分析　analysis of survival group
平均死亡年龄　average death age
静止人口特征　static population characteristic
人口状态　population state
★ 总寿命　total life
★ 总活过年数　total years of living
★ 总剩余寿命　total time-until-death
剩余寿命　time-until-death
静止人口数学问题　static population mathematical problem
★ Lexis 图　Lexis figure
目前平均年龄　current average age
均匀出生　uniform birth

生育函数　fertility function
★ 稳定人口模型　stable population models
★ 人口内在增长率　intrinsic rate of growth
死亡函数　death function
内在增长率　inner growth rate
r_i 的近似求解　approximate solution of r_i
★ 女性生育函数　female childbearing function

★ 平均世代间隔　mean length of a generation
出生率增长　birth rate growth
拟稳定人口　quasi stable population
静止人口　stationary population
初始死力　initial force of mortality
连续死亡函数　continuous death function

例 题　Example

例 1　求 x 岁及以上的 T_x 人的总活过的年数和总寿命。

解　对于 x 岁及以上的人,它们在 x 岁前已活过的年数为 xT_x,而满 x 岁后,总活过的年数为 $\int_x^\infty (y-x)l_x dy = Y_x$,那么总活过的年数为 $xT_x + Y_x$。总寿命＝总活过的年数＋总剩余寿命。

x 岁及以上的 T_x 人总剩余寿命为 Y_x,所以,总寿命 $= xT_x + Y_x + Y_x = xT_x + 2Y_x$。

Ex. 1　Determine x years old and above, T_x people of the total years of living and total life.

Solution　For people is x years old and over x years old, total years of living is xT_x before aged-x. After aged-x, total years of living is $\int_x^\infty (y-x)l_x dy = Y_x$. So total years of living is $xT_x + Y_x$. Total life ＝ Total years of living ＋ Total time-until-death.

For x years old and above, total time-until-death life of T_x people is Y_x, so total life$= xT_x + Y_x + Y_x = xT_x + 2Y_x$.

例 2　一个女性在 20 岁时生了一个女儿,在 40 岁时又生了一个女儿,给定 $S(20) = 0.96, S(40) = 0.9$,当一个人口中所有活着的女性人口都具有这种生育模型时,求该人口内在增长率 r_i。

解　因为所有女人具有同样生育模式,有 $f_{20}^f = f_{40}^f = 1$,而对于其他年龄有 $f_x^f = 0$。由于 $\sum_{x=\alpha}^{\beta} e^{-r_i y} S(y) f_x^f = 1$,所以 $e^{-20r_i} \cdot S(20) + e^{-40r_i} \cdot S(40) = 1$,即 $e^{-20r_i} \cdot (0.96) + e^{-40r_i} \cdot (0.9) = 1$,解得 $r_i = 0.0217$。

EX. 2　A female has given birth to a daughter when she is 20 years old; she has given birth to a daughter when she is 40 years old. Let $S(20) = 0.96, S(40) = 0.9$. When all the female population have this fertility model, please determine the inner growth rate r_i of the population.

Solution　Because all fertility have same fertility pattern, we have $f_{20}^f = f_{40}^f = 1$, for other ages $f_x^f = 0$.

Since $\sum_{x=\alpha}^{\beta} e^{-r_i y} S(y) f_x^f = 1$, we have $e^{-20r_i} \cdot S(20) + e^{-40r_i} \cdot S(40) = 1$, then $r_i = 0.0217$

本章重点

掌握静止人口模型、稳定人口模型、拟稳定人口模型的概念,性质及计算方法。

Key points of this chapter

Grasp the concept, properties and calculation method for stationary population model, the stable population model and quasi-steady population model.

第九章　人口规划及人口普查应用
Chapter 9　Population Planning and Census Application

本章首先介绍了人口数据估计的几个模型概念及计算公式。其次,对人口规划的概念与方法进行了阐述。最后介绍了人口统计数据在社会保障中的作用。

This chapter introduces the concepts and the computational formula for several modes on population data estimation. We then show the concepts and methods for population planning. The role of demographic data in social security is introduced in the end of this chapter.

单词和短语 Words and expressions

★ 插值　interpolation
人口数据估计　estimation of population data
数学模型　mathematical model
线性插值　linear interpolation
多项式插值　polynomial interpolation
线性插值估测　linear interpolation estimation
年增长力　annual growth force
年增长率　annual growth rate
★ 几何模型　geometric model
★ Logistic 模型　Logistic model
单位人口增长力　population growth force of unit
增长上限　growth upper limit
模型参数　model parameter
模型函数　model function
拐点　inflection point
人口规划　population planning
生存因素　survival factor
生育—生存因素　fertility—survival factor
分性别生命表　sex life table
育龄区间　range of fertility age
分年龄—性别生育率集合　age- sex fertility rate collection
成活产儿数　number of live birth
★ Leslie 矩阵　Leslie matrix
极限年龄　limit age
矩阵阶数　order of matrix
预测向量　prediction vector

预测动植物生长　forecast for animal and plant growth
年龄组间距　age group spacing
规划远期人口　Planning long-term population
完全生命表　complete life table
简略生命表　abridged life table
人口统计资料　population statistics data
光滑性　smoothness
国家医疗　national health
健康水准　healthy standard
额外利益　additional benefits
其他社会　other society
心理因素　psychological factor
医疗保险　medical insurance
医院保险数据　hospital insurance data
测试比率　test ratio
退休保险　retirement insurance
筹资方法　financing method
在职劳动者　On-the-job laborer
计划组织者　scheme organizer
★ 完全基金制　completely fund system
资产管理成效　asset management effectiveness
计划赞助者　plan sponsor
养老负担　pension burden, PB
养老负担变化率　rate of change of the pension burden
老年化　aging

例题 Example

例 已知人口普查数据：$P(1950) = 1600000, P(1960) = 2000000, a = 3000000$，利用 Logistic 模型求 $P(1980)$。

解 取 1950 年为计时开始时刻，$A = \dfrac{1}{a} = 0.0033, B = \dfrac{1}{P(1950)} - A = 0.0029$。

则 $P(1970) = \dfrac{1}{0.0033 + 0.0029 e^{-20k}} = 2000000$

导出 $e^{-20k} = 0.5714$。

因为 $e^{-10k} = 0.7559$，

所以 $P(1980) = \dfrac{1}{0.0033 + 0.0029 e^{-30k}} = 2180000$

Ex Suppose census data $P(1950) = 1600000, P(1960) = 2000000, a = 3000000$, Using the Logistic model to determine $P(1980)$

Solution Time beginning moment is 1950, $A = \dfrac{1}{a} = 0.0033$, $B = \dfrac{1}{P(1950)} - A = 0.0029$

So $P(1970) = \dfrac{1}{0.0033 + 0.0029e^{-20k}} = 2000000.$

Get $e^{-20k} = 0.5714$

Since $e^{-10k} = 0.7559$, then $P(1980) = \dfrac{1}{0.0033 + 0.0029e^{-30k}} = 2180000$

本章重点
1. 掌握人口数据估计的几个模型的概念及计算公式。
2. 熟知人口规划的概念与方法。

Key points of this chapter
1. Grasp the concepts and the computational formulas of population data estimation for several models.
2. Know the concepts and methods of population planning.

第三篇　人口统计
Part 3　Population Statistics

第十章　表格数据修匀
Chapter 10　Table Data Smoothing

本章首先给出修匀的基本概念和基本符号，简单阐述光滑性检验和拟合检验的概念及计算公式。其次，对表格数据修匀的方法进行阐述。最后，介绍二维修匀的基本概念、二维 Whittaker 修匀的表达式和计算公式。

This chapter first introduces the basic concepts and symbols of data smoothing. A briefly discussion on the concepts and calculates formulas on the smooth test and fitting test is given. Then the method for table data smoothing is discussed. In the end of this chapter, we present the basic concept of two-dimensional smoothing as well as the expressions and calculation formula of 2D Whittaker smoothing.

单词和短语 Words and expressions

★ 修匀算子　smoothing operator
修匀法　smoothing method
样本信息　sample information
未知参数　unknown parameter
重新抽样　re-sampling
先验观点　priori viewpoint
修匀值　smoothing value
修正初始估计值　amend initial estimated value
编制生命表　prepare life table
初始生命表　initial life table
死亡率序列　mortality sequence
连续变化　continuous change
待估参数　estimating parameter
修匀随机变量　smoothing random variables
估计误差随机变量　evaluated error random vari-able

修匀过程　smoothing process
修匀误差　smoothing error
参数序列　parameter sequence
★ 光滑性检验　test of smooth
★ 拟合检验　fit testing
光滑量　smooth measure
样本容量　sample size
修匀的拟合检验　the fitting test of smoothing
加权偏差　weighted deviation
修匀死亡数　smoothing deaths
观察死亡数　observation deaths
观察死亡年龄　observe death age
修匀死亡年龄　smoothing death age
生命变化规律　variation of life

偏差序列　deviation sequence
修匀序列　smoothing sequence
初始估计序列　initial estimation sequence
★表格数据修匀　table data smoothing
★参数修匀　parameter smoothing
★移动加权平均修匀法（M—W—A）　moving weighted average smoothing
中心对称　centro-symmetric
端值问题　end value problem
★再生性　regeneration
最小指标　minimum indicator
最大指标　maximum index
光滑系数　smoothing factor
★Whittaker 修匀　Whittaker smoothing
拟合度量算子　fitting measure operator
光滑性度量算子　smooth measure operator
观察数据　observation data
修匀数据　smoothing data
标准 Whittaker 型　standard Whittaker type
Makeham 法则　Makeham law
Lowrie 形式　Lowrie form
Schuette 形式　Schuette form
标准死亡表　standard mortality table
Bayes 修匀　Bayes smoothing
Bayes 估计　Bayes estimation

连续型随机变量　continuous random variable
先验密度　priori density
先验分布　priori distribution
后验分布　posteriori distribution
★Kimeldorf—Jones（K—J）方法　Kimeldorf—Jones(K—J)method
a_1 类矩阵　a_1 class matrix
先验最佳估计　priori best estimate
★Dirichlet 修匀　Dirichlet smoothing
图表修匀　diagram smoothing
方差补整修匀　variance fill refurbishment
3 参照标准表修匀　smoothing of Consult standard table
一维修匀　one dimensional smoothing
二维修匀　two dimensional smoothing
投保年龄　insured age
选择期　select period
终极死亡率　ultimate mortality
垂直光滑度量　vertical smooth measure
水平光滑度量　horizontal smooth measure
★垂直方向差分算子　vertical difference operator
★水平方向差分算子　horizontal difference operator

例　题　Example

例 1　已知 $(1) m = \begin{bmatrix} 1 \\ 2 \\ 3 \end{bmatrix}; (2) v = \begin{bmatrix} 2 \\ 2 \\ 2 \end{bmatrix}; (3) A^{-1} = \begin{bmatrix} 1 & 1 & 2 \\ 1 & 3 & 0 \\ 2 & 0 & 2 \end{bmatrix}$. 运用 K—J 修匀方法，求 $(A+B)^{-1}(u-m)$.

解　$v = u + (I + AB^{-1})^{-1}(m-u) = m + (I + BA^{-1})(u-m),$
　　　$v - m = (AA^{-1} + BA^{-1})^{-1}(u-m) = A(A+B)^{-1}(u-m),$

$$(A+B)^{-1}(u-m) = A^{-1}(v-m) = \begin{bmatrix} -1 \\ 1 \\ 0 \end{bmatrix}$$

Ex. 1　Let $(1) m = \begin{bmatrix} 1 \\ 2 \\ 3 \end{bmatrix}; (2) v = \begin{bmatrix} 2 \\ 2 \\ 2 \end{bmatrix}; (3) A^{-1} = \begin{bmatrix} 1 & 1 & 2 \\ 1 & 3 & 0 \\ 2 & 0 & 2 \end{bmatrix}$, Using K—J smoothing method determine $(A+B)^{-1}(u-m)$.

Solution　$v = u + (I + AB^{-1})^{-1}(m-u) = m + (I + BA^{-1})(u-m),$
　　　$v - m = (AA^{-1} + BA^{-1})^{-1}(u-m) = A(A+B)^{-1}(u-m),$

$$(A+B)^{-1}(u-m) = A^{-1}(v-m) = \begin{bmatrix} -1 \\ 1 \\ 0 \end{bmatrix}$$

例 2 已知观察值 U 和修匀值 V 分别如下：
$U = \begin{bmatrix} 1 & 1 & 6 \\ 3 & 4 & 6 \end{bmatrix}, V = \begin{bmatrix} 2 & 3 & 5 \\ 3 & 5 & 6 \end{bmatrix}$；并且我们知道：

(1) 所有的权数都等于 2；

(2) $F = \sum_{i=1}^{2} \sum_{j=1}^{3} w_{ij}(u_{ij} - v_{ij})^2$；

(3) 垂直光滑度量 ^vS 和水平光滑度量 ^hS 都采用一阶差分；

(4) $M = F + 2^vS + ^hS$.

求 M。

解 $^vS = \sum_{j=1}^{3} \sum_{i=1}^{1} (\Delta v_{ij})^2 = (3-2)^2 + (5-3)^2 + (6-5)^2 = 6$

$^hS = \sum_{i=1}^{2} \sum_{j=1}^{2} (\Delta v_{ij})^2 = (3-2)^2 + (5-3)^2 + (5-3)^2 + (6-5)^2 = 10$

$F = \sum_{i=1}^{2} \sum_{j=1}^{3} 2(u_{ij} - v_{ij})^2 = 14$

则 $M = 14 + 12 + 10 = 36$

Ex. 2 Suppose that observation value U and smoothing values V are $U = \begin{bmatrix} 1 & 1 & 6 \\ 3 & 4 & 6 \end{bmatrix}, V = \begin{bmatrix} 2 & 3 & 5 \\ 3 & 5 & 6 \end{bmatrix}$; and we know:

(1) All weights are equal to 2；

(2) $F = \sum_{i=1}^{2} \sum_{j=1}^{3} w_{ij}(u_{ij} - v_{ij})^2$；

(3) Vertical smooth measure ^vS and horizontal smooth measure ^hS use first-order difference；

(4) $M = F + 2^vS + ^hS$.

Determine M.

Solution $^vS = \sum_{j=1}^{3} \sum_{i=1}^{1} (\Delta v_{ij})^2 = (3-2)^2 + (5-3)^2 + (6-5)^2 = 6$

$^hS = \sum_{i=1}^{2} \sum_{j=1}^{2} (\Delta v_{ij})^2 = (3-2)^2 + (5-3)^2 + (5-3)^2 + (6-5)^2 = 10$

$F = \sum_{i=1}^{2} \sum_{j=1}^{3} 2(u_{ij} - v_{ij})^2 = 14$

So $M = 14 + 12 + 10 = 36$

本章重点

1. 掌握表格数据修匀的概念及计算方法。
2. Whittaker 修匀的表达式及计算方法。
3. 理解 Bayes 修匀的基本概念及两类特殊的 Bayes 修匀方法。

Key points of this chapter

1. Grasp the concept and the formula of table data smoothing.
2. Know the expressions and calculation formula of 2D Whittaker smoothing.
3. Understand basic concept of Bayes smoothing and the Bayes smoothing methods for two special cases.

第十一章 参数修匀
Chapter 11 Parameter Smoothing

本章以常用的三种死力的含参函数形式为基础，阐述了参数估计的配置法、最小二乘修匀法及极大似然法；进而介绍了分段参数修匀和光滑连接修匀的概念和方法。

This chapter is based on three kinds of force of mortality function with parameters, described configuration method of parameter estimation, smoothing method of least squares and method of maximum likelihood. And then introduce the concepts and methods of segmentation parameters smoothing and smooth connection smoothing.

单词和短语 Words and expressions

* ★ 分段参数修匀　segmentation parameters smoothing
* ★ 光滑连接修匀　smooth connection smoothing
* ★ Everrett 公式　Everrett formula
* 寿险精算理论　life insurance actuarial theory
* Gompertz 形式　Gompertz form
* 数据合适预检方法　inspection method of data suitable
* Weibull 公式　Weibull formula
* Makeham 形式　Makeham form
* 配置法　configuration method
* 配置技术　configuration technology
* ★ 最小二乘修匀法　smoothing method of least squares
* 最小二乘函数　Least squares function
* ★ 极大似然法　method of maximum likelihood
* 退出　exits
* 总似然函数　total likelihood function
* 分组死亡　group death
* 保险年度　insurance annual
* 投保人　applicant
* 保单周年日　policy anniversary
* 最低年龄　minimum age
* 最高年龄　maximum age
* 年龄区间　age interval
* 似然因子　likelihood factor
* 一次可微　a differentiable
* 最小二乘三次样条　least squares cubic spline
* 两弧三次样条　two arcs cubic spline
* 光滑连接　smooth connection
* 结点　node
* 中心差分算子　central difference operator
* 插值公式　interpolation formula
* 线性插值公式　linear interpolation formula
* 再生多项式函数　regenerate polynomial function
* 最高次数　maximum times
* 四点修匀公式　four-point smoothing formula
* 六点修匀公式　six point smoothing formula

例题 Example

例　已知(1) $v_x = \begin{cases} p_0(x), & 10 \leqslant x \leqslant k, \\ p_0(x) - 0.0064(x-k), & k \leqslant x \leqslant 13, \end{cases}$ $11 < k < 12$

(2) 已知数据如表 11-1

表 11-1

x	10	11	12	13
w_x	4	4	2	3
u_x	0.02	0.03	0.04	0.04
$p_0(x)$	0.020	0.031	0.042	0.053

用最小二乘样条修匀法来拟合观察值 u_x，并求 k。

解

$$SS = \sum_{x=10}^{13} (v_x - u_x)^2 w_x = (0.02-0.02)^2 \times 4 + (0.031-0.03)^2 \times 4$$
$$+ [0.042 - 0.0064(12-k) - 0.04]^2 \times 2$$
$$+ [0.053 - 0.0064(13-k) - 0.04]^2 \times 3$$

$$\frac{\partial SS}{\partial k} = 0$$

即 $k = 11.3$。

Ex Let (1) $v_x = \begin{cases} p_0(x), & 10 \leqslant x \leqslant k, \\ p_0(x) - 0.0064(x-k), & k \leqslant x \leqslant 13, \end{cases}$ $11 < k < 12$, (2) The data is shown in table 11-1.

Table 11-1

x	10	11	12	13
w_x	4	4	2	3
u_x	0.02	0.03	0.04	0.04
$p_0(x)$	0.020	0.031	0.042	0.053

Using smoothing spline method of least squares to imitate the observed values u_x, then determine k.

Solution

$$SS = \sum_{x=10}^{13} (v_x - u_x)^2 w_x = (0.02-0.02)^2 \times 4 + (0.031-0.03)^2 \times 4$$
$$+ [0.042 - 0.0064(12-k) - 0.04]^2 \times 2$$
$$+ [0.053 - 0.0064(13-k) - 0.04]^2 \times 3$$

$$\frac{\partial SS}{\partial k} = 0$$

So $k = 11.3$

本章重点

1 掌握参数修匀，以及表格数据修匀与参数修匀的差别。

Key points of this chapter

1 Grasp the parameter smoothing and know the difference between the table data smoothing and parameter smoothing.

第八部分　抽样调查
Part 7　Sampling Survey

引　言

社会发展离不开统计数据。统计作为一种计数的活动由来已久，我国远在夏商时代就已有人口和土地的记载，在治国中起到了重要的作用。随着社会的发展，就更加需要对各种社会经济现象作定量研究。进入二十世纪以来，科学和技术迅猛发展，人类的经济和社会生活以及其他活动都发生了巨大的变化。大规模的专业化生产、产品的标准化和劳务的综合利用，一切活动都是以追求最大经济效率为目标。从事活动的规模越大，就越需要对发展计划进行周密的安排，而计划的设计、实施以及对未来成效的评估，不论工业、商业还是政府活动，都不可缺少的要以客观资料作为依据。各种决策资料都需要有健全的依据，因此就会用到各种统计信息。抽样调查就是一种重要的统计数据的方法。

抽样调查是按照概率论与数理统计的原理，从研究的总体中按随机原则来抽取样本，通过对样本的调查获取数据，以此来对总体的特征做出估计推断，对推断中可能出现的抽样误差可以从概率的意义上加以控制。

关于抽样调查的历史发展，可分三个不同的时间段进行考察。

1. 1895 年以前，抽样调查处在实践探索过程和萌芽阶段。

大量事实表明，在 1895 年以前，抽样调查的实践应用已经在许多领域展开，这些都为抽样调查的进一步发展创造了条件，积累了宝贵经验。但这时的抽样调查仍处于探索过程之中，没有形成系统的理论，也没有得到普遍的认可。

2. 1895~1925 年，抽样调查逐步得以确认的过程。

从 1895 年到 1925 年，经过 30 年的反复讨论，代表性方法，即抽样方法，才得到人们的最终承认。在这一过程中，凯尔以其坚持不懈的努力，被称为抽样调查的先驱者是当之无愧的。而鲍利则从抽样理论上有力地支持了代表性调查的主张，从而对抽样调查的初步发展做出了贡献。在抽样调查的理论得以逐步公认的同时，抽样的实践活动继续得以深入发展。

3. 1925 年以后，抽样调查进入全面发展阶段、逐步走向成熟。

如今，无论是为了科研目的还是为了管理目的，抽样调查作为一种为种类繁多的主题提供统计数据的手段已经被广泛接受。在社会学、社会心理学、人口学、政治学、经济学、教育以及公共卫生等领域，抽样调查都已得到广泛应用，这些调查被用于形成、检验和改进研究假设。

本部分主要介绍抽样调查的基本常用方法：简单随机抽样、分层抽样、系统抽样、整群抽样、多阶段抽样、不等概率抽样、二重抽样。

Introduction

Social development is inseparable from statistical data. Statistics had been on for a long time as a count of activities, since our country had records of population and lands in the dynasties of Xia and Shang, which played an important role in country management. With the development of society, we need more quantitative research on social economic phenomenon. Since the 20th century, science and technology have been rapidly developed, and human economics and other activities were changed greatly. Large-scale specialized production, standardization of products and the comprehensive utilization of labor, all of these activities are pursuing maximum economic efficiency. It needs a more careful arrangement of development plan when engaging larger activities. No matter in the industrial, commercial or government activities, when designing, resolving and evaluating the effectiveness of the plans, it's indispensable to base on object data. All kinds of decision-making material must have a sound basis, so they will use all sorts of statistical information. Sampling survey is an important method of statistical data.

Sampling survey is in accordance with the theory of probability and mathematical statistics, which sampling from the study population according the random principle, and collecting data through the investigation of the sample, then making estimate for population depend on the data, and controlling the possible sampling error in the sense of probability.

The historical development of sampling survey can be divided into three different times.

1. Before 1895, sampling survey was in the practice exploration and the embryonic stage.

A large number of facts show that the application of sampling survey has been commenced in many fields before 1985. These created the condition and accumulated valuable experience for the further development of sampling survey. But during this period sampling survey was still in the exploration process and there was no systematic theory and generally recognized.

2. 1895~1925, Sampling survey was gradually confirmed.

From 1895 to 1925, after 30 years repeated discussion, sampling method finally got the people's approval. In this process, Kiaer well deserved the title of sampling pioneer by his unremitting efforts, and Bowley made contribution on the development of sampling survey by his sampling theory strongly supported the representative survey claims. While sampling survey had been gradually recognized, sampling practice continued to be further developed.

3. After 1925, sampling survey had entered a stage of comprehensive development, and gradually matured.

Now, no matter for scientific or administrative purpose, sampling survey has been widely accepted as a means of statistical for a large variety of topics. In sociology, social, psychology, demographic, political, science, economics, education, public health and other fields, sampling survey has been widely used, the survey was used to form, inspect and improve the research hypotheses.

This part mainly introduces the basic methods of sampling survey: simple random sampling, stratified sampling, systematic sampling, cluster sampling, multistage sampling, varying probability sampling, double sampling.

第一章 简单随机抽样
Chapter 1 Simple Random Sampling

在有限总体抽样中,简单随机抽样是最基础的抽样方法,其他许多抽样方法都可看作是在该方法基础上的修正,以便在更方便或更精确两个方面得到改进。本章将会介绍简单随机抽样的实现,总体均值和方差的估计,进一步会介绍比例估计和区间估计以及样本容量的确定方法等内容。

In finite population sampling, simple random sampling is the most basic sampling method, and many other sampling methods can be seen as a correction on the basis of this method in order to be more convenient or accurate. This chapter will introduce simple random sampling, estimation of population mean and variance. Furthermore we will introduce ratio estimation, interval estimation and sample size determination method.

单词和短语 Words and expressions

总体	population	随机数	random number
样本	sample	样本量	sample size
总体均值	population mean	有放回抽样	sampling with replacement
样本均值	sample mean	无放回抽样	sampling without replacement
总体方差	population variance	标准误	standard error
样本方差	sample variance	抽样方差	sampling variance

抽样比　sampling fraction
无偏估计　unbiased estimate
比例估计　ratio estimate
区间估计　interval estimate
回归估计　regression estimate

基本概念和性质　Basic concepts and properties

1 简单随机抽样又称纯随机抽样，这是一种最基本的抽样方式。它定义为：设有限总体共有 N 个单元，从中抽取容量为 n 个单元的样本，使得每一个可能的样本被抽中的概率相同，这种抽样方法称为简单随机抽样，所抽到的样本称为简单随机样本。

Simple random sampling, also called pure random sampling, is one of the most basic sampling methods. The definition is: let a finite population contain N units, we select a sample of n units from it, which make sure every unit has the same probability to be selected. This sample method is called simple random sampling, and the sample selected is called simple random sample.

2 在简单随机抽样中样本均值是总体均值的无偏估计，样本方差是总体方差的无偏估计。

In simple random sampling, sample mean is an unbiased estimate of the population mean, and sample variance is an unbiased estimate of the population variance.

3 设 P 是总体中具有某种特性的比例，a 是简单随机样本中具有该特性的单元数，则 $\hat{P}=\dfrac{a}{n}$ 是 P 的无偏估计且 $V(\hat{P})=\dfrac{P(1-P)}{n}\dfrac{N-n}{N-1}$。

Let P be the proportion of units who have some certain features in population, and a be the number of the units with this feature in sample, then $\hat{P}=\dfrac{a}{n}$ is an unbiased estimate of P and $V(\hat{P})=\dfrac{P(1-P)}{n}\dfrac{N-n}{N-1}$.

本章重点

1 掌握简单随机抽样的概念和抽样方法以及如何进行样本量的确定。
2 掌握简单随机抽样总体均值的估计值，区间估计以及比例估计等方法。
3 熟悉简单随机抽样的回归估计。

Key points of this chapter

1 Master the concept of the simple random sampling and sampling method, also know how to determine the sample size.
2 Master the point estimate and the interval estimate of population mean and the method of ratio estimate.
3 Be familiar with simple random sampling regression estimation.

第二章　分层抽样
Chapter 2　Stratified Sampling

抽样调查的一种常见情况是，我们对于要研究的总体元素已经掌握了一定的信息，这类辅助信息既可以用于设计阶段改进抽样设计，也可以用于分析阶段提高样本估计量的精确度，或者兼用于两个阶段。本章讨论利用辅助信息进行分层来改进抽样设计，即分层抽样，而且分层抽样是抽样调查时最常使用的抽样技术。

A common situation of sampling survey is that we have mastered a certain amount of information about the population elements to be studied. Such auxiliary information can be used to improve the sampling in design stage, and can also be used to improve the accuracy of estimator of the sample in analysis stage, or used in both stages. This chapter will discuss using of auxiliary information to stratify to improve sampling survey, and stratified sampling is the most commonly used sampling technique.

单词和短语 Words and expressions

分层抽样	stratified sampling	事后分层	post stratification
层	strata	设计效应	design effect
比例分配	proportional allocation	等额样本	equality sample
奈曼分配	Neyman allocation	层内方差	variance within strata
最优分配	optimum allocation	层间方差	variance between strata
目录抽样	catalogue sampling		

基本概念和性质 Basic concepts and properties

1 分层抽样也称分类抽样或类型抽样。这种抽样方法是在抽样之前将总体的 N 个单位划分为互不交叉重叠的若干层,设为 K 层,每一层包含的单元数分别为 $N_1, N_2, \cdots, N_k$,且 $N = \sum_{j=1}^{k} N_j$,然后再从各层中独立的抽取一定数量的总体单元组成样本,设总的样本量为 n,各层的样本量分别为 $n_1, n_2, \cdots, n_k$,且 $n = \sum_{j=1}^{k} n_j$。由此获得的样本称为分层样本。如果每层中的抽样都是简单随机的,那么这种分层抽样称作分层随机抽样。

Stratified sampling is also called group sampling. This sampling method first divide the N units of population into K non overlapping stratums, and the units number of each stratum were $N_1, N_2, \cdots, N_k$, and $N = \sum_{j=1}^{k} N_j$. Then it selects a certain number of units from every stratum independent comprised the sample. Let the total sample size is n, and the sample size of each stratum were $n_1, n_2, \cdots, n_k$, and $n = \sum_{j=1}^{k} n_j$. Thus the sample obtained is called stratified sample. And if a simple random sample is taken in each stratum, then the stratified sampling is called stratified random sampling.

2 如果忽略抽样比 $f = \frac{n}{N}, f_i = \frac{n_i}{N_i} (i = 1, \cdots, K)$,则

$$V_{opt}(\bar{y}_{st}) \leqslant V_{prop}(\bar{y}_{st}) \leqslant V(\bar{y})$$

其中 $V_{opt}(\bar{y}_{st})$ 为按奈曼最优分配分层抽样,层内简单随机抽样估值的抽样方差;$V_{prop}(\bar{y}_{st})$ 为按比例分配分层抽样,层内简单随机抽样估值的抽样方差;$V(\bar{y})$ 为不分层简单随机抽样估值的抽样方差。

If we ignore the sampling fraction $f = \frac{n}{N}, f_i = \frac{n_i}{N_i} (i = 1, \cdots, K)$, then

$$V_{opt}(\bar{y}_{st}) \leqslant V_{prop}(\bar{y}_{st}) \leqslant V(\bar{y})$$

In which $V_{opt}(\bar{y}_{st})$ is the variance of the estimated means of stratified random sampling with Neyman optimum allocation. And $V_{prop}(\bar{y}_{st})$ is the variance the estimated means of stratified sampling with proportional allocation. $V(\bar{y})$ is the variance the estimated means of non stratified simple random sampling.

本章重点

1 掌握分层抽样的概念和总体估计值的性质。
2 掌握分层抽样中常见的样本量分配方法,了解通过这几种方法如何确定样本容量。
3 熟悉事后分层等一些其他分层技术。

Key points of this chapter

1 Master the concept of stratified sampling and the properties of population estimate.
2 Master the common methods of sample size allocation in stratified sampling, and understand how to determine sample size by these methods.
3 Be familiar with post stratification and other stratified methods.

第三章 系统抽样
Chapter 3 Systematic Sampling

系统抽样是一种比简单随机抽样更方便的抽样方法,而且能保证每个单位拥有相同的概率被抽到样本中。本章将会介绍系统抽样的概念和特点,抽样方差的估计以及与简单随机抽样方法的比较等内容。

Systematic sampling is a sampling scheme, which is more convenient than simple random sampling and which ensures that each unit has equal chance of being included in the sample. This chapter will introduce the concept and characteristic of systematic sampling, the sampling variance and the comparison with simple random sampling will be mentioned.

单词和短语 Words and expressions

系统抽样 systematic sampling	部分区间 fractional interval
抽样区间 sampling interval	端点校正 end corrections
随机起始数 random start	居中样本 centrally located sample
系统样本 systematic sample	平衡系统抽样 balanced systematic sampling
直线系统抽样 linear systematic sampling	周期变化 periodic variation
环状系统抽样 circular systematic sampling	

基本概念和性质 Basic concepts and properties

1 在系统抽样中,先将总体从 $1 \sim N$ 相继编号,并选取抽样区间 K 为最接近 N/n 的整数,其中 N 为总体单位数,n 为样本容量。然后在 $1 \sim K$ 中抽取随机数 r,作为样本的第一个单位,接着取 $r+k, r+2k, \cdots$,直至抽够 n 的单位为止。随机数 r 称为随机起始数,通过这一过程选取的样本称为具有随机起始数的样本。很容易看出,r 值决定了整个样本。

In systematic sampling, we number the population with $1 \sim N$ successively, and choose the integer nearest to N/n as the sampling interval k, where N is the total number of the population, and n is the sample size. Then choose a number r at random from $1 \sim K$ as the first unit of the sample, and extract $r+k, r+2k, \cdots$, until we got n units at the end. The random r is called the random start, and the sample selected by this procedure is called systematic sample with a random start. One can easily see that r determines the entire sample.

2 系统抽样法的一个缺点是,无法只由一个单一有偏差的样本,去无偏的估计总体平均数或总和的方差。

In systematic sampling, it is almost not possible to obtain an unbiased estimate of the variance of either the population mean or population total on the basis of a single sample, and this is one of the disadvantages of systematic sampling.

3 定义系统样本内的方差 S_{usy}^2 为

$$S_{usy}^2 = \frac{1}{k(n-1)} \sum_{i=1}^{k} \sum_{j=1}^{n} (y_{ij} - \bar{y}_i)^2$$

用它表示的方差为

$$var(\bar{y}_a y) = \frac{N-1}{N} S^2 - \frac{k(n-1)}{N} S_{usy}^2$$

所以系统抽样的精度优于简单随机抽样当且仅当 $S_{usy}^2 > S^2$。
We have

$$var(\bar{y}_a y) = \frac{N-1}{N} S^2 - \frac{k(n-1)}{N} S_{usy}^2$$

where

$$S_{uuy}^2 = \frac{1}{k(n-1)} \sum_{i=1}^{k} \sum_{j=1}^{n} (y_{ij} - \bar{y}_i)^2$$

is the variance among units that lie within the same systematic sample.

So the mean of the systematic sample is more precise than the mean of simple random sampling if and only if $S_{uuy}^2 > S^2$.

本章重点

1. 掌握系统抽样的概念以及直线系统抽样和环状系统抽样等方法。
2. 熟悉系统抽样的特点以及与简单随机抽样等方法的比较。
3. 了解样本有直线趋势时系统抽样的改进。

Key points of this chapter

1. Master the concept of the systematic sampling, the linear systematic sampling and circular systematic sampling.
2. Be familiar with the characteristic of systematic sampling and the comparison with simple random sampling.
3. Understand how to improve systematic sampling when the sample has linear trends.

第四章 整群抽样
Chapter 4 Cluster Sampling

在大多数抽样问题中,可以将总体看成是由元素组成的若干个集合所构成的。利用这种集合进行抽样的一种方法是将它们当作层来处理,如分层抽样。另一种是将这些集合作为群处理,也就是我们本章将要介绍的整群抽样,本章主要讲述整群抽样的概念和抽样方法以及它的方差估计。

In most sampling problems, the population can be viewed as a collection of several sets which consist of units. One sampling method at the basis of sets is treated them as stratums, like stratified sampling. Another method treating these sets as clusters is cluster sampling which this chapter will introduce. The concept and sample method, also estimate of cluster sampling will be introduced in this chapter.

单词和短语 Words and expressions

整群抽样	cluster sampling	群间方差	between cluster variance
群	cluster	群内方差	inter cluster variance
抽样框	sampling frame	方差函数	variance function
等整群抽样	sampling of equal cluster	成本函数	cost function
群大小不定	varying cluster size	最佳群大小	optimum cluster size
群内相关系数	intraclass correlation coefficient		

基本概念和性质 Basic concepts and properties

1. 设总体有 N 个群组成,第 i 个群包含 M_i 个单元。从总体中按某种方式抽取 n 个群,观测其中所有的单元。这种抽样就称为整群抽样。在实际抽样调查工作中,整群抽样是一种常用的抽样方法。
当群的大小相等,即 $M_i = M$ 时,整群抽样比较简单,假定群的抽取是无放回的简单随机抽样,总体均值的估计量为

$$\bar{y} = \frac{1}{nM} \sum_{i=1}^{n} \sum_{j=1}^{M} y_{ij} = \frac{1}{n} \sum_{i=1}^{n} \bar{y}_i$$

显然该估计量为一个无偏估计量,估计量的方差为

$$V(\bar{y}) = \frac{1-f}{nM} \frac{1}{N-1} \sum_{j=1}^{N} (\bar{Y}_j - \bar{Y})^2$$

方差的估计量为

$$\widetilde{V}(\bar{y}) = \frac{1-f}{n} \frac{1}{n-1} \sum_{j=1}^{n} (\bar{y}_j - \bar{\bar{y}})^2$$

Let the population consist of N clusters, and the ith cluster contain M_j units. Selecting n clusters from population by some method, and observing all the units of these clusters. This sampling method is called cluster sampling. Cluster sampling has a widespread application in actual sampling survey. When the clusters have the same size, that is $M_j = M$, cluster sampling is simple. Assuming that using the simple random sampling to select clusters, the estimate of population mean is

$$\bar{\bar{y}} = \frac{1}{nM} \sum_{i=1}^{n} \sum_{j=1}^{M} y_{ij} = \frac{1}{n} \sum_{i=1}^{n} \bar{y}_i$$

Obviously this estimate is unbiased, and its variance is

$$V(\bar{\bar{y}}) = \frac{1-f}{nM} \frac{1}{N-1} \sum_{j=1}^{N} (\bar{Y}_j - \bar{\bar{Y}})^2$$

The estimate of this variance is

$$\widetilde{V}(\bar{\bar{y}}) = \frac{1-f}{n} \frac{1}{n-1} \sum_{j=1}^{n} (\bar{y}_j - \bar{\bar{y}})^2$$

本章重点

1. 掌握整群抽样的概念和优点。
2. 掌握等整群抽样的实施和估计方法，熟悉群大小不定时整群抽样的实施办法和估计。

Key points of this chapter

1. Master the concept and advantages of cluster sampling.
2. Master the method and estimate of sampling of equal cluster, and be familiar with the method and the estimate of sampling of varying cluster size.

第五章 多阶段抽样
Chapter 5 Multistage Sampling

假设总体由 N 个初级单位组成，每个初级单位又包含若干个次级单位。如果同一初级单位的次级单位比较相似，按上一章所述方法构成的整群样本代表性较差。这时可以在每个被抽中的初级单位中对次级单位再进行一次抽样，这种抽样就称为二阶抽样。同样可以定义一般的多阶段抽样。这就是本章将要介绍的内容。

Let population consist of N primary stage units, and every primary stage unit contain several secondary stage units. If the secondary stage units of identical primary stage unit are similar, the cluster sample composed by the method introduced in former chapter is poor representation. At this time we can select secondary stage units from every primary stage units which have been selected. This method is called two stage sampling. We can also define multistage sampling like this, which we will introduce in this chapter.

单词和短语 Words and expressions

多阶抽样 multistage sampling	初级单位 primary stage units
二阶抽样 two stage sampling	第二阶段单位 secondary stage units
三阶抽样 three stage sampling	多级抽样 multiphase sampling
第一阶段单位 first stage units	

基本概念和性质 Basic concepts and properties

1. 多阶抽样是实际工作中常用的一种抽样技术，特别是当调查总体规模很大时，几乎都是将总体分成若干小总体,实行多阶抽样。多阶抽样保持了整群抽样样本单元相对集中的特点,因而实施方便、节省费用,它的效率又高于整群抽样,所以应用很广。

Multistage sampling is a sampling technique which used commonly in practical work, especially when the population scale is very large. Mostly the population will be divided into several subpopulations, and multistage sampling will be carried out. Multistage sampling keeps the advantages of cluster sampling, that is sample units relatively concentrated, thus easy to implement and cost saving, and it is more efficient than cluster sampling, so it has widely application.

本章重点

1 掌握多阶抽样的概念和优点。
2 掌握简单随机抽样二阶段抽样的方法和估计,熟悉三阶段抽样设计和复合抽样设计。

Key points of this chapter

1 Master the concept and advantages of multistage sampling.
2 Master the method and estimate of two stage scheme with simple random sampling, familiar with three stage sampling and sampling methods composite sampling designs.

第六章　不等概率抽样
Chapter 6　Varying probability Sampling

前几章介绍的抽样方法都是以简单随机抽样为基础的。在整群抽样和多阶抽样中,也是按照简单随机抽样抽取初级单元,即各个初级单元不论大小如何,被抽到的概率都是相等的。如果初级单元的大小悬殊很大,这样做就不合适了。此时可采用不等概率抽样,即各个初级单元被抽到的机会是不同的。本章将说明不等概率抽样的必要性和它的抽选方法以及在不等概率抽样下的方差估计等问题。

The sampling methods introduced in the first few chapters are all based on simple random sampling. Cluster sampling and multistage sampling also select primary units on the basis of simple random sampling. That is to say, regardless of the size of each primary unit, the probability of being extracted is equal. It is unsuitable when the primary units are great disparity. Varying probability sampling can be used at this time, that is, each primary unit selected with unequal probability. This chapter will introduce the necessity of varying probability sampling and its select method, as well as the estimation under varying probability sampling.

单词和短语 Words and expressions

概率抽样	probability sampling	拉稀里法	Lahiri method
不等概率抽样	varying probability sampling	布鲁尔法	Brewer method
规模度量	measure of size	德宾方法	Durbin method
比例抽样	probability proportional to size sampling	桑普福德方法	Sampford method
放回比例抽样	PPS sampling with replacement	莫蒂方法	Muthy method
不放回抽样	PPS sampling without replacement		

基本概念和性质　Basic concepts and properties

1 不等概率抽样是指在抽取样本之前给总体的每一个单元赋予一定的被抽中概率。不等概率抽样分为放回与不放回两种情况。有放回的不等概率中,最常用的是按总体单元的规模大小来确定抽选的概率。设总体中第 j 个单元的规模度量为 M_j,总体的总规模度量为 $M_0 = \sum_{j=1}^{N} M_j$,则该单元的抽选概率应为 $Z_j = \dfrac{M_j}{M_0}$。这种不等概率抽样称作按规模大小成比例的概率抽样,简称为 PPS 抽样。

Varying probability sampling refers to give every unit a probability to be selected before sampling. Varying probability is divided into two cases of with replacement and without replacement. In varying probability sampling with replacement the most common used method is selecting units according to their size. Let M_j be the measure of size of the jth unit, and the total size of population be $M_0 = \sum_{j=1}^{N} M_j$, thus the probability to be selected of this unit is $Z_j = \frac{M_j}{M_o}$. This method is called probability proportional to size sampling (PPS).

2 不放回的不等概率抽样,是指在抽样的过程中被抽中的单元不能再被抽中,因此在抽取了第一个单元之后,余下的 $N-1$ 的单元中再以什么样的概率抽选就比较复杂,接着抽取第三和第四个单元时就面临更复杂的问题,因此抽样的实施比较困难。这种抽样要求做到第 j 个单元入样概率为 π_j,在样本容量为 n 时所有 N 个单元的入样概率之和就应等于 n。

Varying probability sampling without replacement refers that the units selected can no longer be selected in sampling process. Thus when the first unit was selected, the selected probability of remain $N-1$ units is complicated, and more complicated problem will be faced when selecting 3th unit and 4th unit. So it is difficult to select sample. This method ask that the selected probability of jth is π_j, and when the sample size is n, the total probability of the N units equalsn.

本章重点
1 掌握不等概率抽样的必要性和概念。
2 熟悉有放回不等概率抽样的抽样方法以及估计方法。
3 掌握不放回的不等概率抽样的几种实施方法并了解其估计方法。

Key points of this chapter
1 Master the necessity and concept of the varying probability sampling.
2 Master the sample selection method of varying probability sampling with replacement and the estimate of this method.
3 Master the sample selection methods of varying probability sampling without replacement, and understand estimate of these methods.

第七章 二重抽样
Chapter 7 Double Sampling

前几章作抽样设计或估值时,经常利用一些辅助变量的信息,如分层的辅助变量,PPS 的规模测度等。我们前面总是假定抽样框已经具有这些辅助信息,但实际工作中有时并无现成的信息。当这些变量缺乏现成的资料,而调查这些信息又比较方便和便宜时,我们可以先做一次样本容量相当大的调查来获得这些辅助信息的资料,然后再进行抽样,这便是我们本章将要介绍的二重抽样方法。

When sampling design or estimating in first few chapters, we usually use some auxiliary variables information, such as auxiliary variables of stratifying, measurable size of PPS and so on. We always assume that sampling frame has these auxiliary variables, but in practical works these information don't exist sometimes. When these auxiliary variables are lack of available information and it's convenient and cheap to survey these data, we can take a considerable sample to get data of these auxiliary variables before sampling. This is the double sampling which will be introduced in this chapter.

单词和短语 Words and expressions

二重抽样	double sampling	第一重样本	first sample
双相抽样	two phase sampling	第二重样本	second sample

基本概念和性质 Basic concepts and properties

1 二重抽样又称双相抽样,这种抽样时指在抽样时分两次来抽取样本。设总体的单元数位 N,第一次抽取一个容量 n' 较大的样本,称作第一重样本,其目的是为了获取总体有关辅助信息,为下步的第二重抽样提供条件和提高估计精度。第二重样本的样本量用 n 表示，它的样本量较小，它可以从第一重样本中抽取，也可以从总体中独立的抽取。结合第一重样本提供的信息和第二重样本的数据来对总体做出估计。

Double sampling, also called two phase sampling, means selecting sample in two steps. Let population contain N units. The first selected bigger sample with the size n' is called first sample. The purpose of the first sample is to get the auxiliary information of population, also provide conditions for the second sample and improve the estimate accurate. The size of the second sample is n, which is smaller. The second sample can be selected from the first sample or selected from population randomly. The estimate of population based on the information provided by the first sample and data from the second sample.

本章重点

1. 掌握多阶抽样的概念和特点。
2. 掌握为分层而进行的二重抽样和为 PPS 抽样而进行的二重抽样的实施方法和估计方法。
3. 了解二重抽样样本量的最优分配。

Key points of this chapter

1. Master the concept and characters of multistage sampling.
2. Master the method and estimate of double sampling for stratification and double sampling for PPS.
3. Understand the optimum allocation of the sample size of double sampling.

参考文献

[1] 华东师范大学数学系. 数学分析. 3版. 北京：高等教育出版社, 2001
[2] 马知恩, 王绵森. 工科数学分析基础. 1版. 北京：高等教育出版社, 1998
[3] Frank Ayres, Elliott Mendelson. 微积分. 1版. 北京：高等教育出版社, 2000
[4] Wilfred Kaplan. 高等微积分学. 5版. 北京：电子工业出版社, 2004
[5] Patrick M, Fitzpatrick. 高等微积分. 1版. 北京：机械工业出版社, 2003
[6] Dale Varberg, Edwin J. Prucell, Steven E. Rigdon. 微积分. 8版. 北京：机械工业出版社, 2002
[7] 北京大学数学系. 高等代数. 3版. 北京：高等教育出版社, 2003
[8] 同济大学数学教研室. 线性代数. 4版. 北京：高等教育出版社, 2006
[9] S. K. Jain and A. D. Gunawardena, Linear Algebra: An Interactive Approach. 北京：机械工业出版社, 2003
[10] 吕林根, 许子道. 解析几何. 3版. 北京：高等教育出版社, 2001
[11] A. W. Goodman. Analytic Geometry and The Calculus. New York: The Macmillan Company
[12] Belmont, California. Douglas F. Riddle. Analytic Geometry. Wadsworth Publishing Company
[13] 林世明. 实用英汉数学词汇. 兰州：西北工业大学出版社, 1997
[14] 吴光磊, 田畴. 解析几何简明教程. 北京：高等教育出版社, 2003
[15] 尤承业. 解析几何. 北京：北京大学出版社, 2004
[16] 王敬庚, 傅若男. 空间解析几何. 2版. 北京：北京师范大学出版社, 2003
[17] 丘维声. 解析几何. 2版. 北京：北京大学出版社, 1996
[18] 廖华奎, 王宝富. 解析几何教程（理科类）. 北京：科学出版社, 2000
[19] 杨文茂, 李金英. 空间解析几何（修订版）. 武汉：武汉大学出版社, 2001
[20] 唐焕文, 贺明峰. 数学模型引论. 北京：高等教育出版社, 2001
[21] 姜启源. 数学模型. 2版. 北京：高等教育出版社, 1993
[22] 叶其孝. 大学生数学建模竞赛辅导教材. 长沙：湖南教育出版社, 1993
[23] 杨启帆, 边馥萍. 数学模型. 杭州：浙江大学出版社, 1990
[24] 汪国强. 数学建模优秀案例选编. 广州：华南理工大学出版社, 1998
[25] 华中科技大学数学系. 复变函数与积分变换. 北京：高等教育出版社, 1999
[26] Jerrold E. Marsden. Basic Complex Analysis. W. H. Frecman and Company San Fracisco, 1973
[27] Ahlfors, Lars Valerian. Complex Analysis: an introduction to the theory of analytic function of one complex variable(third edition). New York; Toronto: McGraw-Hill, 1979
[28] Athanasios Papoulis. The fourier integral and its applications. McGraw-Hill book company, Inc, 1962
[29] 南京工学院数学教研组. 积分变换. 北京：高等教育出版社, 1989
[30] 上海交通大学数学系. 积分变换. 上海：上海交大出版社, 1988
[31] M. R 施皮格尔. 拉普拉斯（变换）原理及题解. 张智星, 译. 台北：晓园出版社, 1993
[32] B. L. Van der Waerden. Algebra. New York: Frederick Ungar Publishing Co., 1970
[33] J. Goldhaber and G. Ehrich, Algebra. New York: Macmillan Company, 1979
[34] I. Kaplansky. Commutative Rings. Boston: Bacon, Inc., 1970
[35] 盛骤, 谢式千, 潘承毅. 概率论与数理统计. 北京：高等教育出版社, 2001
[36] 吴亚森, 孙爱霞. 概率论与数理统计. 广州：华南理工大学出版社, 1999
[37] 安希忠, 等. 实用概率统计. 北京：中国农业出版社, 2000

[38] Marray R. Spiegel. Theory and problems of probability and statistics. McGraw-Hill Book Company. 1982
[39] Charles J. Stone. A Course in Probability and Statistics(英文版). 北京：机械工业出版社，2003
[40] J. Laurie Snell. Introduction to Probability. Random House. 1998
[41] 丁同仁，李承治. 常微分方程教程. 北京：高等教育出版社，2001
[42] C. Henry Edwards, David E. Penney. Differential Equations. The University of Georgia, Prentice Hall, Upper Saddle River, NJ 07458, 2000
[43] N. Finizio, G. Ladas. Ordinary Differential Equations with Modern Applications, University of Rhode Island, Belmont, California, Wadsworth Publishing Company. 1981
[44] 顾英利，张析. 新英汉数学词汇. 北京：科学出版社，2003
[45] 东北师范大学数学系. 常微分方程. 北京：高等教育出版社，1998
[46] 王高雄，等. 常微分方程. 北京：高等教育出版社，1994
[47] J. Stoer and R. Bulirsch, Introduction to Numerical Analysis. 2版. 北京：世界图书出版公司，1998
[48] 孙志忠，袁慰平，闻震初. 数值分析. 2版. 南京：东南大学出版社，2002
[49] 史万明，等. 数值分析. 北京：北京理工大学出版社，2002
[50] 王仁宏. 数值逼近. 北京：高等教育出版社，2003
[51] 程正兴，李水根. 数值逼近与常微分方程数值解. 西安：西安交通大学出版社，2001
[52] 蒋尔雄，赵风光. 数值逼近. 上海：复旦大学出版社，2002
[53] 张凯院，徐仲. 数值代数, 北京：科学出版社，2006
[54] George Bachmann. Lawrence Narici. Functional Analysis. New York：Courier Dover Publications，2000
[55] 程其襄，等. 实变函数与泛函分析基础. 北京：高等教育出版社，1999
[56] 夏道行，等. 实变函数与泛函分析. 北京：高等教育出版社，1985
[57] 匡纪昌. 实分析与泛函分析. 北京：高等教育出版社，2002
[58] 唐焕文，秦学志. 实用最优化方法. 大连：大连理工大学出版社，2001
[59] Russell C. Walker. Introduction to Mathematical Programming(英文版). 北京：机械工业出版社，2005
[60] Gass S. I. Linear programming: Methods and Applications. New York：McGraw-Hill，1985
[61] 廖玉麟. 数学物理方程. 武汉：华中理工大学出版社，1995
[62] 梁昆淼. 数学物理方法. 2版. 北京：高等教育出版社，1978
[63] 吴崇试. 数学物理方法. 北京：北京大学出版社，1999
[64] 徐长发，李红. 偏微分方程数值解法. 2版. 武汉：华中理工大学出版社，2000
[65] 严镇军. 数学物理方程，合肥：中国科学技术大学出版社，1989
[66] 崔锦泰. 小波分析导论. 程正兴，译. 西安：西安交通大学出版社，1995
[67] Charles K. Chui. An Introduction to Wavelets. Volumes 1，San Diego：Academic Press，1992
[68] Albert Boggess, Francis J. Narcowich. A First Course in Wavelets with Fourier Analysis. Publishing House of Electronics Industry，2002
[69] 李建平，唐远炎. 小波分析方法的应用. 重庆：重庆大学出版社，2000
[70] 李建平. 小波分析与信号处理——理论、应用及软件实现. 重庆：重庆出版社，2001
[71] 程正兴. 小波分析算法与应用. 西安：西安交通大学出版社，1998
[72] 王慧. a 尺度 Harr 小波与多小波理论与研究. 重庆大学硕士学位论文，2003
[73] Chui C K, Lian J. A Study of Orthonomal Multi-wavelet. J Appox Nomer Math，1996
[74] Mei Xiang Ming. Differential Geometry. High Educational Press，Beijing，1998

[75] M. Do Carmo, Differential Geometry of Curves and Surfaces, Prentice Hall, Englewood Cliffs, NJ, 1976

[76] A. Gray, Modern Differential Geometry of Curves and Surfaces, CRC Press, Boca Raton, 1994

[77] M. Spivak, A Comprehensive Introduction to Differential Geometry, Publish or Perish, Wilmington, Del., 1979

[78] John R. Taylor, An Introduction to Error Analysis-the Study of Uncertainties in Physical Measurements, University Science Books, Unite States of America, 1982

[79] D. P. Laurie, Numerical Solution of Partial Differential Equations: Theory, Tools and Case Studies, Birkhuser Verlag Basel, Switzerland, 1983

[80] Haim Brezis, Felix Browder, Partial Differential Equations in 20th Century. Advances in Mathematics 135, 1998:77-144

[81] 马丁.J. 普林格. 技术分析. 4版. 任若恩,等,译. 北京:中国财政经济出版社,2003

[82] 威廉. F. 夏普,戈登.J. 亚历山大,杰弗里. V. 贝利. 投资学. 北京:人民大学出版社,2002

[83] 马丁.J. 普林格. 技术分析精论(英文版). 北京:经济科学出版社,2000

[84] Joseph Stampfli, Victor Goodman. 金融数学(英文版). 北京:机械工业出版社,2004

[85] Zvi Bondie, Alex Kane, Alan J. Marcus. Essentials of investments. Von Hoffmann Press, Inc, 1998